Molecular Computing and Bioinformatics

Molecular Computing and Bioinformatics

Special Issue Editors

Xiangxiang Zeng
Alfonso Rodríguez-Patón
Quan Zou

MDPI • Basel • Beijing • Wuhan • Barcelona • Belgrade

MDPI

Special Issue Editors
Xiangxiang Zeng
Xiamen University
China

Alfonso Rodríguez-Patón
Universidad Politcnica de Madrid
Spain

Quan Zou
University of Electronic Science and Technology of China
China

Editorial Office
MDPI
St. Alban-Anlage 66
4052 Basel, Switzerland

This is a reprint of articles from the Special Issue published online in the open access journal *Molecules* (ISSN 1420-3049) from 2018 to 2019 (available at: https://www.mdpi.com/journal/molecules/special_issues/Molecular_Computing_Bioinformatics).

For citation purposes, cite each article independently as indicated on the article page online and as indicated below:

LastName, A.A.; LastName, B.B.; LastName, C.C. Article Title. *Journal Name* **Year**, *Article Number*, Page Range.

ISBN 978-3-03921-195-1 (Pbk)
ISBN 978-3-03921-196-8 (PDF)

Contents

About the Special Issue Editors

Xiangxiang Zeng is affiliated with the College of Information Science and Engineering, Hunan University, Changsha, Hunan, China. He received a BSc degree in Automation from Hunan University, China, in 2005 and a PhD degree in System Engineering from the Huazhong University of Science and Technology, China, in 2011. From 2011 to 2018, he was an Assistant Professor and Associate Professor in the Department of Computer Science, Xiamen University. His main research interests include systems biology, natural computing and data mining.

Alfonso Rodríguez-Patón received a BSc degree in Electronic and Computational Specialty from the University of Santiago de Compostela, Spain, in 1992. He received a PhD degree in Computer Science from the Polytechnic University of Madrid, in 1999. He is an Associate Professor in the Faculty of Informatics at the Polytechnic University of Madrid. His main research interests focus on the interplay between computer science, biology, and engineering.

Quan Zou received a PhD degree from the Harbin Institute of Technology, P.R. China, in 2009. From 2015 to 2018, he was a Professor of Computer Science at Tianjin University. From 2009 to 2015, he was an Assistant and Associate Professor at Xiamen University, P.R. China. His research is in the areas of bioinformatics, machine learning, and parallel computing. Several related works have been published by Science, Briefings in Bioinformatics, and Bioinformatics, among others. Google Scholar shows that over 100 of his papers have been cited more than 3,000 times.

Editorial

Molecular Computing and Bioinformatics

Xin Liang [1], Wen Zhu [1], Zhibin Lv [2,*] and Quan Zou [2,3,*]

[1] School of Mathematics and Statistics, Hainan Normal University, Haikou 570100, China
[2] Institute of Fundamental and Frontier Sciences, University of Electronic Science and Technology of China, Chengdu 610054, China
[3] Center for Informational Biology, University of Electronic Science and Technology of China, Chengdu 611731, China
* Correspondence: lvzhibin@pku.edu.cn (Z.L.); zouquan@nclab.net (Q.Z.)

Received: 24 June 2019; Accepted: 25 June 2019; Published: 26 June 2019

Abstract: Molecular computing and bioinformatics are two important interdisciplinary sciences that study molecules and computers. Molecular computing is a branch of computing that uses DNA, biochemistry, and molecular biology hardware, instead of traditional silicon-based computer technologies. Research and development in this area concerns theory, experiments, and applications of molecular computing. The core advantage of molecular computing is its potential to pack vastly more circuitry onto a microchip than silicon will ever be capable of—and to do it cheaply. Molecules are only a few nanometers in size, making it possible to manufacture chips that contain billions—even trillions—of switches and components. To develop molecular computers, computer scientists must draw on expertise in subjects not usually associated with their field, including organic chemistry, molecular biology, bioengineering, and smart materials. Bioinformatics works on the contrary; bioinformatics researchers develop novel algorithms or software tools for computing or predicting the molecular structure or function. Molecular computing and bioinformatics pay attention to the same object, and have close relationships, but work toward different orientations.

Keywords: molecular computing; bioinformatics; machine learning; protein; DNA; RNA; drug; bio-inspired

---

1. Introduction

The origin of molecular computing was as early as 1961, which was conceived by Feynman [1]. Due to the limitations of experimental conditions, materials, and biotechnology at that time, Feynman's idea was not really realized. In the following decades, biological theories have been evolving, and new biotechnology and experimental methods have been constantly emerging, which paved the way for the final reality for molecular computing. In 1994, Adleman [1] put forward a DNA molecular biological calculation method based on the Hamilton graph and successfully achieved molecular computing in DNA solution for the first time. Adleman's pioneering work opened a new field for computational science, which was of great significance and soon gained extensive attention from researchers in the field of mathematics, computer, biology, etc. In addition, other biological computing models, such as membrane computing [2], bacterial computing [3], evolutionary calculation [4,5], and virus calculation [6] have been proposed and implemented.

With the development of new generation sequencing technology, the scale of DNA, RNA, and protein biological database has been increasing dramatically [7]. An era of biological big data set in. How to efficiently analyze biological big data becomes a great challenge. Bioinformatics is an important means to cope with this challenge [8,9]. Bioinformatics combines the tools of mathematics, computer science, and biology to more efficiently elucidate and understand the biological implications and significance for a variety of sequence and structure data as well as other biological data, which has enormously promoted the research and development of many areas relative to biology. For instance, specific biological macromolecules identification and functional analysis could be achieved via bioinformatics [10,11]. By means of bioinformatics, we could uncover the relationship between genes and diseases and analyze the mechanism of diseases, both of which would benefit diseases diagnosis, diseases treatment, and even epidemic prevention [12,13]. Using the relationship between the structure and function of biomolecules gained by bioinformatics, we could analyze the effective composition of complex drugs, discover the target of new drugs, and design new drugs [14]. All of these achievements come with new software, new algorithms, and new tools originated from continuously evolving bioinformatics.

After a rigorous review process, 25 papers submitted from numerous countries including China, Malaysia, South Korea, Poland, Saudi Arabia, and so on are published in the special issue. Twenty-two of these papers are directly related to topics of molecular computation and bioinformatics. Three of them are new areas with overlapping frontiers, which are assigned to bio-inspired research areas. It is hoped that the researchers' results and perspectives in the issue will arouse readers' interest and inspire readers.

2. Molecular Computing

Differing from traditional silicon-based computing, DNA computing is an integrated technology with DNA molecules, biochemical reactions, and molecular biology. As the field has gained insight into the molecular structures, physical–chemical properties and biomechanisms of DNA, DNA computing has been developing rapidly and become an increasingly important branch in the field of computing. The DNA double strands complementary hybridization rule is the cornerstone for DNA computing. Based on this, it uses well-designed DNA sequences with a variety of carefully selected parameters such as the position binding force of the double-strand formation to realize the chemical reaction of the DNA chain system for DNA computing. Two articles in the issue focus on DNA computing. Han et al. [15] designed an 8-bit adder/subtractor with domain tags based on DNA chain displacement. The adder/subtractor used different domains to represent 0 and 1 signals instead of high and low DNA concentration. Their simulation results proved the feasibility and accuracy of the adder/subtractor logic calculation model based on the domain label, which could extend its application for molecular logic circuits. Beak et al. [16] developed an enzyme weight-updating algorithm on the basics of DNA molecular learning for future smart molecular computing systems. The new algorithm used a hypernetwork model, which integrated the internal circulation structure of DNA and ensemble learning to update the enzyme weight. It enabled the enzyme to be used for the large-scale parallel processing of DNA. At the same time, the intuitive method of DNA data construction in Beak's work could significantly reduce the number of unique DNA sequences that are needed for covering the large search space of the feature set. It was an algorithm that realized the combination of molecular computation and machine learning.

Along with DNA computing as one of the biological computing models, there are other forms of biological computing, including membrane calculation [17–19], evolutionary calculation [4,5], virus calculation [6], etc. The purpose of bacterial computing is to build "bacterial computers" to solve complex problems. In this issue, Wang et al. [20] proposed a bacterial and plasmid computing system (BP system). Two bacteria, 34 plasmids, and two genes were used to build two BP systems to demonstrate the possibility of building powerful bacterial computers.

3. Bioinformatics

3.1. Biomolecules Structure and Function Analysis

The analysis of the structure and function of biomolecules is an important area in biology, which involves multiple subjects such as protein secondary structures, protein and gene identification, and the analysis of specific functional binding sites for DNA and proteins, etc. The algorithm tools and software provided by bioinformatics greatly advance the progress in these fields. This special issue contains six related papers to the subtopic. Ping et al. [21] utilized bioinformatics tools and software such as the Basic Local Alignment Search Tool (BLAST), MEGA7.0, GSDS2.0 etc. to identify laccase gene families from three different Brassics. A series of changes under the stress for BnLACs (laccase genes from the Brassica napus genome) expression was investigated by RNA sequencing and quantitative real-time polymerase chain reaction and resulted in better insights for BnLACs' evolutions and functions. Su et.al. [22] used TransportTP, WOLF-PSORT, MEME, and other bioinformatics tools to conduct genome-wide identification and comparison of oligopeptide transporter (OPT) family genes for ginseng and 11 flowering plants. They also analyzed the expression, evolution, and biological function of OPT family genes. Their work improved the interpretation of metabolic transport mechanism and signal transduction during the cultivation of ginseng plants. Miskiewicz et al. [23] applied WebLogo, ContextFold, RNApdbee, RNAComposer and other tools to discover structural motifs in miRNA precursors from the Viridiplantae kingdom, and they revealed the secondary structural pattern of microRNA. Kalidasan et al. [24] studied the iron harvesting system of stenotrophomonas maltophilia using BLAST tools and biological experimental techniques, and proved that stenotrophomonas maltophilia acquired iron during iron starvation and used specific iron sources. Zhang et al. [25] proposed a method called Reprsent Concat, which integrated multiple heterogeneous interactive networks. The method was able to infer gene function. More heterogeneous network methods and applications could be referred to the review [26]. Feng et al. [27] carried out a support vector machine ensemble classifier algorithm to construct a recognition method for D modification site in the saccharomyces cerevisiae transcriptome. They achieved an accuracy of 83.09% with a Matthew correlation coefficient of 0.62. Using machine learning to predict modification sites is currently a hot topic in the field of biological information. Some state-in-art deep learning methods have been developed for predicting N6-methyladenosine(m6A) [28], N4 -methylcytosine (4mC) [29], and so on.

In addition, molecular topological index is defined as the invariant of the distance or degree of the vertex in the molecule, which is used to describe molecules and is useful for predicting the physical and chemical properties of proteins, DNA, and RNA and for verifying macromolecular structural characteristics. In the issue, Zhang et al. [30] employed two classical operations in graph theory, i.e., Cartesian product and graph connection, to construct an edge version topological index for atomic bond connection and geometric frameworks. They gave the proof detail of theory involved.

3.2. Drug Research and Development (R&D)

It is well known that drug R&D is notoriously long and expensive. A study published in Nature Medicine in 2010 found that a drug took an average of 13 years and cost $1.8 billion to develop from its initial laboratory study to its final release [31]. However, bioinformatics enables us to effectively reduce the drug R&D period and expense, thus making it more productive for drug R&D. In the issue, Chen et al. [32] gave a comprehensive overview of machine learning algorithms for drug-target interaction prediction, and also summarized a brief list of frequently used databases. They introduced the principles, pros, and cons of representative methods, especially the latest new algorithms, and expounded the challenges and future trends for drug–target interaction prediction. In response to the challenge regarding the dense protein interaction network identification algorithm not being suitable for sparse protein–protein interaction (PPI) networks, Cao et al. [33] developed a new method for identifying punitive protein complexes based on penalized matrix decomposition (PMD). This method surpassed previously reported methods, and achieved an ideal overall f-measure performance, better accuracy (ACC), and a maximum matching rate. Chen et al. [34] constructed a prediction algorithm for the outflow mechanism of p-glycoprotein compound substrates, which could be used for drug discovery and development. In Chen's work, a new hierarchical support vector regression scheme was built to study the nonlinear quantitative structure–activity relationship (QSAR) and explore the complex relationship between descriptor and outflow rate. With deep learning framework, Hu et al. [35] proposed a general method (SDHINE) for predicting adverse drug reactions by embedding heterogeneous networks, which integrated protein–protein interaction (PPI) information into drug embedding. Indeed, machine learning—including deep learning—is so helpful for drug R&D that quite a mass of works has published in recent years. For example, besides in this issue, Su et al. [36] used different deep learning methods to predict the efficacy and adverse reactions of cancer drugs. Ding et al. predicted the correlation between drug targets [37,38] and drug side effects [39,40] with types of machine learning methods.

Additionally, a review of the use of bioinformatics to identify Chinese herbs is presented in this special issue. Han et al. [41] outlined the two kinds of technology—biochip and DNA barcode—and their application for the identification of Chinese herbal medicine. Chinese herbs generally came from a wide range of sources, and some of them seemed to be so similar that it was hard to distinguish them by shape, color, or other apparent characteristic. However, with bioinformatics strategic methods, the identification of Chinese herbal medicine composition was speedy and accurate, as mentioned by Han et al.

3.3. Disease Analysis and Research

Bioinformatics affords us a feasible and novel means for studying on diseases diagnosis, treatment, and even on transmission mechanism. This special issue includes several related papers. Oh et al. [42] used the TRANSFAC tool and biological experimental technique to study the therapeutic effect of the HIF-1 alpha hypoxia inducer on peri-implant bone formation in diabetic mice, and concluded that the local application of HIF-1 alpha induced gene expression and growth promotion of the bone around the implant. On the basis of amino acid mutation, Qiang et al. [43] established a prediction model of avian influenza transmission from bird to human via using random forest, support vector, and other machine learning methods. Their research concluded that there were three molecular patterns of avian-to-human transmission for avian influenza that existed in nature. Xu et al. [44] exploited a support vector machine (SVM) to discriminate genes of Alzheimer's syndrome (AD) with an accuracy of 85.7%. Zakariah et al. [45] used the new generation sequencing technology, Hum-mPLoc 3.0, and other tools to study the human mycoplasma protein targeting the endoplasmic reticulum and its effect on the causes of prostate cancer. Their prediction found that intercellular infection in host cells was capable of leading to prostate cancer. Abnormal miRNA expression in various environmental factors (such as anxiety, alcoholism, etc.) gives rise to a series of diseases. The identification of the relationship between miRNA and environmental factors would facilitate the curing of diseases. Luo et al. [46] developed a new algorithm that integrated multiple types of biological information to reveal the interaction between miRNA and environmental factors, and the area under curve(AUC) of the algorithm reached 0.8208. Similarly, web-based methods have also been applied to predict the relationship between miRNA and disease [47–50]. The gene fusion structure is a common somatic mutation in cancer genome. The identification of drivers for fusion structures is of great importance for many downstream analyses, and is useful for clinical practice. Xu et al. [51] proposed a new algorithm for the stable identification of fusion structure driver genes. The algorithm took the gene network as a priori information and estimated the driver gene according to the destructive hypothesis.

Beyond the above-mentioned studies, this issue collectsan article on large-scale biomedical text data mining. Xing et al. [52] developed a parallel processing framework called ParaBTM for biomedical text mining on supercomputers. When running on the Tianhe-2 supercomputer, it took less than 12 h to process 60178 PubMed full texts by ParaBTM.

4. Bio-Inspired Research

The remaining three papers are on cross-cutting research and organized as a bio-inspired research area. Inspired by DNA sequences with the biological properties such as parallel computation and low energy consumption, DNA computation and DNA coding are widely used in image encryption [53]. In this issue, Wang et al. [54] introduced their new algorithm for correcting image encryption errors by using DNA coding. Hamming distance was used to reduce the similarity of DNA sequences for error correcting. Image edge detection is a fundamental task in image processing and computer vision. Yuan et al. [55] applied the enzymatic numerical P system (ENPS) to solve image edge detection problems. ENPS was a cell-like P system with a nested membrane structure consisting of four membranes. The calculation of edge detection was carried out in parallel among the three inner membranes. Exploring and examining the causal relationship between variables has shown great practical value in recent years, and could be used for scientific discovery from big data. Hong et al. [43] constructed the so-called K2 and BSO combined causal discovery optimization algorithm, which mimicked the human way of solving problems with brainstorming. Their algorithm took advantage of the K2 mechanism and used BSO to design the optimal topological order of searching nodes instead of the traditional graph space, which was able to solve the problem that the traditional algorithm could not work properly, since the graph space was too large.

5. Conclusions

This special issue covers several emerging topics in the fields of molecular computing and bioinformatics, which is supposed to intrigue a wide variety of readers. It must express gratitude to the Molecules editorial board for offering such a good opportunity to organize such a special issue. It must also appreciate the efforts of the reviewers to ensure the high quality of this special issue. Finally, it is thankful for all those who have contributed to this issue. More authors and readers are expected to contribute to Molecules in the future.

Funding: The work was supported by the National Key R&D Program of China (2018YFC0910405), the Natural Science Foundation of Hainan (Grant No. 119MS036), and the Natural Science Foundation of China (No. 61771331, 61863010, 61572178, 61672214 and 61873076).

Conflicts of Interest: The authors declare no conflict of interest.

References

1. Adleman, L.M. Molecular computation of solutions to combinatorial problems. *Science* **1994**, *266*, 1021. [CrossRef] [PubMed]
2. Păun, G. Computing with Membranes. *J. Comput. Syst. Sci.* **2000**, *61*, 108–143. [CrossRef]
3. Poet, J.L.; Campbell, A.M.; Eckdahl, T.T.; Heyer, L.J. Bacterial computing. *XRDS* **2010**, *17*, 10–15. [CrossRef]
4. Xu, H.; Zeng, W.; Zhang, D.; Zeng, X.X. MOEA/HD: A Multiobjective Evolutionary Algorithm Based on Hierarchical Decomposition. *IEEE Trans. Cybern.* **2019**, *49*, 517–526. [CrossRef] [PubMed]
5. Zhang, X.; Tian, Y.; Jin, Y. A knee point-driven evolutionary algorithm for many-objective optimization. *IEEE Trans. Evol. Comput.* **2015**, *19*, 761–776. [CrossRef]
6. Chen, X.; Pérez-Jiménez, M.J.; Valencia-Cabrera, L.; Wang, B.; Zeng, X. Computing with viruses. *Theor. Comput. Sci.* **2016**, *623*, 146–159. [CrossRef]
7. Mardis, E.R. Next-Generation DNA Sequencing Methods. *Annu. Rev. Genom. Hum. Genet.* **2008**, *9*, 387–402. [CrossRef]
8. Min, S.; Lee, B.; Yoon, S. Deep learning in bioinformatics. *Brief. Bioinform.* **2017**, *18*, 851–869. [CrossRef]
9. Li, Y.; Huang, C.; Ding, L.; Li, Z.; Pan, Y.; Gao, X. Deep learning in bioinformatics: Introduction, application, and perspective in the big data era. *arXiv* **2019**. [CrossRef]
10. Lo Bosco, G.; Di Gangi, M.A. Deep Learning Architectures for DNA Sequence Classification. In *Fuzzy Logic and Soft Computing Applications, Wilf 2016*; Petrosino, A., Loia, V., Pedrycz, W., Eds.; Taylor & Francis Group: Oxfordshire, UK, 2017; Volume 10147, pp. 162–171.
11. Zahiri, J.; Emamjomeh, A.; Bagheri, S.; Ivazeh, A.; Mahdevar, G.; Sepasi Tehrani, H.; Mirzaie, M.; Fakheri, B.A.; Mohammad-Noori, M. Protein complex prediction: A survey. *Genomics* **2019**. [CrossRef]
12. Wang, Y.; Deng, G.; Zeng, N.; Song, X.; Zhuang, Y. Drug-Disease Association Prediction Based on Neighborhood Information Aggregation in Neural Networks. *IEEE Access* **2019**, *7*, 50581–50587. [CrossRef]
13. Tsuji, S.; Aburatani, H. Machine Learning Applications in Cancer Genome Medicine. *Gan to kagaku ryoho. Cancer Chemother.* **2019**, *46*, 423–426.

14. Stephenson, N.; Shane, E.; Chase, J.; Rowland, J.; Ries, D.; Justice, N.; Zhang, J.; Chan, L.; Cao, R. Survey of Machine Learning Techniques in Drug Discovery. *Curr. Drug Metab.* **2019**, *20*, 185–193. [CrossRef] [PubMed]

15. Han, W.; Zhou, C. 8-Bit Adder and Subtractor with Domain Label Based on DNA Strand Displacement. *Molecules* **2018**, *23*, 2989. [CrossRef] [PubMed]

16. Baek, C.; Lee, S.-W.; Lee, B.-J.; Kwak, D.-H.; Zhang, B.-T. Enzymatic Weight Update Algorithm for DNA-Based Molecular Learning. *Molecules* **2019**, *24*, 1409. [CrossRef] [PubMed]

17. Cabarle, F.G.C.; Adorna, H.N.; Jiang, M.; Zeng, X. Spiking Neural P Systems With Scheduled Synapses. *IEEE Trans. Nanobioscience* **2017**, *16*, 792–801. [CrossRef] [PubMed]

18. Song, T.; Rodríguez-Patón, A.; Zheng, P.; Zeng, X. Spiking Neural P Systems with Colored Spikes. *IEEE Trans. Cognitive Dev. Syst.* **2018**, *10*, 1106–1115. [CrossRef]

19. Zhang, X.; Pan, L.; Păun, A. On the universality of axon P systems. *IEEE Trans. Neural Networks Learn. Syst.* **2015**, *26*, 2816–2829. [CrossRef]

20. Wang, X.; Zheng, P.; Ma, T.; Song, T. Small Universal Bacteria and Plasmid Computing Systems. *Molecules* **2018**, *23*, 1307. [CrossRef]

21. Ping, X.; Wang, T.; Lin, N.; Di, F.; Li, Y.; Jian, H.; Wang, H.; Lu, K.; Li, J.; Xu, X.; et al. Genome-Wide Identification of the LAC Gene Family and Its Expression Analysis Under Stress in Brassica napus. *Molecules* **2019**, *24*, 1985. [CrossRef]

22. Su, H.; Chu, Y.; Bai, J.; Gong, L.; Huang, J.; Xu, W.; Zhang, J.; Qiu, X.; Xu, J.; Huang, Z. Genome-Wide Identification and Comparative Analysis for OPT Family Genes in Panax ginseng and Eleven Flowering Plants. *Molecules* **2018**, *24*, 15. [CrossRef] [PubMed]

23. Miskiewicz, J.; Szachniuk, M. Discovering Structural Motifs in miRNA Precursors from the Viridiplantae Kingdom. *Molecules* **2018**, *23*, 1367. [CrossRef] [PubMed]

24. Kalidasan, V.; Azman, A.; Joseph, N.; Kumar, S.; Awang Hamat, R.; Neela, V.K. Putative Iron Acquisition Systems in Stenotrophomonas maltophilia. *Molecules* **2018**, *23*, 2048. [CrossRef] [PubMed]

25. Zhang, J.; Deng, L. Integrating Multiple Interaction Networks for Gene Function Inference. *Molecules* **2018**, *24*, 30. [CrossRef] [PubMed]

26. Liu, X.; Hong, Z.; Liu, J.; Lin, Y.; Alfonso, R.-P.; Zou, Q.; Zeng, X. Computational methods for identifying the critical nodes in biological networks. *Brief. Bioinform.* **2019**. [CrossRef] [PubMed]

27. Feng, P.; Xu, Z.; Yang, H.; Lv, H.; Ding, H.; Liu, L. Identification of D Modification Sites by Integrating Heterogeneous Features in Saccharomyces cerevisiae. *Molecules* **2019**, *24*, 380. [CrossRef] [PubMed]

28. Zou, Q.; Xing, P.; Wei, L.; Liu, B. Gene2vec: Gene Subsequence Embedding for Prediction of Mammalian N6-Methyladenosine Sites from mRNA. *RNA* **2019**, *25*, 205–218. [CrossRef] [PubMed]

29. He, W.; Jia, C.; Zou, Q. 4mCPred: Machine Learning Methods for DNA N4-methylcytosine sites Prediction. *Bioinformatics* **2019**, *35*, 593–601. [CrossRef] [PubMed]

30. Zhang, X.; Jiang, H.; Liu, J.-B.; Shao, Z. The Cartesian Product and Join Graphs on Edge-Version Atom-Bond Connectivity and Geometric Arithmetic Indices. *Molecules* **2018**, *23*, 1731. [CrossRef] [PubMed]

31. Paul, S.M.; Mytelka, D.S.; Dunwiddie, C.T.; Persinger, C.C.; Munos, B.H.; Lindborg, S.R.; Schacht, A.L. How to improve R&D productivity: The pharmaceutical industry's grand challenge. *Nat. Rev. Drug Discov.* **2010**, *9*, 203. [PubMed]

32. Chen, R.; Liu, X.; Jin, S.; Lin, J.; Liu, J. Machine Learning for Drug-Target Interaction Prediction. *Molecules* **2018**, *23*, 2208. [CrossRef] [PubMed]

33. Cao, B.; Deng, S.; Qin, H.; Ding, P.; Chen, S.; Li, G. Detection of Protein Complexes Based on Penalized Matrix Decomposition in a Sparse Protein–Protein Interaction Network. *Molecules* **2018**, *23*, 1460. [CrossRef] [PubMed]

34. Chen, C.; Lee, M.-H.; Weng, C.-F.; Leong, M.K. Theoretical Prediction of the Complex P-Glycoprotein Substrate Efflux Based on the Novel Hierarchical Support Vector Regression Scheme. *Molecules* **2018**, *23*, 1820. [CrossRef] [PubMed]

35. Hu, B.; Wang, H.; Wang, L.; Yuan, W. Adverse Drug Reaction Predictions Using Stacking Deep Heterogeneous Information Network Embedding Approach. *Molecules* **2018**, *23*, 3193. [CrossRef] [PubMed]

36. Su, R.; Liu, X.; Wei, L.; Zou, Q. Deep-Resp-Forest: A deep forest model to predict anti-cancer drug response. *Methods* **2019**. [CrossRef]

37. Shen, C.; Ding, Y.; Tang, J.; Xu, X.; Guo, F. An Ameliorated Prediction of Drug–Target Interactions Based on Multi-Scale Discrete Wavelet Transform and Network Features. *Int. J. Mol. Sci.* **2017**, *18*, 1781. [CrossRef] [PubMed]

38. Ding, Y.; Tang, J.; Guo, F. Identification of drug-target interactions via multiple information integration. *Inf. Sci.* **2017**, *417–419*, 546–560. [CrossRef]

39. Ding, Y.; Tang, J.; Guo, F. Identification of drug-side effect association via multiple information integration with centered kernel alignment. *Neurocomputing* **2019**, *325*, 211–224. [CrossRef]

40. Ding, Y.; Tang, J.; Guo, F. Identification of Drug-side Effect Association via Semi-supervised Model and Multiple Kernel Learning. *IEEE J. Biomed. Health Inform.* **2019**. [CrossRef]

41. Han, K.; Wang, M.; Zhang, L.; Wang, C. Application of Molecular Methods in the Identification of Ingredients in Chinese Herbal Medicines. *Molecules* **2018**, *23*, 2728. [CrossRef]

42. Oh, S.-M.; Shin, J.-S.; Kim, I.-K.; Kim, J.-H.; Moon, J.-S.; Lee, S.-K.; Lee, J.-H. Therapeutic Effects of HIF-1α on Bone Formation around Implants in Diabetic Mice Using Cell-Penetrating DNA-Binding Protein. *Molecules* **2019**, *24*, 760. [CrossRef] [PubMed]

43. Qiang, X.; Kou, Z.; Fang, G.; Wang, Y. Scoring Amino Acid Mutations to Predict Avian-to-Human Transmission of Avian Influenza Viruses. *Molecules* **2018**, *23*, 1584. [CrossRef] [PubMed]

44. Xu, L.; Liang, G.; Liao, C.; Chen, G.-D.; Chang, C.-C. An Efficient Classifier for Alzheimer's Disease Genes Identification. *Molecules* **2018**, *23*, 3140. [CrossRef] [PubMed]

45. Zakariah, M.; Khan, S.; Chaudhary, A.A.; Rolfo, C.; Ben Ismail, M.M.; Alotaibi, Y.A. To Decipher the Mycoplasma hominis Proteins Targeting into the Endoplasmic Reticulum and Their Implications in Prostate Cancer Etiology Using Next-Generation Sequencing Data. *Molecules* **2018**, *23*, 994. [CrossRef] [PubMed]

46. Luo, H.; Lan, W.; Chen, Q.; Wang, Z.; Liu, Z.; Yue, X.; Zhu, L. Inferring microRNA-Environmental Factor Interactions Based on Multiple Biological Information Fusion. *Molecules* **2018**, *23*, 2439. [CrossRef]

47. Jiang, L.; Ding, Y.; Tang, J.; Guo, F. MDA-SKF: Similarity Kernel Fusion for Accurately Discovering miRNA-Disease Association. *Front. Genet.* **2018**, *9*, 1–13. [CrossRef] [PubMed]

48. Jiang, L.; Xiao, Y.; Ding, Y.; Tang, J.; Guo, F. FKL-Spa-LapRLS: An accurate method for identifying human microRNA-disease association. *BMC Genom.* **2019**, *19*, 11–25. [CrossRef]

49. Zeng, X.; Liu, L.; Lü, L.; Zou, Q. Prediction of potential disease-associated microRNAs using structural perturbation method. *Bioinformatics* **2018**, *34*, 2425–2432. [CrossRef]

50. Zhang, X.; Zou, Q.; Rodriguez-Paton, A.; Zeng, X.X. Meta-Path Methods for Prioritizing Candidate Disease miRNAs. *IEEE-ACM Trans. Comput. Biol. Bioinform.* **2019**, *16*, 283–291. [CrossRef]

51. Xu, M.; Zhao, Z.; Zhang, X.; Gao, A.; Wu, S.; Wang, J. Synstable Fusion: A Network-Based Algorithm for Estimating Driver Genes in Fusion Structures. *Molecules* **2018**, *23*, 2055. [CrossRef]

52. Xing, Y.; Wu, C.; Yang, X.; Wang, W.; Zhu, E.; Yin, J. ParaBTM: A Parallel Processing Framework for Biomedical Text Mining on Supercomputers. *Molecules* **2018**, *23*, 1028. [CrossRef] [PubMed]

53. Liu, H.; Wang, X.; Kadir, A. Image encryption using DNA complementary rule and chaotic maps. *Appl. Soft Comput.* **2012**, *12*, 1457–1466. [CrossRef]

54. Wang, B.; Xie, Y.; Zhou, S.; Zheng, X.; Zhou, C. Correcting Errors in Image Encryption Based on DNA Coding. *Molecules* **2018**, *23*, 1878. [CrossRef] [PubMed]

55. Yuan, J.; Guo, D.; Zhang, G.; Paul, P.; Zhu, M.; Yang, Q. A Resolution-Free Parallel Algorithm for Image Edge Detection within the Framework of Enzymatic Numerical P Systems. *Molecules* **2019**, *24*, 1235. [CrossRef] [PubMed]

molecules

MDPI

Review

Application of Molecular Methods in the Identification of Ingredients in Chinese Herbal Medicines

Ke Han [1],*, Miao Wang [2], Lei Zhang [2] and Chunyu Wang [3]

1   School of Computer and Information Engineering, Harbin University of Commerce, Harbin 150028, China
2   Life sciences and Environmental Sciences Development Center, Harbin University of Commerce, Harbin 150010, China; w1993817m@163.com (M.W.); 13212921382@163.com (L.Z.)
3   School of Computer Science and Technology, Harbin Institute of Technology, Harbin 150001, China; chunyu@hit.edu.cn
*   Correspondence: hanke@hrbcu.edu.cn; Tel.: +86-138-3611-6965

Received: 9 September 2018; Accepted: 20 October 2018; Published: 22 October 2018

Abstract: There are several kinds of Chinese herbal medicines originating from diverse sources. However, the rapid taxonomic identification of large quantities of Chinese herbal medicines is difficult using traditional methods, and the process of identification itself is prone to error. Therefore, the traditional methods of Chinese herbal medicine identification must meet higher standards of accuracy. With the rapid development of bioinformatics, methods relying on bioinformatics strategies offer advantages with respect to the speed and accuracy of the identification of Chinese herbal medicine ingredients. This article reviews the applicability and limitations of biochip and DNA barcoding technology in the identification of Chinese herbal medicines. Furthermore, the future development of the two technologies of interest is discussed.

Keywords: bioinformatics; identification of Chinese herbal medicines; biochip technology; DNA barcoding technology

1. Introduction

The traditional method of determining the authenticity of traditional Chinese medicine is to evaluate the color, shape, or nature of the medicinal materials using either physical/chemical methods or microscopy. These methods are still used, are relatively simple to carry out, and have led to substantial advancements in the screening of Chinese herbal medicines. However, these methods have some shortcomings and disadvantages [1]. It is assumed that there is a certain level of contamination of counterfeit and inferior medicinal ingredients in compound medications made by processing drugs of many ingredients, and it is difficult to identify the contaminating materials following the traditional method of identification. After thousands of years of development, traditional Chinese medicine has grown into a vast system, and its medicinal ingredients are numerous and complex [2]. Moreover, there are many disadvantages to the all too common practices of using the same name to describe many different herbs, or in having multiple names for a single medicinal material. This further increases the difficulty in the identification of Chinese herbal medicines [3,4]. In addition, other conventional Chinese herbal screening methods require the operators to have very rich work experience and expertise, or some subtle changes will lead to errors in the identification results [5]. Due to the error-prone nature of the identification process, it is often difficult to meet actual needs.

Bioinformatics is a discipline that integrates the most advanced knowledge of computer information technology and molecular biology [6–8]. This discipline uses the powerful data-processing ability of computers to analyze and compare proteomic, transcriptomic, genomic, and microorganism data, and so on, in order to identify and solve specific problems. At present, this technology has been applied to many fields, including Chinese herbal screening, as described in this review. It has also been applied to other medical fields, such as studying the effect of a drug on the prevention and treatment of diseases, and studying special drugs for cancer cell proliferation and death [9]. Therefore, it is of great significance and value for Chinese medicine modernization to further study the identification of Chinese herbal medicines by bioinformatics.

At present, bioinformatics assists in the identification and assessment of Chinese herbal medicines through the use of DNA barcoding technology and biochip technology. DNA barcoding seeks to identify biological material through sequencing a selected genetic marker, and then comparing that DNA sequence to sequences from the same genetic marker in other species [10,11]. Nowadays, this technology has been applied widely, and can be regarded as a mature tool for biological exploration and research. Not only can it be used to identify organisms that are difficult to distinguish by other methods, it can also be used to discover new organisms that have not been discovered by the biological research community to date [12]. If a specimen of a suspected new organism is collected, this technique can be used to extract its corresponding DNA barcode, and then compare and analyze it with DNA barcodes (nucleotide sequences) that are available in the public nucleotide sequence database, so that the relevant identity information of this organism can be determined step-by-step. On the other hand, biochip technology is a microarray that is composed of some biological components (nucleic acids, proteins, cells, etc.) wrapped on solid supports such as nylon film or a silicon wafer [13]. Through the use of a chip, this technology takes advantage of automation, speed, and big data. This review discusses the promotion and use of bioinformatics in the field of Chinese herbal medicine identification in recent years, and compares the advantages and disadvantages of various methods. With the broad applicability of these techniques, we hope that the identification of Chinese herbal medicines can be further developed.

2. Biochip Technology

Since the 1990s, due to continuous progress and development in science and technology, especially the rapid development of computer and network technology, a series of related disciplines emerged [14–18]. This also led to the development of biological chip technology for microanalysis research [19]. Biochip technology is a comprehensive technology discipline that combines chemical and physical technology. Biochip is the crystallization of DNA hybridization probe technology combined with semiconductor industry technology. The technique is to hybridize a large number of probe molecules with fluorescent-labeled DNA or other sample molecules (e.g., proteins, factors, or small molecules) after immobilizing them on the support. The number and sequence information of the sample molecules can be obtained by detecting the hybridization signal intensity of each probe molecule [20]. Fluorescence-labeled target molecules play a role in a variety of microbodies on the chip. A spectrophotometer is used for analysis of spectral/absorption characteristics, and the results will vary according to the intensity of the material. Next, a special instrument is used to collect and convert the data to be processed by a computer. In this way, the required biological information will be obtained [21,22]. Nowadays, biochips can be divided into tissue chips [23], protein chips, and gene chips. The latter two are widely used in Chinese herbal medicine identification.

2.1. Gene Chip

Gene chip technology is fast, highly efficient, automated, parallel, and economical. It has become an important technical method in the field of screening and the evaluation of inferior and counterfeit drugs of Chinese herbal medicine. The process involves obtaining the standard atlas of positive drugs and the atlas of prepared identification (query) drugs, and then analyzing and comparing the differences between the two. To obtain the two maps, one must first extract DNA from the corresponding samples, and then let them hybridize with DNA chips [24–27]. The detection steps of gene chip technology are shown in Figure 1.

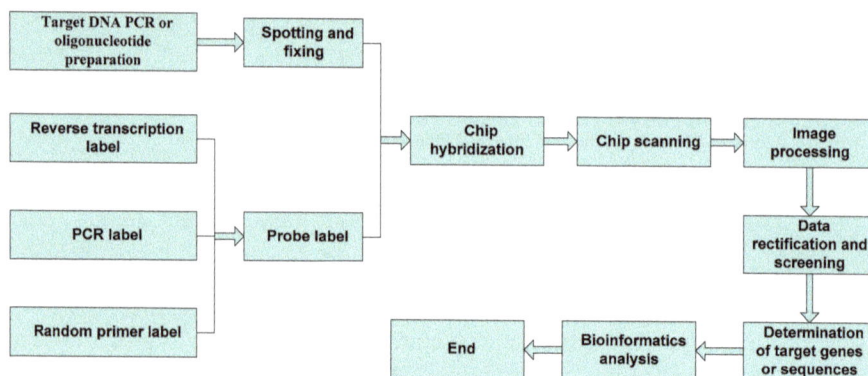

Figure 1. Steps of gene chip technology.

Zhang et al. [28] used gene chip technology to study the *Dendrobium nobile* (*Orchidaceae, Dendrobium Sw.*) species mixed in compound medication, and five species of *Dendrobium nobile* were successfully detected and loaded into China Pharmacopoeia (https://www.sinicave.com/pd_pharmacopoeia.cfm). It has been shown that this technique can be used in the field of Chinese herbal medicine taxonomy. Hao et al. [29] was the first to use AFLP (amplified fragment length polymorphism) technology in the study of the characteristics of Chinese herbal medicines, after which they drew the atlas of genetic diversity and AFLP fingerprints on the target taxa. This method can be used to easily and effectively differentiate between the characteristics of Chinese herbal medicines. The AFLP marker is then turned into a SCAR (sequence characterized amplified regions) marker, and the mounting glue can be recycled, or molecular cloning by PCR can be performed. The obtained PCR can be used to detect the characteristics of the target medicinal ingredient. Among expectorant and antitussive drugs, *Fritillaria cirrhosa* (*Liliaceae, Fritillaria*) has a positive effect, and is favored by many [30]. However, the high price of the highly sought-after drug encouraged sales of counterfeit or low-quality replicas at lower prices. In view of this, some scholars [31] began to sequence and study the 5S rRNA gene of *Fritillaria hupehensis, F. anhuiensis, F. thunbergii*, and *Fritillaria cirrhosa* by gene chip technology, and found that *Fritillaria cirrhosa* has a unique sequence of 5′-CTTTTGTCGATCA-3′, which is absent in other *Fritillaria* species. This sequence was used to make the gene chip. This technique is used to detect and extract the gene sequences of some tested products, and then compare them with the gene sequence of the positive *Fritillaria cirrhosa* control, to determine whether the sample that is being tested contains the expected product.

In another study, Chen et al. [32] isolated the genomic DNA of *Coptis chinensis* (*Ranunculus, Coptis Salisb.*) as a template, and then analyzed the many thermal cycle parameters in the ISSR (inter-simple sequence repeat) reaction system using two factors and single factors, as well as the effect and influence of some major components of amplification to identify the optimal conditions. Finally, the reaction system and amplification procedure that are suitable for the ISSR analysis of *Coptis chinensis* were established under optimum conditions. The establishment of this optimization system provides a standardized procedure for the identification and genetic diversity analysis of *Coptis chinensis* by the ISSR marker technique in the future.

The greatest advantage of a gene chip is high throughput. The process of marking and hybridizing probes for many genes can be completed in a single experiment; the degree of automation is high, and the data are objective and reliable. However, its greatest shortcoming is that the gene chip cannot be used to find new things; it can only be used to find things that have been found before (and have been printed on the chip).

2.2. Protein Chip

Protein chip technology is mainly used for protein analysis and exploration. The methodology involves the use of advanced microelectronics technology to analyze the surface of some carriers in order to establish a system that is suitable for microbiological research [33–35]. First, some known proteins are fixed to the carrier in order; then, the molecules can specifically interact with known molecules based on molecular properties, after which the molecules are ready for purification and subsequent treatment. Finally, the protein components can be more quickly and accurately screened [36,37].

Li et al. [38] used an NP10 chip and a protein chip combined with SELDI-TOF MS (surface-enhanced laser desorption/inionation-time of flight-mass spectra) to analyze the peptide composition and protein of tortoise shell glue. The peptide component/protein quality fingerprint of tortoise shell glue was obtained. This technology can be used for the digital analysis of tortoise shell glue. Wang et al. [39] obtained the protein/peptide from the dried and processed *Pheretima asiatica*, and analyzed the protein/peptide information with the surface-enhanced laser desorption/ionization time-of-flight mass spectrometry technique. Twenty-nine peaks of protein/peptide molecular weight were obtained. Among them, crude and dried *Pheretima asiatica* had 17 peaks of meaningful molecular weight, and processed *Pheretima asiatica* had 12 peaks. There was an amino acid residue among multiple adjacent protein peaks, and multi-groups of molecular weight of crude and processed *Pheretima asiatica* were extremely similar, indicating that these peptides may represent the same peptides. Molecular weight fingerprinting is obtained from the crude and processed *Pheretima asiatica* protein/peptide by laser desorption/ionization time-of-flight mass spectrometry; it can serve as a digitalized quality and control standard of *Pheretima asiatica*, and provide references for the further isolation, purification, and verification of the proteins/peptides associated with *Pheretima asiatica* function.

Protein chip technology has the following advantages: it is low-cost because it requires only a few samples and reagents in the process, it is more accurate and sensitive than conventional ELISA (enzyme linked immune sorbent assay), it is easy to operate due to automation, it is highly accurate, a large number of proteins can be quickly analyzed, and it has few sample requirements. The sample to be tested can be detected and analyzed by a protein chip only by simple processing. However, this technology has some shortcomings compared to gene chip technology: the purification of protein is more difficult, there is a lack of mature protein technology, the technology is inseparable from the function and re-modification of other proteins, and proteins are more variable among themselves. Therefore, protein chips are used less frequently than gene chips.

3. DNA Barcoding

Ever since Canadian scientist Paul Herbert [40] proposed DNA barcoding technology back in 2003, this diagnostic technology has been widely used in the field of species identification. DNA barcoding uses a short, unique DNA sequence to identify a species. Since traditional Chinese medicines are derived from a variety of animals, plants, and minerals, this technology can be used for the accurate identification and evaluation of medicinal ingredients. This technique is widely used in the field of biological and medicinal material identification because it is not influenced by the morphology of the sample or the surrounding environment [41]. The application of this technology has broadened widely, and has reached multiple fields and industries, such as ecology, development, and evolution, Chinese herbal medicine identification, and genetic identification [42]. A non-trivial proportion of the public fungal DNA sequences are compromised in terms of quality and reliability, contributing noise and bias to sequence-borne inferences. R. Henrik Nilsson et al. [43] discussed various aspects and pitfalls of sequence quality assessment. Based on their observations, they provided a set of guidelines to assist in the manual quality management of newly generated, near-full-length (Sanger-derived) fungal internal transcribed spacer (ITS) sequences, and to some extent also sequences of shorter read lengths, other genes or markers, and groups of organisms. A flow chart describing the process of the DNA barcoding molecular identification of Chinese herbal medicines is shown in Figure 2.

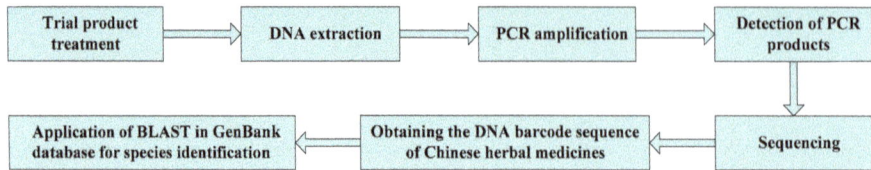

Figure 2. Molecular identification process of DNA barcoding in Chinese medicinal materials.

According to the selection of different DNA sequences, DNA barcoding can be roughly divided into four categories: (1) mitochondrial DNA barcode, (2) ribosomal DNA barcode, (3) chloroplast DNA barcode, and (4) DNA barcode combination identification. A suitable DNA barcode should meet the following conditions: high throughput so that it can be routinely sequenced in plant species, suitable for producing a high-sequence mass coverage of bidirectional sequences with minimal unsequenced bases, high resolution so that most species can be distinguished, and DNA fragments should be short enough so that degraded DNA can be amplified.

3.1. Mitochondrial DNA Barcode

Mitochondrial DNA barcode technology was the first technique to be used in the classification and identification of animal mitochondria. Mitochondrial COI (cytochrome oxidase) gene barcode technology has been an important method in identifying animal Chinese herbal medicine [44]. Hebert [40] eventually selected COI sequences, because COI sequences ensured sufficient variation and were easy to be amplified by universal primers, and there were few deletions and insertions in its own DNA sequence, so it was suitable for the analysis of closely related taxa.

Shi Linchun et al. [45], in order to distinguish *periostracum serpentis*(the skin that the snake shed, PS) from its adulterants, PCR amplified and sequenced COI sequences of 68 samples from 13 species. Furthermore, the DNA barcoding gap and phylogenetic cluster analysis were carried out. The results showed that three specimens of *periostracum serpentis Elaphe taeniura* (Cope), *E. carinata* (Guenther), and *Zaocys humnades* (Cantor)—had DNA barcode gaps, and they were separated into independent branches on the neighbor-joining (NJ) system clustering tree. As a DNA barcode, COI can not only identify three basic genera and species of Chinese herbal snakeskin, it can also distinguish between snakeskin and their easily confused products. This shows that DNA barcoding can be used for the identification of the snake shedding that is found in Chinese medicinal products. Zhang Hongyin et al. used the same method to identify the pseudo products of *centipede* [46], *deer medicine* [47], and *Gekko gecko* [48]. The results showed that the COI gene can be used as an effective marker for identifying at least metazoan Chinese medicinal ingredients.

The COI gene sequence exists in the vast majority of animal cells. There are only one set of genomic chromosomes in a cell, whereas there are hundreds of mitochondria per cell; that is why mitochondrial DNA is more easily recovered. The mitochondrial DNA mutation rate is 10 times greater than that of genomic DNA. According to this characteristic, it is easier for us to accurately differentiate species. However, one disadvantage of this technology is that it requires a lot of tedious work in the identification of species, and its identification results are not produced directly or quickly.

3.2. Ribosomal DNA Barcode

The nuclear ribosomal ITS (internal transcribed spacer) region contains the ITS1 intergenic region, the ITS2 intergenic region, and the 5.8S gene (ITS1-5.8S-ITS2) ranging in size from 400 bp to over 1000 bp. ITS has a high capacity for species identification and technical scalability. Ribosomal DNA is a polygene family, and ITS exists in highly repetitive ribosomal DNA [49,50]. Due to its fast evolution speed and short length, this genetic marker is widely used in the field of angiosperm branching analysis in plant systems and fungal metabarcoding [51].

Li et al. [52] used ITS primers to amplify the ITS sequences of *Hedyotis diffusa Willd* (*Rubiaceae*, *Cerastium*) and *Corymbose Hedyotis Herb* (*Oldenlandia corymbosa* L.), and found that there are obvious differences in the ITS sequences between the two plants. They then used the established phylogenetic tree to analyze the common *Herba Hedyotis* that is available on the market, and found that only the medicinal herbs purchased from Guangzhou were genuine; those purchased from Boston and Hong Kong were actually *Corymbose Hedyotis Herb*. Moreover, there was confusion regarding the samples that were used in previous studies of medicinal materials, indicating that it is difficult to distinguish between the two types of medicinal materials based on their shape and characteristics. Thus, the identification of Chinese medicinal materials assisted by bioinformatics is more accurate than the traditional identification method. Based on ITS2 barcode technology, Yu Junlin et al. [53] explored 17 samples of *Bupleurum longiradiatum Turcz* and 31 samples of *Bupleurum* species. They examined the intraspecific variation of two species by analyzing and studying two categories, and identified an accurate distinction between two major species. In order to ensure the accuracy of the study, the samples used in the study have been validated using BLAST (Basic Local Alignment Search Tool) following the DNA barcoding identification system for Chinese medicinal materials (http://www. tcmbarcode.cn). The ITS2 barcode sequence can accurately identify *Bupleurum* and *B. longiradiatum Turcz*. The minimum kimura 2-parameter(K2P) distance between *Bupleurum* and *B. longiradiatum Turcz* is far greater than the maximum K2P distance within the species of *Bupleurum*, and the NJ tree shows that the *B. longiradiatum Turcz* constitutes a single branch, which can be distinguished from *Bupleurum*. Therefore, *Bupleurum* and *B. longiradiatum Turcz* can be distinguished consistently and accurately using the ITS2 barcode. Shi Yuhu et al. [54] used the ITS2 sequence as a barcode to identify the herbal tea ingredient *Plumeriarubra* and its adulterants. Genomic DNAs from 48 samples were extracted; the ITS2 sequences were amplified and sequenced bidirectionally; and then they were assembled and obtained using CodonCode Aligner (https://www.codoncode.com/). The sequences were aligned using ClustalW, the genetic distances were computed by the K2P model, and the NJ phylogenetic tree was constructed using MEGA5.0. The results showed that the length of the ITS2 sequence of *P. rubra* were 244 bp. The intraspecific genetic distances (0–0.0166) were much smaller than interspecific ones between *P. rubra* and its adulterants (0.3208–0.6504). The NJ tree indicated that *P. rubra* and its adulterants could be distinguished clearly. Therefore, using the ITS2 barcode can accurately and effectively distinguish the herbal tea ingredient *P. rubra* from its adulterants, which provides a new molecular method to identify *P. rubra* and ensure its safety in use.

ITS sequences are usually used to distinguish some species that are closely related to each other. It may be difficult to identify the different geographical distribution and host types of some fungi. Not all 18 S and 28 rRNA databases have been established, and many of the data are still lacking, which results in the inability to rapidly identify the required ITS fragments, thus inevitably affecting the application of ITS methods.

3.3. Chloroplast DNA Barcode

psbA-trnH intervals in common flowering plants are between 340–660 bp. This sequence interval was compared with nine other genetic markers (matK, rbcL, and ITS are included) with a discrimination efficiency of 83% and an amplification efficiency of 100% [55–58]. Zhang Yaqin et al. [59] used psbA-trnH sequence technology to study and explore *Pyrrosia* in order to distinguish the appearance and form of counterfeit herbs that are similar to those of *Pyrrosia*. They collected partial sequences of psbA-trnH in the chloroplast genes of *Pyrrosias* and some counterfeit products. Next, the group used two conventional methods based on the partial sequence of the collected psbA-trnH: the minimum distance method and the similarity search method. It was found that the former method could not distinguish between two *Pyrrosia* species; however, the use of the psbA-trnH sequence clearly distinguishes some other counterfeit products mixed with other kinds of *Pyrrosias*. This study shows that the psbA-trnH sequence can effectively distinguish and identify genuine and counterfeit products of pteridophytes.

As one of the fastest evolving regions in plant chloroplasts, the matK gene is about 1500 bp [60]. Hilu and Lahaye [61] also pointed out that it can be implemented in a single fragment when choosing a barcode of a plant. The fragment is then added according to the complexity of the group. Genievskaya Y et al. [62] identified species defined by morphological traits using sequences of the nuclear ribosomal DNA ITS1-5.8S-ITS2 region and matK. The polymorphic sequence positions in Kazakh populations and GenBank (Benson et al. 2017) references were acquired by comparison with GenBank sequences, which identified a difference between local populations of sand rice (*Agriophyllum squarrosum*). ITS and matK sequence analysis revealed a segregation of *Agriophyllum squarrosum* (L.) *Moq* from *A. minus* into separate branches in maximum-likelihood dendrograms. ITS analysis can be used to characterize the populations of *A. squarrosum* growing far away from each other. The data obtained in this study laid the foundation for the further study of *A. squarrosum* populations, and summed up the advantages and disadvantages of this technology. Wang Xiaoming et al. [63] used the matK sequence method in their analysis and exploration of *Herba Abri* (*Leguminous, Abrus* L.), and summarized the advantages and disadvantages of this technology in the applications of the DNA barcode in this plant. The modified CTAB (Cetyltriethyl Ammnonium Bromide) method was used to extract the total DNA of nine kinds of *Herba Abri* from different regions, and the matK sequence was amplified using the universal primers of leguminous plants. After that, both K2P genetic distance calculation and the creation of the NJ tree showed that the ITS2 sequences can be used as the DNA barcode sequence of the *Fabaceae* plants. The results showed that the total length of the matK sequence was 889–895 bp; the genetic distance between different plants was far greater than the genetic distance within the populations of the same plant, and even the smallest interspecific genetic distance still exceeded the maximum intraspecific genetic distance. Therefore, matK sequence technology can serve as a DNA barcode for leguminous plants.

rbcL is a fragment of the coding region of a chloroplast gene. The rbcL fragment has a low species resolution, but it is of relatively high species resolution in angiosperms [64–66]. Chen Jianxiong et al. [67] collected nine samples of Chinese lobelia of different origins and seven samples of *Mazus japonicus*, and extracted the total DNA from all of the samples collected. The rbcL fragment in the chloroplast DNA of the sample was sequenced, and Clustal X 2.1 software (University College Dublin, Dublin, Ireland) was used for multiple sequence alignments. The NJ clustering feature of MEGA 5.0 software (Center of Evolutionary Functional Genomics Biodesign Institute Arizona State University, Phoenix, AZ, USA) was then used for cluster analysis. They designed specific primers identified by SNP micropoints for two groups of samples, established a specific PCR identification method, and used SYBR Green I (Molecular Probes) dye to establish a rapid detection method for two kinds of Chinese herbs. They successfully identified the Chinese lobelia and *Mazus japonicus*. The results showed that rbcL has strong molecular identification ability and rbcL is easy to amplify and compare. So, rbcL was selected as the DNA barcode for species identification. Huang Qionglin et al. [68] identified *Nervilia fordii (Hance) Schltr.* and its adulterants byrbcLsequencing. They used a commercial kit to extract genomic DNA from fresh leaves; PCR amplification and sequencing were conducted with a pair of universal primers. DNAMAN (https://www.lynnon.com/); Clustal X (University College Dublin, Dublin, Ireland) and MEGA 4.0 software (Center of Evolutionary Functional Genomics Biodesign Institute Arizona State University, Phoenix, AZ, USA) were used for sequence alignment, genetic distance analysis, and clustering analysis. They acquired 502 bp sequences of the rbcL gene from *N. fordii* and its adulterants. Three types of *N. fordii* showed completely consistent sequence data, and differences in five sites were shown between *N. fordii* and *N. plicata*. The interspecific variations were larger than the intraspecific ones. In the cluster dendrogram, all of the species were monophyletic and distinguished from the others. The results showed that the rbcL gene can be used as a DNA barcode to identify *N. fordii* and its counterfeit.

3.4. DNA Barcode Combination Identification

Studies have found that it is difficult to identify some species accurately if only one genetic marker is used in the identification of Chinese herbal medicines. Therefore, it is not possible to rely on a single genetic marker to identify all of the species, especially for those species with complex genetic backgrounds [69–72]. Therefore, a combination of DNA barcodes can be used to accurately identify species.

The CBOL Plant Working Group [73] found that rbcL and matK were suitable barcodes for plants, because these two barcodes were used to successfully identify 550 species from 907 samples, with a success rate of roughly 72%. Moreover, the barcode can become a basic criterion for the molecular identification of species, and it can also provide references for discovering hidden species. Yong et al. [74] tested the universality of rbcL, matK, trnH-psbA, and ITS as DNA barcodes for tree species, and then examined the accuracy of the phylogenetic inference and species identification in three tropical cloud forests (Table 1). Their results suggested that rbcL and trnH-psbA should be adopted as the standard DNA barcode for tree species in tropical cloud forests. The success rates of identifying four fragments were all higher than 41.00%, demonstrating that these fragments are candidates for use in species identification. They used random fragment combinations of rbcL, matK, and trnH-psbA to infer phylogenetic relationships, and established the optimal evolutionary tree with high supporting values in tropical cloud forests.

Table 1. Results of PCR amplification success rate and DNA sequencing rate for rbcL, matK, trnH-psbA, and internal transcribed spacer (ITS), respectively.

	RbcL	MatK	TrnH-psbA	ITS
Success rates of PCR amplification	75.26% ± 3.65%	57.24% ± 4.42%	79.28% ± 7.08%	50.31% ± 6.64%
Rates of DNA sequencing	63.84% ± 4.32%	50.82% ± 4.36%	72.87% ± 11.37%	45.15% ± 8.91%

Priyanka et al. [75] studied the steno-endemic species of the genus *Decalepis* (*Decalepisar ayalpathra Venter*, which is locally known as Amirthapala, is a steno-endemic species in the eastern and western ghats of peninsular India), and found the corresponding DNA barcode, which can be used to monitor and stop the illegal trade of these endangered species. They used rbcL, matK, psbA-trnH, ITS, and ITS2 as DNA barcode candidates. The average intraspecific variation was 0–0.27%, which was less than the distance to the nearest neighbor (0.4–11.67%) with ITS and matK. Finally, they combined rbcL, matK, andITS to produce 100% species resolution using the PAUP (http://paup.phylosolutions.com/, Phylogenetic Analysis Using PAUP) and BOLD (http://v3.boldsystems.org/, Bold systems) methods with the least number of marker combinations to support a character-based approach. They found that the most advantageous barcode datasets were achieved by combining rbcL, matK, ITS, mat, ITS, and ITS2, with a consistency index (CI) of 85% and 90%, respectively. They included 2106 characters in the former dataset for parsimony analysis, among which 103 were parsimony informative, and 18 variable characters were parsimony-uninformative. The 1836 total characters of the latter dataset contributed 146 informative characters. Therefore, the rbcL, matK, and ITS combination is considered to be the best choice for species resolution in the genus *Decalepis*. DNA barcoding has greatly improved species identification and resolution.

Compared with use of single DNA barcodes for identifying species, combining DNA barcodes can greatly improve species resolution. However, researchers need to find the right combination of DNA barcodes. The use of DNA barcode combinations for species identification is more complicated and costlier than use of a single barcode. Therefore, different identification methods should be selected for different species, so as to achieve quick and convenient identification results.

3.5. The Limitations of DNA Barcoding

Although DNA barcoding has many advantages in the identification of Chinese herbal medicine, it still has some limitations. In fungi, the ITS region has been sequenced for something similar to

1% of the estimated number of extant species. This makes it tricky to compare newly generated sequences to the entries in GenBank. First, traditional Chinese herbal medicines are usually made of dead animals and plants. Due to collection, processing, storage time, and storage conditions, the DNA macromolecules in the test sample may have been destroyed and degraded prior to analysis. Second, it is difficult to obtain the template that is needed to identify DNA molecules when only small fragments are retained in the sample. Third, although DNA barcoding technology can effectively identify and evaluate plant and animal material in medicinal ingredients, the technology cannot identify the mineral components. Finally, DNA barcoding technology requires technical expertise and is costly.

In order to solve these problems, we should promote basic research, enrich the genome sequencing data, and find small fragments of molecular markers. When dealing with medicinal materials, we should strictly follow standard methods and conditions to avoid damaging the DNA. Results can be validated by using multiple fragments or molecular markers to obtain more accurate DNA fragments, and we should simply test the plant or animal tissues or extracts upfront, before they are used to make medicines. Previous researchers used ClustalW to multiple sequence alignment, but this method is a tool that is long-since obsolete. The user should go for a recent (and readily updated) one, rather than relying on old programs. We recommend any of MAFFT (Katoh et al. 2010), Muscle (Edgar et al. 2004), and PRANK (Löytynoja et al. 2005) for large or otherwise non-trivial sequence datasets [76]. For phylogenetic analysis, NJ is a long-since obsolete tool; Bayesian inference in MrBayes is a whole lot more powerful. In future research, researchers can use these better tools to improve identification.

4. Future Perspectives

Chinese medicine reveals a rich history of our people's long-term struggle against diseases, and has made great contributions to the prosperity and rebirth of the Chinese nation. Starting from Shennong's Herbal Classic (https://en.wikipedia.org/wiki/Shennong_Ben_Cao_Jing), Chinese medicine plays a very important role in our historical society. However, after a period of instability, the development of traditional Chinese medicine is regrettably at a standstill. With improvements in people's standards of living and access to technology, medical practitioners and researchers in China and abroad have taken a new interest in Chinese herbal medicine. However, this increased interest in traditional Chinese medicine has led to the emergence of counterfeit and low-quality Chinese herbal medicine in the market, which has greatly reduced the efficacy of Chinese herbal medicine. Therefore, it is essential that we develop a fast and accurate method to identify counterfeit and low-quality Chinese herbal medicines.

Chinese herbal medicines can be identified based on three aspects: appearance, chemical composition, and molecular characteristics. Many medicinal herbs are similar in appearance and are not easy to differentiate, which increases the risk of false identification. Since many Chinese herbal medicines are made up of many kinds of materials that are subsequently ground into a powder, it is very difficult to identify Chinese medicinal ingredients according to chemical constituents. Bioinformatics has brought new opportunities for the identification of traditional Chinese herbal medicine ingredients using high throughput and big data. At present, there are mainly two methods of bioinformatics identification of Chinese herbal medicine: biochip technology and DNA barcoding. The use of DNA barcoding is more extensive, because this method has different identification methods for different kinds of Chinese herbal medicines, making it a more targeted approach. Moreover, the combination identification method of DNA barcoding allows for more species to be identified accurately. The first step toward the DNA barcoding of Chinese herbal medicine is to extract different DNA fragments from different species and select the most suitable DNA fragment, followed by PCR amplification to detect PCR products. Finally, the target bands should be bidirectionally sequenced by DNA sequencing. This method is not without its limitations, and cannot be used to identify all of the species. Therefore, future studies should combine traditional Chinese medicine identification methods, biochip technology, DNA barcoding technology, and high-throughput sequencing-driven metabarcoding to

Molecules **2018**, *23*, 2728

identify Chinese medicinal ingredients and obtain more accurate and rapid identification results. We believe that through the continuous efforts of researchers, the bioinformatics-assisted identification of Chinese herbal medicines will make substantial advancements. It will one day be possible to quickly identify counterfeit and inferior medicinal ingredients from a large number of mixed Chinese medicinal materials, thus standardizing the market of Chinese herbal medicines.

Funding: The work was supported by the Young Reserve Talents Research Foundation of Harbin Science and Technology Bureau, grant number2015RQQXJ082.

Conflicts of Interest: The authors declare no conflict of interest. The founding sponsors had no role in the design of the study; in the collection, analyses, or interpretation of data; in the writing of the manuscript, and in the decision to publish the results.

References

1. Zhao, Y.; Hellum, B.H.; Liang, A.H.; Nilsen, O.G. The in vitro inhibition of human cyp1a2, cyp2d6 and cyp3a4 by tetrahydropalmatine, neferine and berberine. *Phytother. Res.* **2012**, *26*, 277–283. [CrossRef] [PubMed]
2. Chen, R.; Dong, J.; Cui, X.; Wang, W.; Yasmeen, A.; Deng, Y.; Zeng, X.M.; Tang, Z. DNA based identification of medicinal materials in chinese patent medicines. *Sci. Rep.* **2012**, *2*, 958. [CrossRef] [PubMed]
3. Han, J.; Song, J.; Yao, H.; Chen, X.; Chen, S. Comparison of DNA barcoders in identifying medicinal materials. *China J. Chin. Mater. Med.* **2012**, *37*, 1056–1061.
4. Qian, Z.; Lin, C.R.; Xiao, P.G. Identification and standard of traditional Chinese Medicine. *China J. Chin. Mater. Med.* **2014**, *39*, 2153–2154.
5. Read, T.D.; Salzberg, S.L.; Pop, M.; Shumway, M.; Umayam, L.; Jiang, L.X.; Holtzapple, E.; Busch, J.D.; Smith, K.L.; Schupp, J.M.; et al. Comparative genome sequencing for discovery of novel polymorphisms in bacillus anthracis. *Science* **2002**, *296*, 2028–2033. [CrossRef] [PubMed]
6. Giacomelli, L.; Covani, U. Bioinformatics and data mining studies in oral genomics and proteomics: New trends and challenges. *Open Dent. J.* **2010**, *4*, 67–71. [CrossRef] [PubMed]
7. Wang, J.X.; Li, M.; Yu, Z.G. Nonlinear science and network methods for prediction problems in bioinformatics and systems biology. *Curr. Bioinform.* **2016**, *11*, 154–155. [CrossRef]
8. Prejzendanc, T.; Wasik, S.; Blazewicz, J. Computer representations of bioinformatics models. *Curr. Bioinform.* **2016**, *11*, 551–560. [CrossRef]
9. Gao, J.Z.; Cui, W.; Sheng, Y.J.; Ruan, J.S.; Kurgan, L. Psionplus: Accurate sequence-based predictor of ion channels and their types. *PLoS ONE* **2016**, *11*, e0152964. [CrossRef] [PubMed]
10. Hou, D.; Song, J.; Shi, L.; Ma, X.; Xin, T.; Han, J.; Xiao, W.; Sun, Z.; Cheng, R.; Yao, H. Stability and accuracy assessment of identification of traditional chinese materia medica using DNA barcoding: A case study on flos lonicerae japonicae. *BioMed. Res. Int.* **2013**, *2013*, 549037. [CrossRef] [PubMed]
11. Sun, Z.Y.; Chen, S.L. Identification of cortex herbs using the DNA barcode nrits2. *J. Nat. Med.* **2013**, *67*, 296–302. [CrossRef] [PubMed]
12. Cao, S.P.; Guo, L.N.; Luo, H.M.; Yuan, H.; Chen, S.Y.; Zheng, J.; Lin, R.C. Application of coi barcode sequence for the identification of snake medicine (zaocys). *Mitochondrial DNA Part A* **2016**, *27*, 483–489. [CrossRef] [PubMed]
13. Ma, J.W.; Gu, H. A novel method for predicting protein subcellular localization based on pseudo amino acid composition. *BMB Rep.* **2010**, *43*, 670–676. [CrossRef] [PubMed]
14. Avin, F.A.; Bhassu, S.; Tan, Y.S.; Shahbazi, P.; Vikineswary, S. Molecular divergence and species delimitation of the cultivated oyster mushrooms: Integration of igs1 and its. *Sci. World J.* **2014**, *10*, 793414. [CrossRef] [PubMed]
15. Liu, Y.; Zeng, X.; He, Z.; Zou, Q. Inferring microrna-disease associations by random walk on a heterogeneous network with multiple data sources. *Trans. Comput. Biol. Bioinform.* **2017**, *14*, 905–915. [CrossRef] [PubMed]
16. Wang, S.; Huang, G.; Hu, Q.; Zou, Q. A network-based method for the identification of putative genes related to infertility. *Biochim. Biophys. Acta Gen. Subj.* **2016**, *1860*, 2716–2724. [CrossRef] [PubMed]
17. Zou, Q.; Li, J.; Hong, Q.; Lin, Z.; Wu, Y.; Shi, H.; Ju, Y. Prediction of microrna-disease associations based on social network analysis methods. *BioMed Res. Int.* **2015**, *2015*, 810514. [CrossRef] [PubMed]

18. Zou, Q.; Li, J.; Wang, C.; Zeng, X. Approaches for recognizing disease genes based on network. *BioMed. Res. Int.* **2014**, *2014*, 416323. [CrossRef] [PubMed]
19. Chen, G.X.; Zhao, J.C.; Zhao, X.; Zhao, P.S.; Duan, R.J.; Nevo, E.; Ma, X.F. A psammophyte *Agriophyllum squarrosum* (L.) moq.: A potential food crop. *Genet. Resour. Crop Evol.* **2014**, *61*, 669–676. [CrossRef]
20. Tamura, K.; Stecher, G.; Peterson, D.; Filipski, A.; Kumar, S. Mega6: Molecular evolutionary genetics analysis version 6.0. *Mol. Biol. Evol.* **2013**, *30*, 2725–2729. [CrossRef] [PubMed]
21. Smith, D.R. Mutation rates in plastid genomes: They are lower than you might think. *Genome Biol. Evol.* **2015**, *7*, 1227–1234. [CrossRef] [PubMed]
22. Darriba, D.; Taboada, G.L.; Doallo, R.; Posada, D. Jmodeltest 2: More models, new heuristics and parallel computing. *Nat. Methods* **2012**, *9*, 772. [CrossRef] [PubMed]
23. Bucher, C.; Torhorst, J.; Bubendorf, L.; Schraml, P.; Kononen, J.; Moch, H.; Mihatsch, M.; Kallioniemi, O.; Sauter, G. Tissue microarrays (tissue chips) for high-throughput cancer genetics: Linking molecular changes to clinical endpoints. *Am. J. Hum. Genet.* **1999**, *65*, A10.
24. Li, P.; Guo, M.; Wang, C.; Liu, X.; Zou, Q. An overview of snp interactions in genome-wide association studies. *Brief. Funct. Genom.* **2015**, *14*, 143–155. [CrossRef] [PubMed]
25. Yao, H.; Song, J.Y.; Liu, C.; Luo, K.; Han, J.P.; Li, Y.; Pang, X.H.; Xu, H.X.; Zhu, Y.J.; Xiao, P.G.; et al. Use of its2 region as the universal DNA barcode for plants and animals. *PLoS ONE* **2010**, *5*, e13102. [CrossRef] [PubMed]
26. Mei, Q.L.; Zhang, H.X.; Liang, C. A discriminative feature extraction approach for tumor classification using gene expression data. *Curr. Bioinform.* **2016**, *11*, 561–570. [CrossRef]
27. Hossen, M.B.; Siraj-Ud-Doulah, M. Identification of robust clustering methods in gene expression data analysis. *Curr. Bioinform.* **2017**, *12*, 558–562. [CrossRef]
28. Zhang, Y.B.; Wang, J.; Wang, Z.T.; But, P.P.H.; Shaw, P.C. DNA microarray for identification of the herb of dendrobium species from chinese medicinal formulations. *Planta Med.* **2003**, *69*, 1172–1174. [PubMed]
29. Liu, Z.Q.; Hao, M.G. Application of the AFLP combining with SCAR to identify the geoherbalism of traditional Chinese medicines. *Chin. J. Mod. Chin. Med.* **2005**, *1*, 1.
30. Li, H.; Li, Z.; Song, H. Study on quality standard of biyanling tablet. *China Pharm.* **2012**, *21*, 31–32.
31. Gong, B.Q.; Luo, G.H.; Wang, W. Research Progress on the Identification Methods of Fritillaria. *J. Anhui Agric. Sci.* **2009**, *37*, 1603–1604.
32. Chen, D.; Li, L.; Lu, C.; Zhong, G.; Qu, X.; Peng, R. Study on optimization of issr reaction conditions for coptis chinensis franch. *Bull. Bot. Res.* **2007**, *27*, 77–81.
33. Xin, T.Y.; Li, X.J.; Yao, H.; Lin, Y.L.; Ma, X.C.; Cheng, R.Y.; Song, J.Y.; Ni, L.H.; Fan, C.Z.; Chen, S.L. Survey of commercial rhodiola products revealed species diversity and potential safety issues. *Sci. Rep.* **2015**, *5*, 8337. [CrossRef] [PubMed]
34. Miao, C.P.; Li, X.H.; Jiang, D.M. Spatial variability of agriophyllum squarrosum across scales and along the slope on an active sand dune in semi-arid china. *Arid Land Res. Manag.* **2013**, *27*, 231–244. [CrossRef]
35. Zhao, P.S.; Zhang, J.W.; Qian, C.J.; Zhou, Q.; Zhao, X.; Chen, G.X.; Ma, X.F. Snp discovery and genetic variation of candidate genes relevant to heat tolerance and agronomic traits in natural populations of sand rice (agriophyllum squarrosum). *Front. Plant Sci.* **2017**, *8*, 536. [CrossRef] [PubMed]
36. Wang, J.; Xia, N.H. Quantitative Analysis of Morphological Characters of Ardisia crenata Complex (Primulaceae). *J. Trop. Subtrop. Bot.* **2013**, *21*, 543–548.
37. Mishra, P.; Kumar, A.; Nagireddy, A.; Mani, D.N.; Shukla, A.K.; Tiwari, R.; Sundaresan, V. DNA barcoding: An efficient tool to overcome authentication challenges in the herbal market. *Plant Biotechnol. J.* **2016**, *14*, 8–21. [CrossRef] [PubMed]
38. Li, D.Z.; Gao, L.M.; Li, H.T.; Wang, H.; Ge, X.J.; Liu, J.Q.; Chen, Z.D.; Zhou, S.L.; Chen, S.L.; Yang, J.B.; et al. Comparative analysis of a large dataset indicates that internal transcribed spacer (its) should be incorporated into the core barcode for seed plants. *Proc. Natl. Acad. Sci. USA* **2011**, *108*, 19641–19646. [PubMed]
39. Wang, R.G.; Li, C.M.; Wang, L.Y.; Liu, X.L.; Liu, H.P.; You, Z.L. Biological characteristics of dried and processed pheretima asiatica by the interaction of protein/peptide in traditional chinese medicine with protein chips as carriers. *J. Clin. Rehabil. Tissue Eng. Res.* **2008**, *12*, 2489–2492.
40. Hebert, P.D.N.; Cywinska, A.; Ball, S.L.; DeWaard, J.R. Biological identifications through DNA barcodes. *Proc. R. Soc. B Biol. Sci.* **2003**, *270*, 313–321. [CrossRef] [PubMed]

41. Miao, M.; Warren, A.; Song, W.B.; Wang, S.; Shang, H.M.; Chen, Z.G. Analysis of the internal transcribed spacer 2 (its2) region of scuticociliates and related taxa (ciliophora, oligohymenophorea) to infer their evolution and phylogeny. *Protist* **2008**, *159*, 519–533. [CrossRef] [PubMed]

42. Coleman, A.W. Pan-eukaryote its2 homologies revealed by rna secondary structure. *Nucleic Acids Res.* **2007**, *35*, 3322–3329. [CrossRef] [PubMed]

43. Nilsson, R.H.; Tedersoo, L.; Abarenkov, K.; Ryberg, M.; Kristiansson, E.; Hartmann, M.; Schoch, C.L.; Nylander, J.A.A.; Bergsten, J.; Porter, T.M. Five simple guidelines for establishing basic authenticity and reliability of newly generated fungal its sequences. *Mycokeys* **2012**, *4*, 37–63. [CrossRef]

44. Trias-Blasi, A.; Vorontsova, M. Plant identification is key to conservation. *Nature* **2015**, *521*, 161. [CrossRef] [PubMed]

45. Shi, L.C.; Chen, J.; Liu, D.; Zhang, Y.H.; Jia, J.; Zhang, H.; Yao, H. Molecular Identification of Serpentis Periostracum and Its Adulterants Based on COI Sequence. *World Sci. Technol. Modemization Tradit. Chin. Med.* **2014**, *2*, 284–287.

46. Zhang, H.; Chen, J.; Jia, J.; Liu, D.; Shi, L.; Zhang, H.; Song, J.; Yao, H. Identification of scolopendra subspinipes mutilans and its adulterants using DNA barcode. *China J. Chin. Mater. Med.* **2014**, *39*, 2208–2211.

47. Liu, D.; Qian, Q.; Zhang, H.; Zeng, D.; Jia, J.; Zhang, H. Molecular Identification of the Traditional Chinese Medicine of the Deers Using COI Barcode Sequence. *World Sci. Technol. Modemization Tradit. Chin. Med.* **2014**, *2*, 274–278.

48. Zhang, H.; Shi, L.; Liu, D.; Jia, J.; Chen, J.; Zhang, H. Identification of Gekko geeko Linnaeus and Adulterants Using the COI Barcode. *World Sci. Technol.-Mod. Tradit. Chin. Med.* **2014**, *2*, 269–273.

49. Chen, S.L.; Yao, H.; Han, J.P.; Liu, C.; Song, J.Y.; Shi, L.C.; Zhu, Y.J.; Ma, X.Y.; Gao, T.; Pang, X.H.; et al. Validation of the its2 region as a novel DNA barcode for identifying medicinal plant species. *PLoS ONE* **2010**, *5*, e8613. [CrossRef] [PubMed]

50. Holtken, A.M.; Schroder, H.; Wischnewski, N.; Degen, B.; Magel, E.; Fladung, M. Development of DNA-based methods to identify cites-protected timber species: A case study in the meliaceae family. *Holzforschung* **2012**, *66*, 97–104.

51. Erickson, D.L.; Jones, F.A.; Swenson, N.G.; Pei, N.; Bourg, N.A.; Chen, W.; Davies, S.J.; Ge, X.J.; Hao, Z.; Howe, R.W.; et al. Comparative evolutionary diversity and phylogenetic structure across multiple forest dynamics plots: A mega-phylogeny approach. *Front. Genet.* **2014**, *5*, 358. [CrossRef] [PubMed]

52. Li, M.; Jiang, R.W.; Hon, P.M.; Cheng, L.; Li, L.L.; Zhou, J.R.; Shaw, P.C.; But, P.P.H. Authentication of the anti-tumor herb baihuasheshecao with bioactive marker compounds and molecular sequences. *Food Chem.* **2010**, *119*, 1239–1245. [CrossRef]

53. Yu, J.; Zhao, S.; Ren, M.; Qian, Q.; Pang, X. Identification of bupleurum chinense and b. Longiradiatum based on its2 barcode. *China J. Chin. Mater. Med.* **2014**, *39*, 2160–2163.

54. Shi, Y.H.; Sun, W.; Fang, G.H.; Deng, R.B.; Xu, W.L.; Huang, X.D.; Weng, S.Q.; Li, C.Y.; Chen, S.L. Identification of herbal tea ingredient Plumeria rubra and its adulterants using DNA barcoding. *China J. Chin. Mater. Med.* **2014**, *39*, 2199–2203.

55. Liu, J.; Yan, H.F.; Newmaster, S.G.; Pei, N.C.; Ragupathy, S.; Ge, X.J. The use of DNA barcoding as a tool for the conservation biogeography of subtropical forests in china. *Divers. Distrib.* **2015**, *21*, 188–199. [CrossRef]

56. Pei, N.C.; Erickson, D.L.; Chen, B.F.; Ge, X.J.; Mi, X.C.; Swenson, N.G.; Zhang, J.L.; Jones, F.A.; Huang, C.L.; Ye, W.H.; et al. Closely-related taxa influence woody species discrimination via DNA barcoding: Evidence from global forest dynamics plots. *Sci. Rep.* **2015**, *5*, 15127. [CrossRef] [PubMed]

57. Wang, X.; Long, W.; Yang, X.; Xiong, M.; Kang, Y.; Huang, J.; Wang, X.; Hong, X.; Zhou, Z.; Lu, Y.; et al. Patterns of plant diversity within and among three tropical cloud forest communities in hainan island. *Chin. J. Plant Ecol.* **2016**, *40*, 469–479.

58. Burgess, K.S.; Fazekas, A.J.; Kesanakurti, P.R.; Graham, S.W.; Husband, B.C.; Newmaster, S.G.; Percy, D.M.; Hajibabaei, M.; Barrett, S.C.H. Discriminating plant species in a local temperate flora using the rbcl plus matk DNA barcode. *Methods Ecol. Evol.* **2011**, *2*, 333–340. [CrossRef]

59. Zhang, Y.; Shi, Y.; Song, M.; Lin, Y.; Ma, X.; Sun, W.; Xiang, L.; Liu, X. Identification of pyrrosiae folium and its adulterants based on psba-trnh sequence. *China J. Chin. Mater. Med.* **2014**, *39*, 2222–2226.

60. Lu, M.M.; Ci, X.Q.; Yang, G.P.; Li, J. DNA barcoding of subtropical forest trees—A study from ailao mountains nature reserve, Yunnan, China. *Plant Divers. Resour.* **2013**, *35*, 733–741.

61. Lahaye, R.; Van der Bank, M.; Bogarin, D.; Warner, J.; Pupulin, F.; Gigot, G.; Maurin, O.; Duthoit, S.; Barraclough, T.G.; Savolainen, V. DNA barcoding the floras of biodiversity hotspots. *Proc. Natl. Acad. Sci. USA* **2008**, *105*, 2923–2928. [CrossRef] [PubMed]

62. Genievskaya, Y.; Abugalieva, S.; Zhubanysheva, A.; Turuspekov, Y. Morphological description and DNA barcoding study of sand rice (agriophyllum squarrosum, chenopodiaceae) collected in kazakhstan. *BMC Plant Biol.* **2017**, *17*, 177. [CrossRef] [PubMed]

63. Wang, X.; Niu, X.; Wei, N.; Ji, K. Its sequence used as DNA barcode applied in chinese herb abrus cantoniensis. *Genom. Appl. Biol.* **2012**, *31*, 603–608.

64. Huang, X.C.; Ci, X.Q.; Conran, J.G.; Li, J. Application of DNA barcodes in asian tropical trees—A case study from xishuangbanna nature reserve, Southwest China. *PLoS ONE* **2015**, *10*, e0129295. [CrossRef] [PubMed]

65. Aldrich, J.; Cherney, B.W.; Merlin, E.; Christopherson, L. The role of insertions/deletions in the evolution of the intergenic region between psba and trnh in the chloroplast genome. *Curr. Genet.* **1988**, *14*, 137–146. [CrossRef] [PubMed]

66. Fazekas, A.J.; Burgess, K.S.; Kesanakurti, P.R.; Graham, S.W.; Newmaster, S.G.; Husband, B.C.; Percy, D.M.; Hajibabaei, M.; Barrett, S.C.H. Multiple multilocus DNA barcodes from the plastid genome discriminate plant species equally well. *PLoS ONE* **2008**, *3*, e2802. [CrossRef] [PubMed]

67. Chen, J.X.; Wei, Y.C.; Huang, Z.H.; Xu, H.L.; Lu, W.; Liang, Y.C. Rapid PCR Differentiation between Lobelia chinensis and Mzaus pumilus. *Fujian J. Agric. Sci.* **2017**, *32*, 730–733.

68. Huang, Q.; Liang, L.; Rui, H.E.; Zhan, R.; Chen, W. Molecular identification of nervilia fordii (hance) schltr. And its adulterants by rbcl gene. *Chin. J. Trop. Crop.* **2012**, *33*, 1630–1634.

69. Devey, D.S.; Chase, M.W.; Clarkson, J.J. A stuttering start to plant DNA barcoding: Microsatellites present a previously overlooked problem in non-coding plastid regions. *Taxon* **2009**, *58*, 7–15.

70. Simeone, M.C.; Piredda, R.; Papini, A.; Vessella, F.; Schirone, B. Application of plastid and nuclear markers to DNA barcoding of euro-mediterranean oaks (quercus, fagaceae): Problems, prospects and phylogenetic implications. *Bot. J. Linn. Soc.* **2013**, *172*, 478–499. [CrossRef]

71. Saarela, J.M.; Sokoloff, P.C.; Gillespie, L.J.; Consaul, L.L.; Bull, R.D. DNA barcoding the canadian arctic flora: Core plastid barcodes (rbcl plus matk) for 490 vascular plant species. *PLoS ONE* **2013**, *8*, e77982. [CrossRef] [PubMed]

72. Letcher, S.G. Phylogenetic structure of angiosperm communities during tropical forest succession. *Proc. R. Soc. B Biol. Sci.* **2010**, *277*, 97–104. [CrossRef] [PubMed]

73. Hollingsworth, P.M.; Forrest, L.L.; Spouge, J.L.; Hajibabaei, M.; Ratnasingham, S.; van der Bank, M.; Chase, M.W.; Cowan, R.S.; Erickson, D.L.; Fazekas, A.J.; et al. A DNA barcode for land plants. *Proc. Natl. Acad. Sci. USA* **2009**, *106*, 12794–12797.

74. Kang, Y.; Deng, Z.Y.; Zang, R.G.; Long, W.X. DNA barcoding analysis and phylogenetic relationships of tree species in tropical cloud forests. *Sci. Rep.* **2017**, *7*, 12564. [CrossRef] [PubMed]

75. Mishra, P.; Kumar, A.; Sivaraman, G.; Shukla, A.K.; Kaliamoorthy, R.; Slater, A.; Velusamy, S. Character-based DNA barcoding for authentication and conservation of iucn red listed threatened species of genus decalepis (apocynaceae). *Sci. Rep.* **2017**, *7*, 14910. [CrossRef] [PubMed]

76. Hyde, K.D.; Udayanga, D.; Manamgoda, D.S.; Tedersoo, L.; Larsson, E.; Abarenkov, K.; Bertrand, Y.J.K.; Oxelman, B.; Hartmann, M.; Kauserud, H.; et al. Incorporating molecular data in fungal systematics: A guide for aspiring researchers. *Quant. Biol.* **2013**, *3*, 1–32. [CrossRef]

molecules

MDPI

Review

Machine Learning for Drug-Target Interaction Prediction

Ruolan Chen [1], Xiangrong Liu [1], Shuting Jin [1], Jiawei Lin [1] and Juan Liu [2,*]

[1] Department of Computer Science, School of Information Science and Technology, Xiamen University, Xiamen 361005, China; chenruolan@stu.xmu.edu.cn (R.C.); xrliu@xmu.edu.cn (X.L.); stjin.xmu@gmail.com (S.J.); 23020161153321@stu.xmu.edu.cn (J.L.)

[2] Department of Instrumental and Electrical Engineering, School of Aerospace Engineering, Xiamen University, Xiamen 361005, China

* Correspondence: cecyliu@xmu.edu.cn

Received: 5 August 2018; Accepted: 27 August 2018; Published: 31 August 2018

Abstract: Identifying drug-target interactions will greatly narrow down the scope of search of candidate medications, and thus can serve as the vital first step in drug discovery. Considering that in vitro experiments are extremely costly and time-consuming, high efficiency computational prediction methods could serve as promising strategies for drug-target interaction (DTI) prediction. In this review, our goal is to focus on machine learning approaches and provide a comprehensive overview. First, we summarize a brief list of databases frequently used in drug discovery. Next, we adopt a hierarchical classification scheme and introduce several representative methods of each category, especially the recent state-of-the-art methods. In addition, we compare the advantages and limitations of methods in each category. Lastly, we discuss the remaining challenges and future outlook of machine learning in DTI prediction. This article may provide a reference and tutorial insights on machine learning-based DTI prediction for future researchers.

Keywords: drug-target interaction prediction; machine learning; drug discovery

---

1. Introduction

Most drugs demonstrate efficacy via the in-vivo interactions with their target molecules such as enzymes, ion channels, nuclear receptors and G protein-coupled receptors (GPCRs). Therefore, identifying drug-target interactions (DTIs) has become a vital precondition in cognate areas including poly-pharmacology, drug repositioning, drug discovery, side-effect prediction and drug resistance [1]. The experimentation and confirmation of drug-target pairs have been great hindrances to many drug researches. On top of that biochemical experiments for undiscovered drug-target interactions involve significantly costly, time-consuming and challenging work. For instance, it takes around 1.8 billion dollars for each new molecular entity (NME) [2] as well as an average time span of 9 to 12 years for the approval of a new drug application (NDA) [3].

Besides the known interactions already stored in various databases, there exist countless unpaired small molecule compounds that could potentially be discovered and developed into new medications. Only a small number of drug-target pairs have been experimentally validated in the current data set. In fact, although there are more than 90 million compounds described in the PubChem database, a large proportion of interactions still remain to be discovered [4]. Furthermore, the number of truly innovative drugs approved by regulatory agencies has decreased in recent years, despite the progress in biotechnology. For instance, it is reported that US Food and Drug Administration (FDA) only approves approximately 20 novel drugs every year with high investment costs [5]. These large time, money and resource costs, both human and material, have motivated researchers to constantly develop innovative technology for the exploitation of new drugs. Interaction prediction helps to screen new drugs candidates effectively and efficiently.

Identifying new targets for existing or abandoned drugs, namely drug repositioning, is another important part in drug discovery. The "multi-target, multi-drug" in place of "one target, one drug" model has been widely accepted as our understanding of pharmacology deepens [1]. The important fact is that drugs typically target multiple proteins rather than only one. The anticancer drugs sunitinib (Sutent) and imatinib (Gleevec) are both concrete evidence. What's more, drugs may interact with other proteins in addition to the primary therapeutic targets, namely off-target effects. Off-target effects are typically considered harmful side effects. However, in some cases, they may be beneficial since they could lead to unexpected therapeutic effects and provide a new perspective on the molecular mechanisms of drug side effects. The purpose of drug repositioning is the detection for new clinical uses for existing drugs. An obvious benefit of drug repositioning is that existing drugs have already been strictly verified for their safety and bioavailability. Omitting some previously completed steps can greatly speed up the drug development process. Governments, academic institutions and non-trading organizations around the world have made more effort into drug repositioning recently which will effectively facilitate the repositioning research [6].

For all the reasons mentioned above, detecting drug-target interactions is fundamental to both new drug discovery and old drug repositioning. The known drug-target interactions based on wet-lab experiments are limited to a very small number. The huge gap between known and unknown drug-target pairs has prompted interest in DTI prediction. Traditional prediction strategies in vitro have faced the limitations of time and monetary costs, while recently developed computational or in silico methods can more efficiently predict potential interaction candidates. Computational methods have achieved favorable performance in many related bioinformatics fields, such as disease-related miRNA prediction [7–9], disease genes prediction [10], protein-protein interaction prediction [11] and protein subcellular location prediction [12]. They greatly narrow the broad scope of research of experimental DTI validation. Therefore, there is a continuous and urgent demand for the development of computational techniques on DTI predictions.

Currently, the ligand-based, docking simulation, and chemogenomic approaches are the three main classes of computational methods for predicting DTIs. Ligand-based methods [13] like Quantitative Structure Activity Relationship (QSAR) utilize the idea that similar molecules usually bind to similar proteins. Specifically, these methods predict interactions by comparing a new ligand to known proteins ligands. However, ligand-based methods perform poorly when the number of known ligands is insufficient.

As for docking simulation methods [14], the three-dimensional (3D) structures of proteins are required for simulation hence becoming inapplicable when there are numerous proteins with unavailable 3D structures. Moreover they cannot be applied to membrane proteins like ion channel and G-Protein Coupled Receptors (GPCRs) whose structures are too complex to obtain. Docking simulations usually take significant time and thus it can be especially inefficient.

To address the difficulties of traditional methods, chemogenomic approaches [15] have recently been performed successfully in drug discovery and repositioning on a large scale. There are four main types of target frequently involved in DTI prediction, namely protein, disease, gene and side effect. For the purpose of drug-target pair prediction, these methods integrate both the chemical space of compounds and the genomic space of target proteins into a unified space: pharmacological space. Hence, chemogenomic approaches can make full use of abundant biological data that is favorable for prediction. In such a DTI prediction problem, the major challenge is the scarcity of known drug-protein interactions and unverified negative drug-target interaction samples. These chemogenomic approaches can be classified into different categories, such as machine learning-based methods, graph-based methods and network-based methods [16]. Among all the chemogenomic approaches, machine learning-based methods have gained the most attention for their reliable prediction results. Most of these methods generally utilize the chemical and biological features of drugs and targets, and adopt various machine learning techniques to predict interactions between drugs and targets. Figure 1 is a branch diagram of recent computational methods for DTI prediction.

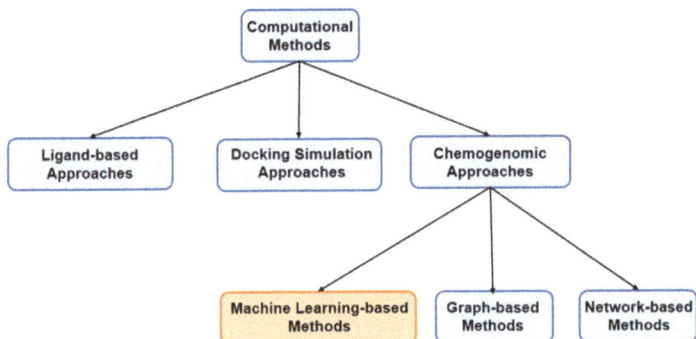

Figure 1. Branch diagram of recent computational methods for DTI prediction.

In this review, we focus on machine learning methods applied to DTI prediction. To be specific, we aim to provide a comprehensive overview on a subclass of chemogenomic approaches exploiting machine learning frameworks. Compared with those ligand-based methods that also apply machine learning strategies, the methods discussed in this review can be applicable to target proteins with insufficient known ligands. Firstly, we summarize a brief list of databases frequently used in drug discovery. Next, we adopt a hierarchical classification scheme. In particular, we classify the machine learning methods into two major categories i.e., supervised and semi-supervised methods, and provide more subclasses. We attempt to introduce several representative methods of each category, respectively. Furthermore, we present the advantages and disadvantages for methods of each category. Finally, we will discuss the challenges and further outlook for current machine learning methods in DTI prediction domain from our point of view.

1. Supervised Learning Methods Both positive labels and negative labels are required in the training set. Then these labeled samples are used to train the learning models for subsequent DTI prediction.

   - Similarity-based methods The similarities among drugs or among targets are calculated via various similarity measurement strategies. Similarity matrices can be utilized in various types of kernel functions:

     (i) The nearest neighbor methods: The nearest neighbor methods make predictions based on the information of the nearest neighbors.

     (ii) Bipartite local models: Two local models are firstly trained for drugs and targets respectively. The final prediction result for each drug-target pair is computed based on the operation of the two independent prediction scores.

     (iii) Matrix factorization methods: Drug-target interaction matrix is factorized into two latent feature matrices that when multiplied together can approximate the original matrix.

   - Feature vector-based methods The training data is represented as feature vectors. Then some machine learning models, like Random Forest, can be utilized for prediction based on these vectors.

2. Semi-Supervised Learning Methods Semi-supervised learning methods make predictions only based on a small amount of labeled data and a large amount of unlabeled data. To our best

knowledge, there are already some excellent reviews on chemogenomic approaches for DTI prediction [6,15–19]. Compared to previous works, we focus on the special topic of machine learning methods used in DTI prediction. Besides, we utilize a hierarchical classification scheme and summarize several latest prediction methods such as [20–23] which are hardly mentioned in any previous review. In particular, review [17] is written only from a narrow viewpoint, namely similarity-based approaches, which are a subclass of machine learning methods. Surveys [6,15,18,19] all provide a more general and comprehensive overview of chemogenomic approaches rather than emphasizing machine learning. In recent years, machine learning has made breakthroughs and attracted a lot of public attention. Discussing state-of-the-art DTI prediction strategies from this special perspective can demonstrate more methodology details. Although review [16] also focuses on learning-based methods, its emphasis is only on supervised learning. In comparison, we provide more detailed sub-classes and introduce newly developed methods after review [16] was published. The rest of this article is organized as follows: The "Databases" section describes current available data sources for DTI prediction research. The "Methods" section briefly introduces several representative machine learning methods via a hierarchical classification scheme. Then we discuss advantages and limitations of methods in each category as well as remaining challenges. Finally, the "Conclusions and Outlook" section makes a future perspective for machine leaning in DTI prediction.

2. Databases

Data mining and utilization based on the existing bioinformatics databases is a significant methodology for drug discovery. With the development of molecular biology, abundant information about drugs and targets has accumulated. Thus, it is necessary to establish databases for managing and maintaining the data. There exist a number of different professional databases involving potential cellular targets for various families of chemical compounds up to now. A large portion of them are publicly available. Moreover, the data size is increasing owing to the contributions of researchers from around the world. As more information about drugs and targets is collected, there are more opportunities for drug discovery research. To a certain degree, these databases have promoted the development of latest methodologies for drug discovery. In Table 1, we list frequently used databases, their web servers and brief descriptions. Table 2 shows the statistics of the number of compounds, targets and compound-target interactions in these databases. Note that not all databases provide complete information in their databases and published papers.

Some of these databases are being updated frequently, such as DrugBank, KEGG, and STITCH and so on, while the data in other databases has remained almost the same for several years, such as SuperPred which was last updated in April 2014. It is, however, encouraging that more new databases and easy-to-use web servers have been recently established. On one hand, the existing databases provide plentiful data sources of drug space and target space. It is time for the researchers to make efforts to integrate more different types of heterogeneous data. On the other hand, current databases do not involve any non-interaction information. This common drawback has limited the prediction result of supervised learning methods. Thus it would be meaningful to make public both interactions and non-interactions between drugs and targets in the future.

Table 1. Databases supporting drug discovery methods.

Database and URL	Brief Descriptions
KEGG [29] http://www.genome.jp/kegg	An encyclopedia of genes and genomes for both functional interpretation and practical application of genomic information.
BRENDA [30] http://www.brenda-enzymes.org/	The main enzyme and enzyme-ligand information system.
PubChem [31] https://pubchem.ncbi.nlm.nih.gov/	A database for information on chemical substances and their biological activities involving three inter-linked databases, i.e., Substance, Compound and BioAssay.
TTD [32] http://bidd.nus.edu.sg/group/ttd/ttd.asp	Therapeutic Target Database providing comprehensive information about the drug resistance mutations, gene expressions and target combinations data.
DrugBank [33] http://www.drugbank.ca	Consisting of two parts information involving detailed drug data (i.e., chemical, pharmacological and pharmaceutical) and drug target information (i.e., sequence, structure, and pathway) respectively.
SuperTarget [34] http://bioinf-apache.charite.de/supertarget	A database integrating drug-related information with more than 330,000 compound-target protein relations.
ChEMBL [35] https://www.ebi.ac.uk/chembldb	Data resource for molecule structures and molecule-protein interactions collected from the primary published literature on a regular basis.
STITCH [36] http://stitch.embl.de/	Repository of known and predicted chemical-protein interactions.
MATADOR [37] http://matador.embl.de/	A database of protein-chemical interactions including as many direct and indirect interactions as possible.
BindingDB [38] http://www.bindingdb.org/bind	A public database of protein-ligand binding affinities.
TDR targets [39] http://tdrtargets.org/	A chemogenomics resource for neglected tropical diseases.
SIDER [40] http://sideeffects.embl.de/	Serving information on marketed medicines and their recorded adverse drug reactions.
ChemBank [41] http://chembank.broad.harvard.edu/	Collections of available data derived from small molecules and small-molecule screens and resources for studying their properties.
DCDB [42] http://www.cls.zju.edu.cn/dcdb/	The Drug Combination Database for collecting and organizing known examples of drug combinations.
CancerDR [43] http://crdd.osdd.net/raghava/cancerdr/	Cancer Drug Resistance Database of 148 anticancer drugs and their effectiveness against around 1000 cancer cell lines.
ASDCD [44] http://asdcd.amss.ac.cn/	The first Antifungal Synergistic Drug Combination Database including published synergistic antifungal drug combinations, targets, indications, and other pertinent data.
SuperPred [45] http://prediction.charite.de/	Resource of compound-target interactions.

Table 2. The statistics of the number of compounds, targets and compound-target interactions in the databases covered in the review.

Databases	The Number of Compounds	The Number of Targets	The Number of Compound-Target Interactions
KEGG	18,380	26,885,475	
BRENDA		7341	
PubChem	96,479,316	68,868	
TTD	34,019	3101	
DrugBank	11,682	26,889	131,724
SuperTarget	195,770	6219	332,828
ChEMBL	2,275,906	12,091	
STITCH	500,000	9,600,000	1,600,000,000
MATADOR	775		
BindingDB	652,068	7082	1,454,892
TDR targets	2,000,000	5300	
SIDER	5868	1430	139,756
ChemBank	1,700,000		
DCDB	904	805	
CancerDR	148	116	
ASDCD	105	1225	210
SuperPred	341,000	1800	665,000

3. Methods

In the era of big data, machine learning methods are designed to generate predictive models based on some underlying algorithm and a given big data set. For biological and biomedical research, machine learning plays a pivotal role in filtering large amounts of data into patterns [24–27]. The general machine learning workflow in DTI prediction can be divided into three steps. First, preprocessing the input data of the drug and the target; second, training the underlying model based on a set of learning rules; third, utilizing the predictive model to make predictions for a test data set.

From our research, study [28] is the first work that applies machine learning to protein-chemical interaction prediction. This work establishes a SVM analysis framework of amino acid sequence data, chemical structure data and mass spectrometry data. This pioneering study has inspired subsequent studies. Machine learning for drug discovery has become a field of long-standing and growing interest since then.

For simplicity, we classify machine learning methods for drug-target interaction prediction into two major categories, i.e., supervised learning and semi-supervised methods. Specifically, the supervised learning methods can be further classified into two sub-classes including similarity-based methods and feature-based methods.

3.1. Supervised Learning Methods

Supervised learning methods are applied to train the learning model and identify patterns when labels are available. For the DIT prediction problem, known drug-target interactions are labeled as positive samples and the rest are labeled as negative ones. Next, these labels are used to train the model for subsequent interaction predicting. In fact, those drug-target pairs without explicit interaction information may correspond to unknown or missing interactions rather than non-interactions. In general results of non-interactions between drugs and targets are not published. Methods of this category regard all the unknown drug-target interactions as non-interaction despite inaccuracy. In the section, we will review the supervised methods proposed so far in two categories, i.e., similarity-based methods and feature-based methods.

3.1.1. Similarity-Based Methods

A key underlying assumption of similarity-based machine learning methods is the "guilt-by-association" assumption, that is, similar drugs tend to share similar targets and vice versa. In this kind of approach, the similarity among drugs or among targets is computed by various similarity measures. The constructed similarity matrices define several types of kernel functions.

• The Nearest Neighbor Methods

The nearest neighbor methods generally adopt relatively simple similarity functions. Researchers often integrate these methods with some other approaches to help predict new drugs or targets, such as models in paper [46,47]. In the early stage, study [48] proposed two exploratory approaches, namely the nearest profile method (NN) and the weighted profile method. The nearest profile method follows the key concept that similar drugs or targets tend to be close in the network. This method was used in [49] as the baseline. In contrast, the weighted profile method utilizes the similarities of all the other drugs and targets and then adopts a weighted average. However, these methods show poor performance in the case when targets bound to similar drug share low sequence similarity or vice versa.

In the studies [23,50] by Zhang et al., methods that make drug-drug pair predictions based on neighbors were developed. These studies further extended the classic neighbor recommender method to the integrated neighborhood-based method (INBM). In simple terms, neighbor recommender method generally uses the weighted average information of neighbors for prediction. INBM is an ensemble model that integrates several neighborhood-based models for a robust prediction. For each drug-drug pair, three commonly used formulas, namely Jaccard similarity, Cosine similarity and Pearson correlation similarity, are used to calculate similarity score.

Another novel methodology in this category is Similarity-Rank-based predictor (SRP) [51]. Two indices, i.e., tendency index and inverse tendency index, are computed to construct a SRP. To be specific, the former represents the likelihood that each drug–target pair tends to interact, while the latter measures the tendency that each drug–target pair does not interact. The calculation formulas involve both similarity and similarity rank. Then an interaction likelihood score is computed as the likelihood ratio of the two indices. This method can generate two interaction likelihood scores, one from the drug side and the other from the target side. The final prediction score is the average of the two scores. The clear advantage of SRP is that it is a lazy and non-parametric model without the requirements of an optimization solver, prior statistical knowledge as well as tunable parameters.

In recent years, other new similarity-based methods have been proposed one after another, such as rule-based inference. Due to the limitation of the previous topology-based methods, a similarity-based deep learning method [52] merges the similarity measure with two rule-based inference methods. In other words, drug-based similarity inference (DBSI) and target-based similarity inference (TBSI) [48,53] are adopted to discover the drug-target interactions with the similarities. Though it is flexible to assemble any kernel functions, the method cannot predict new drugs or targets.

Note that most of similarity measures only utilize some important drug-related or disease-related properties to perform drug-disease prediction and ignore the known drug-disease interaction information [54]. Some researchers have proposed new similarity measures. Luo et al. [54] have designed a comprehensive similarity measure. In order to improve traditional similarity measures for drug-disease prediction, the comprehensive similarity measure has integrated drug or disease feature information with known drug–disease interactions. The similarity measure can be broken down into three steps. In the first step, drug similarity and disease similarity are calculated based on drug-related properties or disease-related properties respectively. In the second step, these similarity values are adjusted by a logistic function based on the analysis and evaluation results. In the last step, a weighted drug network can be established for the drug similarity. The edge weight represents the number of common diseases between corresponding drugs. Then a cluster method, ClusterONE, is applied to identify potential drug clusters. Similarity between drugs belonging to the same cluster is enhanced and thus comprehensive drug similarity is obtained. Disease similarity can be improved in the same way as for drugs.

• Bipartite Local Models

Bipartite local models (BLMs) firstly generate two independent prediction for drugs and targets respectively. The final prediction result is then obtained by aggregating the two prediction scores.

The concept of BLM was first introduced in the pioneering work by Bleakley and Yamanishi [49]. This method can transform the drug-target interaction prediction problem into a binary classification problem. More specifically, a local model is trained for drugs based on chemical similarity. Another one is trained for proteins based on sequence structure. Therefore, two SVM classifiers can generate two independent prediction results from the drug or target side respectively. Final prediction result for each drug-target pair is computed based on the average of these two independent prediction scores.

Analogously, another method [55] developed a regularized least square classifier introducing two algorithms, called RLS-avg and RLS-kron. In particular, Regularized Least Squares (RLS-avg) utilizes kernel ridge regression to perform prediction. While in RLS-kron, all pairs of drugs and targets are combined into one to make Kronecker product, bringing the runtime down greatly.

Considering the limitation of the BLM-based methods above of predicting new drug or target without any known interactions available, Mei et al. [46] extended existing BLM by adding a preprocessing to infer training data from neighbors' interaction profiles. The method is called Bipartite Local Models with Neighbor-based Interaction Profile Inferring (BLM-NII). BLM-NII involves RLS-avg algorithm and is proven to be effective in new candidate problem.

• Matrix Factorization Methods

Matrix factorization methods are typically used in recommendation systems to find potential user-item interactions. The DTI prediction can be regarded as a matrix completion problem that aims to look for missing interactions. Therefore, drug-target interaction matrix can be factorized into two other matrices that when multiplied together can approximate the original matrix.

Kernelized Bayesian Matrix Factorization with Twin Kernels (KBMF2K) [56] is the original method that introduced matrix factorization to DTI prediction. Following some previous approaches, KBMF2K defines two kernel matrices only based on chemical similarity between drug compounds and genomic similarity between target proteins. It combines Bayesian probabilistic formulation, matrix factorization and binary classification for prediction problem.

Another study adopting probabilistic formulations is Probabilistic Matrix Factorization (PMF) [57]. PMF is distinguished greatly from KBMF2K by its independence of drug or target similarity matrices. Furthermore, the study presented the active learning (AL) strategy along with probabilistic matrix factorization.

Zheng et al. [58] proposed an extension of weighted low-rank approximation from one-class collaborative filtering (CMF), namely Multiple Similarities Collaborative Matrix Factorization (MSCMF). MSCMF integrates multiple similarity matrices, including chemical structure similarity, genomic sequence similarity, ATC similarity, GO similarity and PPI network similarity. Weights over the matrices are estimated to select similarities automatically. This strategy improves predictive performance in the experiment. Drugs and targets are projected into low-rank matrices. Then weights over similarity matrices are estimated using an alternating least squares algorithm. However, regardless of its performance, under this data integration strategy, a large amount of information may be lost, thus leading to sub-optimal solution.

The method developed by Ezzat et al. [59], employed two matrix factorization methods (i.e., GRMF and WGRMF). It was revealed in previous work [60] that data usually lies on or nears to the low-dimensional and non-linear manifold. Therefore, GRMF and WGRMF perform manifold learning implicitly by means of graph regularization. In addition, a preprocessing step (WKNKN) was applied to new drug or target prediction by transforming all the 0's in the original drug-target matrix into interaction likelihood values. This important step distinguishes this method from other work that regards all the 0's of given drug-target matrix as non-interaction roughly, and thus enhances the prediction results.

3.1.2. Feature Vector-Based Methods

Generally, similarity-based prediction algorithms do not take heterogeneous types and interactions defined in semantic networks into consideration. In addition, it may be difficult to add the long indirect connections between two nodes. Therefore, feature vector-based methods have been utilized for DTI prediction. The input of feature vector-based methods is drug-target pairs represented by fixed-length feature vectors. The feature vectors are encoded by various properties of drugs and targets.

In the systematic approach [61], chemical descriptors are calculated using DRAGON program (http://www.talete.mi.it/index.htm). Finally, each drug is represented as a set of 1080 descriptors, including constitutional descriptors, topological descriptors, 2D autocorrelations, eigenvalue-based indices and so on. Likewise, each protein is represented by a set of structural and physicochemical descriptors via PROFEAT WEBSEVER (http://jing.cz3.nus.edu.sg/cgi-bin/prof/prof.cgi). The descriptors involve Amino acid composition descriptors, Dipeptide composition descriptors, and Autocorrelation descriptors and so on. Then each protein sequence with changeable length can be transformed into a standard feature vector of 1080 dimensions. Hence, a set of 2160-dimensional feature vectors for each drug-target pair can be constructed. Subsequent prediction step performs Random Forest (RF) algorithm which introduces random training set (bootstrap) and random input vectors into the trees. The comprehensive framework shows its robustness against the over fitting problem and performs more efficiently for a large-scale data set in experiments.

In order to integrate diverse information from heterogeneous data sources, a method named DTINet was proposed by Luo et al. [20]. Through DTINet, a low dimensional feature vector that accurately explains the topological properties of each node in the heterogeneous network is first learned. In the further step, DTINet applies inductive matrix completion to best project drug space onto protein space.

Due to the fact that DTINet separates features and may result in loss of the optimal solution, Wan et al. [21] created a new framework called neural integration of neighbor information for DTI prediction (NeoDTI). The inspiration of NeoDTI came from convolution neural networks (CNNs). It integrates the neighbor information in heterogeneous network. After extracting the complex hidden features vectors of drugs and targets, NeoDTI automatically learns topology-preserving representations to achieve superior prediction performance.

The pioneering effort in [62] introduced a two-layer undirected graphical model, namely restricted Boltzmann machine (RBM), into a large-scale drug-target interaction prediction. There are no intra-layer connections in these layers. What's more, RBM model is trained via a practical learning algorithm, i.e., Contrastive Divergence (CD). Where the method significantly outperforms other existing approaches is in that it can predict different types of DTIs on a multidimensional network. In other words, the method can identify binary DTIs as well as their corresponding types of interactions, including relationships and drug modes of action.

In the paper published by Fu and cooperators [63], a state-of-the-art machine learning model was constructed based on meta-path-based topological features. Two measures of topological features are calculated, including the number of path instances between nodes and a normalization process to it. Given features, a Random Forest algorithm is used as supervised classification. Furthermore, intrinsic feature ranking algorithm embedded in Random Forest selects the important topological features for better prediction. This framework has shown precise predictability.

3.2. Semi-Supervised Learning Methods

Considering the negative sample selection has a great influence on the accuracy of DTI prediction results, some researchers have proposed semi-supervised methods to address the problem. These methods use only a small amount of labeled data and a large amount of unlabeled data. Semi-supervised methods typically use the labeled data to infer labels for unlabeled data. On the other hand, the unlabeled data can also help provide insights into the structure of training set.

Having no use of negative samples, study [64] first employed a manifold Laplacian regularized least square (LapRLS) based on the BLM concept. Furthermore, an extension of the standard LapRLS, namely NetLapRLS, was proposed. NetLapRLS integrates information from chemical space, genomic space and drug-protein interaction for a new kernel. These semi-supervised methods have achieved encouraging results than using the labeled data alone. However, it is time-consuming when implementing them on a large scale.

Another method is designed for both semi-supervised and unsupervised settings. Ma et al. [22] presented a new framework to learn accurate and interpretable similarity measures when labels are scarce. This framework constructs a set of Graph Auto-Encoder (GAE)-based models and integrates multi-view drug similarities. Besides, an attentive mechanism is used for view selection and better interpretability.

3.3. Discussion

Each machine learning model possesses its unique advantages as well as disadvantages. Note that just as the popular concept in computer science, namely "no free lunch theorem" [65], machine learning methods are context-specific. Therefore, in this review we can only evaluate the advantages and disadvantages of each method category based on DTI prediction context.

A number of supervised models have been already proven feasible for DTI prediction. However, most supervised methods simply regard all the unlabeled drug-target pairs as negative samples and thus generate inaccurate predictive results. What's more, each similarity-based method has its limitation when extending to large a data set because of high complexity of similarity matrices computation.

Consider the three sub-classes of similarity-based methods respectively. Although the nearest neighbor methods generally apply relatively simple similarity functions, most of them construct neighborhoods only based on first-order similarity and do not involve the transitivity of similarity [66]. A key advantage of bipartite local models is that they process much fewer drug-target pairs, and thus they have much lower complexity than global models. Nevertheless, bipartite local models cannot handle the scenario that both drugs and targets are not involved in the training set unless combined with other methods. According to the experiment result in [19], matrix factorization methods generally have more superior performance than other methods including the nearest neighbor models and bipartite local models.

A small number of known drug-target interactions results in an imbalanced dataset. As an effective solution for imbalanced datasets, semi-supervised learning uses only a small amount of labeled data with a large amount of unlabeled data and generates more reliable prediction than supervised one.

In addition to the aforementioned single machine learning methods, we also have introduced several ensemble methods [61,63]. A better and robust prediction generally results from the biases trade-off of each single method. Generally, ensemble methods can combine different learning models. For more ensemble methods applied to drug-target interaction prediction task, please refer to [67–69].

Generally, machine learning has achieved favorable performance in DTI prediction. Nonetheless, a number of challenges still remain. Above all, recently, some researchers have emphasized that predictive models based on machine learning are usually established and evaluated with overly simplified settings. Prediction results under such experiment settings may be over optimistic and deviate from the real case. Particularly, most of machine learning methods simply regard drug-target interaction as an on-off relationship and ignore other vital factors like molecule concentrations and quantitative affinities. Pahikkala et al. [24] have pointed out four factors having significant impact on prediction results, including problem formulation, evaluation data set, evaluation procedure and experimental setting. Considering the binding affinities and dose-dependence of drug-target pairs, the DTI prediction problem should be formulated as a regression or rank prediction problem rather than a standard binary classification problem. The second challenge is the imbalanced dataset problem. Due to the small number of known drug-target pairs, the current dataset is imbalanced. Some models like decision trees and SVMs, have a great bias for recognizing the majority class and thus result in poor performance [16]. Thirdly, most machine learning models possess "poor interpretability" properties. In other words, it is difficult to understand the underlying drug mechanism of action from a biological perspective. Note that in most case, it is easier to explain relatively simple models. This case is consistent with one of the "rules of thumb" [70], that is "simple is often better". Nonetheless, for most current state-of-the-art approaches achieving high DTI prediction accuracy, such as deep learning methods, it is difficult to interpret them from a pharmacology perspective. Last but not least, there are still no uniform evaluation metrics special for DTI prediction. Previous studies have adopted some common evaluation metrics in bioinformatics [71], such as sensitivity, specificity, Area Under the Precision-Recall (AUPR) curve and Area under the ROC curve (AUC). The fact is that if the sensitivity increases, the specificity decreases. Considering the limitation of using sensitivity or specificity alone, AUPR and AUC may be better choices in evaluation tasks. In the currently accessible datasets, the number of unknown samples is much more than the known ones, and thus false positives should be weighed more. AUPR can reduce the impact of false positive data on evaluation results as possible [72], and AUC is insensitive to imbalance dataset [73]. Thus both AUPR and AUC are generally adequate metrics for evaluating the performance of machine learning-based methods.

4. Conclusions and Outlook

DTIs contribute to the selection of potential drugs and thus effectively reduce the scope of research for biochemical experiments. Besides, they can provide deep insights into the side effects and the mechanism(s) of action of drugs. Hence, DTI prediction is a vital prerequisite for drug discovery. In fact, a number of public available databases have been established and promoted the development of innovatory DTI prediction strategies.

In this review, we focus on machine learning-based methods integrating chemical space and genomic space. We summarize the databases and machine learning methods frequently used in DTI prediction. In particular, we focus on several state-of-the-art predictive models appearing in recent years. We adopt a hierarchical classification scheme. We classify machine learning methods into two major categories: supervised and semi-supervised methods, and provide more subclasses.

Machine learning will be promising in DTI prediction for the next several years. However, there is still much room for improvement. Hence, we conclude with some advice as a reference for the future researchers.

Firstly, ensemble approaches combine multiple independent classifiers into one model and typically achieve a better prediction results. Next, semi-supervised learning is a powerful tool for addressing the imbalanced dataset problem. However, only a small number of semi-supervised learning methods have been proposed recently. Hence, the research on semi-supervised learning methods needs more attention. Furthermore, note the fact that drug-target pairs involve binding affinities and dose-dependence. It is more practical and meaningful to study new regression methods for DTI prediction problem. The using of quantitative bioactivity data will lead to a more accurate and reliable predictive result. Finally, with the development of high throughput biotechnology, the available data has been growing quickly recently. It is time for further machine learning technology to take full advantage of more different types of heterogeneous data.

5. Key Points

1. Identifying drug-target interactions is the vital first step in drug discovery research.
2. A number of existing professional databases serve known data resources for DTI prediction and thus promote the drug discovery.
3. Machine learning-base methods are generally effective and reliable for DTI prediction.
4. Different machine learning methods have their merits and demerits. Hence, it is essential to choose appropriate methods or assemble models for special prediction tasks.
5. A more effective prediction model can be established by integrating more heterogeneous data sources of drugs and targets.
6. In reality, DTI prediction is a regression problem with quantitative bioactivity data.

Author Contributions: Conceptualization, R.C.; Writing-Original Draft Preparation, R.C.; Writing-Review & Editing, R.C., X.L., S.J. and J.L. (Jiawei Lin); Funding Acquisition, X.L.; Supervision, J.L. (Juan Liu).

Funding: This research was funded by the National Natural Science Foundation of China grant numbers [61472333, 61772441, 61472335, 61425002], Project of marine economic innovation and development in Xiamen grant number [16PFW034SF02], Natural Science Foundation of the Higher Education Institutions of Fujian Province grant number [JZ160400], Natural Science Foundation of Fujian Province grant number [2017J01099], President Fund of Xiamen University grant number [20720170054], and the National Natural Science Foundation of China grant number [81300632].

Acknowledgments: We would like to thank all authors of the cited references.

Conflicts of Interest: The authors declare no conflicts of interest.

References

1. Masoudi-Nejad, A.; Mousavian, Z.; Bozorgmehr, J.H. Drug-target and disease networks: Polypharmacology in the post-genomic era. *In Silico Pharmacol.* **2013**, *1*, 17. [CrossRef] [PubMed]
2. Paul, S.M.; Mytelka, D.S.; Dunwiddie, C.T.; Persinger, C.C.; Munos, B.H.; Lindborg, S.R.; Schacht, A.L. How to improve R&D productivity: The pharmaceutical industry's grand challenge. *Nat. Rev. Drug Discov.* **2010**, *9*, 203–214. [CrossRef] [PubMed]
3. Dickson, M.; Gagnon, J.P. Key factors in the rising cost of new drug discovery and development. *Nat. Rev. Drug Discov.* **2004**, *3*, 417–429. [CrossRef] [PubMed]
4. Wang, Y.; Bryant, S.H.; Cheng, T.; Wang, J.; Gindulyte, A.; Shoemaker, B.A.; Thiessen, P.A.; He, S.; Zhang, J. Pubchem bioassay: 2017 update. *Nucleic Acids Res.* **2017**, *45*, D955–D963. [CrossRef] [PubMed]
5. Chen, H.; Zhang, Z. A semi-supervised method for drug-target interaction prediction with consistency in networks. *PLoS ONE* **2013**, *8*, e62975. [CrossRef] [PubMed]
6. Li, J.; Zheng, S.; Chen, B.; Butte, A.J.; Swamidass, S.J.; Lu, Z. A survey of current trends in computational drug repositioning. *Brief. Bioinform.* **2016**, *17*, 2–12. [CrossRef] [PubMed]
7. Zeng, X.; Liu, L.; Lu, L.; Zou, Q. Prediction of potential disease-associated micrornas using structural perturbation method. *Bioinformatics* **2018**, *34*, 2425–2432. [CrossRef] [PubMed]
8. Zhang, X.; Zou, Q.; Rodríguez-Patón, A.; Zeng, X. Meta-path methods for prioritizing candidate disease mirnas. *IEEE/ACM Trans. Comput. Biol. Bioinform.* **2017**. [CrossRef]
9. Hua, S.; Yun, W.; Zhiqiang, Z.; Zou, Q. A discussion of micrornas in cancers. *Curr. Bioinform.* **2014**, *9*, 453–462. [CrossRef]
10. Zeng, X.; Liao, Y.; Liu, Y.; Zou, Q. Prediction and validation of disease genes using hetesim scores. *IEEE/ACM Trans. Comput. Biol. Bioinform.* **2017**, *14*, 687–695. [CrossRef] [PubMed]
11. Zeng, J.; Li, D.; Wu, Y.; Zou, Q.; Liu, X. An empirical study of features fusion techniques for protein-protein interaction prediction. *Curr. Bioinform.* **2016**, *11*, 4–12. [CrossRef]
12. Wang, Z.; Zou, Q.; Jiang, Y.; Ju, Y.; Zeng, X. Review of protein subcellular localization prediction. *Curr. Bioinform.* **2014**, *9*, 331–342. [CrossRef]

13. Keiser, M.J.; Roth, B.L.; Armbruster, B.N.; Ernsberger, P.; Irwin, J.J.; Shoichet, B.K. Relating protein pharmacology by ligand chemistry. *Nat. Biotechnol.* **2007**, *25*, 197–206. [CrossRef] [PubMed]
14. Arola, L.; Fernandez-Larrea, J.; Blay, M.; Salvado, M.J.; Blade, C.; Ardevol, A.; Vaque, M.; Pujadas, G. Protein-ligand docking: A review of recent advances and future perspectives. *Curr. Pharm. Anal.* **2008**, *4*, 1–19. [CrossRef]
15. Yamanishi, Y. Chemogenomic approaches to infer drug–target interaction networks. In *Data Mining for Systems Biology: Methods and Protocols*; Mamitsuka, H., DeLisi, C., Kanehisa, M., Eds.; Humana Press: Totowa, NJ, USA, 2013; Volume 939, pp. 97–113. ISBN 978-1-62703-107-3.
16. Mousavian, Z.; Masoudi-Nejad, A. Drug-target interaction prediction via chemogenomic space: Learning-based methods. *Expert Opin. Drug Metab. Toxicol.* **2014**, *10*, 1273–1287. [CrossRef] [PubMed]
17. Ding, H.; Takigawa, I.; Mamitsuka, H.; Zhu, S. Similarity-based machine learning methods for predicting drug-target interactions: A brief review. *Brief. Bioinform.* **2014**, *15*, 734–747. [CrossRef] [PubMed]
18. Chen, X.; Yan, C.C.; Zhang, X.; Zhang, X.; Dai, F.; Yin, J.; Zhang, Y. Drug-target interaction prediction: Databases, web servers and computational models. *Brief. Bioinform.* **2016**, *17*, 696–712. [CrossRef] [PubMed]
19. Ezzat, A.; Wu, M.; Li, X.L.; Kwoh, C.K. Computational prediction of drug-target interactions using chemogenomic approaches: An empirical survey. *Brief. Bioinform.* **2018**. [CrossRef] [PubMed]
20. Luo, Y.; Zhao, X.; Zhou, J.; Yang, J.; Zhang, Y.; Kuang, W.; Peng, J.; Chen, L.; Zeng, J. A network integration approach for drug-target interaction prediction and computational drug repositioning from heterogeneous information. *Nat. Commun.* **2017**, *8*, 573. [CrossRef] [PubMed]
21. Wan, F.; Hong, L.; Xiao, A.; Jiang, T.; Zeng, J. Neodti: Neural integration of neighbor information from a heterogeneous network for discovering new drug-target interactions. *Bioinformatics* **2018**. [CrossRef]
22. Ma, T.; Xiao, C.; Zhou, J.; Wang, F. Drug similarity integration through attentive multi-view graph auto-encoders. *arXiv*, **2018**; arXiv:1804.10850.
23. Zhang, W.; Chen, Y.; Liu, F.; Luo, F.; Tian, G.; Li, X. Predicting potential drug-drug interactions by integrating chemical, biological, phenotypic and network data. *BMC Bioinform.* **2017**, *18*, 18. [CrossRef] [PubMed]
24. Pahikkala, T.; Airola, A.; Pietila, S.; Shakyawar, S.; Szwajda, A.; Tang, J.; Aittokallio, T. Toward more realistic drug-target interaction predictions. *Brief. Bioinform.* **2015**, *16*, 325–337. [CrossRef] [PubMed]
25. Zeng, X.; Zhang, X.; Zou, Q. Integrative approaches for predicting microrna function and prioritizing disease-related microrna using biological interaction networks. *Brief. Bioinform.* **2016**, *17*, 193–203. [CrossRef] [PubMed]
26. Zou, Q.; Ju, Y.; Li, D. Protein folds prediction with hierarchical structured SVM. *Curr. Proteom.* **2016**, *13*, 79–85. [CrossRef]
27. Wang, X.; Zeng, X.; Ju, Y.; Jiang, Y.; Zhang, Z.; Chen, W. A classification method for microarrays based on diversity. *Curr. Bioinform.* **2016**, *11*, 590–597. [CrossRef]
28. Nagamine, N.; Sakakibara, Y. Statistical prediction of protein chemical interactions based on chemical structure and mass spectrometry data. *Bioinformatics* **2007**, *23*, 2004–2012. [CrossRef] [PubMed]
29. Kanehisa, M.; Furumichi, M.; Mao, T.; Sato, Y.; Morishima, K. Kegg: New perspectives on genomes, pathways, diseases and drugs. *Nucleic Acids Res.* **2017**, *45*, D353–D361. [CrossRef] [PubMed]
30. Placzek, S.; Schomburg, I.; Chang, A.; Jeske, L.; Ulbrich, M.; Tillack, J.; Schomburg, D. Brenda in 2017: New perspectives and new tools in brenda. *Nucleic Acids Res.* **2017**, *45*, D380–D388. [CrossRef] [PubMed]
31. Kim, S.; Thiessen, P.A.; Bolton, E.E.; Chen, J.; Fu, G.; Gindulyte, A.; Han, L.; He, J.; He, S.; Shoemaker, B.A. Pubchem substance and compound databases. *Nucleic Acids Res.* **2016**, *44*, D1202–D1213. [CrossRef] [PubMed]
32. Qin, C.; Zhang, C.; Zhu, F.; Xu, F.; Chen, S.Y.; Zhang, P.; Li, Y.H.; Yang, S.Y.; Wei, Y.Q.; Tao, L. Therapeutic target database update 2014: A resource for targeted therapeutics. *Nucleic Acids Res.* **2014**, *42*, D1118–D1123. [CrossRef] [PubMed]
33. Wishart, D.S.; Feunang, Y.D.; Guo, A.C.; Lo, E.J.; Marcu, A.; Grant, J.R.; Sajed, T.; Johnson, D.; Li, C.; Sayeeda, Z. Drugbank 5.0: A major update to the drugbank database for 2018. *Nucleic Acids Res.* **2017**, *46*, D1074–D1082. [CrossRef] [PubMed]
34. Hecker, N.; Ahmed, J.; Von, E.J.; Dunkel, M.; Macha, K.; Eckert, A.; Gilson, M.K.; Bourne, P.E.; Preissner, R. Supertarget goes quantitative: Update on drug-target interactions. *Nucleic Acids Res.* **2012**, *40*, D1113–D1117. [CrossRef] [PubMed]

35. Gaulton, A.; Bellis, L.J.; Bento, A.P.; Chambers, J.; Davies, M.; Hersey, A.; Light, Y.; Mcglinchey, S.; Michalovich, D.; Allazikani, B. ChEMBL: A large-scale bioactivity database for drug discovery. *Nucleic Acids Res.* **2012**, *40*, D1100–D1107. [CrossRef] [PubMed]
36. Szklarczyk, D.; Santos, A.; Von, M.C.; Jensen, L.J.; Bork, P.; Kuhn, M. STITCH 5: Augmenting protein-chemical interaction networks with tissue and affinity data. *Nucleic Acids Res.* **2016**, *44*, D380–D384. [CrossRef] [PubMed]
37. Günther, S.; Kuhn, M.; Dunkel, M.; Campillos, M.; Senger, C.; Petsalaki, E.; Ahmed, J.; Urdiales, E.G.; Gewiess, A.; Jensen, L.J. Supertarget and matador: Resources for exploring drug-target relationships. *Nucleic Acids Res.* **2008**, *36*, D919–D922. [CrossRef] [PubMed]
38. Liu, T.; Lin, Y.; Wen, X.; Jorissen, R.N.; Gilson, M.K. Bindingdb: A web-accessible database of experimentally determined protein–ligand binding affinities. *Nucleic Acids Res.* **2007**, *35*, D198–D201. [CrossRef] [PubMed]
39. Magariños, M.P.; Carmona, S.J.; Crowther, G.J.; Ralph, S.A.; Roos, D.S.; Shanmugam, D.; Voorhis, W.C.V.; Agüero, F. TDR targets: A chemogenomics resource for neglected diseases. *Nucleic Acids Res.* **2012**, *40*, D1118–D1127. [CrossRef] [PubMed]
40. Kuhn, M.; Campillos, M.; Letunic, I.; Jensen, L.J.; Bork, P. A side effect resource to capture phenotypic effects of drugs. *Mol. Syst. Biol.* **2010**, *6*, 343–348. [CrossRef] [PubMed]
41. Seiler, K.P.; George, G.A.; Happ, M.P.; Bodycombe, N.E.; Carrinski, H.A.; Norton, S.; Brudz, S.; Sullivan, J.P.; Muhlich, J.; Serrano, M. Chembank: A small-molecule screening and cheminformatics resource database. *Nucleic Acids Res.* **2008**, *36*, D351–D359. [CrossRef] [PubMed]
42. Liu, Y.; Wei, Q.; Yu, G.; Gai, W.; Li, Y.; Chen, X. DCDB 2.0: A major update of the drug combination database. *Database* **2014**, *2014*. [CrossRef] [PubMed]
43. Kumar, R.; Chaudhary, K.; Gupta, S.; Singh, H.; Kumar, S.; Gautam, A.; Kapoor, P.; Raghava, G.P.S. CancerDR: Cancer drug resistance database. *Sci. Rep.* **2013**, *3*, 1445. [CrossRef] [PubMed]
44. Chen, X.; Ren, B.; Chen, M.; Liu, M.X.; Ren, W.; Wang, Q.X.; Zhang, L.X.; Yan, G.Y. ASDCD: Antifungal synergistic drug combination database. *PLoS ONE* **2014**, *9*, e86499. [CrossRef] [PubMed]
45. Nickel, J.; Gohlke, B.O.; Erehman, J.; Banerjee, P.; Rong, W.W.; Goede, A.; Dunkel, M.; Preissner, R. SuperPred: Update on drug classification and target prediction. *Nucleic Acids Res.* **2014**, *42*, W26–W31. [CrossRef] [PubMed]
46. Mei, J.P.; Kwoh, C.K.; Yang, P.; Li, X.L.; Zheng, J. Drug-target interaction prediction by learning from local information and neighbors. *Bioinformatics* **2013**, *29*, 238–245. [CrossRef] [PubMed]
47. Van Laarhoven, T.; Marchiori, E. Predicting drug-target interactions for new drug compounds using a weighted nearest neighbor profile. *PLoS ONE* **2013**, *8*, e66952. [CrossRef] [PubMed]
48. Yamanishi, Y.; Araki, M.; Gutteridge, A.; Honda, W.; Kanehisa, M. Prediction of drug–target interaction networks from the integration of chemical and genomic spaces. *Bioinformatics* **2008**, *24*, i232–i240. [CrossRef] [PubMed]
49. Bleakley, K.; Yamanishi, Y. Supervised prediction of drug–target interactions using bipartite local models. *Bioinformatics* **2009**, *25*, 2397–2403. [CrossRef] [PubMed]
50. Zhang, W.; Zou, H.; Luo, L.; Liu, Q.; Wu, W.; Xiao, W. Predicting potential side effects of drugs by recommender methods and ensemble learning. *Neurocomputing* **2016**, *173*, 979–987. [CrossRef]
51. Shi, J.Y.; Yiu, S.M. SRP: A concise non-parametric similarity-rank-based model for predicting drug-target interactions. In Proceedings of the 2015 IEEE International Conference on Bioinformatics and Biomedicine (BIBM), Washington, DC, USA, 9–12 November 2015; IEEE: New York, NY, USA, 2015; pp. 1636–1641.
52. Zong, N.; Kim, H.; Ngo, V.; Harismendy, O. Deep mining heterogeneous networks of biomedical linked data to predict novel drug-target associations. *Bioinformatics* **2017**, *33*, 2337–2344. [CrossRef] [PubMed]
53. Cheng, F.; Liu, C.; Jiang, J.; Lu, W.; Li, W.; Liu, G.; Zhou, W.; Huang, J.; Tang, Y. Prediction of drug-target interactions and drug repositioning via network-based inference. *PLoS Comput. Biol.* **2012**, *8*, e1002503. [CrossRef] [PubMed]
54. Luo, H.; Wang, J.; Li, M.; Luo, J.; Peng, X.; Wu, F.X.; Pan, Y. Drug repositioning based on comprehensive similarity measures and bi-random walk algorithm. *Bioinformatics* **2016**, *32*, 2664–2671. [CrossRef] [PubMed]
55. Van Laarhoven, T.; Nabuurs, S.B.; Marchiori, E. Gaussian interaction profile kernels for predicting drug–target interaction. *Bioinformatics* **2011**, *27*, 3036–3043. [CrossRef] [PubMed]
56. Gönen, M. Predicting drug–target interactions from chemical and genomic kernels using bayesian matrix factorization. *Bioinformatics* **2012**, *28*, 2304–2310. [CrossRef] [PubMed]

57. Cobanoglu, M.C.; Liu, C.; Hu, F.; Oltvai, Z.N.; Bahar, I. Predicting drug–target interactions using probabilistic matrix factorization. *J. Chem. Inf. Model.* **2013**, *53*, 3399–3409. [CrossRef] [PubMed]

58. Zheng, X.; Ding, H.; Mamitsuka, H.; Zhu, S. Collaborative matrix factorization with multiple similarities for predicting drug-target interactions. In Proceedings of the ACM SIGKDD International Conference on Knowledge Discovery and Data Mining, Chicago, IL, USA, 11–14 August 2013; ACM: New York, NY, USA, 2013; pp. 1025–1033.

59. Ezzat, A.; Zhao, P.; Wu, M.; Li, X.L.; Kwoh, C.K. Drug-target interaction prediction with graph regularized matrix factorization. *IEEE/ACM Trans. Comput. Biol. Bioinform.* **2016**, *14*, 646–656. [CrossRef] [PubMed]

60. Tenenbaum, J.B.; Silva, V.D.; Langford, J.C. A global geometric framework for nonlinear dimensionality reduction. *Science* **2000**, *290*, 2319–2323. [CrossRef] [PubMed]

61. Yu, H.; Chen, J.; Xu, X.; Li, Y.; Zhao, H.; Fang, Y.; Li, X.; Zhou, W.; Wang, W.; Wang, Y. A systematic prediction of multiple drug-target interactions from chemical, genomic, and pharmacological data. *PLoS ONE* **2012**, *7*, e37608. [CrossRef] [PubMed]

62. Wang, Y.; Zeng, J. Predicting drug-target interactions using restricted boltzmann machines. *Bioinformatics* **2013**, *29*, i126–i134. [CrossRef] [PubMed]

63. Fu, G.; Ding, Y.; Seal, A.; Chen, B.; Sun, Y.; Bolton, E. Predicting drug target interactions using meta-path-based semantic network analysis. *BMC Bioinform.* **2016**, *17*, 160. [CrossRef] [PubMed]

64. Xia, Z.; Wu, L.Y.; Zhou, X.; Wong, S.T. Semi-supervised drug-protein interaction prediction from heterogeneous biological spaces. *BMC Syst. Biol.* **2010**, *4*, S6. [CrossRef] [PubMed]

65. Wolpert, D.H.; Macready, W.G. No free lunch theorems for optimization. *IEEE Trans Evol. Comput.* **1997**, *1*, 67–82. [CrossRef]

66. Zhang, P.; Wang, F.; Hu, J.; Sorrentino, R. Label propagation prediction of drug-drug interactions based on clinical side effects. *Sci. Rep.* **2015**, *5*, 12339. [CrossRef] [PubMed]

67. Ezzat, A.; Wu, M.; Li, X.L.; Kwoh, C.K. Drug-target interaction prediction using ensemble learning and dimensionality reduction. *Methods* **2017**, *129*, 81–88. [CrossRef] [PubMed]

68. Ezzat, A.; Wu, M.; Li, X.L.; Kwoh, C.K. Drug-target interaction prediction via class imbalance-aware ensemble learning. *BMC Bioinform.* **2016**, *17*, 267–276. [CrossRef] [PubMed]

69. Zhang, R. An ensemble learning approach for improving drug–target interactions prediction. In Proceedings of the 4th International Conference on Computer Engineering and Networks, Shanghai, China, 19–20 July 2015; Wong, W.E., Ed.; Springer International Publishing: Cham, Switzerland, 2015; pp. 433–442.

70. Camacho, D.M.; Collins, K.M.; Powers, R.K.; Costello, J.C.; Collins, J.J. Next-generation machine learning for biological networks. *Cell* **2018**, *173*, 1581–1592. [CrossRef] [PubMed]

71. Zeng, X.; Lin, W.; Guo, M.; Zou, Q. A comprehensive overview and evaluation of circular rna detection tools. *PLoS Comput. Biol.* **2017**, *13*, e1005420. [CrossRef] [PubMed]

72. Davis, J.; Goadrich, M. The relationship between Precision-Recall and ROC curves. In Proceedings of the 23rd International Conference on Machine Learning (ICML '06), Pittsburgh, PA, USA, 25–29 June 2006; ACM Press: New York, NY, USA, 2006; pp. 233–240.

73. Fawcett, T. An introduction to ROC analysis. *Pattern Recognit. Lett.* **2006**, *27*, 861–874. [CrossRef]

molecules

MDPI

Article

Genome-Wide Identification of the LAC Gene Family and Its Expression Analysis Under Stress in *Brassica napus*

Xiaoke Ping [1], Tengyue Wang [1], Na Lin [1], Feifei Di [1], Yangyang Li [1], Hongju Jian [1], Hao Wang [2], Kun Lu [1], Jiana Li [1], Xinfu Xu [1] and Liezhao Liu [1,*]

[1]   College of Agronomy and Biotechnology, Chongqing Engineering Research Center for Rapeseed, Academy of Agricultural Sciences, State Cultivation Base of Crop Stress Biology for Southern Mountainous Land, Southwest University, Chongqing 400715, China; xiaokeping1995@163.com (X.P.); tengyue1992@126.com (T.W.); linna123@yeah.net (N.L.); sddifeifei@163.com (F.D.); liyangyangswu@163.com (Y.L.); jianhongju1989@126.com (H.J.); drlukun@swu.edu.cn (K.L.); ljn1950@swu.edu.cn (J.L.); xinfuxu@126.com (X.X.)
[2]   Hybrid Rapeseed Research Center of Shanxi Province, Shanxi Rapeseed Branch of National Centre for Oil Crops Genetic Improvement, Yangling 712100, China; wangzy846@sohu.com
*   Correspondence: liezhao@swu.edu.cn; Tel.: +86-023-6825-1264

Received: 15 April 2019; Accepted: 17 May 2019; Published: 23 May 2019

Abstract: Lignin is an important biological polymer in plants that is necessary for plant secondary cell wall ontogenesis. The laccase (*LAC*) gene family catalyzes lignification and has been suggested to play a vital role in the plant kingdom. In this study, we identified 45 *LAC* genes from the *Brassica napus* genome (*BnLACs*), 25 *LAC* genes from the *Brassica rapa* genome (*BrLACs*) and 8 *LAC* genes from the *Brassica oleracea* genome (*BoLACs*). These *LAC* genes could be divided into five groups in a cladogram and members in same group had similar structures and conserved motifs. All *BnLACs* contained hormone- and stress- related elements determined by cis-element analysis. The expression of *BnLACs* was relatively higher in the root, seed coat and stem than in other tissues. Furthermore, *BnLAC4* and its predicted downstream genes showed earlier expression in the silique pericarps of short silique lines than long silique lines. Three miRNAs (miR397a, miR397b and miR6034) target 11 *BnLACs* were also predicted. The expression changes of *BnLACs* under series of stresses were further investigated by RNA sequencing (RNA-seq) and quantitative real-time polymerase chain reaction (qRT-PCR). The study will give a deeper understanding of the *LAC* gene family evolution and functions in *B. napus*.

Keywords: laccase; *Brassica napus*; lignification; stress

1. Introduction

B. napus originated from either the Mediterranean or Northern Europe and was formed by chromosome doubling after an interspecific natural cross between *B. rapa* (AA, 2n = 20) and *B. oleracea* (CC, 2n = 18) [1]. Rapeseed oil was once considered as a bad food choice because the seeds contain erucic acid and cholesterol, but with breeding selection and industrial improvement, *B. napus* has nowadays become the third largest source of vegetable oil. Unfortunately, *B. napus* is susceptible to various biotic and abiotic stresses, such as drought, heat, low temperature and fungi infection.

Lignin widely existed and composed of three monomers: coniferyl (G), sinapyl (S), and *p*-coumaryl (H) alcohols. Lignin in plants is involved in the formation of cell walls and together with cellulose increases cellular hardness. Studies have proved that lignin is related to drought stress [2] and a high content can improve the resistance to lodging and *Sclerotinia sclerotiorum* (*S. sclerotiorum*) [3,4]. Laccases are widely distributed with obvious functional differences in plants and fungi [5,6]. It can degrade lignin in *Pleurotus ostreatus* [7] and is expressed in lignifying cells in many plant species [8,9]. LACs are also named multicopper enzymes and supposed to catalyze lignin formation by polymerizing monolignols in plants [10].

To date, *LACs* have been characterized in many species. Lacquer tree contains LAC in the resin ducts and secreted resin [11]. In cotton, an ex-planta phytoremediation system was built based on the overexpression of *LACs* [12]. Forty-four, 46 and 84 *LACs* were identified from *Gossypium arboretum (G. arboretum), Gossypium raimondii (G. raimondii)* and *Gossypium hirsutum (G. hirsutum)*, respectively [13]. A total of 27 laccase candidates (*SbLAC1-SbLAC27*) were identified in *Sorghum bicolor* [14]. *LACs* are continually being detected in other species. An acidic *LAC* gene was found through cDNA cloning in sycamore maple and tobacco [15,16]. Five different LAC-encoding cDNA sequences were identified from ryegrass, with four from the stem and one from the meristematic tissue [17]. *Acer pseudoplatanus* has been found to produce and excrete LAC under cell culture [18,19]. In *Populus euramericana*, five distinct *LACs* were found in xylem tissue [20].

Many studies have indicated the relationship between LAC and lignification. In loblolly pine, LAC was purified from different xylem and shown to coincide with lignin formation in time and place. In *A. thaliana*, gene structure and molecular analysis of the laccase-like multicopper oxidase (*LMCO*) gene family noted that LAC genes (*AtLACs*), *AtLAC4, AtLAC7, AtLAC8* and *AtLAC15* were mainly expressed in the seed coat, root, pollen grains and cell walls, respectively, and all these tissues present high lignification [21,22]. In maize (*Zea mays*), the *ZmLAC2, ZmLAC3, ZmLAC4*, and *ZmLAC5* coincided with the tissues undergoing lignification [23]. Northern blot analysis indicated that five *LACs* (*LAC1, LAC2, LAC3, LAC90* and *LAC110*) in poplar were highly expressed in stems, although their sequences vary greatly [20]. *SofLAC* was reported as a new *LAC* gene and proven to participate in lignification in sugarcane [24]. In *B. napus* and *Brachypodium distachyon* LAC has been shown to affect the accumulation of lignin [25,26]. In addition to oxidative lignin polymerization, LAC can also protect plants from biotic stresses and abiotic stresses such as the toxic phytoalexins and tannins in the host environment. In another study about maize, *ZmLAC3* was induced by wound, whereas *ZmLAC2* and *ZmLAC5* were repressed and *ZmLAC4* gene expression was unaffected [23]. The *OsChI1* gene encodes a putative *LAC* precursor protein in rice (*Oryza sativa*), overexpression of *OsChI1* in *A. thaliana* improved plants drought and salt tolerance [27]. Compared with the numerous reports about *LAC* gene in other species, few reports are available for *B. napus* especially at the genome level and the influence of stress on *BnLACs* [28,29].

In our study, we identified *LAC* gene family in *B. napus* and characterized them by gene structure, motif and cis-element analysis. The expression patterns of all *BnLACs* were identified by RNA-seq and some of them were analyzed under different stresses by RNA-seq or qRT-PCR. The research not only uncovers the evolutionary relationship of *LAC* gene family but also provides information about LAC respond to biotic and abiotic stresses.

2. Results

2.1. Characterization of the 45 BnLACs

Basic Local Alignment Search Tool Protein (BLASTp) was performed and confirmed 45 *BnLACs* in the *B. napus* genome by using 17 *AtLACs* protein sequences as queries (Table 1). Except for *AtLAC2, AtLAC8* and *AtLAC16*, the remaining *AtLACs* had more than one homologous gene in the *B. napu* genome. *AtLAC3, AtLAC4, AtLAC5, AtLAC11, AtLAC12* and *AtLAC17* had four homologs genes were the most. The genomic sequences lengths of *BnLACs* had a wide range from 1937 (*BnLAC13-1*) to 7114 bp (*BnLAC5-4*). The average MW is 63.02 kDa. The pI values of these proteins varied from 6.10 (*BnLAC9-1*) to 9.74 (*BnLAC14-2*). Subcellular localization predicted results showed all the 45 proteins are located in secretory except for *BnLAC11-4*, which is predicted in mitochondrion. Thirty-nine *BnLACs* are accurately, unevenly mapped on the 12 *B. napus* chromosomes and no tandem duplication. The remaining six *BnLACs* are located on the unmapped scaffolds in the Ann_random and Cnn_ random genome.

Table 1. Characterization of *BnLACs* identified in *B. napus* genome.

LAC	Gene ID	Predicted Subcellular Location	Chr	Position	Genomic Sequences Length (bp)	cDNA Length (bp)	Protein Sequences Length (aa)	MW (kDa)	pI
BnLAC1-1	*BnaC08g17780D*	Secretory	C08	21234694:21237623	2929	1737	578	65.20	9.09
BnLAC1-2	*BnaA08g22770D*	Secretory	A08	16433641:16436613	2972	1740	579	65.23	9.09
BnLAC2	*BnaC04g54790D*	Secretory	C04	2100118: 2102571	2453	1737	578	64.46	9.53
BnLAC3-1	*BnaA04g17380D*	Secretory	A04	14120708:14122939	2231	1713	570	64.07	9.48
BnLAC3-2	*BnaC04g41010D*	Secretory	C04	41756886:41759146	2260	1713	570	64.12	9.59
BnLAC3-3	*BnaA05g12170D*	Secretory	A05	6998148: 7000182	2034	1632	543	60.91	9.41
BnLAC3-4	*BnaC04g14580D*	Secretory	C04	11956954:11959010	2056	1551	516	57.83	9.63
BnLAC4-1	*BnaA05g06610D*	Secretory	A05	3604302: 3607497	3195	1677	558	61.54	9.36
BnLAC4-2	*BnaC04g07220D*	Secretory	C04	5406354: 5409465	3111	1677	558	61.54	9.36
BnLAC4-3	*BnaA04g21810D*	Secretory	A04	16546132:16549220	3088	1680	559	61.79	9.41
BnLAC4-4	*BnaC04g45660D*	Secretory	C04	45245585:45248722	3137	1683	560	61.88	9.41
BnLAC5-1	*BnaC04g47080D*	Secretory	C04	46166799:46169078	2279	1710	569	63.00	8.57
BnLAC5-2	*BnaA05g05410D*	Secretory	A05	2791689: 2794263	2574	1725	574	63.65	8.92
BnLAC5-3	*BnaC04g04810D*	Secretory	C04	3510057: 3512615	2558	1725	574	63.52	8.85
BnLAC5-4	*BnaA04g29320D*	Secretory	A04	1313319: 1320433	7114	1722	573	63.47	8.57
BnLAC6-1	*BnaA04g27180D*	Secretory	A04	19140832:19142936	2104	1710	569	63.62	8.19
BnLAC6-2	*BnaC04g50890D*	Secretory	C04	48319588:48321669	2081	1743	580	65.01	8.7
BnLAC7-1	*BnaC05g29170D*	Secretory	A05	20475705:20479975	4270	1668	555	61.16	9.11
BnLAC7-2	*BnaCnng24340D*	Secretory	Cnn	22790373:22794992	4619	1707	568	62.68	9.1
BnLAC8	*BnaC02g03650D*	Secretory	C02	1747091: 1749618	2527	1542	513	57.09	8.88
BnLAC9-1	*BnaC03g00490D*	Secretory	C03	235782: 239591	3809	1704	567	62.86	6.1
BnLAC9-2	*BnaC03g00480D*	Secretory	C03	229247: 232215	2968	1779	592	65.59	6.67
BnLAC9-3	*BnaA03g00500D*	Secretory	A03	177682: 180567	2885	1734	577	63.99	6.83
BnLAC10-1	*BnaC02g03710D*	Secretory	C02	1776151: 1778498	2347	1680	559	61.18	9.46
BnLAC10-2	*BnaAnng13970D*	Secretory	Ann	15018444:15020836	2392	1692	563	61.62	9.46
BnLAC11-1	*BnaA10g26680D*	Secretory	A10	16966690:16969252	2562	1686	561	62.54	8.99
BnLAC11-2	*BnaC02g03260D*	Secretory	C02	1549295: 1551691	2396	1683	560	62.25	8.49
BnLAC11-3	*BnaAnng18410D*	Secretory	Ann	19628923:19631267	2344	1683	560	62.14	8.75
BnLAC11-4	*BnaCnng02950D*	Mitochondrion	Cnn	2437726: 2440207	2481	1731	576	64.31	8.88
BnLAC12-1	*BnaAnng01310D*	Secretory	Ann	804829: 807062	2233	1698	565	62.58	9.52
BnLAC12-2	*BnaA10g25010D*	Secretory	A10	16204889:16207264	2375	1698	565	62.65	9.1
BnLAC12-3	*BnaC02g02320D*	Secretory	C02	1036921: 1039163	2242	1695	564	62.50	9.49
BnLAC12-4	*BnaC09g49940D*	Secretory	C09	48045040:48047342	2302	1698	565	62.58	9.23
BnLAC13-1	*BnaA10g23590D*	Secretory	A10	15550049:15551986	1937	1704	567	63.05	6.65
BnLAC13-2	*BnaC09g48310D*	Secretory	C09	47172850:47174803	1953	1701	566	62.86	6.65
BnLAC14-1	*BnaC09g47160D*	Secretory	C09	46664083:46666581	2498	1731	576	65.13	9.67
BnLAC14-2	*BnaA10g22590D*	Secretory	A10	15188457:15190992	2535	1743	580	65.74	9.74
BnLAC15-1	*BnaC02g38340D*	Secretory	C02	41316880:41322634	5754	1692	563	63.60	9.01
BnLAC15-2	*BnaAnng08030D*	Secretory	Ann	8068829: 8073341	4512	1680	559	63.42	9.04
BnLAC15-3	*BnaA06g30430D*	Secretory	A06	20553666:20556168	2502	1683	560	63.26	9.09
BnLAC16	*BnaC09g34170D*	Secretory	C09	37560731:37563386	2655	1713	570	63.07	9.18
BnLAC17-1	*BnaA03g09140D*	Secretory	A03	4114497: 4116527	2030	1722	573	63.71	9.28
BnLAC17-2	*BnaC03g11450D*	Secretory	C03	5566603: 5568690	2087	1722	573	63.82	9.32
BnLAC17-3	*BnaA02g06580D*	Secretory	A02	3141348: 3143631	2283	1719	572	63.47	9.27
BnLAC17-4	*BnaA10g12900D*	Secretory	A10	10509237:10513003	3766	1722	573	63.59	9.28

Chromosome C04 has the most *LACs* (eight) and A06, C08 only have one *LAC* gene (Figure 1). To further infer the phylogenetic mechanisms of *BnLACs*, a comparative syntenic map of *B. napus* associated with *A. thaliana* was constructed. Thirteen *BnLACs* show syntenic relationship with those in *A. thaliana* and focus on chromosome 02 and chromosome 05 (Figure 2).

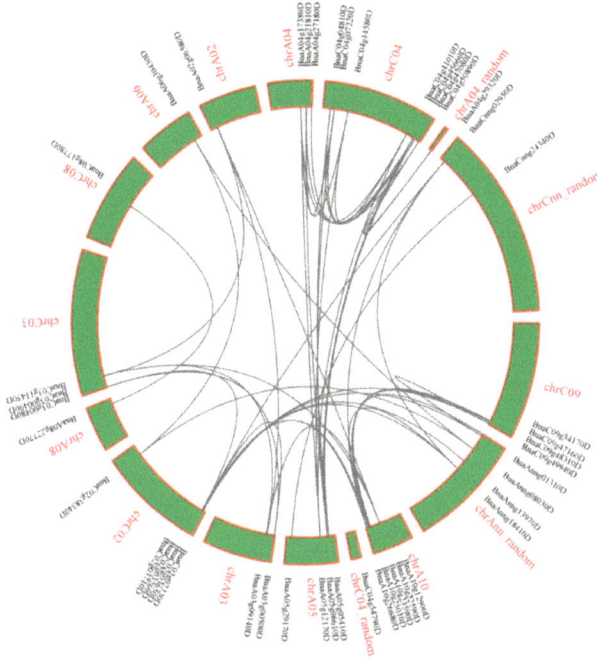

Figure 1. Interchromosomal relationships of *BnLACs* and gray lines indicate duplicated *LAC* gene pairs. The chromosome number and gene ID are indicated with red and black, respectively. The figure was generated by MCScanX with the default parameters [30].

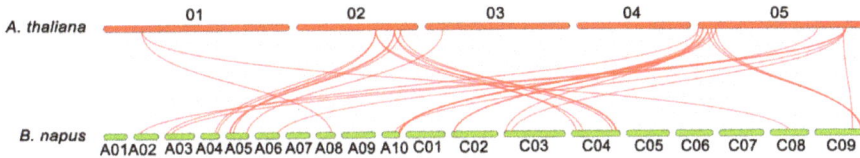

Figure 2. Synteny analysis of *LACs* between *A. thaliana* and *B. napus*. Gray lines in the background indicate all the collinear blocks, red lines highlight the syntenic *LAC* gene pairs. The figure was constructed by TBtools 0.66444553 (https://github.com/CJ-Chen/TBtools).

2.2. Phylogenetic Analysis of LACs in A. thaliana, B. napus, B. rapa and B. oleracea

To study the evolutionary relationship of *LACs* in *A. thaliana, B. napus, B. rapa* and *B. oleracea*, a cladogram containing 45 *BnLACs*, 25 *BrLACs*, eight *BoLACs* and 17 *AtLACs* was constructed and divided into five groups with well-supported bootstrap values (Figure 3). Groups I, II, III, IV and V had 35, 26, 5, 15 and 14 members, respectively. Forty-five *BnLACs* were unevenly divided into five groups, the most being in Group I which contained 16 *BnLACs* the least being in Group III that only contained one. According to the bootstrap value in the tree, genes in same group were closely related but in different groups were far apart. Genes divided into the same group are thought to have similar functions and number of *LACs* in *B. oleracea* was far less than in *B. rapa* and *B. napus*, however, every group contained at least one *BoLAC*, which is essential to keep complete gene function of *LAC* in *B. oleracea*.

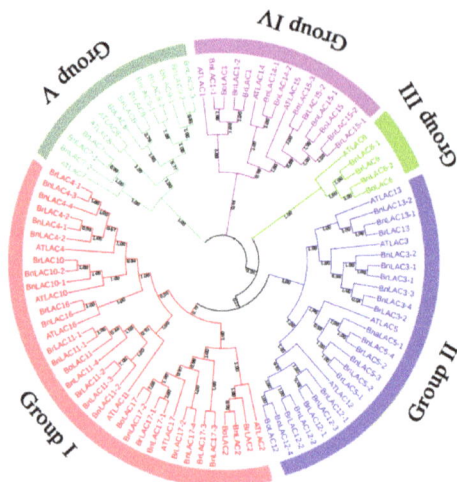

Figure 3. The cladogram of *LAC* proteins from *A. thaliana* (17), *B. napus* (45), *B. rapa* (25) and *B. oleracea* (8). The cladogram was constructed by the MEGA 7.0 software [31] using the neighbour-joining option with 1000 bootstrap replicates and pairwise deletion. Distinct colour segment represents different groups, bootstrap value are shown near nodes.

2.3. Gene Structure and Conservative Domain Analysis of BnLACs and AtLACs

As shown in the cladogram, members in the same group had highly similar gene structures, and the number of exons in the 45 *LACs* ranged from 4 to 7 (Figure 4). Compared with introns, exons were more stable in length. For example, *BnLAC5-4* had a separate longer intron but other members in the Group II didn't contain one. *BnLACs* homologous with *AtLAC4*, *AtLAC7*, *AtLAC8* and *AtLAC9* showed diversity with other *BnLACs* because of one or two long introns existed. For exploring more characteristics about *LACs*, introns number of *LACs* from *A. thaliana*, *S. bicolor*, *G. arboretum*, *G. raimondii* and *G. hirsutum* were compared. In *A. thaliana*, there was no *LAC* gene had more six introns. Most *LACs* in *S. bicolor* have one to three introns less than other species with five introns (Figure 5).

Full-length protein sequences of 17 *AtLACs* and 45 *BnLACs* were analyzed to identify their conserved motifs and further understand their functions (Figure 6). The length of the 20 motifs ranged from 6 to 50 amino acids and motif sequences are provided in Table S1. *AtLAC4*, *BnLAC8* had the fewest motifs (eight), *BnLAC11-4* had the most motifs (twenty). In contrast to others *LACs*, *AtLAC2* and *BnLAC11-4* had an extra motif 18, *BnLAC3-3*; *BnLAC3-4* and *BnLAC7-2* lose the motif 2. *LACs* divided into the same groups also had different motifs, *BnLAC5-4* and *AtLAC16* lost the motif 6 compared with other genes in their corresponding group. *AtLAC8*, *AtLAC9* and their homologous gene in *B. napus* genome lacked the motif 5 when compared with others in Group V. Motif 1 and motif 3 are highly conserved and can be found in all the protein sequences of 62 *LACs*.

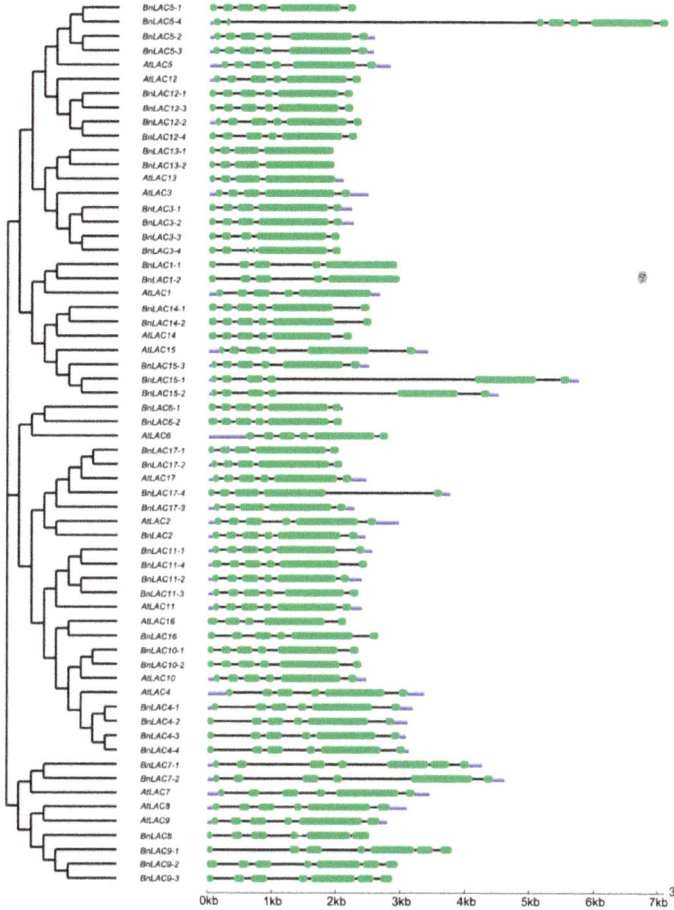

Figure 4. Exon-intron structures of *BnLACs* and *AtLACs* with a cladogram. The result generated by GSDS 2.0 (http://gsds.cbi.pku.edu.cn/). The green boxes, black lines and blue box indicate exons, introns, untranslated region, respectively.

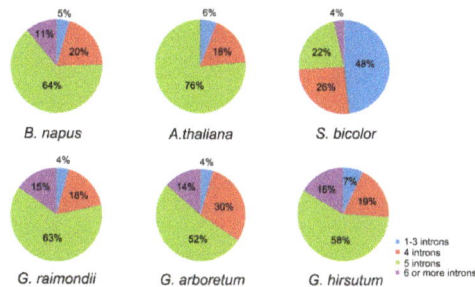

Figure 5. Number of introns in *A. thaliana*, *S. bicolor*, *G. arboretum*, *G. raimondii* and *G. hirsutum*. The figure was constructed by GraphPad Prism 6 (Graphpad Software Inc., La Jolla, CA, USA, www.graphpad.com).

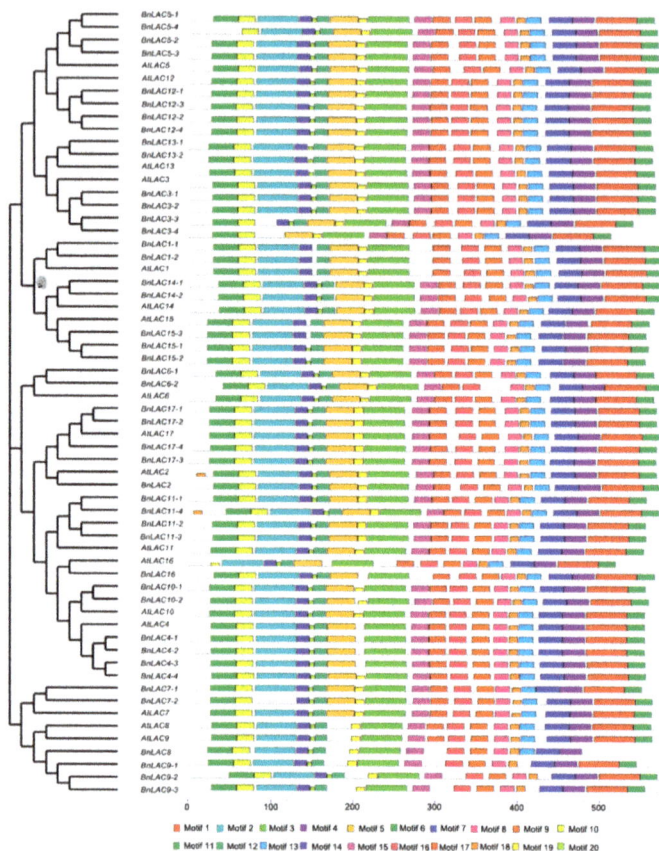

Figure 6. Conserved motif analysis of *BnLACs and AtLACs* proteins presented with a cladogram. Conserved motifs were generated by MEME 5.0.4 (http://meme-suite.org/tools/meme) and boxes of different colors represent motifs 1-20. Motif sequences are provided in Table S1.

2.4. Diverse cis Regulatory Elements and miRNAs are Predicted

To investigate what condition could influence *BnLACs* expression, 1500 bp upstream of the initiation codons were analysed for cis-elements. Eleven kinds of elements including light responsive, six hormone-related, four stress-related elements were searched (Table S2). All the *BnLACs* are light and hormone responsive. Twenty-six *BnLACs* have a gibberellin element, 25 *BnLACs* have an ethylene element, which are the two most elements. Seven members contained a wound element and thirty *BnLACs* were influenced under drought stress. Fourteen members had low temperature elements and 34 members may have been influenced by heat stress.

Eleven *BnLACs* were predicted to have their expression regulated by miRNAs (Table S3, expectation number under 3.0 were selected). All the predicted *BnLACs* were regulated by miR397a and miR397b; *BnLAC4-1* and *BnLAC4-2* were also regulated by miR6034. As the expectation number showed, *BnLAC17-3* was the most likely targeted gene by miR397a and miR397b.

2.5. Expression Pattern Analysis of BnLACs

To investigate the expression patterns of 45 *BnLACs*, a heatmap was built based on the RNA-seq (BioProject ID PRJNA358784) using 32 different tissues and stages of *B. napus* as samples (Table S4, Figure 7). Results suggested that *BnLACs* have different expression patterns across tissues and stages. Strong expression occurred in highly lignified tissues such as roots and stems. *BnLAC15-1* and *BnLAC15-2* had the highest expression in the seeds and seed coats. *BnLAC4-1, BnLAC4-2, BnLAC4-3* and *BnALC4-4* had high expression in silique pericarps. *BnLAC5-2, BnLAC5-3, BnLAC15-1, BnLAC15-2* and *BnLAC15-3* highly expressed in seed coats. No *BnLACs* highly expressed in leaves. *BnLAC9-1* and *BnLAC14-2* rarely expressed in any tissue and stage. These results showed *BnLACs* functioned differently and some members are redundant.

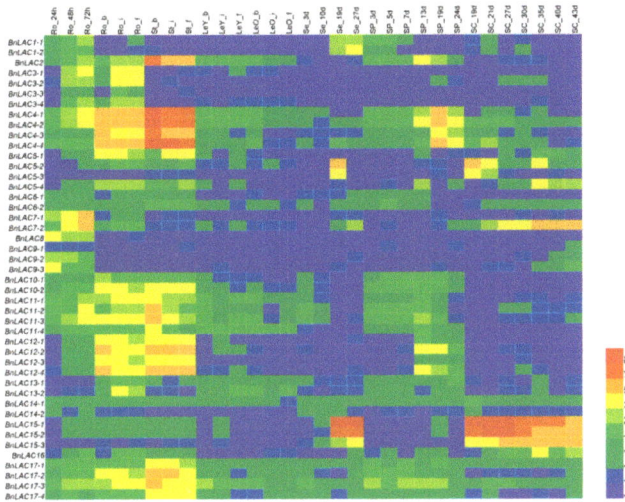

Figure 7. RNA-seq of *BnLACs* in *B. napus* 32 different tissues and stages. Ro, root; St, stem; LeY, young leaf; LeO, old leaf; Se, seed; SP, silique pericarp; SC, seed coat; s, seedling stage; b, bud stage; i, initial flowering stage; and, f, full-bloom stage. The 24, 48, and 72 h means the time that had passed after seed germination. The 3, 10, 27 d and other indicate the number of DAF. The bar on the lower right corner represents the Fragments Per Kilobase of Transcript Per Million Fragments Mapped (FPKM) values and different colors represent different expression levels. Heat map was generated by HemI 1.0 software (http://hemi.biocuckoo.org/faq.php).

2.6. Responses of BnLACs upon Abiotic Stress

As many studies described lignification can response to stresses, RNA-seq and qRT-RCR were used to analyze the expression patterns of several *BnLACs* under different stresses. RNA-seq showed that under Cd^{2+} stress, *BnLAC14-2, BnLAC15-1* and *BnLAC15-2* were up-regulated, *BnLAC12-1* and *BnLAC12-4* were down-regulated at 24 hours after stress then up-regulated at 72 hours after stressed. *BnLAC6-1, BnLAC9-1* and *BnLAC15-3* had no change and other *BnLACs* were down-regulated at the analysed time points. Most *BnLACs* had low expression after 72 hours of Cd^{2+} stress and only 7 *BnLACs* (*BnLAC4-1, BnLAC4-2, BnLAC7-2, BnLAC11-2, BnLAC11-3, BnLAC11-4* and *BnLAC17-3*) still highly expressed (Table S5, Figure 8a). Under NH_4^+ toxicity, Expression levels of *BnLAC6-2, BnLAC13-1* and *BnLAC13-2* were low and not influenced by NH_4^+ toxicity. *BnLAC4-1, BnLAC10-2, BnLAC11-1* and *BnLAC17-2* were up-regulated at 3 or 12 hours after stress and changed to normal level at 48 hours. Some members responsed NH_4^+ toxicity until 48 hours after stressed, *BnLAC2, BnLAC4-2* and *BnLAC14-2* were down-regulated and members such as *BnLAC3-2, BnLAC3-3, BnLAC3-4, BnLAC5-2, BnLAC7-2, BnLAC12-2, BnLAC12-4* and *BnLAC15-1* show significant upregulation (Table S5, Figure 8b). Under drought and wound stress, expression levels of 14 *BnLACs* in stems and leaves of three lines (ZS11, 7191 and D2) were analyzed by qRT-RCR.

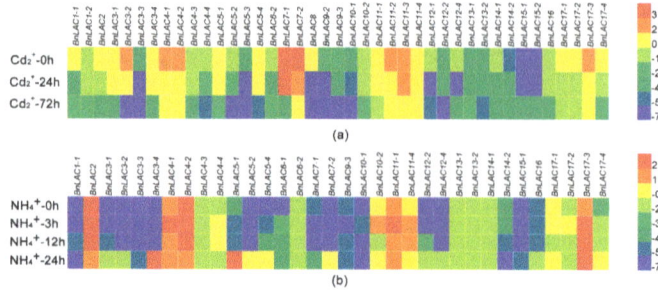

Figure 8. RNA-seq of *BnLACs* in *B. napus* tissues in response to Cd^{2+} stress (**a**), NH_4^+ stress (**b**). The 0, 3, 12 and other time points in (**a**,**b**) denotes the time after Cd^{2+}, NH_4^+ stress respectively. The bar on the lower right corner represents the FPKM values and different colors represent different expression levels. Heat map was generated by HemI 1.0 software (http://hemi.biocuckoo.org/faq.php).

High expression concentrated at stems, which were consistent with the RNA-seq analysis (Table S6). In leaves of three lines under drought stress, the 14 *BnLACs* had similar expression pattern (Figure 9). *BnLAC2*, *BnLAC4-1*, *BnLAC4-2*, *BnLAC4-4*, *BnLAC11-3*, *BnLAC12-1*, *BnLAC12-2* and *BnLAC17-3* have stable expression and close to zero. *BnLAC6-1*, *BnLAC6-2*, *BnLAC11-2*, *BnLAC11-4*, *BnLAC14-1* and *BnLAC17-1* have relatively high expression. *BnLAC6-1*, *BnLAC11-2* and *BnLAC11-4* were down-regulated, *BnLAC6-2* and *BnLAC17-1* were slightly up-regulated under drought stress.

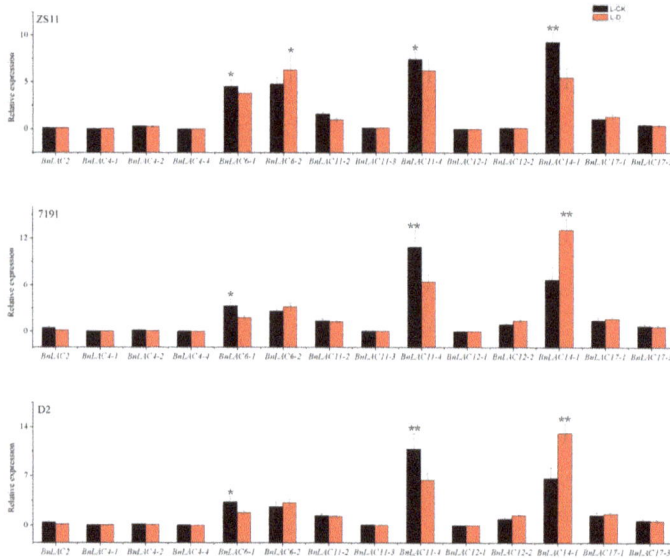

Figure 9. Quantitative RT-PCR of 14 *BnLACs* in *B. napus* leaves in response to drought stress. Expression of *Actin* and *UBC21* were used to normalize the expression level of each *LAC* gene. Bars represent means ± SEM of three biological replicates. Bars marked with asterisks indicate significant differences (Student's t-test) to corresponding control samples for the same time point, *$P < 0.05$, **$P < 0.01$. The results were presented by OriginPro 8 (OriginLab Corporation, Northampton, MA, USA). L-CK and L-D means no drought treated and drought treated, respectively.

In wounded leaves of line ZS11, the expression of *BnLAC2*, *BnLAC11-2*, *BnLAC11-3*, *BnLAC17-1* and *BnLAC17-3* rose firstly then returned to normal levels. *BnLAC6-1*, *BnLAC11-4*, *BnLAC12-2* and

BnLAC14-1 had similar expression patterns and showed high expression at 0.5 and 3 hours after wounding. *BnLAC4-1, BnLAC4-2, BnLAC4-4, BnLAC6-2* and *BnLAC12-1* showed no expression in any samples. In wounded leaves of line 7191, *BnLAC4-1, BnLAC4-2, BnLAC4-4, BnLAC11-4* and *BnLAC12-1* were up-regulated followed by down-regulation and showed the highest expression at different time points. *BnLAC4-1, BnLAC4-2, BnLAC4-4* and *BnLAC12-1* showed the highest expression at 1.5 hours after wounding and *BnLAC11-4* showed the highest expression at 0.5 hours after wounding. *BnLAC6-2, BnLAC12-2* and *BnLAC14-1* were down-regulated for all the analyzed time points and *BnLAC11-3* had no change after wounding. In wounded leaves of line D2, *BnLAC2, BnLAC6-2 BnLAC11-2, BnLAC12-2, BnLAC17-1* and *BnLAC17-3* had high expression at two time points, 0.5 and 1.5 hours after wounding, respectively. *BnLAC4-1, BnLAC4-2, BnLAC4-4, BnLAC11-3* and *BnLAC12-1* nearly have no expression in the control sample and kept stable under wounding stress. Expression levels of *BnLAC6-1, BnLAC11-4,* and *BnLAC14-1* responded to wounds slowly and showed high expression at 3 and 6 hours (Figure 10).

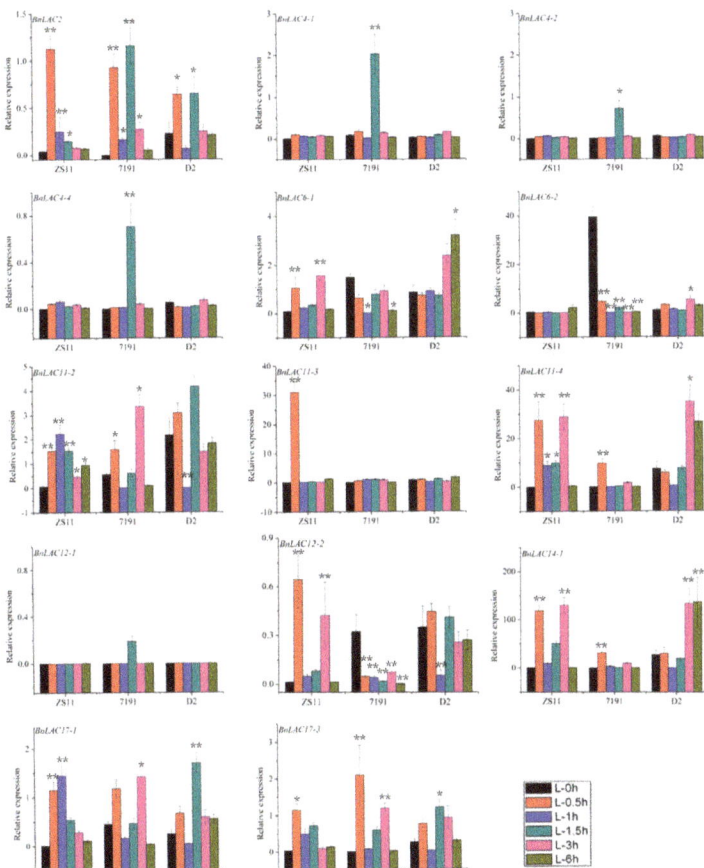

Figure 10. Quantitative RT-PCR of 14 *BnLACs* in *B. napus* leaves in response to wounding stress. Expression of *Actin* and *UBC21* were used to normalize the expression level of each *LAC* gene at each time point. Bars represent means ± SEM of three biological replicates. Bars marked with asterisks indicate significant differences (Student's t-test) to corresponding control samples for the same time point, *$P < 0.05$, **$P < 0.01$. The results were presented by OriginPro 8. Different color means different time after wound shown as bar at lower right corner.

In wounded stems of line ZS11, *BnLAC2*, *BnLAC4-1*, *BnLAC4-4*, *BnLAC11-3*, *BnLAC11-4* and *BnLAC12-1* had the highest expression at 1.5 hours after wounding and changed to zero at 6 hours. *BnLAC6-2*, *BnLAC11-2*, *BnLAC12-2*, *BnLAC17-1* and *BnLAC17-3* were up-regulated followed by down-regulation. *BnLAC6-1* and *BnLAC14-1* were down-regulated all the time after wounding. In wounded stems of 7191 line, *BnLAC4-1*, *BnLAC4-4*, *BnLAC11-2* and *BnLAC11-3* were down-regulated followed by up-regulation and showed high expression at 6 hours after wounding. *BnLAC12-1*, *BnLAC12-2* and *BnLAC17-1* were down-regulated at all the analyzed time points. *BnLAC6-1*, *BnLAC6-2*, *BnLAC11-4* and *BnLAC14-1* showed similar expression patterns and had their highest expression at 6 hours after wounding. In wounded stems of D2 line, *BnLAC2* and *BnLAC4-4* had the highest expression at 0.5 hour after wounding. *BnLAC4-1*, *BnLAC4-2*, *BnLAC11-3*, *BnLAC12-1*, *BnLAC17-1* and *BnLAC17-3* showed high expression at 0.5 and 1 hours after wounding. Expression levels of *BnLAC6-1*, *BnLAC12-2* and *BnLAC14-1* almost no change and were close to zero for the analyzed time points. Both *BnLAC11-2* and *BnLAC11-4* had the highest expression at 1 hour and lowest expression at 1.5 hours after wounding (Figure 11).

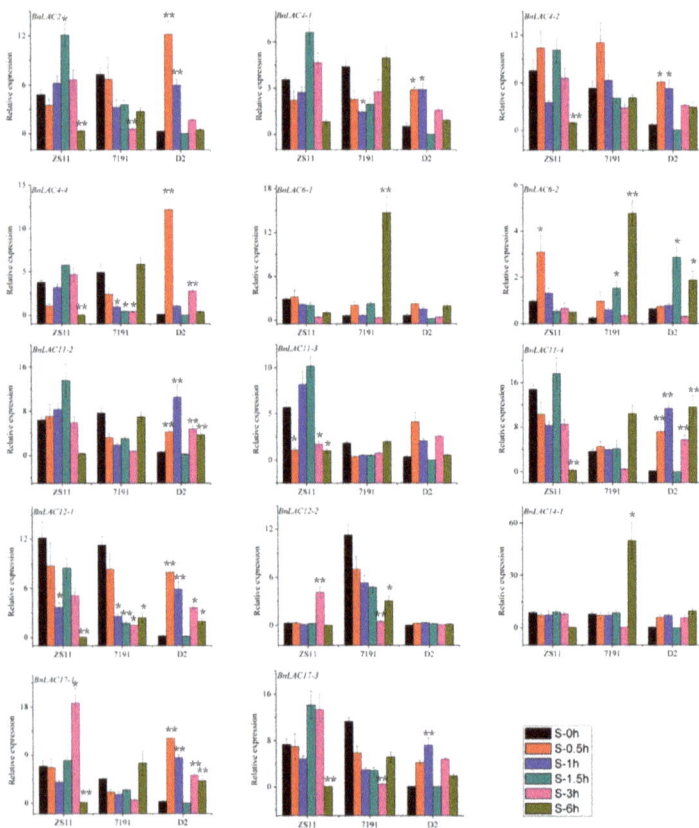

Figure 11. Quantitative RT-PCR of 14 *BnLACs* in *B. napus* stems in response to wounding stress. Expression of *Actin* and *UBC21* were used to normalize the expression level of each *LAC* gene at each time point. Bars represent means ± SEM of three biological replicates. Bars marked with asterisks indicate significant differences (Student's t-test) to corresponding control samples for the same time point, *$P < 0.05$, **$P < 0.01$. The results were presented by OriginPro 8. Different color means different time after wound shown as bar at lower right corner.

2.7. BnLAC4 and its Predicted Downstream Genes are Differentially Expressed in the Silique Pericarp between Long and Short Silique Lines

In STRING platform, CTL2, IRX3, CESA4, IRX1, LAC17, GAUT12, IRX6, PGSIP1, GLP10 and FLA11 were predicted to interact with protein AtLAC4. A total 70 homologous genes were identified in the B. napus genome and expression levels of them in silique pericarps of long and short siliques lines were showed by a heatmap (Table S7, Figure 12). Most of the identified genes showed higher expression in silique pericarps of short silique lines on the 16th Days After Flower (DAF) and almost equal expression in two kinds of silique pericarps on the 25th DAF. On the contrary, many genes showed a higher expression in long silique lines on the 35th DAF. Those results showed BnLAC4 and its' predicted downstream genes expressed earlier in silique pericarp of short siliques lines.

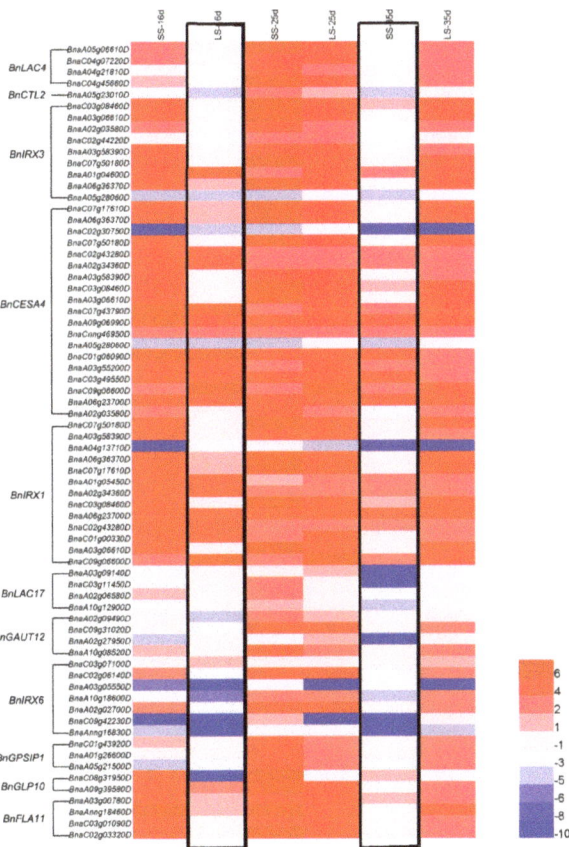

Figure 12. RNA-seq of BnLAC4 BnCTL2, BnIRX3, BnCESA4, BnIRX1, BnLAC17, BnGAUT12, BnIRX6, BnPGSIP1, BnGLP10 and BnFLA11 genes in six kinds of B. napus silique pericarps. LS, long silique lines; SS, short silique lines. The 16, 25, and 35 d means DAF. The bar on the lower right corner represents the FPKM values and different colors represent different expression levels. Heat map was generated by HemI 1.0 software (http://hemi.biocuckoo.org/faq.php). Lower expression between long and short silique lines at 16th and 35th DAF were indicated by black box.

3. Discussion

B. napus is the third largest source of vegetable oil worldwide and plays an important role in national economies and food industries. The rapeseed yield decreases frequently because of lodging and other biotic and abiotic stresses. Many studies have shown that lignin aids in the resistance to fungi, stress and the LAC acts as an enzyme related to lignification in plants. Previous studies on the *LAC* gene family have been performed, but no related reports on the analysis of this family in *B. napus* exist until now. We identified and analyzed the LAC gene family with the aim of providing a reference at the genome level and deeper understanding of lignification.

3.1. Loss Events Occurred in the LAC gene family Along with the Evolution

Loss events occur frequently during evolution because of hybridization and chromosome doubling [32]. As a result of whole-genome triplication (WGT), the genes in *A. thaliana* should have three homologs in *B. rapa* and *B. oleracea*. In the study, only *AtLAC1* had three orthologous genes in *B. rapa* and no *AtLACs* corresponding to three orthologous genes were found in *B. oleracea*. Some *AtLACs* even have no orthologous genes in *B. rapa* and *B. oleracea* like *AtLAC14* in *B. rapa*, *AtLAC3* and *AtLAC4* in *B. oleracea*. Furthermore, *B. napus* formed by natural hybridization and polyploidization of *B. rapa* and *B. oleracea*, but no one *AtLAC* corresponding to six *BnLACs* was seen in this study and the most is four. Hence, the conclusion can be drawn that not only during the whole-genome triplication but also the formation of *B. napus*, gene loss events existed in *LAC* gene family universally.

Twenty four *BoLACs* were identified in *B. oleracea*, but only eight of them contained four essential conserved domains as described in the Materials and Methods section. Compared with *B. rapa* that contained 25 *BrLACs*, faster or broader gene loss happened in *B. oleracea*.

3.2. Regulation of BnLAC Genes

Cis-element analysis showed the expression of *BnLACs* was regulated at transcriptional level (Table S2). Research has shown *LAC* genes were also regulated at the post-tanscriptional level or through post-translational modifications [14]. G-box can be found in some *BnLACs* and related to light-response and salt tolerance in rice flag leaf by combinating with bZIP, bHLH, and NAC TFs [33–35]. The *AtLAC4* gene has been confirmed to be up-regulated after MYB58 binds to AC elements [36]. Our results show diverse cis-elements in promoters of *BnLACs*, 11 kinds of promoters were selected for analysis including light responsive element, six hormone-related elements and four stress-related elements (Table S2). These findings revealed that *BnLACs* are also regulated by a series of factors. Related studies have reported that Ptr-miR397a is a negative regulator of *LACs* in *Populus trichocarpa* [37]. Several *LACs* are targeted by miR408, miR397, and miR857 in *A. thaliana* when Cu is absent [38]. Ptr-miR397a and Os-miR397 are involved in negative regulation of *PtrLACs* and *OsLACs* [38,39]. Seven *SbLACs* have also been predicted to be sbi-miRNA targets [14]. In the study, 11 *BnLACs* were predicted to be regulated by miR397a, miR397b and miR6034 (Table S3). All the 11 predicted *BnLACs* were regulated by miR397a and miR397b, and *BnLAC4-1* and *BnLAC4-2* were also regulated by miR6034. Research about miR6034 is very few and the process it participates in is not clear. Results have proved *BnLAC17-3* was the most likely targeted gene by miR397a and miR397b in our study and in *A. thaliana* [40]. The findings of the study and combined with previous researches suggest *LAC* genes are truly regulated by miRNAs, and miRNA397 likely plays a very important role in the regulatory network.

3.3. Expression Patterns and Response to Stress

Abundant expression focuses on the roots, stems, and seed coats, whereas the expression in leaves, petals, pistils, and stamens are very low. The expression coincides with lignification in different parts of the plant. *BnLAC13-1* and *BnLAC13-2* showed different expression pattern though contain same cis-elements (Table S2, Figure 7). The reason would be a network containing other factors exists and regulates expression patterns of *BnLACs* such as miRNA and epigenetic modifications [41]. Some members, such as *BnLAC9-1* and *BnLAC14-1*, were never highly expressed in any tissues or stages. It might be that their function was not required in biological processes or was only induced. by certain environmental factors, similar genes can also be found in cotton and Sorghum bicolor [13,14].

Many studies have illustrated that *LACs* are influenced by different kinds of stress and our study also proved that. *OsChL1* as a putative laccase precursor, its expression was increased under drought stress and overexpressed the gene in *A. thaliana* can increase drought and salt tolerance [27]. In the study, six out of 14 *BnLACs* were influenced by drought stress. *BnLAC6-2*, *BnLAC11-2* and *BnLAC11-4* were up-regulated, *BnLAC6-1* and *BnLAC11-4* were down-regulated. *BnLAC14-1* was up-regulated in line 7191 and D2 but down-regulated in line ZS11. Metal ions would influence the expression of *LACs* and miRNAs directly or indirectly. In *Citrus*, the expression of *LAC7* was up-regulated by boron toxicity [42]. miR397 has been confirmed as a regulatory factor of *LACs* and its expression were influenced by Cd^{2+} [43]. In our study, most *BnLACs* were down-regulated after Cd^{2+} treatment and only four *BnLACs* showed upregulation. Another study in our lab has indicated NH_4^+ enrichment treatment would increase the lignin content in stem and root. Consistent with phenotype, the expression of most *BnLACs* were up-regulated at different time after NH_4^+ enrichment treatment by RNA-seq. According to the results of the promoter analysis, some members like *BnLAC5-1*, *BnLAC6-1*, *BnLAC6-2* contain cis element about wound and a study has been reported *LACs* were influenced by wound [23]. qRT-RCR results showed the expression of the selected 14 *BnLACs* in leaves and sterms changed intricately after wounded. We also found gene in different lines responded to stress differently, *BnLAC11-4* was up-regulated in lines 7191 and D2 but down-regulated in line ZS11. Further works are needed to find out the relation between wound healing and *LACs*.

3.4. BnLAC4 and its Downstream Genes May Participate in Silique Elongation in B. napus

Studies have reported that miR397 (both miR397a and b) regulate lignin content and yield traits in *Rice* and *Populus trichocarpa* via modulating *LACs* [39]. In *A. thaliana*, overexpression of miR397b-resistant *AtLAC4* results in an increased silique length and decreased lignin content [40]. In our research, homologous genes of *AtLAC4*, *AtCTL2*, *AtIRX3*, *AtCESA4*, *AtIRX1*, *AtLAC17*, *AtGAUT12*, *AtIRX6*, *AtPGSIP1*, *AtGLP10* and *AtFLA11* in *B. napus* show earlier expression in silique pericarp of short silique lines than long silique lines. Some of those genes predicted like *AtIRX1* and *AtIRX6* are related to secondary cell wall biosynthesis. Combined with the reports, it could be that the period of *BnLAC4* and its downstream genes expression may regulate silique length in *B. napus*.

4. Materials and Methods

4.1. Plant Materials and Stress Treatment

Inbred line Zhongshuang11 (ZS11), 7191, and D2 were sown in humus and grown to the four-leaf-stage. Half of the plants in each line were transplanted for drought stress and the remaining were left for the control. After 25 days without irrigation, the lines in the drought stress treatment showed a wilted phenotype and young leaves from control and stressed lines were frozen immediately in liquid nitrogen and stored at −80 °C.

Leaves and stems were wounded at the bolt stage. Leaves were wounded by a plastic comb-like brush, which was 8.5 cm long and had 42 spikes with a diameter of 1 mm that were equally arrange. Every leaf received three rows of wounds on each side of the midrib and parallel with it; the total number of punctures in each leaf was 252. Stems were wounded to a centimeter depth by scalpel blades. Samples were harvested around the cut at 0.5, 1, 1.5, 3, and 6 hours after wound [44]. The collected samples were frozen immediately in liquid nitrogen and stored at −80 °C for RNA isolation.

The seeds were grown at 22 °C with a light intensity of 200 mol/m²/s and a photoperiod of 16 hours for 7 days in hydroponic culture. Subsequently, the plants were collected after exposure to 1 mM Cd^{2+} ($CdCl_2$) at 0, 24, and 72 h immediately frozen in liquid nitrogen for RNA sequencing.

Seeds of the *B. napus* line ZS11 were surface-sterilized with 1.2% sodium hypochlorite and germinated in a chamber room (16 hours light 15000Lx/8 hours dark at temperature 25 °C) with Hoagland solution. At four-leaf-stage, a portion of the seedlings were cultivated with modified Hoagland solution (0 mM NO_3^-, 10 mM NH_4^+, pH 6.0; other ions were not changed) for NH_4^+ toxicity treatment. After 3, 12 and 48 hours, the third and fourth true leaves were immediately frozen in liquid nitrogen and stored at −80 °C for RNA sequencing.

Long and short silique lines were selected from a recombinant inbred line (RIL) population constructed from a cross between GH06 (female parent) and P174 (male parent). Lines were planted in open field and grew under normal condition, silique pericarps of two kinds lines were collected respectively at 16th, 25th and 35th DAF and frozen in liquid nitrogen for RNA sequencing.

4.2. Characterisation of the LAC Gene Family

To date, 17 *LACs* have been reported in *A. thaliana* [35]. BLASTp was performed in the *B. napus, B. rapa* and *B. oleracea* genome using the *AtLACs* protein sequences as queries and sequences with E-value less than 1×10^{-20} were selected [45]. Some repeated sequences were manually deleted according to the E-value. All the remaining genes were checked by InterProScan (http://www.ebi.ac.uk/interpro) [46], and the sequences with four essential conserved domains of multicopper oxidase type 1 (IPR001117), multicopper oxidase type 2 (IPR011706), multicopper oxidase type 3 (IPR011707) and laccase (IPR017761) were deemed as candidate *LACs* [13]. Another BLASTp was performed in *A. thaliana* genome using candidate *BnLACs* protein sequences as queries and hold those genes that corresponded to *AtLACs*. *LACs* identified from *B. napus, B. rapa* and *B. oleracea* were named according to the orthologous sequence in *A. thaliana*. Information about putative sequences were searched in the date bases of BRAD (http://brassicadb.org/) and *B. napus* Genome Browser (http://www.genoscope.cns.fr/brassicanapus/). The chromosomal locations were shown by MapChart software [47], and the number of amino acids, isoelectric point (pI) and molecular weight (MW) of the protein sequences were searched using the ExPASy website (http://web.expasy.org/). The subcellular localization pattern of *LAC* genes were predicted using the web-based tool TargetP1.1 server (http://www.cbs.dtu.dk/services/TargetP/) [48]. Multiple Collinearity Scan toolkit (MCScanX) was adopted to analyze the gene duplication events with the default parameters [30]. The synteny relationship of the *LACs* in *B. napus* and *A. thaliana* were constructed using the Dual Systeny Plotter software (https://github.com/CJ-Chen/TBtools).

4.3. Evolutionary relationship of the LAC Genes Family in A. thaliana, B. napus, B. rapa, and B. oleracea

A cladogram containing the sequences identified from the four species was built using MEGA 7.0 software [31], with 1000 bootstrap replicates performed, and it was then modified by iTOL (http://itol.embl.de/) and Photoshop CS 5 to further visualize evolutionary relationship.

4.4. Gene Structure and Conserved Motif Analysis

The cDNA sequences, genomic sequences and full-length protein sequences of *AtLACs* and *BnLACs* were obtained from the *A. thaliana* genome (http://www.arabidopsis.org/) and *B. napus* Genome Browser (http://www.genoscope.cns.fr/brassicanapus/) respectively. Gene structures were analysed by Gene Structure Display Server (GSDS2.0, http://gsds.cbi.pku.edu.cn/) [49], conserved motifs were tested by and Multiple EM for Motif Elicitation version 5.0.4 (MEME, http://meme-suite.org/tools/meme) with a limit of 20 motifs and any number of repetitions deemed as motif sites [50].

4.5. Cis-Elements Analysis and Prediction of miRNA Target BnLACs

One thousand and five hundred base pairs (bp) upstream of the initiation codons (ATG) were searched in the *B. napus* Genome Browser and cis-elements were analysed using the PlantCARE database (http://bioinformatics.psb.ugent.be/webtools/plantcare/html/). The genome sequences of the 45 *BnLACs* were submitted to the psRNATarget Server (http://plantgrn.noble.org/psRNATarget/) with default parameters to predicte the miRNAs with a target site on *BnLACs*. MiRNAs from the *B. napus* genome were selected and expectation number under 3 were selected [51].

4.6. Expression Patterns Analysis of B. napus LAC Genes

The expression patterns of *BnLACs* were based on RNA-seq, using data from the BioProject ID PRJNA358784. The data included the expression in different tissues in different stages of the *B. napus* cultivar ZS11. The clean reads were aligned to the *B. napus* reference genome and these sequence data and corresponding gene annotation files were downloaded from the genome website (http://www.genoscope.cns.fr/brassicanapus). The BWA and Bowtie softwares were used to map the reads to a reference genome and the reference genes, respectively [52]. The alignment results were visualized by IGV (Integrative Genomics Viewer) and genes expression levels were quantified on the basis of their FPKM values using Cufflinks with default parameters. For RNA-seq data about Cd^{2+} and NH_4^+ stresses, HISAT2 was used to map the reads to a reference genome and genes [53]. The alignment results were also visualized by IGV and the level of each gene expression was measured as FPKM by StringTie [54]. The heatmaps were built to represent the expression level of the *BnLACs* using HemI 1.0 software (http://hemi.biocuckoo.org/faq.php) [55].

4.7. RNA Extraction, Reverse Transcription and qRT-PCR

Total RNA was extracted using the EZ-10 DNAaway RNA Mini-prep Kit (Sangon Biotech, Shanghai, China). NanoDrop 2000 (Thermo Fisher Scientific, Worcester, MA, USA) and electrophoresis were used to measure concentrations and RNA integrity. Complementary DNA was obtained using the iScriptTM cDNA Synthesis Kit (Bio-Rad, Hercules, CA, USA) and diluted 15 times with distilled deionized water for qRT-PCR. The composition of qRT-PCR contained 2 μL of 15-fold diluted cDNA solution, 10 μL of SYBR®Green Supermix (Bio-Rad), 0.4 μL of 10 mM forward and reverse primers and 7.2 μL of distilled deionized water. Primers were designed on Primer Premier Software (version 5.0) (Table S8) [56] and qRT-PCR was performed on a CFX96 Real-time System (Bio-Rad) with the following conditions: 98 °C for 30 s, then 40 cycles of 98 °C for 15 s, 55 °C for 30 s, and an increase from 65–95 °C at increments of 0.5°C every 0.05 s. Three biological replicates and three technical replications were used for qRT-PCR. According to the $2^{-\Delta\Delta Ct}$ method using Actin7 and UBC21 as internal controls, the gene expression levels were determined and displayed by OriginPro 8 (OriginLab Corporation, Northampton, MA, USA).

4.8. Proteins Interaction with AtLAC4 and Identified their Homologous Genes in the B. napus genome

MiR397b regulated both lignin content and silique length via modulating *AtLAC4* has been identified [39]. To understand whether the similar interaction exit in *B. napus*, the protein sequences of *AtLAC4* was obtained from the *A. thaliana* genome (https://www.arabidopsis.org/) and used to predict the interacting proteins in STRING platform (https://string-db.org/?tdsourcetag=s_pctim_aiomsg). Homologous genes of predicted protein sequences were searched in the *B. napus* genome as the method of identifying *BnLACs*.

5. Conclusions

A total of 45 putative *BnLACs* were identified in the *B. napus* genome and unevenly mapped on the 12 *B. napus* chromosomes with no tandem duplication. *BnLACs* were divided into five groups in the cladogram and members in same group had similar structures and motifs. *BnLACs* had high expression in lignified tissues such as roots, stems and seed coats. After high concentration of NH_4^+ toxicity, most *BnLACs* were up-regulated and lignin more and faster deposited. Expression of many *BnLACs* were close to zero in leaves and uninfluenced by drought stress. Some *BnLACs* were down- regulated and individual gene showed different responses in different lines. Many members were intricately influenced by wounding stress, and significantly regulated members may take part in the healing process. By RNA-seq between long and short silique lines, we forecasted that earlier lignification may be a reason for the short siliques. The results in the study give a chance to further study the functions of *BnLACs* in lignification and the interaction with other biological processes.

Supplementary Materials: The following are available online at http://www.mdpi.com/1420-3049/24/10/1985/s1, **Table S1:** Motif sequences of conserved motif analysis, **Table S2:** Putative cis-elements in 45 *BnLACs* promoters, **Table S3:** List of *BnLACs* with putative miRNA target sites, **Table S4:** Expression levels of the *BnLACs* in different tissues and stages of *B. napus*. The values represent the FPKM values, **Table S5:** Expression levels of the *BnLACs* under Cd^{2+}, NH_4^+ enrichment treatment. The values represent the FPKM values, **Table S6:** Expression levels of 14 selected *BnLACs* in the L(leaf) and S(stem) of the three lines under drought and wound stress, **Table S7:** Expression levels of *BnLAC4*, *BnCTL2*, *BnIRX3*, *BnCESA4*, *BnIRX1*, *BnLAC17*, *BnGAUT12*, *BnIRX6*, *BnPGSIP1*, *BnGLP10* and *BnFLA11* genes in 6 kinds of silique pericarps. The values represent the FPKM values, **Table S8:** Primers used for qRT-PCR analysis.

Author Contributions: L.L., H.W. and J.L. designed the experiments. X.P., F.D., Y.L., T.W. and H.J. performed the experiments. X.P., T.W., N.L., X.X. and K.L. analyzed the data. X.P. wrote the paper.

Funding: This work was supported by grants from the National Key Research and Development Program of China (2016YFD0100202), National Natural Science Foundation of China (31771830, 31701335), the Fundamental Research Funds for Central Universities (XDJK2017A009) and the "111" Project (B12006).

Conflicts of Interest: The authors declared no conflict of interest.

References

1. Chalhoub, B.; Denoeud, F.; Liu, S.; Parkin, I.A.P.; Tang, H.; Wang, X.; Chiquet, J.; Belcram, H.; Tong, C.; Samans, B. Early Allopolyploid Evolution in the Post-Neolithic Brassica Napus Oilseed Genome. *Science* **2014**, *6199*, 950–953. [CrossRef] [PubMed]

2. Mourasobczak, J.; Souza, U.; Mazzafera, P. Drought Stress and Changes in the Lignin Content and Composition in Eucalyptus. *BMC Proc.* **2011**, *7*, 103.

3. Peng, D.; Chen, X.; Yin, Y.; Lu, K.; Yang, W.; Tang, Y.; Wang, Z. Lodging Resistance of Winter Wheat (*Triticum aestivum* L.): Lignin Accumulation and Its Related Enzymes Activities Due to the Application of Paclobutrazol or Gibberellin Acid. *Field Crops Res.* **2014**, *2*, 1–7. [CrossRef]

4. Cruickshank, A.W.; Cooper, M.; Ryley, M.J.; Cruickshank, A.W.; Cooper, M.; Ryley, M.J. Peanut Resistance to Sclerotinia Minor and S. Sclerotiorum. *Aust. J. Agric. Res.* **2002**, *10*, 1105–1110. [CrossRef]

5. Mayer, A.M.; Harel, E. Polyphenol Oxidase in Plants. *Phytochemistry* **1979**, *18*, 193–195. [CrossRef]

6. Solomon, E.I.; Sundaram, U.M.; Machonkin, T.E. Multicopper Oxidases and Oxygenases. *Chem. Rev.* **1996**, *7*, 2563. [CrossRef]

7. Cohen, R.; Persky, L.; Hadar, Y. Biotechnological Applications and Potential of Wood-Degrading Mushrooms of the Genus Pleurotus. *Appl. Microbiol. Biotechnol.* **2002**, *5*, 582–594. [CrossRef]

8. Bao, W.; O'malley, D.M.; Whetten, R.; Sederoff, R.R. A Laccase Associated with Lignification in Loblolly Pine Xylem. *Science* **1993**, *5108*, 672–674. [CrossRef] [PubMed]

9. Sterjiades, R.; Dean, J.F.; Eriksson, K.E. Laccase from Sycamore Maple (*Acer pseudoplatanus*) Polymerizes Monolignols. *Plant Physiol.* **1992**, *3*, 1162–1168.

10. Liang, M.; Haroldsen, V.; Cai, X.; Wu, Y. Expression of a Putative Laccase Gene, Zmlac1, in Maize Primary Roots under Stress. *Plant Cell Environ.* **2006**, *5*, 746. [CrossRef]

11. Hüttermann, A.; Mai, C.; Kharazipour, A. Modification of Lignin for the Production of New Compounded Materials. *Appl. Microbiol. Biotechnol.* **2001**, *4*, 387–394. [CrossRef]

12. Wang, G.D.; Li, Q.J.; Luo, B.; Chen, X.Y. Ex Planta Phytoremediation of Trichlorophenol and Phenolic Allelochemicals Via an Engineered Secretory Laccase. *Nat. Biotechnol.* **2004**, *7*, 893. [CrossRef]

13. Balasubramanian, V.K.; Rai, K.M.; Thu, S.W.; Mei, M.H.; Mendu, V. Genome-Wide Identification of Multifunctional Laccase Gene Family in Cotton (*Gossypium* spp.); Expression and Biochemical Analysis During Fiber Development. *Sci. Rep.* **2016**, *6*, 34309. [CrossRef]

14. Wang, J.; Feng, J.; Jia, W.; Fan, P.; Bao, H.; Li, S.; Li, Y. Genome-Wide Identification of Sorghum Bicolor Laccases Reveals Potential Targets for Lignin Modification. *Front. Plant Sci.* **2017**, *8*, 714. [CrossRef]

15. Lafayette, P.R.; Eriksson, K.E.; Dean, J.F. Nucleotide Sequence of a Cdna Clone Encoding an Acidic Laccase from Sycamore Maple (*Acer pseudoplatanus* L.). *Plant Physiol.* **1995**, *2*, 667–668. [CrossRef]

16. Kiefermeyer, M.C.; Gomord, V.; O'Connell, A.; Halpin, C.; Faye, L. Cloning and Sequence Analysis of Laccase-Encoding Cdna Clones from Tobacco. *Gene* **1996**, *1–2*, 205–207. [CrossRef]

17. Gavnholt, B.; Larsen, K.; Rasmussen, S.K. Isolation and Characterisation of Laccase Cdnas from Meristematic and Stem Tissues of Ryegrass (Lolium Perenne). *Plant Sci.* **2002**, *6*, 873–885. [CrossRef]

18. Bligny, R.; Gaillard, J.; Douce, R. Excretion of Laccase by Sycamore (*Acer pseudoplatanus* L.) Cells. Effects of a Copper Deficiency. *Biochemical Journal*. **1983**, *2*, 583–588.

19. Tezuka, K.; Hayashi, M.; Ishihara, H.; Onozaki, K.; Nishimura, M.; Takahashi, N. Occurrence of Heterogeneity of N-Linked Oligosaccharides Attached to Sycamore (*Acer pseudoplatanus* L.) Laccase after Excretion. *Biochem. Mol. Biol. Int.* **1993**, *3*, 395–402.

20. Ranocha, P.; Mcdougall, G.; Hawkins, S.; Sterjiades, R.; Borderies, G.; Stewart, D.; Cabanesmacheteau, M.; Boudet, A.M.; Goffner, D. Biochemical Characterization, Molecular Cloning and Expression of Laccases—A Divergent Gene Family—in Poplar. *FEBS J.* **1999**, *259*, 485–495. [CrossRef]

21. Turlapati, P.V.; Kim, K.W.; Davin, L.B.; Lewis, N.G. The Laccase Multigene Family in Arabidopsis Thaliana: Towards Addressing the Mystery of Their Gene Function(S). *Planta* **2011**, *3*, 439–470. [CrossRef] [PubMed]

22. Mccaig, B.C.; Meagher, R.B.; Dean, J.F.D. Gene Structure and Molecular Analysis of the Laccase-Like Multicopper Oxidase (Lmco) Gene Family in Arabidopsis Thaliana. *Planta* **2005**, *5*, 619–636. [CrossRef] [PubMed]

23. Caparrós-Ruiz, D.; Fornalé, S.; Civardi, L.; Puigdomènech, P.; Rigau, J. Isolation and Characterisation of a Family of Laccases in Maize. *Plant Sci.* **2006**, *2*, 217–225. [CrossRef]

24. Cesarino, I.; Araújo, P.; Mayer, J.L.S.; Vicentini, R.; Berthet, S.; Demedts, B.; Vanholme, B.; Boerjan, W.; Mazzafera, P. Expression of Soflac, a New Laccase in Sugarcane, Restores Lignin Content but Not S:G Ratio of Arabidopsis Lac17 Mutant. *J. Exp. Bot.* **2013**, *6*, 1769–1781. [CrossRef]

25. Zhang, K.; Lu, K.; Qu, C.; Liang, Y.; Wang, R.; Chai, Y.; Li, J. Gene Silencing of Bntt10 Family Genes Causes Retarded Pigmentation and Lignin Reduction in the Seed Coat of Brassica Napus. *PLoS ONE* **2013**, *4*, e61247. [CrossRef]

26. Wang, Y.; Le, B.P.; Antelme, S.; Soulhat, C.; Gineau, E.; Dalmais, M.; Bendahmane, A.; Morin, H.; Mouille, G.; Lapierre, C. Laccase 5 Is Required for Lignification of the Brachypodium Distachyon Culm. *Plant Physiol.* **2015**, *1*, 192–204. [CrossRef] [PubMed]

27. Cho, H.Y.; Lee, C.; Hwang, S.G.; Park, Y.C.; Lim, H.L.; Jang, C.S. Overexpression of the Oschi1 Gene, Encoding a Putative Laccase Precursor, Increases Tolerance to Drought and Salinity Stress in Transgenic Arabidopsis. *Gene* **2014**, *1*, 98–105. [CrossRef]

28. Sato, Y.; Bao, W.; Sederoff, R.; Whetten, R. Molecular Cloning and Expression of Eight Laccase Cdnas in Loblolly Pine (*Pinus taeda*). *J. Plant Res.* **2001**, *2*, 147–155. [CrossRef]

29. O'Malley, D.M.; Ross, W.; Bao, W.; Chen, C.; Sederoff, R.R. The Role of Laccase in Lignification. *Plant J.* **1993**, *5*, 751–757. [CrossRef]

30. Wang, Y.; Tang, H.; DeBarry, J.D.; Tan, X.; Li, J.; Wang, X.; Lee, T.-h.; Jin, H.; Marler, B.; Guo, H.; et al. Mcscanx: A Toolkit for Detection and Evolutionary Analysis of Gene Synteny and Collinearity. *Nucleic Acids Res.* **2012**, *7*, e49. [CrossRef]

31. Kumar, S.; Stecher, G.; Tamura, K. MEGA7: Molecular Evolutionary Genetics Analysis Version 7.0 for Bigger Datasets. *Mol. Biol. Evol.* **2016**, *33*, 1870–1874. [CrossRef]

32. Paterson, A.H.; Bowers, J.E.; Chapman, B.A. Ancient Polyploidization Predating Divergence of the Cereals, and Its Consequences for Comparative Genomics. *Proc. Natl. Acad. Sci. USA* **2004**, *26*, 9903–9908. [CrossRef]

33. Toledoortiz, G.; Huq, E.; Quail, P.H. The Arabidopsis Basic/Helix-Loop-Helix Transcription Factor Family. *Plant Cell* **2003**, *8*, 1749–1770. [CrossRef]

34. Guo, Y.; Gan, S. Atnap, a Nac Family Transcription Factor, Has an Important Role in Leaf Senescence. *Plant J.* **2006**, *4*, 601–612. [CrossRef]

35. Shen, H.; Cao, K.; Wang, X. Atbzip16 and Atbzip68, Two New Members of Gbfs, Can Interact with Other G Group Bzips in Arabidopsis Thaliana. *BMB Rep.* **2008**, *2*, 132–138. [CrossRef]

36. Zhou, J.; Lee, C.; Zhong, R.; Ye, Z.H. Myb58 and Myb63 Are Transcriptional Activators of the Lignin Biosynthetic Pathway During Secondary Cell Wall Formation in Arabidopsis. *Plant Cell* **2009**, *1*, 248–266. [CrossRef]

37. Lu, S.; Li, Q.; Wei, H.; Chang, M.J.; Tunlayaanukit, S.; Kim, H.; Liu, J.; Song, J.; Sun, Y.H.; Yuan, L. Ptr-Mir397a Is a Negative Regulator of Laccase Genes Affecting Lignin Content in Populus Trichocarpa. *Proc. Natl. Acad. Sci. USA* **2013**, *26*, 10848–10853. [CrossRef]

38. Abdelghany, S.E.; Pilon, M. Microrna-Mediated Systemic Down-Regulation of Copper Protein Expression in Response to Low Copper Availability in Arabidopsis. *J. Biol. Chem.* **2008**, *23*, 15932–15945. [CrossRef]

39. Zhang, Y.C.; Yu, Y.; Wang, C.Y.; Li, Z.Y.; Liu, Q.; Xu, J.; Liao, J.Y.; Wang, X.J.; Qu, L.H.; Chen, F. Overexpression of Microrna Osmir397 Improves Rice Yield by Increasing Grain Size and Promoting Panicle Branching. *Nat. Biotechnol.* **2013**, *9*, 848. [CrossRef]

40. Wang, C.-Y.; Zhang, S.; Yu, Y.; Luo, Y.-C.; Liu, Q.; Ju, C.; Zhang, Y.-C.; Qu, L.-H.; Lucas, W.J.; Wang, X. Mir397b Regulates Both Lignin Content and Seed Number in Arabidopsis Via Modulating a Laccase Involved in Lignin Biosynthesis. *Plant Biotechnol. J.* **2015**, *8*, 1132–1142. [CrossRef]

41. Bottcher, A.; Cesarino, I.; Santos, A.B.; Vicentini, R.; Mayer, J.L.; Vanholme, R.; Morreel, K.; Goeminne, G.; Moura, J.C.; Nobile, P.M. Lignification in Sugarcane: Biochemical Characterization, Gene Discovery, and Expression Analysis in Two Genotypes Contrasting for Lignin Content. *Plant Physiol.* **2013**, *4*, 1539. [CrossRef]

42. Jin, L.F.; Liu, Y.Z.; Yin, X.X.; Peng, S.A. Transcript Analysis of Citrus Mirna397 and Its Target Lac7 Reveals a Possible Role in Response to Boron Toxicity. *Acta Physiol. Plant.* **2016**, *1*, 18. [CrossRef]

43. Shen, C.; Huang, Y.Y.; He, C.T.; Zhou, Q.; Chen, J.X.; Tan, X.; Mubeen, S.; Yuan, J.G.; Yang, Z.Y. Comparative Analysis of Cadmium Responsive Micrornas in Roots of Two Ipomoea Aquatica Forsk. Cultivars with Different Cadmium Accumulation Capacities. *Plant Physiol. Biochem.* **2016**, *111*, 329–339. [CrossRef]

44. Bo, P.; Hopkins, R.; Rask, L.; Meijer, J. Differential Wound Induction of the Myrosinase System in Oilseed Rape (*Brassica napus*): Contrasting Insect Damage with Mechanical Damage. *Plant Sci.* **2005**, *3*, 715–722.

45. Altschul, S.F.; Madden, T.L.; Schäffer, A.A.; Zhang, J.; Zhang, Z.; Miller, W.; Lipman, D.J. Gapped Blast and Psi-Blast: A New Generation of Protein Database Search. *Nucleic Acids Res.* **1997**, *25*, 3389–3402. [CrossRef]

46. Mitchell, A.; Chang, H.Y.; Daugherty, L.; Fraser, M.; Hunter, S.; Lopez, R.; Mcanulla, C.; Mcmenamin, C.; Nuka, G.; Pesseat, S. The Interpro Protein Families Database: The Classification Resource after 15 Years. *Nucleic Acids Res.* **2015**, *43*, 213–221. [CrossRef]

47. Voorrips, R.E. Mapchart: Software for the Graphical Presentation of Linkage Maps and Qtls. *J. Hered.* **2002**, *1*, 77–78. [CrossRef]

48. Emanuelsson, O.; Nielsen, H.; Brunak, S.; Von Heijne, G. Predicting Subcellular Localization of Proteins Based on Their N-Terminal Amino Acid Sequence. *J. Mol. Biol.* **2000**, *4*, 1005–1016. [CrossRef]

49. Hu, B.; Jin, J.; Guo, A.Y.; Zhang, H.; Luo, J.; Gao, G. Gsds 2.0: An Upgraded Gene Feature Visualization Server. *Bioinformatics* **2015**, *8*, 1296. [CrossRef]

50. Bailey, T.L.; Boden, M.; Buske, F.A.; Frith, M.; Grant, C.E.; Clementi, L.; Ren, J.; Li, W.W.; Noble, W.S. Meme Suite: Tools for Motif Discovery and Searching. *Nucleic Acids Res.* **2009**, *37*, 202–208. [CrossRef]

51. Dai, X.; Zhao, P.X. Pssrnaminer: A Plant Short Small Rna Regulatory Cascade Analysis Server. *Nucleic Acids Res.* **2008**, *36*, W114–W118. [CrossRef]

52. Li, H.; Durbin, R. Fast and Accurate Short Read Alignment with Burrows–Wheeler Transform. *Bioinformatics* **2009**, *25*, 1754–1760. [CrossRef]

53. Daehwan, K.; Langmead, B.; Salzberg, S.L. Hisat: A Fast Spliced Aligner with Low Memory Requirements. *Nat. Methods* **2015**, *4*, 357–360.

54. Pertea, M.; Pertea, G.M.; Antonescu, C.M.; Chang, T.C.; Mendell, J.T.; Salzberg, S.L. Stringtie EnablesImproved Reconstruction of a Transcriptome from Rna-Seq Reads. *Nat. Biotechnol.* **2015**, *3*, 290–295. [CrossRef]

55. Deng, W.; Wang, Y.; Liu, Z.; Cheng, H.; Xue, Y. Hemi: A Toolkit for Illustrating Heatmaps. *PLoS ONE* **2014**, *11*, e111988. [CrossRef]

56. Lalitha, S. Primer Premier 5. *Biotech Softw. Internet Rep.* **2000**, *1*, 270–272. [CrossRef]

Sample Availability: Samples of the compounds are available from the authors.

![molecules logo] *molecules*

MDPI

Article

Enzymatic Weight Update Algorithm for DNA-Based Molecular Learning

Christina Baek [1,†], Sang-Woo Lee [2], Beom-Jin Lee [2], Dong-Hyun Kwak [1] and Byoung-Tak Zhang [1,2,3,*]

[1] Interdisciplinary Program in Neuroscience, Seoul National University, Seoul 08826, Korea; dsbaek@bi.snu.ac.kr (C.B.); dhkwak@bi.snu.ac.kr (D.-H.K.)
[2] School of Computer Science and Engineering, Seoul National University, Seoul 08826, Korea; slee@bi.snu.ac.kr (S.-W.L.); bjlee@bi.snu.ac.kr (B.-J.L.)
[3] Interdisciplinary Program in Cognitive Science, Seoul National University, Seoul 08826, Korea
[*] Correspondence: btzhang@bi.snu.ac.kr; Tel.: +82-2-880-1847
[†] Current address: Building 138-115, 1 Gwanak-ro, Gwanak-gu, Seoul 08826, Korea.

Academic Editors: Xiangxiang Zeng, Alfonso Rodríguez-Patón and Quan Zou
Received: 19 February 2019; Accepted: 4 April 2019; Published: 10 April 2019

Abstract: Recent research in DNA nanotechnology has demonstrated that biological substrates can be used for computing at a molecular level. However, in vitro demonstrations of DNA computations use preprogrammed, rule-based methods which lack the adaptability that may be essential in developing molecular systems that function in dynamic environments. Here, we introduce an in vitro molecular algorithm that 'learns' molecular models from training data, opening the possibility of 'machine learning' in wet molecular systems. Our algorithm enables enzymatic weight update by targeting internal loop structures in DNA and ensemble learning, based on the hypernetwork model. This novel approach allows massively parallel processing of DNA with enzymes for specific structural selection for learning in an iterative manner. We also introduce an intuitive method of DNA data construction to dramatically reduce the number of unique DNA sequences needed to cover the large search space of feature sets. By combining molecular computing and machine learning the proposed algorithm makes a step closer to developing molecular computing technologies for future access to more intelligent molecular systems.

Keywords: molecular computing; molecular learning; DNA computing; self-organizing systems; pattern classification; machine learning

---

1. Introduction

Molecular computing is a fast-developing interdisciplinary field which uses molecules to perform computations rather than traditional silicon chips. DNA is one such biomolecule which has complementary base pairing properties that allow for both specificity in molecular recognition and self-assembly and for massively parallel reactions which can take place in minute volumes of DNA samples. Pioneering work by Adleman demonstrate solutions to combinatorial problems using molecular computing [1].

Since then, research exploring and exploiting these DNA properties in DNA computing provide the core nanotechnologies required to build DNA devices that are capable of decision-making at a molecular level. Such examples are the implementation of logic gates [2–4], storing and retrieving information [5,6], simple computations for differentiating biological information [7], classifying analogue patterns [8], training molecular automatons [9] and even playing games [10]. In particular, molecular computing based on enzymes has also attracted much attention as enzymes can respond to a range of small molecule inputs, have advantage in signal amplification, and are highly specific in recognition capabilities [11].

As DNA computing has the advantage of biocompatibility with living tissue or organisms, implementing molecular computations has also led to promising biomedical applications. An example of this is the logic gates used to specifically target cells or tissue types, thereby minimizing side effects and releasing chemicals at a specific location [12]. Nanostructures have also been used to efficiently deliver cytotoxic drugs to targeted cancerous cells as a therapeutic drug delivery system for various cancers [13,14].

However, these molecular computation approaches were generally preprogrammed, rule-based, or used logic gates for simple forms of computations which may not exceed the ability of reflex action from the perspective of intelligence. Such as in the work of [15,16] where a perceptron algorithm was designed with a weighted sum operation and [17] where a feedforward and recurrent neural network was constructed with cascading nodes using DNA hybridization; although these studies realized pre-defined perceptrons, the idea of learning, where computational weight parameters were updated to train the model was lacking.

Another state-of-the-art molecular pattern recognition work using the winner-take-all model has been recently published, demonstrating molecular recognition using MNIST digit data and DNA-strand-displacement [18]. This work recognizes patterns into defined categories of handwritten digits '1' to '9' using a simple neural network model called the winner-take-all model. Though similar to our study, a key difference is that this work focuses on 'remembering' patterns during training for recognition and our study focuses on online learning of patterns for classification, where learning refers to the generalization of data following multiple iterations of update during molecular learning. Another key difference is the focus of our work to implement a complete in vitro molecular learning experiment, in wet lab conditions. This is further discussed in the results section as a comparative study with our work (Section 3.4).

Another related area of research includes the implementation of dynamic reaction networks. in vitro biochemical systems, transcriptional circuits have been used to form complex networks by modifying the regulatory and coding sequence domains of DNA templates [19]. A combination of switches with inhibitory and excitatory regulation are used as oscillators similar to that which are found as natural oscillators. Another study also use chemical reactions inspired from living organisms to demonstrate assembling of a de novo chemical oscillator, where the wiring of the corresponding network is encoded in a sequence of DNA templates [20]. These studies use the synthetic systems to further understand the complex chemical reactions found in nature to deepen our understanding of the principle of biological dynamics. A key similarity to our work is the use of modular circuits to model more complex networks. However, it is important to note that these studies are all demonstrated in silico, although it illustrates the potential of in vitro transcriptional circuitry. Computational tools are also being developed, one example being the EXPonential Amplification Reaction (EXPAR), to facilitate the assay design of isothermal nucleic acid amplification [21]. This method helps accelerate DNA assay design, identifying template performance links to specific sequence motifs.

These dynamic system programming paradigms could be valid approaches to implement machine learning algorithms, as programmable chemical synthesis and the instruction strands of DNA dictate which reaction sequence to perform. We ponder that this kind of powerful information-based DNA system technology could also be manipulated to perform defined reactions in specific orders similar to what our study strives to do, thus, implementation operations in vitro to demonstrate molecular learning with the hypernetwork or other machine learning algorithms [22].

Recent work by [23], implement mathematical functions using DNA strand displacement reactions. This study demonstrates considerably more complex mathematical functions to date, can be designed through chemical reaction networks in a systematic manner. It is similar to our work in that it strives to compute complex functions using DNA though a key difference is that the design and validation of this work were presented in silico whereas our work focuses on in vitro implementation of molecular learning. However, the mass-action simulations of the chemical kinetics of DNA strand displacement reactions may be key in developing in vitro learning implementations, as learning consists of target mathematical operations which need to be performed with DNA in a systematic manner to manipulate DNA datasets. Consequently, operations or computational constructs are crucial in implementing machine learning algorithms, from simple perceptrons to neural networks, and this is proposed by this system and thus shares our interests in building systemic molecular implementations of chemical reactions for molecular machine learning. Further examples include a study where an architecture of three autocatalytic amplifiers interacts together to perform computations [24]. The square root, natural logarithm and exponential functions for x in tunable ranges are computed with DNA circuits.

Molecular learning algorithms for pattern recognition in vitro with DNA molecules may be a step towards more advanced systems with higher complexity, adaptability, robustness, and scalability. This could be useful for solving more advanced problems, and be more applicable to use with more intelligent molecular learning devices in order to function in dynamic in vitro and in vivo environments. Some in vitro and in silico studies which aim to create more complex molecular computing systems include associative recall and supervised learning frameworks using strand-displacement [25–29].

There are many difficulties in implementing molecular learning in wet lab experimental settings. DNA may be a more stable biomolecule compared to others such as RNA, however it still requires storage in appropriate solutions and temperature and it is prone to contamination, manipulation techniques often result in heavily reduced yield, and performing and analyzing the molecular biology results can be tedious and time consuming. Furthermore, applying learning algorithms familiar in machine learning bears critical differences to the current demonstration of DNA computing, such as predefined or rule-based models and logic gates.

Our previous study displayed in vitro DNA classification results [30] by retrieving information from a model that was trained with a pseudo-update like operation of increasing the concentration of the matched DNA strands. However, the adaptability and scalability of the model was limited, due in part to the restrictions in the representation of the model by creating single strand DNA (features) with a fixed variable length. Here, this refers to a fixed length of DNA sequence which encodes the variables. Additionally, from the machine learning perspective, updating only with the positive term has critical limitations not common in conventional machine learning methods [31], such as in the log-linear models or neural networks, which require both the positive and the negative terms to perform classification. The accuracy was also somewhat guaranteed because the training and test set were not divided and the features (pool of DNA) were manually designed to have small errors.

In this paper, we introduce a self-improving molecular machine learning algorithm, the hypernetwork [5,27,32], and a novel experimental scheme to implement in vitro molecular machine learning with enzymatic weight update. We present the preliminary in vitro experimental results of proof-of-concept experiments for constructing two types of DNA datasets, the training and test data, with the self-assembling processes of DNA with hybridization and ligation, and DNA cleavage with a nuclease enzyme for the weight update stage of the molecular learning. Our study provides a new method of implementing molecular machine learning with weight update including a negative term.

First, we consider a natural experimental situation typical to machine learning, where we separate the training and test data to evaluate the model. Secondly, by adopting a high order feature extraction method when creating single stranded DNA, a higher-order hypothesis space may be explored which allows for discovery of better solutions even with simple linear classifiers. Thirdly, unlike previous methods which only increased the weight of the model, our proposed method considers both the positive and negative terms of the weight update in the model for learning using an enzymatic weight update operation in vitro. This method is inspired by the energy-based models which use the energy-based objective function to solve problems, and is represented with exponentially proportional probabilities and consists of positive and negative terms to calculate the gradient [33]. Lastly, by encompassing the concept of ensemble learning, the model uses its full feature space for the classification task and also guarantees best performance by voting the best classified labels between each ensemble model. These four aspects are the crucial differences that distinguish our study from previous demonstrations of molecular learning, where without these assumptions of machine learning based aspects, learning, adaptability, and scalability of the model is limited. We show in the results section that the performance of our model gradually increases with the continual addition of training data.

2. Materials and Methods

2.1. The Molecular Learning Model

The hypernetwork is a graphical model with nodes and connections between two or more nodes called hyperedges (Figures 1 and 2) [27,32]. The connections between these nodes are strengthened or weakened through the process of weight update or error correction during learning [32]. We use the term 'hypernetwork' as we refer to hypergraphs which is a generalization of a graph where an edge can join any number of vertices. The hypergraph generally contains nodes and vertices and is a set of non-empty subsets termed hyperedges.

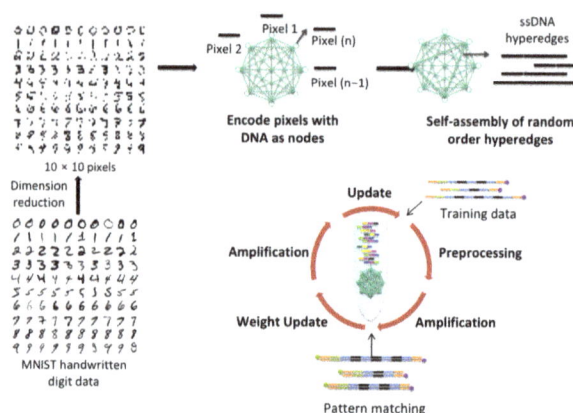

Figure 1. Hypernetwork with nodes and hyperedges with a conceptual overview of molecular machine learning with DNA processes. Each node represents a pixel which is encoded to a unique DNA sequence. These pixel DNA are self-assembled to form random order hyperedges.

This model was inspired by the idea of in vitro evolution, and provides a clear framework for molecular computing to be realized for molecular learning in a test tube. The probability of hyperedges, or weights, are represented by the concentration of DNA species in the tube. In addition, the idea of weight update is implemented here with specific enzymes, which use the gradient descent in a natural way for autonomous weight update or error correction. Gradient descent is an optimizing procedure commonly used in many machine algorithms to calculate the derivative from the training data before calculating update. All of these reactions occur in a massively parallel manner. Related models have also been previously discussed from the constructive machine learning perspective [34,35].

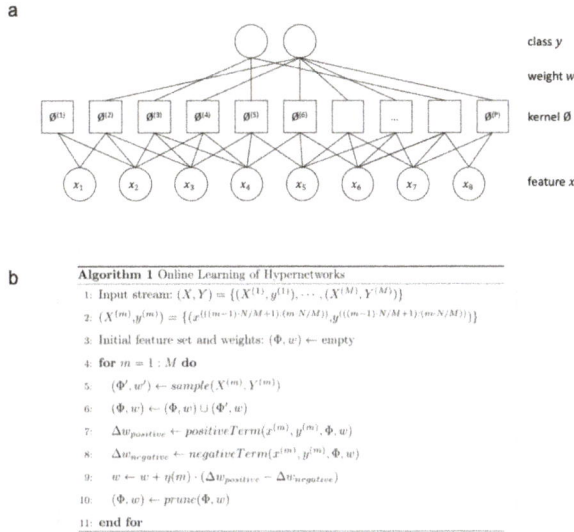

Figure 2. Graphical and algorithmic representation of the hypernetwork model (**a**). The graphical structure of the hypernetwork model. This is the factor graph representation of the model in Figure 1. Note, one kernel corresponds to one factor. (**b**) Algorithm of online learning of hypernetworks.

In this paper, the hypernetwork is interpreted as a maximum entropy classifier with an exponential number of hyperedges as input [36].

$$p(y|x; w) = \frac{1}{Z} \exp(\sum_i w_i \phi^{(i)}(x, y)) \tag{1}$$

$$\phi^{(i)}(x, y) = \delta(y, \tilde{y}^{(i)}) \prod_{j \in C^{(i)}} \delta(x_j, \tilde{x}_j^{(i)}), \tag{2}$$

where Z is the normalization term, weight $w^{(i)}$ is the parameter corresponding to $\phi^{(i)}$, and $C^{(i)}$ is the set of indices of input features corresponding to ith hyperedge. The ith hyperedge consists of the set of input variables $\{\tilde{x}_j^{(i)}\}_{j \in C^{(i)}}$, the output variable $\tilde{y}^{(i)}$, and weight $w^{(i)}$. δ is the identity function. If the whole predefined variables of the ith hyperedge are matched to the corresponding variables of an instance, $\phi^{(i)}$ becomes 1 (Figure 2a).

Our DNA dynamics can be described from the machine learning perspective as presented in Algorithm 1 in Figure 2b. The DNA processes in the paper and Algorithm 1 to the molecular experimental scheme (Figure 3) is matched as following ways:

1. Initializing hyperedge in each epoch corresponds to line 4
2. Figure 3 hybridization corresponds to line 7–8
3. Figure 3 nuclease and amplification corresponds to line 9
4. Merging hyperedges in each epoch corresponds to line 10

In the in vitro implementation of Algorithm 1, the updating of calculated positive or negative term occurs in a slightly different order. In the case of negative weight update, the nuclease cleaves the perfectly matched DNA strands, which occur from the hybridization of complementary DNA hyperedges from training data for '6' and '7', the hybridization being when the negative weight term is calculated. In the case of the positive weight update, the resulting DNA concentrations of each hyperedge from cleavage and purification is amplified, where the positive term was also calculated from the initial hybridization process.

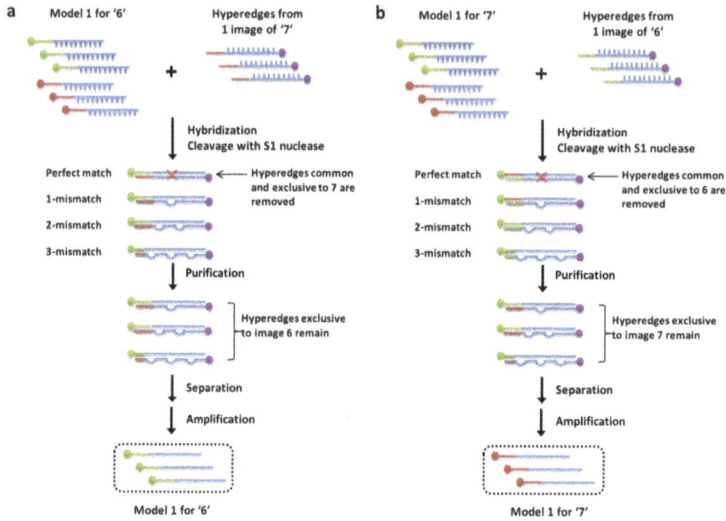

Figure 3. Overview of experimental scheme implementing molecular learning of hypernetworks. (a) Experimental steps for training the image '6' (b) Experimental steps for training the image '7'.

The hybridization rate of DNA datasets to a) construct hyperedges, theta being the hyperdges made from the data and b) to calculate the positive and negative term of Equation (3), is much faster due to the massively parallel nature of DNA computing, compared to the sequential matching of data in silico (Sections 3.1 and 3.2). DNA data representation through the use of sticky ends and ligation enzyme is almost instantaneous too, due to the use of common complementary strands used to ligate single variable DNA to form free-order hyperedges. This step approximates the kernel function in Equation (1). The weight of hyperedges in silico is approximated by the relative concentrations of DNA hyperedges, the relative weights of DNA hyperedges being the probabilistic weight calculated in silico, and the updating of weights in Equation (3) occurs through the PCR amplification and S1 nuclease enzyme cleavage of DNA, thereby increasing and decreasing the concentration of best matched DNA hyperedges respectively (Section 3.3).

The hypernetwork is a suitable model employing molecular machine learning for the following reasons. First, it is a non-linear classifier, which can search the exponential search space of the combination of hyperedges [32,37], unlike the maximum entropy classifier, which has a single variable as input and is a linear classifier.

Secondly, the hypernetwork can be relatively easily implemented in DNA computing as the model utilizes constructive DNA properties such as self assembly and molecular recognition for the generation of hyperedges, and to perform learning operations [30]. Massively-parallel processes can also be exploited with DNA, which means that the search of a large search space is much faster and applicable to the experimental setting.

2.2. DNA Dataset Construction

As an example of pattern classification in a test tube, we use the handwritten digit images from the MNIST database [38], which is commonly used to test machine learning algorithms [39]. In our case this is used to test a two-class classification problem with digits '6' and '7' (Figure 4a). The dimensions of the digit images are reduced to 10×10 which are then used as the input data.

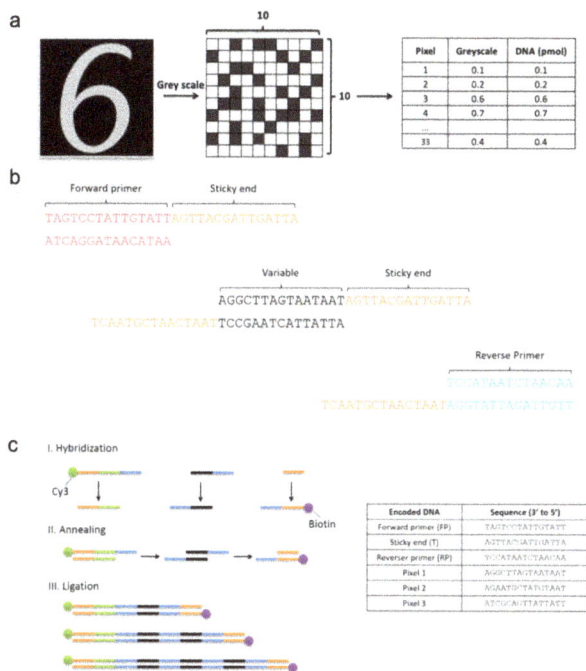

Figure 4. Encoding images to DNA hyperedges (**a**). Encoding of image data to unique DNA sequences (**b**). Complementary DNA sequences of primers, sticky ends and variable DNA that are ligated to form hyperedges (**c**). Self-assembling of free-order hyperedges through three processes, hybridization, annealing, and ligation. Key: Primers (orange), class label (green), tag or sticky end (blue), pixel (black). Please note, the pixel DNA colored black represent unique DNA sequences of various pixels, but for simplicity have been colored black to group them as variable DNA. Table shows DNA sequences for encoded pixels, primers and sticky ends. The final double strands in the sample are separated to single stranded DNA for use as the random DNA library set.

From each image 33 pixels are randomly selected in a non-replacement manner to form a hypernetwork model. 33 pixels were used as we wanted to use the least amount of pixels to represent the largest search space of 10 by 10 pixel images. In other words, only 33 pixels were used in each ensemble of each iteration. So 33 unique DNA sequences could be used to represent 33×3, so 99 pixels in total for each ensemble of learning. In our experiment, we produce three ensembles for each image (Figure 5). The 33 unique pixels from each ensemble are encoded to DNA by allocating 33 unique DNA sequences consisting of 15 base pairs each. Unique DNA oligomers are sequenced with an exhaustive DNA sequence design algorithm, EGNAS [40]. EGNAS, stands for Exhaustive Generation of Nucleic Acid Sequence, and is a software tool used to control both interstrand and intrastrand properties of DNA to generate sets with maximum number of sequence designs with defined properties such as the guanine-cytosine content. This tool is available online for noncommercial use at http://www.chm.tu-dresden.de/pc6/EGNAS. Once each DNA oligomer is assigned to a pixel, that is, one unique DNA to each pixel (Figure 4b), it is the grey scale value (between 0 and 1) for each pixel that determines the amount of DNA to be added to make the DNA dataset (Figure 4a). Each class is labeled with a different fluorescent protein to allow visualization of classes, allowing for the learning of two classes in one tube.

TRAINING DATA

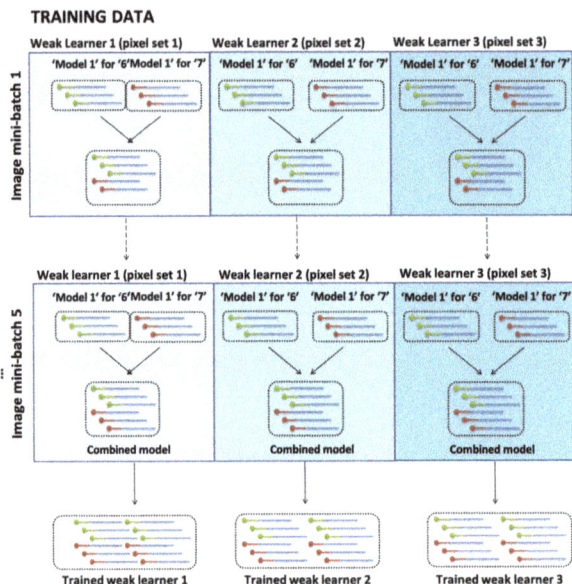

Figure 5. Ensemble of hypernetworks. Three ensembles of both trained models for digit images '6' and '7' are added in an online manner (combined model). The ensembles are added together for ensemble prediction in digit classification in the test stage.

Following the addition of relative amounts of DNA oligomers according to the pixel value, the sequences are joined together to produce free-order hyperedges for the initialized hypernetwork, and training and validation datasets (Figure 4b,c). Here, free-order refers to any number of linked variables in the DNA sequence, for example, 1-order hyperedge consists of one variable, 2-order hyperedge with two variables and 3-order hyperedge with three variables and so on (Figure 3c). Using PCR, the variable DNA, in this instance the pixel, the forward and reverse primers are hybridized to their respective complementary strands to form double stranded DNA. These three units act as building blocks for constructing free-order hyperedges as they are annealed at the tag or sticky end regions with enzyme ligase. It is worth noting that free-order hyperedges enhance the robustness of the model, and it is not only the variables that are learned through the self-organizing hypernetwork, but also the order of hyperedges.

2.3. Learning with Enzymatic Weight Update

Our main idea is that the dual hybridization-separation-amplification process with enzymatic weight update can be interpreted as an approximation of the stochastic gradient descent of hypernetworks.

In the test tube, enzymatic weight update is realized with enzymes which target specific DNA structures. Molecular recognition through hybridization of complementary base pairs allows matching of data to form symmetrical internal loops if incorrectly matched. Symmetric internal loops of DNA [41–43] are used to correlate the differences in training instances. This physical DNA structure is used to determine the degree of matching between two complementary strands for pattern matching. It is these DNA structures which are cleaved by specially chosen enzymes to perform the enzymatic weight update stage of the learning process. Consequently, the cleaving results in decrease in concentration of DNA with symmetric internal loops in the test tube.

First, the training data for digits '6' and '7' is hybridized with the training data for '7' (Figure 3a). Then, S1 nuclease, an enzyme which cleaves the perfectly matched DNA sequences (completely hybridized hyperedges) is added. This allows for the selection of only the perfectly matched hyperedges, leaving any degree of mismatched hyperedges in the tube. This is the enzymatic weight update operation that is demonstrated in vitro. DNA is then purified, separated using biotin, and amplified with PCR resulting in a mixture of single-stranded DNA hyperedges exclusive to '6'. This is repeated for the training data for '6' to train '7' (Figure 3b).

With the enzymatic weight update, through the decrease weight function of S1 nuclease, we eliminate the hyperedges common to both '6' and '7' resulting in only exclusive hyperedges characterizing '6' and '7' for successful digit classification. This corresponds to the negative weight update idea. The discrepancy between the two classes of digit data is represented by the remaining pool of DNA sequences, which are then added to one tube. The addition of the remaining sequences symbolizes the positive weight function. The trained hypernetwork from mini-batch 1 is added to the next minibatch and so on. Here the concept of online learning is applied. The weak learner 1, 2, 3 for ensembles 1, 2, and 3 respectively is added at each iteration to form the final trained weak learner after mini-batch 5 (Figure 5). The ensemble method is discussed in the next section. Repetition of these learning steps is predicted to construct an ensemble of three molecular classifiers which can be used for ensemble prediction in the test stage by measuring the final ratio of fluorescence for Cy3 (Label for 6) and Cy5 (Label for 7). To perform online learning, after the model is trained with each mini-batch, the DNA pool is combined to create a final hypernetwork model for given classification tasks.

The above process can be described theoretically, where the set of hyperedges is determined or the connections between the nodes are strengthened or weakened through the process of weight update or error correction during learning. Equation (3) is the gradient of the log-likelihood of Equation (1).

$$\Delta w_i = \frac{1}{N} \sum_{n=1}^{N} \phi^{(i)}(x^{(n)}, y^{(n)}) \left[1 - \frac{1}{Z} \exp(\sum_{i'} w_{i'} \phi^{(i')}(x^{(n)}, y^{(n)})) \right] \tag{3}$$

The next step of the algorithm consists of pattern matching and weight update (Algorithm 1, line 7, 8). Equation (3) shows the gradient of the log-likelihood of Equation (1). Our algorithm illustrates our learning process of hypernetworks, which can be naturally applied to online learning. The algorithm consists of both parameter learning and structure learning. In parameter learning, Equation (3) is used for stochastic gradient descent. Equation (3) consists of the positive term $\phi^{(i)}(x, y)$ and the negative term $-\frac{1}{Z} \cdot \phi^{(i)}(x, y) \cdot exp(\sum_{i'} w_{i'} \phi^{(i')}(x, y))$.

Without the terms of matching instance $\phi^{(i)}(x, y)$, the positive term is 1 and the negative term is $\frac{1}{Z} \cdot \phi^{(i)}(x, y) \cdot exp(\sum_{i'} w_{i'} \phi^{(i')}(x, y))$. In structure learning, the feature set of hyperedges Φ is updated. The number of possible kernel functions $\phi^{(i)}$ is exponential to the order of the hyperedge. This is required to select the subset of hyperedges, which consist of separable patterns. The candidate hyperedges are sampled from the data instance, where values of the partial input of an instance are used as the features. The hyperedges which are not important are pruned. Large absolute weight values or non-negative weight values can be used as the measure of importance; we use the latter case.

2.4. Ensemble Learning of Hypernetworks

Figure 5 shows that three ensembles were used to train images '6' and '7' in an online manner. We apply the ensemble method to our model for the following reasons. First, the ensemble method guarantees maximum performance within the ensemble models. When different types of models are being ensembled together, the overall performance increases as each model's characteristics of representation and search spaces are different from one another [44]. Another reason for creating a three ensemble model is due to the limitation of interpretability. Since our designed model created free-ordered hyperedges, using 100 pixel produces 100! (factorial of 100) different hyperedges which is almost impossible to visualize with current electrophoresis techniques. Moreover, the formation of the hyperedges is unpredictable since it is affected by a range of external stimuli (temperature, time, concentration etc.). Therefore, we divide the image into three sets and perform the voting method with the final produced results by each of three ensemble models.

The inference procedure is almost the same as the learning process. However, before the S1 nuclease is applied, the concentration of the perfectly matched DNA is measured to decide whether the test data is a digit '6' or '7'.

3. Results

To ensure that the experimental protocol is implemented to the highest degree of efficiency and accuracy, a series of preliminary experiments were undertaken to validate the experimental steps involved in demonstrating the molecular hypernetwork.

3.1. DNA Quantification of 3-Order Hyperedges

The formation of a random single-stranded library was critical in verifying the success of the full experimental scheme. The experimental steps and results are as follows:

1. Hybridization of upper and bottom strands of variable units.
2. Ligation of these variables in a random fashion, all in one tube to create a double-stranded DNA random library.
3. Purification of the sample from ligase.
4. Separation of the double-stranded library to a single-stranded random library using Streptavidin and Biotin.
5. Centrifugal filtering of DNA for concentration.
6. Verification of library formation with the use of complementary strands.

Each step listed above was carried out using wet DNA in a test tube, and at each step, the DNA concentration was measured using a NanoDrop Nucleic Acid Quantification machine, which is a common lab spectrophotometer (Figure 6). Hybridization was performed on the PCR machine with a decrement of 2° from 95° to 10° 100 pmol of each upper and lower strand was used.

Ligation was carried out using Thermo Fisher's T4 ligase enzyme. This enzyme joins DNA fragments with staggered end and blunt ends and repairs nicks in double stranded DNA with 3'-OH and 5'-P ends. Three units of T4 ligase was used with 1 μL of ligation buffer.

Annealing of the ligated DNA strands are then put into PCR conditions with a decrement of 1 degree from 30 to 4 degrees. Purification and extraction of DNA from the ligase inclusive sample was carried out using the QIAEX II protocol for desalting and concentrating DNA solutions. The standard procedures for this were used [45–47]. This procedure is commonly used to remove impurities (phosphatases, endonucleases, dNTPs etc.) from DNA, and to concentrate DNA. The QIAEX desalting and concentration protocol gives quite a detailed description of the procedure.

While there was a significant loss of DNA content after ligation, a sufficient concentration was recovered from the centrifugal filtering step allowing identification of the nine complementary strands possible from the combinations of the three different variables initially used. Bands at the 70 bp marker were present for all nine types of sequences which confirm that all possible sequences were successfully constructed and retrieved during the experimental process. The results shown in Figure 6 present DNA concentrations at various stages of the learning process, and the final confirmation of the success in making a random double-stranded library.

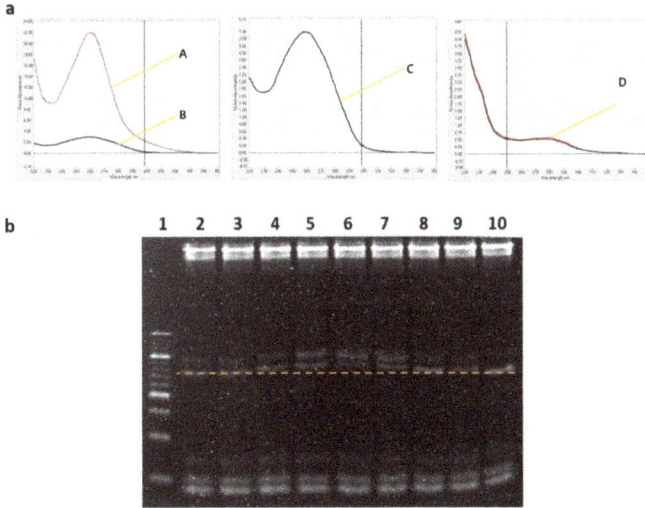

Figure 6. DNA quantification and hybridization of final complementary strands for verifying the presence of a 3-order double-stranded random DNA library. (**a**) Nanodrop analysis for quantification of DNA amount at each step of the protocol; pre-hybridization of variables pre-ligase (A), post-ligase (B), post-purification (C) post-separation (D) (**b**) gel electrophoresis of 3-order random library hybridized with each type of complementary strands (C1-C9 in lanes 2–10), so lane 2 contains the first complementary possible complementary sequence and so on to a total of nine possible hyperedges that could have been produced. Marked line shows 70-mer sequences.

3.2. Creating Free-Order DNA Hypernetworks

For the creation of the free-order hyperedges or different lengths of hyperedges, the concentrations of DNA sequences according to its corresponding pixel greyscale value, was added to a tube and ligation performed as described above. The PAGE electrophoresis gel illustrates the free-order hyperedges which were produced from the ligation procedure (Figures 4c and 7a). Against the 10 bp ladder, many bands of varying intensities can be seen, representing different lengths of DNA present in the sample. From the original tube of 15 bp single stranded DNA following the ligation protocol, the formation of random hyperedges from 30 bp to 300 bp are visible (Figure 7a). This correlates to 0-order to 15-order hyperedges. The same procedure was carried out to produce free-order hyperedges for every dataset: Training data and test data.

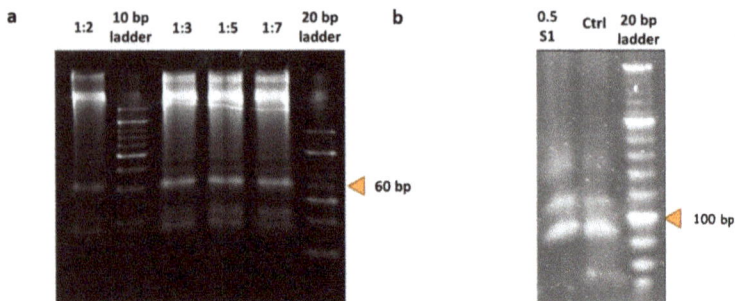

Figure 7. In vitro experimental results (**a**) Free-order hyperedge production. The 10 bp and 20 bp ladders were added in lanes 2 and 6. The ratio between primers and variable DNA sequences were varied to see if there was any effect on the length of the hyperedges produced. (**b**) The control lane shows from 70 base pairs up, perfect, 1, 2, 3-mismatched sequences. The perfectly matched DNA at 70 bp, 1-mismatch at 90 bp, 2-mismatched at 110 bp and 3-mismatched at 130 bp. Lane 1 contains the sample with S1 nuclease treatment. The perfectly matched DNA strands in the lower most band was cleaved.

3.3. Weight Update Feedback Loop for DNA Hypernetwork

Figure 7b shows the result of enzyme treatment for S1 nuclease. DNA was incubated with the enzyme for 30 min at room temperature than enzyme inactivated with 0.2 M of EDTA at heating at 70° for 10 min. The control lane shows 4 bands, the perfectly matched strand, 1-mismatched, 2-mismatched, and 3-mismatched strands from the bottom to the top of the gel. The function of the S1 nuclease is investigated for decreasing weights or in this case DNA concentration, where perfectly matched DNA sequences are cleaved and mismatched sequences remain. In the 0.5 S1 lane, it is evident that only the perfectly matched DNA strands are cleaved and the mismatched strands remain in the mixture. This represents the sequences only present exclusively in the data for digits '6' or '7' can be reproduced with the use of S1 nuclease in the weight update algorithm.

It is interesting to note that the issue of scalability may be addressed through our design which allows 10-class digit classification within the same number of experimental steps. More classes of training data could be added in the hybridization stage to all but the one class of training data for which the label is being learned. This provides a larger scale of digit classification without drastically increasing the workload, time or the need to order new sequences. This novel method of implementing digit classification and experimental results demonstrate the enzymatic reactions which is prerequisite to making this experimentally plausible.

3.4. Performance of In Silico Experiment of DNA Hypernetwork

As described in the Materials and Methods section, we used the MNIST dataset to measure the classifier accuracy [48], which is defined as the estimation of number of correctly classified objects over the total number of objects.

To verify the learning capability of our proposed model, both incremental and online aspects, we compared our model to two existing models; the perceptron model described in [18] and the conventional neural network [49] as a representative example of non-linear classification.

In [18], a basic perceptron model outputs the weighted sum for each class and selects the maximum value as their winning final output. 2-class classification between digits '6' and '7' is demonstrated and nine label 3 grouped class-classifier is described, where all methods first eliminate the outlier and the performance achieved by providing probabilistically calculated weights of the 10 most characteristic features to the designer as a prerequisite. However, in our study, we do not eliminate outliers or give prior weights and use the MNIST dataset as it is for our performance. We exploit the learning ability of a DNA computing model without the need for the designer to previously define weights. Not only do we reduce the labor required by the designer to define weights for selected features but we exploit the massively parallel processing capability of DNA computing whilst demonstrating molecular learning which improves performance with experience as our model is designed for implementation in vitro through molecular biology technique with wet DNA.

Two types of initialization of weights are introduced in our simulation results, 0 weight initialization which is easily implemented in DNA experiments, and random weight initialization which is harder to be conducted in vitro but is more conventionally used in perceptron and neural network models. The perceptron and neural networks convergence of performance are dependent on their initialized weights [50]. We conducted these two methods of initializing the weights, first starting with 0 weight, and second providing random values to the weights.

For the 2-class classification, 1127 and 11,184 images were used for the test and training data respectively. For the 10-class classification 10,000 and 60,000 images were used for the test and training data respectively. As the MNIST dataset is balanced over all classes and not skewed to any class, the accuracy measurement is sufficient to evaluate the classifier's performance [51]. For all cases of learning, we randomly sampled five images. We did this to demonstrate our model's capacity to implement online learning in only a few iterations and more significantly for our work, for the correlation to our wet lab molecular learning protocol, where only five iterations of molecular learning experiments need to be carried out for learning to produce classification performance.

For both the perceptron and neural network model, the learning rate was set to 0.1. For the perceptron model, of the 10 output values from the given input, the output with the biggest value was selected in a winner-take-all method. As the hyperedges produced from the hypernetwork in the 10-class classification (all with weights) was 284, we chose to use 300 hidden nodes for the neural network. All the source codes and relevant results can be found on github repository (https://github.com/drafity/dnaHN).

Figure 8 shows the results of our in silico classification results. As the number of epochs increased, the test accuracy of the hypernetwork also increased. The accuracy of the hypernetwork was also higher than the comparative models of the perceptron and neural network. We note here the significance of having used an accuracy measure to evaluate our classifier. The DNA learning models implemented in vitro in the mentioned related works often lacked appropriate measures to evaluate the classifier's performance. Furthermore, though recognition abilities have been reported through in vitro molecular learning, classification of data through learning and testing this, to consequently group or label the unknown test data to a category by a molecular learning model is to the authors' knowledge a novelty in itself.

A key feature of our model in comparison to the perceptron or neural network models is the minimum number of iterations required to observe significant performance. Our proposed model only needs five iterations of learning to achieve significant classification performance. However, as the results show in Figures 8 and 9, initializing the weight to 0 or giving random weights to the compared models still resulted in low accuracy in small epoch sizes. The perceptron and the neural network require a much larger epoch size for significant classification performance to be achieved.

This is crucial as in vitro experiments to perform molecular learning not only require time-consuming laborious work, but issues with contamination and denaturation can affect the quality of the experimental results. It is only more suitable for molecular learning experiments performed in wet lab conditions to be efficient, exploiting the massively parallel computing possible with DNA but also minimizing the protocol required to perform molecular learning. Our model is designed to do this by autonomously constructing higher order representations, using massively parallel DNA processes to create and update weights in minimal iterations. Furthermore, compared to state-of-the-art studies in molecular recognition, we were able to achieve over 90% accuracy and 60% accuracy in 2-class and 10-class classification respectively, through a molecular learning algorithm in five iterations. Thus, this result present that our model is a novel molecular learning model which learns in an online manner through minimal iterations of learning, suitable for wet lab implementations using DNA.

2-class classification

Full scale

Zoomed view

10-class classification

Full scale

Zoomed view

| DNA Hypernetwork | Perceptron | Neural Network |

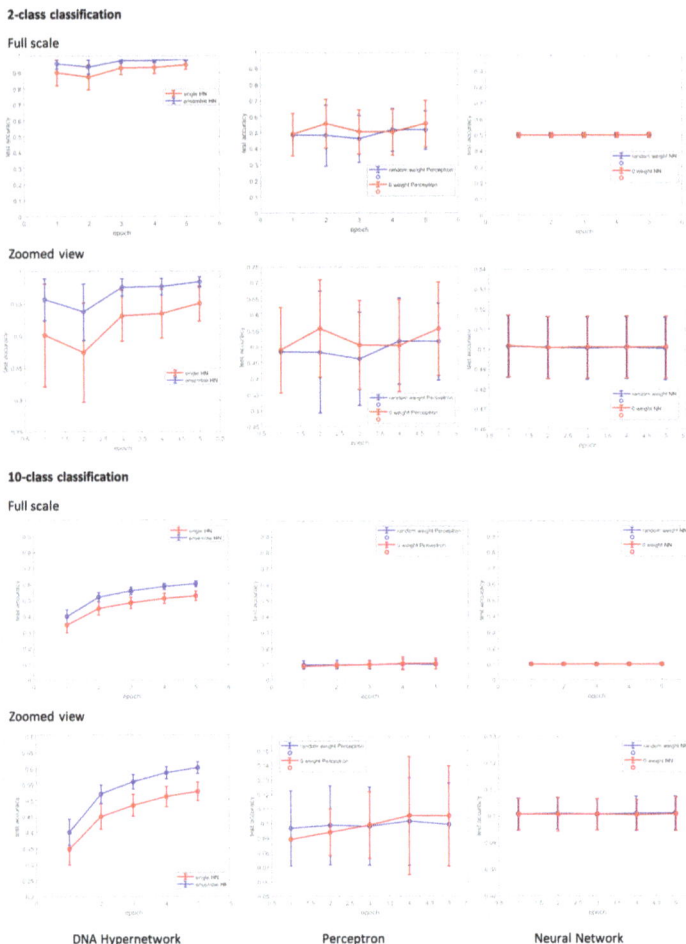

Figure 8. In silico experimental results. Computer simulation of 2-class classification with hypernetwork d. Computer simulation of 10-class classification with hypernetwork.

The hypernetwork, inspired by DNA chemical reactions, when computed in silico, clearly showed the disadvantage of sequential computing in silico and the massively parallel processing advantage of DNA computing in vitro. In an instant, DNA molecules hybridize when complementary strands are added together in an appropriate buffer and thus almost immediately the computing in that tube comes to an end. However, implementation of the hypernetwork in silico is iterative, sequential. For each training and test data, the number of matches and mismatches need to be calculated sequentially, and as the order of hyperedges increases, computational time complexity increase exponentially. As a result, with our computing power, empirically, 1000 iterations require 1000×20 min, a total of approximately 10 days to complete. Therefore it is important to note that there is a sheer advantage in DNA implementation of the hypernetwork compared to in silico. For the same reason, the neural network requires around 1000 iterations to converge and in the case of non-linearly separable data when using the perceptron model, it fails to converge. Thus, the proposed hypernetwork may also be introducing the possibility of a new computing method.

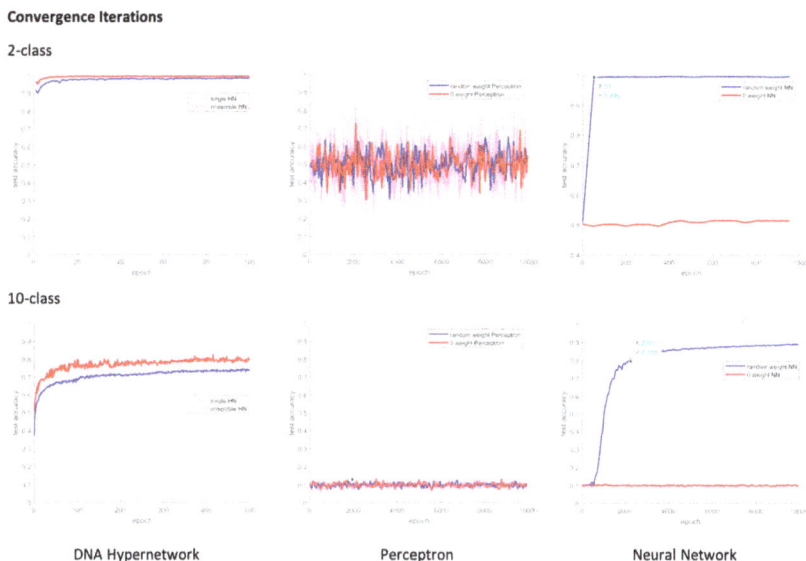

Figure 9. In silico experimental results. Computer simulation of 10-class classification with hypernetwork.

We also discuss the reasons why the perceptron does not perform as well. Due to the nature of perceptron models, as a representative linear classifier, it is difficult for it to solve linearly inseparable problems (XOR problems) without any preprocessing or adding layers to the perceptron to deal with non-linearity problems [52,53]. As illustrated in Figure 8, the perceptron model shows performances close to what would be achieved from random picking for both small and large number of iterations. Depending on how the data is fed, 2-class classification performance levels show major fluctuations, where up to 80% performance is achieved at times and others much lower performance. This phenomenon is typically representative of unsuccessful generalization of the data also called overfitting. For example, in the case of the perceptron, as described in the reference, the performance is achieved only for the data that can be fitted linearly. To learn linearly inseparable data, the model needs a feature reduction or extraction preprocessing methods [54] or a nonlinear kernel to model (e.g., Support Vector Machine [55], Neural Network [53] the high dimension input dataset. As this paper focuses on the implementation of a learning model in vitro only using easy, basic and fundamental learning processes, we believe this is out of the scope of our paper and omit further discussion.

As a support to such arguments, as shown in Figures 8 and 9, both in our results and in Cherry and Qian's work, there are cases where a variety of elimination conditions and previously providing the optimal weights of batch data by the designer can achieve significant performance in 2-class classification i.e., overfitted results (the maximum value of the error bars). However, as in the case of 10-class classification tasks, where the data is not linearly separable where it exceeds the model's capacity, the range of performance levels are smaller and, as acknowledged by Cherry and Qian in their paper, it is difficult for the designer to find the optimal weights for the model.

4. Discussion

We have proposed a novel molecular machine learning algorithm with a validated experimental scheme for in vitro demonstration of molecular learning with enzymatic weight update. The experiments are designed for plausible pattern recognition with DNA through iterative processes of self-organization, hybridization, and enzymatic weight update through the hypernetwork algorithm. Natural DNA processes act in unison with the proposed molecular learning algorithm using appropriate enzymes which allowed updating of weights to be realized in vitro. Unlike in previous studies, a molecular learning algorithm with enzymatic weight update is proposed, where the positive and negative terms of weight update are considered in the model for learning. Using the validated experimental steps, the model can be used for repeated learning iterations for the selection of relevant DNA to cause the DNA pool to continuously change and optimize, allowing large instance spaces to reveal a mixture of molecules most optimized to function as a DNA pattern recognition classifier.

Our experiments showed a higher order feature extraction method was possible in vitro using higher-order DNA hyperedges which was demonstrated by constructing longer DNA sequence datasets. This method of DNA data construction dramatically reduced the number of unique DNA sequences required to cover the large search space of image feature sets. Finally, DNA ensemble learning is introduced for use of the full feature space in the classification tasks.

Although the complete iterations of learning are yet to be carried out, the aim of this paper was to provide a framework, with a synergistic approach between theoretical and experimental designs of molecular learning algorithm. In future experiments we will carry out the iterative molecular learning scheme wet laboratory conditions.

By harnessing the strength of using biomolecules as building blocks for basic computations, new and exciting concepts of information processing have the potential to be discovered through more molecular computing methods. In turn, the implementation of machine learning algorithms through DNA could also act as a starting point for emerging technologies of computational molecular devices, implicated in a diverse range of fields such as intelligent medical diagnostics or therapeutics, drug delivery, tissue engineering, and assembly of nanodevices. As more advanced applications are explored, more intelligent molecular computing systems, with suitable intelligence to navigate and function in dynamic in vivo environments, may bridge gaps in current molecular computing technologies, so that DNA systems can function in uncontrolled, natural environments containing countless unforeseeable variables.

Author Contributions: Authors contributed in the following manner: Conceptualization, C.B., S.-W.L., D.-H.K., B.-J.L., B.-T.Z.; methodology, C.B., S.-W.L.; simulation, S.-W.L., B.-J.L.; data curation, D.-H.K.; validation, C.B.; investigation, C.B., S.-W.L., B.-J.L.; resources, C.B., S.-W.L., B.-J.L.; writing—original draft preparation, C.B.; writing—review and editing, C.B., S.-W.L., D.-H.K., B.-J.L., B.-T.Z.; supervision, B.-T.Z.; project administration, C.B.; funding acquisition, C.B., B.-T.Z.

Funding: This work was supported by Samsung Research Funding Center of Samsung Electronics (SRFC-IT1401-12), the Institute for Information & Communications Technology Promotion (R0126-16-1072-SW.StarLab, 2017-0-01772-VTT), the Agency for Defense Development (ADD-UD130070ID-BMRR), Korea Evaluation Institute of Industrial Technology (10060086-RISF), and the Air Force Office of Scientific Research under award number (FA2386-17-1-4128).

Conflicts of Interest: The funders had no role in the design of the study; in the collection, analyses, or interpretation of data; in the writing of the manuscript, or in the decision to publish the results.

References

1. Adleman, L.M. Molecular computation of solutions to combinatorial problems. *Science* **1994**, *266*, 1021–1024. [CrossRef] [PubMed]
2. Brown, C.W.; Lakin, M.R.; Stefanovic, D.; Graves, S.W. Catalytic molecular logic devices by DNAzyme displacement. *ChemBioChem* **2014**, *15*, 950–954. [CrossRef] [PubMed]
3. Mao, C.; LaBean, T.H.; Reif, J.H.; Seeman, N.C. Logical computation using algorithmic self-assembly of DNA triple-crossover molecules. *Nature* **2000**, *407*, 493–496. [CrossRef]
4. Seelig, G.; Soloveichik, D.; Zhang, D.Y.; Winfree, E. Enzyme-free nucleic acid logic circuits. *Science* **2006**, *314*, 1585–1588. [CrossRef]
5. Zhang, B.T.; Kim, J.K. DNA hypernetworks for information storage and retrieval. *Lect. Notes Comput. Sci.* **2006**, *4287*, 298.
6. Zhang, B.T.; Jang, H.Y. Molecular programming: Evolving genetic programs in a test tube. In Proceedings of the 7th Annual Conference on Genetic and Evolutionary Computation, Washington, DC, USA, 25–29 June 2005; ACM: New York, NY, USA, 2005; pp. 1761–1768.

7. Benenson, Y.; Paz-Elizur, T.; Adar, R.; Keinan, E.; Livneh, Z.; Shapiro, E. Programmable and autonomous computing machine made of biomolecules. *Nature* **2001**, *414*, 430–434. [CrossRef]

8. Yurke, B.; Turberfield, A.J.; Mills, A.P.; Simmel, F.C.; Neumann, J.L. A DNA-fuelled molecular machine made of DNA. *Nature* **2000**, *406*, 605–608. [CrossRef]

9. Stojanovic, M.N.; Stefanovic, D. A deoxyribozyme-based molecular automaton. *Nat. Biotechnol.* **2003**, *21*, 1069–1074. [CrossRef]

10. Pei, R.; Matamoros, E.; Liu, M.; Stefanovic, D.; Stojanovic, M.N. Training a molecular automaton to play a game. *Nat. Nanotechnol.* **2010**, *5*, 773–777. [CrossRef]

11. Katz, E.; Privman, V. Enzyme-based logic systems for information processing. *Chem. Soc. Rev.* **2010**, *39*, 1835–1857. [CrossRef] [PubMed]

12. Douglas, S.M.; Bachelet, I.; Church, G.M. A logic-gated nanorobot for targeted transport of molecular payloads. *Science* **2012**, *335*, 831–834. [CrossRef] [PubMed]

13. Chang, M.; Yang, C.S.; Huang, D.M. Aptamer-conjugated DNA icosahedral nanoparticles as a carrier of doxorubicin for cancer therapy. *ACS Nano* **2011**, *5*, 6156–6163. [CrossRef] [PubMed]

14. Zhang, Q.; Jiang, Q.; Li, N.; Dai, L.; Liu, Q.; Song, L.; Wang, J.; Li, Y.; Tian, J.; Ding, B.; et al. DNA origami as an in vivo drug delivery vehicle for cancer therapy. *ACS Nano* **2014**, *8*, 6633–6643. [CrossRef] [PubMed]

15. Mills, A., Jr.; Turberfield, M.; Turberfield, A.J.; Yurke, B.; Platzman, P.M. Experimental aspects of DNA neural network computation. *Soft Comput.* **2001**, *5*, 10–18. [CrossRef]

16. Lim, H.W.; Lee, S.H.; Yang, K.A.; Lee, J.Y.; Yoo, S.I.; Park, T.H.; Zhang, B.T. In vitro molecular pattern classification via DNA-based weighted-sum operation. *Biosystems* **2010**, *100*, 1–7. [CrossRef] [PubMed]

17. Qian, L.; Winfree, E.; Bruck, J. Neural network computation with DNA strand displacement cascades. *Nature* **2011**, *475*, 368–372. [CrossRef] [PubMed]

18. Cherry, K.M.; Qian, L. Scaling up molecular pattern recognition with DNA-based winner-take-all neural networks. *Nature* **2018**, *559*, 370. [CrossRef] [PubMed]

19. Kim, J.; Winfree, E. Synthetic in vitro transcriptional oscillators. *Mol. Syst. Biol.* **2011**, *7*, 465. [CrossRef]

20. Montagne, K.; Plasson, R.; Sakai, Y.; Fujii, T.; Rondelez, Y. Programming an in vitro DNA oscillator using a molecular networking strategy. *Mol. Syst. Biol.* **2011**, *7*, 466. [CrossRef] [PubMed]

21. Qian, J.; Ferguson, T.M.; Shinde, D.N.; Ramírez-Borrero, A.J.; Hintze, A.; Adami, C.; Niemz, A. Sequence dependence of isothermal DNA amplification via EXPAR. *Nucleic Acids Res.* **2012**, *40*, e87. [CrossRef] [PubMed]

22. Fu, T.; Lyu, Y.; Liu, H.; Peng, R.; Zhang, X.; Ye, M.; Tan, W. DNA-based dynamic reaction networks. *Trends Biochem. Sci.* **2018**, *43*, 547–560. [CrossRef]

23. Salehi, S.A.; Liu, X.; Riedel, M.D.; Parhi, K.K. Computing mathematical functions using DNA via fractional coding. *Sci. Rep.* **2018**, *8*, 8312. [CrossRef]

24. Song, T.; Garg, S.; Mokhtar, R.; Bui, H.; Reif, J. Design and analysis of compact DNA strand displacement circuits for analog computation using autocatalytic amplifiers. *ACS Synth. Biol.* **2017**, *7*, 46–53. [CrossRef] [PubMed]

25. Chen, J.; Deaton, R.; Wang, Y.Z. *A DNA-Based Memory with In Vitro Learning and Associative Recall*; Springer: Berlin, Germany, 2003; pp. 145–156.

26. Lakin, M.; Minnich, A.; Lane, T.; Stefanovic, D. Towards a biomolecular learning machine. In *Unconventional Computation and Natural Computation*; Springer: Berlin, Germany, 2012; pp. 152–163.

27. Lee, J.H.; Lee, B.; Kim, J.S.; Deaton, R.; Zhang, B.T. A molecular evolutionary algorithm for learning hypernetworks on simulated DNA computers. In Proceedings of the 2011 IEEE Congress on Evolutionary Computation (CEC), New Orleans, LA, USA, 5–8 June 2011; pp. 2735–2742.

28. Lim, H.W.; Yun, J.E.; Jang, H.M.; Chai, Y.G.; Yoo, S.I.; Zhang, B.T. Version space learning with DNA molecules. In *International Workshop on DNA-Based Computers*; Springer: Berlin, Germany, 2002; pp. 143–155.

29. Lakin, M.R.; Stefanovic, D. Supervised learning in adaptive DNA strand displacement networks. *ACS Synth. Biol.* **2016**, *5*, 885–897. [CrossRef] [PubMed]

30. Lee, J.H.; Lee, S.H.; Baek, C.; Chun, H.; Ryu, J.h.; Kim, J.W.; Deaton, R.; Zhang, B.T. In vitro molecular machine learning algorithm via symmetric internal loops of DNA. *Biosystems* **2017**, *158*, 1–9. [CrossRef] [PubMed]

31. Bishop, C.M. Machine learning and pattern recognition. In *Information Science and Statistics*; Springer: Heidelberg, Germany, 2006.

32. Zhang, B.T. Hypernetworks: A molecular evolutionary architecture for cognitive learning and memory. *IEEE Comput. Intell. Mag.* **2008**, *3*, 49–63. [CrossRef]

33. Zhou, G.; Sohn, K.; Lee, H. Online incremental feature learning with denoising autoencoders. *Artif. Intell. Stat.* **2012**, *22*, 1453–1461.

34. Heo, M.O.; Lee, S.W.; Lee, J.; Zhang, B.T. Learning global-to-local discrete components with nonparametric bayesian feature construction. In Proceedings of the NIPS Workshop on Constructive Machine Learning, Lake Tahoe, NV, USA, 10 December 2013.

35. Sakellariou, J.; Tria, F.; Loreto, V.; Pachet, F. Maximum entropy model for melodic patterns. In Proceedings of the ICML Workshop on Constructive Machine Learning, Lille, France, 10 July 2015.

36. Nigam, K.; Lafferty, J.; McCallum, A. Using maximum entropy for text classification. In Proceedings of the IJCAI-99 Workshop on Machine Learning for Information Filtering, Stockholm, Sweden, 1 August 1999; Volume 1, pp. 61–67.

37. Zhang, B.T.; Ha, J.W.; Kang, M. Sparse population code models of word learning in concept drift. In Proceedings of the Annual Meeting of the Cognitive Science Society, Sapporo, Japan, 1–4 August 2012; Volume 34.

38. LeCun, Y. The MNIST Database of Handwritten Digits. Available online: http://yann.lecun.com/exdb/mnist/ (accessed on 22 December 2018).

39. Deng, L. The MNIST database of handwritten digit images for machine learning research [best of the web]. *IEEE Signal Process. Mag.* **2012**, *29*, 141–142. [CrossRef]

40. Kick, A.; Bönsch, M.; Mertig, M. EGNAS: An exhaustive DNA sequence design algorithm. *BMC Bioinf.* **2012**, *13*, 138. [CrossRef]

41. Zeng, Y.; Zocchi, G. Mismatches and bubbles in DNA. *Biophys. J.* **2006**, *90*, 4522–4529. [CrossRef]

42. Zacharias, M.; Hagerman, P.J. The influence of symmetric internal loops on the flexibility of RNA. *J. Mol. Biol.* **1996**, *257*, 276–289. [CrossRef] [PubMed]

43. Peritz, A.E.; Kierzek, R.; Sugimoto, N.; Turner, D.H. Thermodynamic study of internal loops in oligoribonucleotides: Symmetric loops are more stable than asymmetric loops. *Biochemistry* **1991**, *30*, 6428–6436. [CrossRef] [PubMed]

44. Oza, N.C. Online bagging and boosting. In Proceedings of the 2005 IEEE International Conference on Systems, Man and Cybernetics, Waikoloa, HI, USA, 10–12 October 2005; Volume 3, pp. 2340–2345.

45. Sambrook, J.; Fritsch, E.F.; Maniatis, T. *Molecular Cloning: A Laboratory Manual*, 2nd ed.; Cold Spring Harbor Laboratory Press: Cold Spring Harbor, NY, USA, 1989.

46. Vogelstein, B.; Gillespie, D. Preparative and analytical purification of DNA from agarose. *Proc. Natl. Acad. Sci. USA* **1979**, *76*, 615–619. [CrossRef] [PubMed]

47. Hamaguchi, K.; Geiduschek, E.P. The effect of electrolytes on the stability of the deoxyribonucleate helix. *J. Am. Chem. Soc.* **1962**, *84*, 1329–1338. [CrossRef]

48. Sammut, C.; Webb, G.I. *Encyclopedia of Machine Learning*; Springer Science & Business Media: Berlin, Germany, 2011.

49. LeCun, Y.; Bottou, L.; Bengio, Y.; Haffner, P. Gradient-based learning applied to document recognition. *Proc. IEEE* **1998**, *11*, 2278–2324. [CrossRef]

50. Nguyen, D.; Widrow, B. Improving the learning speed of 2-layer neural networks by choosing initial values of the adaptive weights. In Proceedings of the 1990 IJCNN International Joint Conference on Neural Networks, San Diego, CA, USA, 17–21 June 1990; pp. 21–26.

51. Powers, D.M. Evaluation: From precision, recall and F-measure to ROC, informedness, markedness and correlation. *J. Mach. Learn. Technol.* **2011**, *2*, 37–63.

52. Rosenblatt, F. *The Perceptron, a Perceiving and Recognizing Automaton Project Para*; Cornell Aeronautical Laboratory: Buffalo, NY, USA, 1957.

53. Russell, S.J.; Norvig, P. *Artificial Intelligence: A Modern Approach*; Pearson Education Limited: Kuala Lumpur, Malaysia, 2016.

54. Liu, C.L.; Nakashima, K.; Sako, H.; Fujisawa, H. Handwritten digit recognition: Benchmarking of state-of-the-art techniques. *Pattern Recognit.* **2003**, *36*, 2271–2285. [CrossRef]

55. Scholkopf, B.; Smola, A.J. *Learning with Kernels: Support Vector Machines, Regularization, Optimization, and Beyond*; MIT Press: Cambridge, MA, USA, 2001.

molecules

MDPI

Article

Therapeutic Effects of HIF-1α on Bone Formation around Implants in Diabetic Mice Using Cell-Penetrating DNA-Binding Protein

Sang-Min Oh [1], Jin-Su Shin [2], Il-Koo Kim [2], Jung-Ho Kim [3], Jae-Seung Moon [2], Sang-Kyou Lee [2,3] and Jae-Hoon Lee [1,*]

[1] Department of Prosthodontics, College of Dentistry, Yonsei University, 134 Shinchon-dong, Seodaemoon-gu, Seoul 03722, Korea; smtop38@naver.com
[2] Department of Biotechnology, College of Life Science and Biotechnology, Yonsei University, 134 Shinchon-dong, Seodaemoon-gu, Seoul 03722, Korea; jinsuand@naver.com (J.-S.S.); tasada19@hanmail.net (I.-K.K.); jsmoon4@hanmail.net (J.-S.M.); sjrlee@yonsei.ac.kr (S.-K.L.)
[3] Research Institute for Precision Immuno-medicine, Good T Cells Incorporated, 134 Shinchon-dong, Seodaemoon-gu, Seoul 03722, Korea; jhokim@goodtcells.co.kr
* Correspondence: jaehoon115@yuhs.ac; Tel.: +82-10-4139-1491

Academic Editors: Xiangxiang Zeng, Alfonso Rodríguez-Patón and Quan Zou
Received: 8 January 2019; Accepted: 18 February 2019; Published: 20 February 2019

Abstract: Patients with uncontrolled diabetes are susceptible to implant failure due to impaired bone metabolism. Hypoxia-inducible factor 1α (HIF-1α), a transcription factor that is up-regulated in response to reduced oxygen during bone repair, is known to mediate angiogenesis and osteogenesis. However, its function is inhibited under hyperglycemic conditions in diabetic patients. This study thus evaluates the effects of exogenous HIF-1α on bone formation around implants by applying HIF-1α to diabetic mice and normal mice via a protein transduction domain (PTD)-mediated DNA delivery system. Implants were placed in the both femurs of diabetic and normal mice. HIF-1α and placebo gels were injected to implant sites of the right and left femurs, respectively. We found that bone-to-implant contact (BIC) and bone volume (BV) were significantly greater in the HIF-1α treated group than placebo in diabetic mice ($p < 0.05$). Bioinformatic analysis showed that diabetic mice had 216 differentially expressed genes (DEGs) and 21 target genes. Among the target genes, NOS2, GPNMB, CCL2, CCL5, CXCL16, and TRIM63 were found to be associated with bone formation. Based on these results, we conclude that local administration of HIF-1α via PTD may boost bone formation around the implant and induce gene expression more favorable to bone formation in diabetic mice.

Keywords: diabetes mellitus; hypoxia-inducible factor-1α; angiogenesis; bone formation; osteogenesis; protein transduction domain

1. Introduction

Dental implants have become an efficient and predictable treatment for replacing missing teeth. The number of implants placed in the United States has been steadily increasing at 12% annually, with improvements in implant materials, designs, and surgical techniques [1]. Despite an implant success rate of 95% in the general population [2], certain risk factors may predispose individuals to implant failure [3]. Among various patient-related risk factors, poorly controlled diabetes mellitus, a chronic metabolic disease characterized by hyperglycemia, has been considered a relative contraindication to dental implant [4–6].

Implant success is highly dependent on osseointegration, the process in which bone and implant surface become structurally and functionally integrated without interposition of the non-bone tissue layer [7]. Osseointegration, which involves bone repair and remodeling, critically affects implant stability [8]. However, the hyperglycemic condition of diabetes inhibits osteoblastic differentiation, mineralization, and adherence of the extracellular matrix and stimulates bone resorption, all consequently interfering with wound healing and bone regeneration [9,10]. Previous experimental studies have reported decreased bone-to-implant contact (BIC) and delayed new bone formation around the implant in diabetic animal models, proving that hyperglycemia impairs osseointegration [11].

Numerous studies have demonstrated that a chronic high glucose level results in defective responses of tissues to hypoxic conditions by impairing the function of hypoxia-inducible factor 1α (HIF-1α) [12–14]. Transcription factor HIF-1α is up-regulated in response to reduced oxygen conditions and influences numerous target genes, such as vascular endothelial growth factor (VEGF) and runt-related transcription factor 2 (RUNX2), which are known to be associated with angiogenesis and osteogenesis [15–17]. HIF-1α, which is well known to play a pivotal role in wound healing, is stabilized against degradation and transactivates under hypoxia [15]. A study carried out by Zou et al. demonstrated that osteogenesis and angiogenesis were enhanced around implants by the up-regulation of HIF-1α in rat bone mesenchymal stem cells (BMSCs) in animal models [18]. In addition, previous studies investigating the effects of HIF-1α on bone regeneration showed that the functions of osteoblasts and chondrocytes are directly regulated by HIF-1α during bone fracture healing in animal models [19].

As many studies have associated the malfunction of HIF-1α in diabetic animal models with delayed bone recovery, attempts have been made to improve bone healing by applying HIF-1α. However, the application of HIF-1α using mesenchymal stem cells to increase its expression is inefficient and time-consuming. To maximize the efficiency of delivery to the implant site, a protein transduction domain (PTD)-mediated DNA delivery system was used in this study. PTDs are short peptides that efficiently transport various proteins, nucleic acids, and nanoparticles into cells across the plasma membranes. The low toxicity and high transduction efficiency of this protein-based strategy constitutes a beneficial method for delivering target DNA to the nucleus [20]. Indeed, our recent study showed that the overexpression of HIF-1α induced by the PTD-mediated DNA delivery system resulted in an increased expression of VEGF and angiogenesis in vitro and in vivo [21].

In this study, taking advantage of the fact that PTD can deliver HIF-1α into cell nuclei, we designed an experiment to determine whether local application of HIF-1α into the implanted sites by using PTD in the femur of diabetic mice enhances osseointegration compared with placebo controls. Using RNA sequencing and histomorphometric analysis, we observed new bone formation and significant changes in the expression of genes associated with wound healing.

2. Results

2.1. RNA Sequencing and Differentially Expressed Genes (DEGs)

Different combinations of groups were designed and RNA sequencing was performed to identify DEGs. Group NH, normal mice with HIF-1α gel; group NP consisted of normal mice with placebo gel; group DH, diabetic mice with HIF-1α gel; group DP, diabetic mice with placebo gel.

The number of up- and down-regulated genes with a certain cutoff (2-fold; p-value < 0.05; FDR < 0.1) for all combinations are described in Table 1. A total of 216 genes were differentially expressed in the DH group compared to the DP group. On the other hand, there were 95 DEGs in the case of normal mice.

Table 1. The number of differentially expressed genes (DEGs) in each combination.

2-Fold; *p*-Value < 0.05; FDR < 0.1	Up	Down
Group NH and NP	94	1
Group DH and DP	201	15

These genes were selected according to the cut-off (2-fold; *p*-value < 0.05; FDR < 0.1), HIF-1α treated group was compared to the placebo group.

2.2. Target Genes of HIF-1α in Bioinformatic Analysis

The software program, TRANSFAC® (Qiagen N.V., Valencia, CA, USA), was used to select the target genes of HIF-1α. Twenty-one genes were identified as target genes of HIF-1α in diabetic mice (Table 2). Among the 21 detected genes, NOS2, GPNMB, CCL2, CCL5, CXCL16, and TRIM63 were found to be associated with wound healing or bone healing-related genes. The functions of these genes are described in Table 3 [22–27]. In normal mice, five genes (NOS2, CCL2, CCL5, CD274, TNF) were identified as target genes of HIF-1α.

Table 2. 21 target genes of Hypoxia-inducible factor 1α (HIF-1α) out of 216 DEGs in diabetic mice through TRANSFAC®.

Gene Symbol	Fold Change (log2X)	Molecule Type
CACNA1S	1.07	Calcium channel
CCL2	1.31	Chemokine
CCL5	1.49	Chemokine
CD274	1.70	Ligand
CXCL16	1.11	Chemokine
COBL	1.46	Cordon bleu
DES	1.20	Enzyme
GPNMB	1.06	ECM
IL2RA	1.76	Binding protein
JSRP1	1.05	Membrane protein
MARCO	2.88	Binding protein
MIA	−1.03	Structural protein
MURC	1.12	Structural protein
MYH14	1.14	Enzyme
MYL3	−1.54	Structural protein
NOS2	1.41	Enzyme
OAS2	1.21	Enzyme
PRKAG3	1.50	Protein kinase
TGTP1	2.16	GTPase
TRIM63	1.93	Ubiquitin protein ligase
TTN	1.05	ECM

With fold change, 1 indicates a two-fold increase in expression and −1 indicates a two-fold decrease in expression. Genes related to tissue healing or bone regeneration are in red.

Table 3. Target genes related to tissue healing or bone regeneration.

Gene Symbol	Functions of Target Gene
NOS2	Mediate increased blood flow Reparative collagen accumulation
GPNMB	Inducing differentiation and mineralization of hBMSCs into osteoblasts Increasing endothelial cell proliferation and migration, resulting in capillary tube formation
CXCL16	Recruitment of osteoclasts to restore the bone lost during the resorptive phase of bone turnover
CCL2	Consistent up-regulation during implant healing
CCL5	Promoting neovascularization and eventual wound repair
TRIM63	Mediating the glucocorticoid-induced promotion of osteoblastic differentiation

These genes were selected out of 21 HIF-1α target genes in diabetic mice.

2.3. Histologic Analysis

In the NH group (Figure 1a,e), abundant and smooth-lined mature bone formation was observed. Mature and smooth-lined bone was also observed in the NP group (Figure 1b,f), but in a lesser amount than in the NH group. In contrast, most of the implant surface in the DP (Figure 1d,h) group showed soft tissue attachment and abundant adipose tissue in surrounding areas. Moreover, bone formation was irregular. In the DH group (Figure 1c,g), many vascular sinusoids with red blood cells were located around implants, and more bone formation and attachment around implants were observed than in the DP group. In addition, there was a tendency toward increased bone formation at the HIF-1α application site around implants. Bone marrow was filled with adipose tissue in areas distant from the application site.

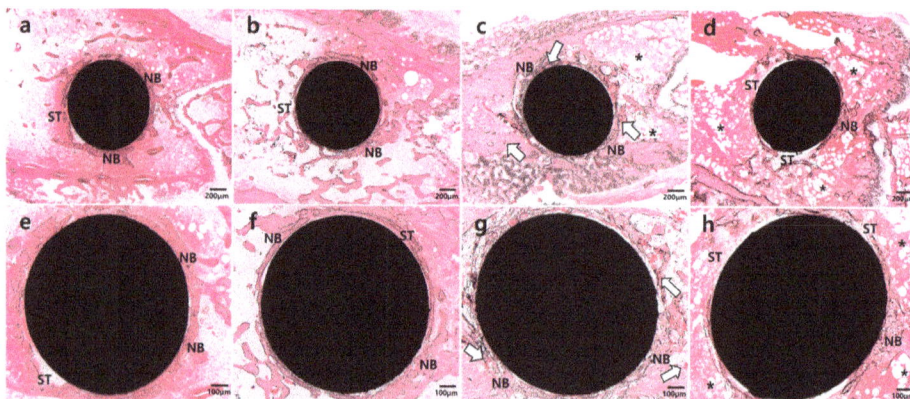

Figure 1. Representative images of undecalcified specimens of four groups. (**a**) and (**e**): Abundant and well-developed new bone (NB) formation observed around the implant in the normal mice with HIF-1α gel (NH) group, with some soft tissue (ST) engagement observed; (**b**) and (**f**): Thin and well-defined new bone formation observed in the normal mice with placebo gel (NP) group; (**c**) and (**g**): Plentiful vascular sinusoids with red blood cells (white arrow) and newly-formed bone surrounding the implant in the diabetic mice with HIF-1α gel (DH) group; (**d**) and (**h**): Fibrotic and adipose tissue (asterisk) surrounding the implant in the diabetic mice with placebo gel (DP) group.

2.4. Histomorphometric and Statistical Analysis

BIC was observed in all specimens. All groups in this study demonstrated normality in the parametric test (Shapiro-Wilk test). The BIC of the HIF-1α treated groups was significantly higher than that of placebo groups in both normal and diabetic mice. There was no significant BIC difference between the NP and DH groups (Figure 2a). Among the diabetic mice, the DH group showed significantly greater BV than the DP group while the groups of normal mice did not show any significant differences in BV. Only the DP group showed significantly lower BV among the four groups (Figure 2b).

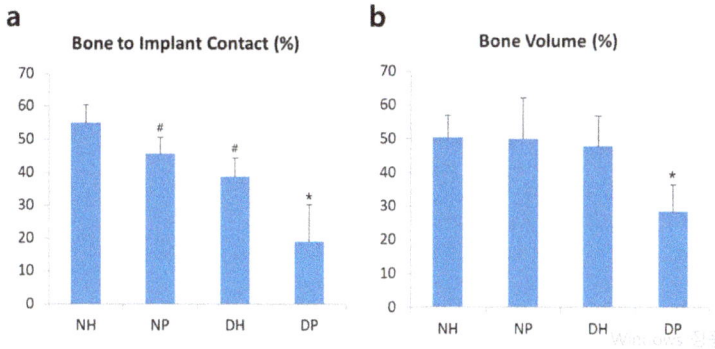

Figure 2. Histomorphometric analysis. (**a**): Linear percentage of direct BIC in the total surface of implants. (**b**): Percentage of newly formed bone area in the circumferential zone within 100 μm of the implant surface. Data represent mean ± SD. *: p-value < 0.05 vs all other groups; #: $p < 0.05$ vs NH group.

3. Discussion

Previous studies reported that hyperglycemia, even in hypoxic conditions, negatively affects the stability and activation of HIF-1α and inhibits the expression of target genes of HIF-1α, which are critical to wound healing [10]. On the other hand, it has been reported that the exogenous increase of HIF-1α resulted in improved bone regeneration and osseointegration around the implant in normoglycemic conditions [18,19]. This study was designed to test the hypothesis that the exogenous increase of HIF-1α would improve bone regeneration in diabetic mice because, based on previous studies, endogenous HIF-1α was suppressed in a hyperglycemic environment and failed to function.

In this study, we found that: (1) HIF-1α improves bone formation around the implant in diabetic mice; (2) HIF-1α induces gene expression that is more favorable to bone regeneration; and (3) exogenous HIF-1α has a greater effect on diabetic mice than normal mice.

Histologic results show that adipose and soft tissue were more engaged in diabetic mice femur bone marrow than in normal ones. Remarkably, most of the bone marrow in diabetic mice was composed of adipose tissue. In addition, the thickness and amount of regenerated bone was thinner and less in diabetic mice. The bone shape was highly irregular and fragile. The histological state of diabetic mice specimens was unfavorable for implant maintenance. These results were coincident with previous studies, which reported slower bone healing in diabetic mice than normal mice as well as poorer biomechanical and histologic bone quality after initial healing [10,28].

Histomorphometric results showed that the DH group had greater bone contact and volume than the DP group. Based on histologic specimens, more vascular sinusoids were generated in those groups with HIF-1a application, demonstrating that HIF-1α increased the expression of VEGF and improved angiogenesis as in our previous study, in which HIF-1a was applied in the same way as this study [21]. It is speculated that HIF-1α, which enhanced angiogenesis, also enhanced bone regeneration and increased BIC and BV levels. In addition, distant sites were full of adipose tissue compared to HIF-1α application sites near the implant, which consisted mainly of dense bone. The difference was even more evident in comparison with the DP group, in which tissue around the implant was filled with adipose tissue. There was no significant difference in BIC and BV when comparing the DH and NP groups. We thus expect diabetic mice to have as much bone formation as normal mice when HIF-1α is applied.

Bioinformatic analysis showed that diabetic mice had 216 DEGs and 21 target genes whereas normal mice had 95 DEGs and 5 target genes. Moreover, the DEGs and target genes in normal mice were mostly included in those of diabetic mice. These results suggest that the application of HIF-1α, suppressed in hyperglycemic conditions, activated the expression of HIF-1α target genes. On the other hand, HIF-1α target genes were less activated in normal conditions because endogenous HIF-1α was already sufficiently expressed around the injured tissue. It is therefore assumed that genes downstream of HIF-1α were more actively expressed in diabetic mice when exogenous HIF-1α was provided, which may account for the histomorphometric results of improved BIC and BV.

In normal conditions, HIF-1α also improved BIC, although not significantly, consistent with previous studies [15,18]. However, HIF-1α was more effective in diabetic mice in that the DH group showed significantly increased BIC and BV than did DP while the NH group showed no significant difference in BIC and BV compared to the NP group. Based on this result, it is assumed that exogenous HIF-1α worked more effectively in hyperglycemic conditions than normoglycemic conditions.

Previous studies used mesenchymal stem cells to increase expression of HIF-1α, a process considered inefficient due to the long and complicated preparation [18,29]. On the other hand, the PTD-mediated DAN delivery system used in this study is a simple and efficient method for the application of HIF-1α expression plasmid, which can easily be produced in large quantities and injected. Moreover, the molecular complex containing Hph-1-GAL4-DBD and HIF-1α-UAS can be solidified into gel form for applications to a local site without diffusion. The efficacy of HIF-1α delivery with a PTD-mediated system in vivo and in vitro was shown in our previous study [21], wherein PTD-mediated HIF-1α delivery increased HIF-1α, VEGF, and other HIF-1α target genes in vitro and in vivo. However, additional studies are needed to determine what the target cells are, how osteoblast progenitors and osteoclasts respond, and whether the gel form releases HIF-1α ideally or not. Moreover, future studies on higher mammals, such as dogs and pigs, may clarify the effects of HIF-1α.

Based on this study, we would like to suggest that the use of angiogenic growth factors, such as HIF-1α, rather than osteogenic growth factors, like BMP, could improve bone quality and quantity around the implant. Bone regeneration can be enhanced by applying bone morphogenetic factor (BMP), an osteogenic growth factor, or vascular endothelial growth factor (VEGF), an angiogenic growth factor [30]. The administration of BMP has been reported to improve bone regeneration in numerous dental studies and is practiced in the dental field [31]. However, several complications associated with BMP treatment have been reported, such as uncontrolled release rates, a short period of BMP release, and a high initial burst of release [32]. BMP also causes an unexpected immune reaction, spontaneous swelling of soft tissues, and difficulties in controlling the diffusion [33]. We think that HIF-1α may serve as an alternative to BMP, which presents several disadvantages.

4. Materials and Methods

4.1. Ethics Statements

This study was carried out in accordance with the guidelines established by the Laboratory Animal Care and Use Committee at Yonsei University Biomedical Research Institute (2014-0032). All surgical procedures were performed via intraperitoneal injection, analgesia, and antibiotics being administered at appropriate time points to minimize suffering and pain. The ARRIVE Guidelines for reporting animal research were abided by in all sections of this report [34].

4.2. Animal Models

Thirteen 8-week-old male C57BL/6 mice (21 g) from Charles River (Orientbio, Gapyeong-gun, Korea) and 13 8-week-old male C57BLKS/J-db/db mice (38 g, Leptin-receptor deficient type 2 diabetes mice) from Charles River (Hinobreeding Center, Tokyo, Japan) were used for the experiments. They were maintained in the Avison Biomedical Research Center at Yonsei University College of Medicine at $23 \pm 2\,°C$ and $50 \pm 10\%$ humidity under 12 h of light alternating with 12 h of darkness.

4.3. Preparation of HIF-1α Gene Construct and Hph-1-GAL4 DNA Binding Domain Protein

HIF-1α encoded plasmid and Hph-1-GAL4 DNA binding protein were provided by Sang-Kyou Lee's Laboratory at Yonsei University, Department of Biotechnology. The HIF-1α gene was inserted into pEGFP-N1 UAS plasmid containing five consensus GAL4 binding sites (UAS: CGGAGGACAGTACTCCG) (HIF-1α-UAS). The GAL4 DNA binding domain that encodes the DNA-interactive domain of yeast transcription factor GAL4 was cloned into pRSETB plasmid (Clonetech) expression vector containing Hph-1-PTD sequence (YARVRRRGPRP) at the N-terminus (Hph-1-GAL4-DBD). pRSETB plasmid with the Hph-1-GAL4 DNA binding domain was transformed into Escherichia coli BL21 star (DE3) pLysS strain (Invitrogen). Protein expression and purification were performed as described previously [20].

4.4. Preparation of HIF-1α Gel

One microgram of HIF-1α-UAS plasmid was mixed with 50 μg of Hph-1-GAL4-DBD at room temperature for 15 min right before surgery, as previously described [21]. The liquid form of Matrigel® (BD Biosciences, San Jose, CA, USA) and the mixture were blended at a 1:1 ratio just before the application of HIF-1α gel during surgery (Figure 3). Pure Matrigel® was used as a placebo gel. Matrigel® was stored in a liquid state at a temperature of $-72\,°C$ in the freezer because it solidifies at $4\,°C$.

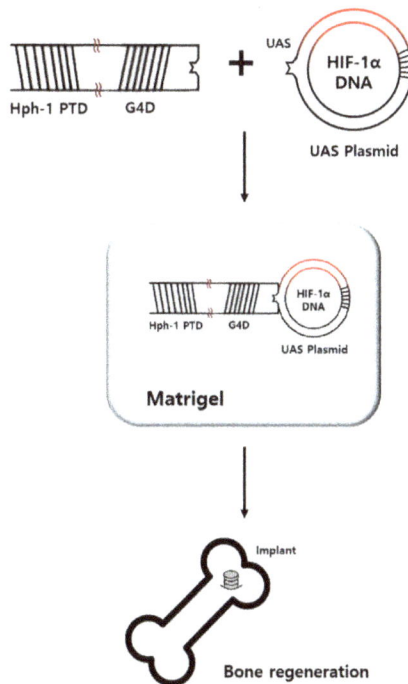

Figure 3. GAL4 DNA binding domain (G4D) was cloned into pRSETB plasmid expression vector containing Hph-1-PTD sequence at the N-terminus (Hph-1-G4D). The HIF-1α gene was inserted into pEGFP-N1 UAS plasmid containing five consensus GAL4 binding sites (HIF-1α-UAS plasmid). HIF-1α-UAS plasmid and Hph-1-G4D were mixed at a 1:50 mass ratio. The mixture and liquid form of Matrigel® were blended at a 1:1 ratio just before application for bone regeneration.

4.5. Surgical Procedure

Thirteen C57BL/6 mice (21 g) and 13 C57BLKS/J-db/db mice (38 g) were given two weeks of acclimatization before surgery. Implant placing methods followed Xu et al. [35]. The mice were anesthetized by intraperitoneal injection of a mixture of Zoletil 50 (30 mg/kg, Vibac Laboratories, Carros, France) and Rompun (10 mg/kg, Bayer Korea, Seoul, Korea) (Figure 4a), the surgical site being shaved (Figure 4b) and disinfected with 10% polyvinylpyrrolidone iodine. An incision was made above both knee joints and the anterior-distal aspect of the femur was accessed using medial parapatellar arthrotomy (Figure 4c). Implant sites were prepared on the anterior-distal surface of the femur through sequential drilling with 0.5 mm and 0.9 mm round burs and 0.7 mm stainless steel twist drills at 1500 rpm with cooled sterile saline irrigation (Figure 4d). To effectively deliver HIF-1α to the implant site via local injection, gel phase materials were prepared as described. HIF-1α gel was injected to the preparation site and cancellous bone of the right femur, placebo gel being injected to the same areas for the left femur (Figure 4e). When the gel hardened, pure titanium implants with a machined surface (1 mm in diameter; 2 mm in length; Shinhung, Seoul, Korea) were inserted into the undersized hole with mild pressure (Figure 4f). The muscles and skin were sutured independently to cover and stabilize the implant (Figure 4g,h). Antibiotics were injected at fixed times daily for 3 days (Enrofloxacin 5 mg/kg, twice a day; Meloxicam, 1 mg/kg, once a day) [36,37]. Three C57BL/6 (21 g) and three C57BLKS/J-db/db (38 g) mice were sacrificed 4 days after the surgery for RNA sequencing, and ten mice from each strain were sacrificed two weeks after the surgery for histologic and histomorphometric analysis.

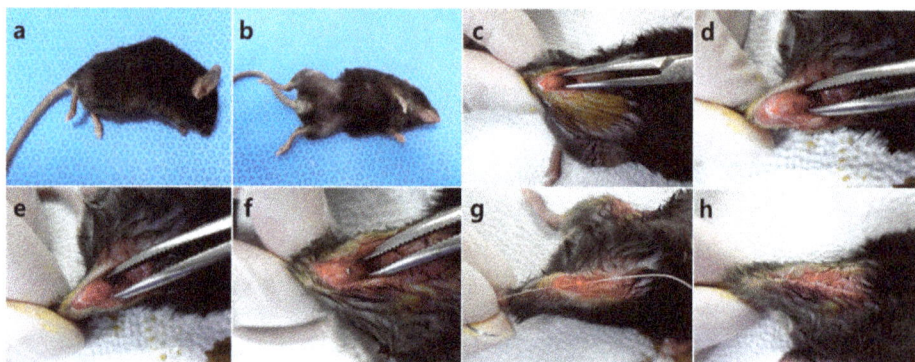

Figure 4. Surgical procedure of the implant. (**a**) Anesthetized mouse, (**b**) skin preparation, (**c**) incision, (**d**) preparation of the implant site, (**e**) gel injection, (**f**) placement of the implant, (**g**) closure of the surgical site layer by layer, (**h**) post surgery.

4.6. RNA Sequencing

Three mice from each strain (C57BL/6 (21 g) and C57BLKS/J-db/db (38 g)) were sacrificed and bone within 1 mm of the implant was taken for RNA sequencing analysis. Because factors related to bone formation are mostly expressed 4 days after implant surgery, RNA sequencing was performed at that time point [38].

RNA purity was determined by assaying 1 μL of total RNA extract on a NanoDrop8000 spectrophotometer (Thermo Fisher Scientific, Wilmington, DE, USA). Total RNA integrity was checked using an Agilent Technologies 2100 Bioanalyzer (Agilent Technologies, Foster City, CA, USA) with an RNA Integrity Number (RIN) value greater than 8. mRNA sequencing libraries were prepared according to the manufacturer's instructions (Illumina TruSeq RNA Prep Kit v2, Illumina, San Diego, CA, USA). mRNA was purified and fragmented from total RNA (1 μg) using poly-T oligo-attached magnetic beads using two rounds of purification. Cleaved RNA fragments primed with random hexamers were reverse transcribed into first strand cDNA using reverse transcriptase and random primers. The RNA template was removed and a replacement strand was synthesized to generate double-stranded (ds) cDNA. End repair, A-tailing, adaptor ligation, cDNA template purification, and enrichment of the purified cDNA templates using PCR were then performed. The quality of the amplified libraries was verified by capillary electrophoresis (Bioanalyzer, Agilent Technologies, Foster City, CA, USA). After performing qPCR using SYBR Green PCR Master Mix (Applied Biosystems, Thermo Fisher Scientific, Foster City, CA, USA), we combined libraries that were index tagged in equimolar amounts in the pool. RNA sequencing was performed using the Illumina NextSeq 500 system (Illumina, San Diego, CA, USA) following the protocols provided for 2 × 75 sequencing.

Reads for each sample were mapped to the reference genome (mouse mm10) by Tophat (v2.0.13). The aligned results were added to Cuffdiff (v2.2.0) to report differentially expressed genes. Geometric and pooled methods were applied for library normalization and dispersion estimation.

4.7. Identification of DEGs

Of the various Cuffdiff output files, "gene_exp.diff" was used to identify DEGs. Two filtering processes were applied to detect DEGs between control and case groups. First, only genes having Cuffdiff status code "OK" were extracted. The status code indicates whether each condition contains enough reads in a locus for a reliable calculation of the expression level, "OK" indicating that the test was successful in calculating the gene expression level. For the second filtering, the 2-fold change was calculated and only genes belonging to the following range were selected:

Up-regulated:
$$\log2[\text{case}] - \log2[\text{control}] > \log2(2) = 1 \tag{1}$$

Down-regulated:
$$\log2[\text{case}] - \log2[\text{control}] < \log2(1/2) = -1 \tag{2}$$

4.8. Identification of Target Genes of HIF-1α

The software program, TRANSFAC® (Qiagen N.V., USA), was used to select the target genes of HIF-1α. TRANSFAC® provides not only a database of eukaryotic transcription factors, but also an analysis of transcription factor binding sites. MATCH analysis was performed with TRANSFAC® using DEGs and an HIF-1α related matrix was selected from the results [39].

4.9. Histologic and Histomorphometric Analysis

Ten mice from each strain (C57BL/6 (21 g) and C57BLKS/J-db/db (38 g)) were sacrificed at 2 weeks after implant surgery to histologically evaluate mature bone in the healing process and estimate the implant stability in each group [37]. Femurs from both sides were obtained and fixed at 10% buffered formalin. After a week of fixation, they were embedded in light curing epoxy resin. The specimens were prepared with a cutting distance of 0.5 mm from the apical end of the implant. Sections were cut with a thickness of 50 μm via a grinding system, stained with hematoxylin and eosin (H&E), then observed using light microscopy (Leica DM LB, Wetzlar, Germany). IMT iSolution Lite ver 8.1® (IMT i-Solution Inc., Vancouver, BC, Canada) was used for histomorphometric measurement. The BIC ratio was calculated as the linear percentage of direct BIC to the total surface of implants. The BV ratio was calculated as the percentage of newly formed bone area to a circumferential zone within 100 μm of the implant surface.

4.10. Statistical Analysis

All statistical procedures were performed using IBM SPSS 23.0 (IBM Corp., Armonk, NY, USA). Raw histomorphometric measurement data were used to calculate the mean ± SD. The Shapiro-Wilk test was used to test normality and one-way analysis of variance (ANOVA) was used to compare groups, which were considered independent. Post hoc was performed with Scheffe's method. The value of $p < 0.05$ was considered statistically significant.

5. Conclusions

PTD-mediated delivery of HIF-1α into implant sites increases local HIF-1α levels, giving rise to a hyperglycemic environment that favors bone regeneration. This method holds tremendous potential and merits further study to determine its effectiveness as a local delivery system.

Author Contributions: J.-H.L. conceived and designed the experiments; J.-H.L. and S.-K.L. guided the research and conducted the research of previous studies; S.-M.O., J.-S.S., I.-K.K., J.-H.K., J.-S.M. and S.-K.L. contributed materials; S.-M.O. performed the experiments and wrote the paper. All authors read and approved the final manuscript.

Funding: This study was supported by the Basic Science Research Program through the National Research Foundation of Korea (NRF) funded by the Ministry of Education (2014R1A1A2055755) to J.-H.L. It was also supported by a National Research Foundation of Korea (NRF) grant funded by the Korea government (MSIP) (NRF-2014R1A2A1A10052466 and NRF-2016K1A1A2912755), the Yonsei University Future-leading Research Initiative of 2015 (2016-22-0053), and the Brain Korea 21 (BK21) PLUS Program, J.-S.S. and J.-S.M. are fellowship awardees by BK21 PLUS program, Republic of Korea to S.-K.L.

Acknowledgments: We thank DNA Link for helping conduct DNA extraction and sample analysis. Furthermore, we want to thank Se-young Kang and Yeun-Ju Kim for their assistance.

Conflicts of Interest: The authors declare no conflict of interest.

References

1. Millennium Research Group. *U.S. Markets for Dental Implants*; Millennium Research Group: Toronto, ON, Canada, 2006.
2. Papaspyridakos, P.; Chen, C.-J.; Singh, M.; Weber, H.-P.; Gallucci, G.O. Success criteria in implant dentistry. *J. Dent. Res.* **2012**, *91*, 242–248. [CrossRef] [PubMed]
3. Moy, P.K.; Medina, D.; Shetty, V.; Aghaloo, T.L. Dental implant failure rates and associated risk factors. *Int. J. Oral Maxillofac. Implants* **2005**, *20*, 569–577. [PubMed]
4. Alsaadi, G.; Quirynen, M.; Komarek, A.; van Steenberghe, D. Impact of local and systemic factors on the incidence of oral implant failures, up to abutment connection. *J. Clin. Periodontol.* **2007**, *34*, 610–617. [CrossRef] [PubMed]
5. Kopman, J.A.; Kim, D.M.; Rahman, S.S.; Arandia, J.A.; Karimbux, N.Y.; Fiorellini, J.P. Modulating the effects of diabetes on osseointegration with aminoguanidine and doxycycline. *J. Periodontol.* **2005**, *76*, 614–620. [CrossRef]
6. Kwon, P.T.; Rahman, S.S.; Kim, D.M.; Kopman, J.A.; Karimbux, N.Y.; Fiorellini, J.P. Maintenance of osseointegration utilizing insulin therapy in a diabetic rat model. *J. Periodontol.* **2005**, *76*, 621–626. [CrossRef] [PubMed]
7. Branemark, P.I. Osseointegrated implants in the treatment of edentulous jaw, experience from a 10-year period. Scand. *J. Plast. Reconstr. Surg.* **1977**, *1*, 1–132.
8. Garetto, L.P.; Chen, J.; Parr, J.A.; Roberts, W.E. Remodeling dynamics of bone supporting rigidly fixed titanium implants: A histomorphometric comparison in four species including humans. *Implant Dent.* **1995**, *4*, 235–243. [CrossRef]
9. Wang, F.; Song, Y.-L.; Li, D.-H.; Li, C.-X.; Wang, Y.; Zhang, N.; Wang, B.-G. Type 2 diabetes mellitus impairs bone healing of dental implants in gk rats. *Diabetes Res. Clin. Pract.* **2010**, *88*, e7–e9. [CrossRef] [PubMed]
10. Funk, J.R.; Hale, J.E.; David Carmines, H.L.G.; Hurwitz, S.R. Biomechanical evaluation of early fracture healing in normal and diabetic rats. *J. Orthop. Res.* **2000**, *18*, 126–132. [CrossRef]
11. Schlegel, K.; Prechtl, C.; Möst, T.; Seidl, C.; Lutz, R.; Wilmowsky, C. Osseointegration of slactive implants in diabetic pigs. *Clin. Oral Implants Res.* **2013**, *24*, 128–134. [CrossRef]
12. Botusan, I.R.; Sunkari, V.G.; Savu, O.; Catrina, A.I.; Grünler, J.; Lindberg, S.; Pereira, T.; Ylä-Herttuala, S.; Poellinger, L.; Brismar, K. Stabilization of hif-1α is critical to improve wound healing in diabetic mice. *Proc. Natl. Acad. Sci. USA* **2008**, *105*, 19426–19431. [CrossRef] [PubMed]

13. Catrina, S.-B.; Okamoto, K.; Pereira, T.; Brismar, K.; Poellinger, L. Hyperglycemia regulates hypoxia-inducible factor-1α protein stability and function. *Diabetes* **2004**, *53*, 3226–3232. [CrossRef] [PubMed]

14. Gao, W.; Ferguson, G.; Connell, P.; Walshe, T.; Murphy, R.; Birney, Y.A.; O'Brien, C.; Cahill, P.A. High glucose concentrations alter hypoxia-induced control of vascular smooth muscle cell growth via a hif-1α-dependent pathway. *J. Mol. Cell. Cardiol.* **2007**, *42*, 609–619. [CrossRef] [PubMed]

15. Ahluwalia, A.; S. Tarnawski, A. Critical role of hypoxia sensor-HIF-1α in VEGF gene activation. Implications for angiogenesis and tissue injury healing. *Curr. Med. Chem.* **2012**, *19*, 90–97. [CrossRef] [PubMed]

16. Kwon, T.G.; Zhao, X.; Yang, Q.; Li, Y.; Ge, C.; Zhao, G.; Franceschi, R.T. Physical and functional interactions between runx2 and hif-1α induce vascular endothelial growth factor gene expression. *J. Cell. Biochem.* **2011**, *112*, 3582–3593. [CrossRef] [PubMed]

17. Tamiya, H.; Ikeda, T.; Jeong, J.-H.; Saito, T.; Yano, F.; Jung, Y.-K.; Ohba, S.; Kawaguchi, H.; Chung, U.-I.; Choi, J.-Y. Analysis of the runx2 promoter in osseous and non-osseous cells and identification of hif2a as a potent transcription activator. *Gene* **2008**, *416*, 53–60. [CrossRef] [PubMed]

18. Zou, D.; Zhang, Z.; He, J.; Zhu, S.; Wang, S.; Zhang, W.; Zhou, J.; Xu, Y.; Huang, Y.; Wang, Y.; et al. Repairing critical-sized calvarial defects with bmscs modified by a constitutively active form of hypoxia-inducible factor-1α and a phosphate cement scaffold. *Biomaterials* **2011**, *32*, 9707–9718. [CrossRef] [PubMed]

19. Zou, D.; He, J.; Zhang, K.; Dai, J.; Zhang, W.; Wang, S.; Zhou, J.; Huang, Y.; Zhang, Z.; Jiang, X. The bone-forming effects of hif-1α-transduced bmscs promote osseointegration with dental implant in canine mandible. *PLoS ONE* **2012**, *7*, e32355. [CrossRef]

20. Kim, E.-S.; Yang, S.-W.; Hong, D.-K.; Kim, W.-T.; Kim, H.-G.; Lee, S.-K. Cell-penetrating DNA-binding protein as a safe and efficient naked DNA delivery carrier in vitro and in vivo. *Biochem. Biophys. Res. Commun.* **2010**, *392*, 9–15. [CrossRef]

21. Jeon, M.; Shin, Y.; Jung, J.; Jung, U.-W.; Lee, J.-H.; Moon, J.-S.; Kim, I.; Shin, J.-S.; Lee, S.-K.; Song, J.S. Hif1a overexpression using cell-penetrating DNA-binding protein induces angiogenesis in vitro and in vivo. *Mol. Cell. Biochem.* **2018**, *437*, 99–107. [CrossRef]

22. Corbett, S.; Hukkanen, M.; Batten, J.; McCarthy, I.; Polak, J.; Hughes, S. Nitric oxide in fracture repair: Differential localisation, expression and activity of nitric oxide synthases. *J. Bone Joint Surg. Br.* **1999**, *81*, 531–537. [CrossRef]

23. Li, B.; Castano, A.P.; Hudson, T.E.; Nowlin, B.T.; Lin, S.-L.; Bonventre, J.V.; Swanson, K.D.; Duffield, J.S. The melanoma-associated transmembrane glycoprotein gpnmb controls trafficking of cellular debris for degradation and is essential for tissue repair. *FASEB J.* **2010**, *24*, 4767–4781. [CrossRef] [PubMed]

24. Azuma, K.; Urano, T.; Ouchi, Y.; Inoue, S. Glucocorticoid-induced gene tripartite motif-containing 63 (trim63) promotes differentiation of osteoblastic cells. *Endocr. J.* **2010**, *57*, 455–462. [CrossRef] [PubMed]

25. Ota, K.; Quint, P.; Weivoda, M.M.; Ruan, M.; Pederson, L.; Westendorf, J.J.; Khosla, S.; Oursler, M.J. Transforming growth factor beta 1 induces cxcl16 and leukemia inhibitory factor expression in osteoclasts to modulate migration of osteoblast progenitors. *Bone* **2013**, *57*, 68–75. [CrossRef] [PubMed]

26. Zins, S.R.; Amare, M.F.; Tadaki, D.K.; Elster, E.A.; Davis, T.A. Comparative analysis of angiogenic gene expression in normal and impaired wound healing in diabetic mice: Effects of extracorporeal shock wave therapy. *Angiogenesis* **2010**, *13*, 293–304. [CrossRef] [PubMed]

27. Lin, Z.; Rios, H.F.; Volk, S.L.; Sugai, J.V.; Jin, Q.; Giannobile, W.V. Gene expression dynamics during bone healing and osseointegration. *J. Periodontol.* **2011**, *82*, 1007–1017. [CrossRef] [PubMed]

28. Macey, L.R.; Kana, S.; Jingushi, S.; Terek, R.; Borretos, J.; Bolander, M. Defects of early fracture-healing in experimental diabetes. *J. Bone Joint Surg. Am.* **1989**, *71*, 722–733. [CrossRef] [PubMed]

29. Seo, B.M.; Miura, M.; Gronthos, S.; Bartold, P.M.; Batouli, S.; Brahim, J.; Young, M.; Robey, P.G.; Wang, C.Y.; Shi, S. Investigation of multipotent postnatal stem cells from human periodontal ligament. *Lancet* **2004**, *364*, 149–155. [CrossRef]

30. Zhang, W.; Zhu, C.; Wu, Y.; Ye, D.; Wang, S.; Zou, D.; Zhang, X.; Kaplan, D.L.; Jiang, X. Vegf and bmp-2 promote bone regeneration by facilitating bone marrow stem cell homing and differentiation. *Eur. Cell. Mater.* **2014**, *27*, 1–12. [CrossRef]

31. Lin, G.H.; Lim, G.; Chan, H.L.; Giannobile, W.V.; Wang, H.L. Recombinant human bone morphogenetic protein 2 outcomes for maxillary sinus floor augmentation: A systematic review and meta-analysis. *Clin. Oral Implants Res.* **2016**, *27*, 1349–1359. [CrossRef]

32. Jeon, O.; Song, S.J.; Yang, H.S.; Bhang, S.-H.; Kang, S.-W.; Sung, M.A.; Lee, J.H.; Kim, B.-S. Long-term delivery enhances in vivo osteogenic efficacy of bone morphogenetic protein-2 compared to short-term delivery. *Biochem. Biophys. Res. Commun.* **2008**, *369*, 774–780. [CrossRef] [PubMed]

33. Shields, L.B.E.; Raque, G.H.; Glassman, S.D.; Campbell, M.; Vitaz, T.; Harpring, J.; Shields, C.B. Adverse effects associated with high-dose recombinant human bone morphogenetic protein-2 use in anterior cervical spine fusion. *Spine* **2006**, *31*, 542–547. [CrossRef] [PubMed]

34. Kilkenny, C.; Browne, W.J.; Cuthill, I.C.; Emerson, M.; Altman, D.G. Improving bioscience research reporting: The arrive guidelines for reporting animal research. *PLoS Biol.* **2010**, *8*, e1000412. [CrossRef] [PubMed]

35. Xu, B.; Zhang, J.; Brewer, E.; Tu, Q.; Yu, L.; Tang, J.; Krebsbach, P.; Wieland, M.; Chen, J. Osterix enhances bmsc-associated osseointegration of implants. *J. Dent. Res.* **2009**, *88*, 1003–1007. [CrossRef] [PubMed]

36. Slate, A.R.; Bandyopadhyay, S.; Francis, K.P.; Papich, M.G.; Karolewski, B.; Hod, E.A.; Prestia, K.A. Efficacy of enrofloxacin in a mouse model of sepsis. *J. Am. Assoc. Lab. Anim. Sci.* **2014**, *53*, 381–386. [PubMed]

37. Tubbs, J.T.; Kissling, G.E.; Travlos, G.S.; Goulding, D.R.; Clark, J.A.; King-Herbert, A.P.; Blankenship-Paris, T.L. Effects of buprenorphine, meloxicam, and flunixin meglumine as postoperative analgesia in mice. *J. Am. Assoc. Lab. Anim. Sci.* **2011**, *50*, 185–191. [PubMed]

38. Lu, H.; Kraut, D.; Gerstenfeld, L.C.; Graves, D.T. Diabetes interferes with the bone formation by affecting the expression of transcription factors that regulate osteoblast differentiation. *Endocrinology* **2003**, *144*, 346–352. [CrossRef]

39. Wingender, E. The transfac project as an example of framework technology that supports the analysis of genomic regulation. *Brief. Bioinform.* **2008**, *9*, 326–332. [CrossRef]

Sample Availability: Samples of the compounds are not available from the authors.

molecules

MDPI

Article

Identification of D Modification Sites by Integrating Heterogeneous Features in *Saccharomyces cerevisiae*

Pengmian Feng [1,5,*,†], Zhaochun Xu [2,3,†], Hui Yang [2], Hao Lv [2], Hui Ding [2] and Li Liu [4,*]

[1]   Innovative Institute of Chinese Medicine and Pharmacy, Chengdu University of Traditional Chinese Medicine, Chengdu 611730, China
[2]   Key Laboratory for Neuro-Information of Ministry of Education, School of Life Science and Technology, Center for Informational Biology, University of Electronic Science and Technology of China, Chengdu 610054, China; jdzxuzhaochun@163.com (Z.X.); huiyang0325@163.com (H.Y.); 13208188368@163.com (H.L.); hding@uestc.edu.cn (H.D.)
[3]   Computer Department, Jingdezhen Ceramic Institute, Jingdezhen 333403, China
[4]   Laboratory of Theoretical Biophysics, School of Physical Science and Technology, Inner Mongolia University, Hohhot 010021, China
[5]   School of Public Health, North China University of Science and Technology, Tangshan 063000, China
*   Correspondence: fengpengmian@gmail.com (P.F.); liliu2010imu@163.com (L.L.); Tel.: +86-315-3725715 (P.F. & L.L.)
†   These authors contributed equally to this work.

Academic Editors: Xiangxiang Zeng, Alfonso Rodríguez-Patón and Quan Zou
Received: 2 December 2018; Accepted: 17 December 2018; Published: 22 January 2019

Abstract: As an abundant post-transcriptional modification, dihydrouridine (D) has been found in transfer RNA (tRNA) from bacteria, eukaryotes, and archaea. Nonetheless, knowledge of the exact biochemical roles of dihydrouridine in mediating tRNA function is still limited. Accurate identification of the position of D sites is essential for understanding their functions. Therefore, it is desirable to develop novel methods to identify D sites. In this study, an ensemble classifier was proposed for the detection of D modification sites in the *Saccharomyces cerevisiae* transcriptome by using heterogeneous features. The jackknife test results demonstrate that the proposed predictor is promising for the identification of D modification sites. It is anticipated that the proposed method can be widely used for identifying D modification sites in tRNA.

Keywords: dihydrouridine; nucleotide physicochemical property; pseudo dinucleotide composition; RNA secondary structure; ensemble classifier

1. Introduction

To date, more than 100 kinds of post-transcriptional modifications have been identified in transfer RNAs (tRNAs). It has been demonstrated that these modifications are involved in all core aspects of tRNA function [1]. Among them, dihydrouridine (D) is a prevalent tRNA modification, which has been found in the three domains of life [2].

The D modification is formed by a dihydrouridine synthase [3]. Unlike uridine (U), the ring of D is not aromatic, which precludes its interactions with other bases in tRNA by stacking interactions [4,5]. By destabilizing the tRNA structure, D can enhance the conformational flexibility of tRNA [6]. Therefore, it is concluded that the flexibility and even the folding of tRNA could be affected by D modification [4,7].

Recent studies have also shown that tRNA lacking D degrades significantly faster, suggesting that D modification can protect tRNAs from degradation [1,8]. Despite the abundant occurrence of D modification, our knowledge about its roles in mediating tRNA biological functions is still limited.

Therefore, it is urgent to develop novel methods to describe the distribution of D modification sites. Since it is cost ineffective and labor intensive to detect D modification sites by using experimental techniques, it is necessary to develop theoretical methods for the detection of D modification.

Therefore, in the present study, an ensemble classifier was proposed for the detection of D modification sites in the *Saccharomyces cerevisiae* transcriptome, in which the nucleotide physicochemical property, pseudo dinucleotide composition, and secondary structure component were employed to train the basic predictors, respectively. In the jackknife test, the ensemble classifier obtained an accuracy of 83.09% for identifying D modification sites. This result demonstrated the superiority of the proposed method for identifying D modification sites in the *S. cerevisiae* transcriptome.

2. Results

2.1. Performances of Different Features

In order to demonstrate the effectiveness of the different kinds of features for identifying D sites, we first built support vector machine (SVM) predictors based on each kind of sequence encoding schemes (i.e., nucleotide physicochemical property, pseudo dinucleotide composition, or secondary structure component). Their jackknife test results for identifying D sites in the *S. cerevisiae* transcriptome are reported in Table 1. Although the nucleotide-physicochemical-property-based predictor (NPCP-SVM) obtained the highest accuracy (Acc) for identifying D sites, its sensitivity (Sn) was only 67.65%, indicating that it still could not accurately identify the real D sites. For the predictors based on pseudo dinucleotide composition and secondary structure component (namely PseDNC-SVM and SSC-SVM), their accuracies (Acc) were only 75.74% and 72.79% with the atthews correlation coefficients (MCC) of 0.5 and 0.45, respectively. Taken together, these results indicate that the performances of the aforementioned three predictors were not fully satisfactory. Therefore, there is still scope to improve the performance for identifying D sites.

Table 1. Performances of different methods for identifying dihydrouridine (D) sites.

Methods	Sn (%)	Sp (%)	Acc (%)	MCC
NPCP-SVM	67.65	100	83.82	0.59
PseDNC-SVM	73.53	77.94	75.74	0.50
SSC-SVM	70.59	75.00	72.79	0.45
Ensemble SVM	76.47	89.71	83.09	0.62

2.2. Improving Predictive Performance Using Ensemble Learning

Several recent works have demonstrated that the ensemble learning scheme can improve the performance of predictors [9–13]. In order to improve the performance of identifying D sites, we constructed an ensemble predictor based on SVM by using different kinds of features. Therefore, three basic SVM-based predictors were built by using nucleotide physicochemical property, pseudo dinucleotide composition, and secondary structure component, respectively. Figure 1 shows the prediction process with the ensemble classifier. The three predictors were integrated as an ensemble predictor via a voting strategy (see Materials and Methods). By combining the results of the three predictors together, a sequence in the benchmark dataset was predicted as a D-site-containing sequence if its prediction probabilities yielded by more than two predictors were all greater than 0.5.

The jackknife test results of the ensemble predictor for identifying D sites in *S. cerevisiae* transcriptome are also listed in Table 1. It was found that the sensitivity of the ensemble predictor was improved to 76.47%. Although its specificity and accuracy was a little lower than NPCP-SVM, the MCC of the ensemble predictor was 0.62, which was higher than that of any single SVM-based predictor, indicating the ensemble predictor was much more stable than NPCP-SVM, PseDNC-SVM, and SSC-SVM for the detection of D modification sites.

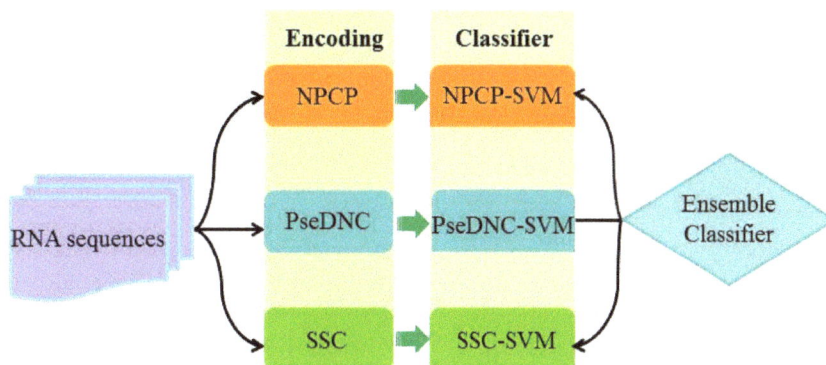

Figure 1. The flaw chart of the ensemble classifiers. NPCP-SVM stands for nucleotide-physicochemical-property-based predictor; PseDNC-SVM stands for pseudo-dinucleotide-composition-based predictor; SSC-SVM stands for secondary-structure-based predictor.

3. Materials and Methods

3.1. Benchmark Dataset

The original 208 positive samples (D-site-containing sequences) were fetched from the RMBase database [14]. All of these sequences in RMBase were 41 nt long with the D site in the center. Preliminary tests indicated that the best prediction results were achieved when the sequence was 41 nt long. In order to avoid redundancy, sequences with more than 80% sequence similarity were removed using the CD-HIT program [15]. Accordingly, we obtained 68 D-site-containing sequences from the *S. cerevisiae* transcriptome.

Negative samples were obtained by selecting 41-nt-long sequences that satisfied the following rules: (1) uridine is the center of the sequence, and (2) no dihydrouridine modification of the centered uridine has been identified experimentally. Accordingly, we could obtain a huge number of negative samples, from which we randomly picked 68 samples to form the negative subset for the purpose of using a balance benchmark dataset to train the model. In summary, our benchmark dataset comprised 68 D-site-containing sequences and 68 false D-site-containing sequences from the *S. cerevisiae* transcriptome, which is available at https://github.com/chenweiimu/D-Pred.

3.2. Sequence Encoding Scheme

3.2.1. Nucleotide Physicochemical Property (NPCP)

Adenosine (A), cytosine (C), guanine (G), and uridine (U) have different chemical properties [16,17]. In terms of ring structures, A and G are purines containing two rings, whereas C and U are pyrimidines containing one ring. When forming secondary structures, C and G form strong hydrogen bonds, whereas A and U form weak hydrogen bonds. In terms of amino/keto bases, A and C belong to the amino group, while G and U belong to the keto group [16,17].

In order to encode RNA sequences using these properties, the (x, y, z) coordinates were used to describe the chemical properties of the four nucleotides, and a value of 0 or 1 was assigned to (x, y, z), respectively. If x, y, and z coordinates stand for the ring structure, the hydrogen bond, and the amino/keto bases, A, C, G, and U can be represented by (1, 1, 1), (0, 0, 1), (1, 0, 0), and (0, 1, 0), respectively.

Accordingly, by using nucleotide chemical properties, each sequence could be encoded by a 123 (3 × 41)-dimensional vector, as given bellow:

$$\mathbf{R}_1 = \begin{bmatrix} \varepsilon_1 & \varepsilon_2 & \varepsilon_3 & \cdots & \varepsilon_i & \cdots & \varepsilon_{123} \end{bmatrix}^{\mathbf{T}} \tag{1}$$

where ε_i indicates the abovementioned nucleotide chemical properties, and its value is 0 or 1.

3.2.2. Pseudo Dinucleotide Composition

The pseudo *k*-tuple nucleotide composition (PseKNC), proposed by Chen et al. [18,19], has been successfully and widely applied in computational genomics [20–22]. PseKNC not only includes local sequence order information but also the global sequence pattern [23]. In the current study, the pseudo dinucleotide composition (PseDNC) was used to encode the RNA sequences and is defined as follows [18,19]:

$$\mathbf{R} = \begin{bmatrix} d_1 & d_2 & \cdots & d_{16} & d_{16+1} & \cdots & d_{16+\lambda} \end{bmatrix}^{\mathbf{T}} \tag{2}$$

where

$$d_u = \begin{cases} \dfrac{f_u}{\sum_{i=1}^{16} f_i + w\sum_{j=1}^{\lambda} \theta_j} & (1 \le u \le 16) \\[4mm] \dfrac{w\theta_{u-16}}{\sum_{i=1}^{16} f_i + w\sum_{j=1}^{\lambda} \theta_j} & (16 < u \le 16+\lambda) \end{cases}. \tag{3}$$

In Equation (3), f_u $(u = 1, 2, \cdots, 16)$ is the normalized occurrence frequency of the *u*-th nonoverlapping dinucleotide in the RNA sequence, and

$$\theta_j = \frac{1}{L-j-1} \sum_{i=1}^{L-j-1} C_{i,\,i+j} \quad (j = 1, 2, \cdots, \lambda; \lambda < L) \tag{4}$$

where θ_j is the *j*-tier correlation factor that reflects the sequence order correlation between all the *j*-th most contiguous dinucleotides. The coupling factor $C_{i,\,i+j}$ is defined as

$$C_{i,\,i+j} = \frac{1}{\mu} \sum_{g=1}^{\mu} \left[P_g(D_i) - P_g(D_{i+j}) \right]^2 \tag{5}$$

where μ is the number of RNA physicochemical properties considered, $P_g(D_i)$ is the normalized numerical value of the *g*-th ($g = 1, 2, 3, \ldots, \mu$) RNA local structural property for the dinucleotide $R_i R_{i+1}$ at position *i*, and $P_g(D_{i+j})$ is the corresponding value for the dinucleotide $R_{i+j}R_{i+j+1}$ at position $i + j$.

Inspired by a recent study [24], the three RNA physicochemical properties, namely, enthalpy [25], entropy [25], and free energy [26], were used to define PseDNC. Thus, in Equation (4), μ is equal to 3. The normalized numerical values of the three physicochemical properties of the 16 different RNA dinucleotides were obtained from our previous work [24].

The two parameters w and λ were optimized in the following ranges [0, 1] and [1, 10] with steps of 0.1 and 1, respectively. In the current work, the optimal values for w and λ were 0.5 and 4, respectively. Hence, the RNA sequence can be formulated by a (16 + 4) = 20-dimensional vector as given below:

$$\mathbf{R}_2 = \begin{bmatrix} d_1 & d_2 & \cdots & d_{16} & d_{17} & \cdots & d_{20} \end{bmatrix}^{\mathbf{T}} \tag{6}$$

3.2.3. Secondary Structure Component (SSC)

Considering the fact that RNA modification is affected by its structures [27], the RNA sequences were also encoded using the RNA secondary structures. By using the RNAfold tool (version 2.1.9) in ViennaRNA package with default parameters [28], we obtained the secondary structure status at each position, which was represented by brackets ("(" or ")") indicating paired nucleotides and by dots (".") indicating unpaired nucleotides. In the current study, we did not distinguish "(" and ")" and used "(" for both situations. For a given trinucleotide, there were eight (2^3) possible structure statuses (i.e., "(((", "((.", "(..", "(.(", ".((", ".(.", "..(", and " ... "). If the first nucleotide in the trinucleotide was further considered, there would be 32 (4×8) possible sequence-structure modes, which were denoted as "A-(((", "A-((.", "A-(..", ... , and "U- ... ". Therefore, a given sequence could be represented by using the following sequence-structure:

$$\mathbf{R}_3 = \left[f^A_{(((}, f^A_{((.}, f^A_{(..}, \cdots , f^A_{...}, f^C_{(((}, \cdots , f^U_{...} \right]^{\mathrm{T}}. \tag{7}$$

The elements in the vector of R_3 indicate the frequency of the 32 sequence-structure modes.

3.3. Support Vector Machine

SVM is a well-known machine learning method for pattern recognition and has been widely used in bioinformatics [29–35]. In the current study, the LibSVM package 3.18 (http://www.csie.ntu.edu.tw/~cjlin/libsvm/) was used to perform SVM. Due to its effectiveness and speed in training process, the radial basis kernel function (RBF) of SVM was often used to find the classification hyperplane. The regularization parameter C and kernel parameter γ of the SVM operation engine was optimized in the ranges of $[2^{-5}, 2^{15}]$ and $[2^{-15}, 2^{-5}]$ with steps of 2 and 2^{-1}, respectively. The prediction was made according to the probability score yielded from SVM. If its probability score was greater than 0.5, a uridine would be predicted as a D site, otherwise, a non-D-site.

3.4. Ensemble Classifiers

By using the NPCP, PseKNC, and SSC features, three basic classifiers were built, which voted for the final result according to the following rule [9]:

$$V_i = \sum_{k=1}^{3} f(pre(C_k), Class_i) \quad (i = 1, 2) \tag{8}$$

where V_i is the voting score for the sequence belonging to the $Class_i$. $f(pre(C_k),Class_i)$ is defined as

$$f(pre(C_k), Class_i) = \begin{cases} 1 & if \ pre(C_k) \in Class_i \\ 0 & if \ pre(C_k) \notin Class_i \end{cases} (i = 1, 2; k = 1, 2, 3). \tag{9}$$

The final prediction is determined by

$$Sgn(i) = \mathrm{argmax}_i\{V_i\} \quad (i = 1, 2) . \tag{10}$$

Sgn(i) is the argument that maximizes the voting score V_i.

3.5. Performance Evaluation

The performance of the method were evaluated by using sensitivity (Sn), specificity (Sp), accuracy (Acc), and the Matthews correlation coefficient (MCC), as given below [36–40]:

$$\begin{cases} Sn = 1 - \dfrac{N^+_-}{N^+} & 0 \leq Sn \leq 1 \\[2mm] Sp = 1 - \dfrac{N^-_+}{N^-} & 0 \leq Sp \leq 1 \\[2mm] Acc = 1 - \dfrac{N^+_- + N^-_+}{N^+ + N^-} & 0 \leq Acc \leq 1 \\[2mm] MCC = \dfrac{1 - \left(\dfrac{N^+_-}{N^+} + \dfrac{N^-_+}{N^-} \right)}{\sqrt{\left(1 + \dfrac{N^-_+ - N^+_-}{N^+}\right)\left(1 + \dfrac{N^+_- - N^-_+}{N^-}\right)}} & -1 \leq MCC \leq 1 \end{cases} \tag{11}$$

where N^+ represents the total number of D-site-containing sequences, while N^+_- is the number of D-site-containing sequences incorrectly predicted to be of false D-site-containing sequences. N^- is the total number of false D-site-containing sequences, while N^-_+ the number of the false D-site-containing sequences incorrectly predicted to be of D-site-containing sequences.

3.6. Jackknife Cross-Validation

Among the three methods (i.e., independent dataset test, K-fold cross-validation test, and jackknife cross-validation), the jackknife cross-validation is deemed to be the least arbitrary, as demonstrated by in a recent review paper [41]. In the jackknife cross-validation, each sample in the training dataset is in turn singled out as an independent test sample and all the rule parameters are calculated without including the one being identified [42–46]. Accordingly, jackknife cross-validation was also used to examine the performance of the method proposed in the current study.

4. Conclusions

In this study, by integrating heterogeneous sequence-based features, a SVM-based ensemble classifier was proposed to identify D modification sites in the *S. cerevisiae* transcriptome. In this predictor, not only was the local and global sequence information included by encoding RNA sequences using PseDNC, but the nucleotide chemical properties and structures were also considered by representing RNA sequences using nucleotide physicochemical properties and predicted RNA secondary structures. The jackknife test results demonstrate that the proposed predictor is promising for the identification of D modification sites. It is anticipated that the proposed method will become an essential computational tool for identifying D modification sites in tRNA.

However, the proposed method has two flaws. The limited number of experimentally verified D modification data hindered us from extracting effective features to describe the D modification sites containing sequences. The other shortcoming is that the present method directly uses the entirety of the features, which may reduce the generalization capacity of the model and increase the computational time. Therefore, in future work, we shall make efforts to collect more D modification data and also employ the feature selection method to winnow out the optimal features.

Author Contributions: P.F. and L.L. conceived and designed the experiments; P.F., Z.X., H.Y., H.L., and H.D. performed the experiments; P.F. and L.L. wrote the paper.

Funding: This work was supported by the National Nature Scientific Foundation of China (31771471, 61772119) and the Natural Science Foundation for Distinguished Young Scholar of Hebei Province (No. C2017209244).

Conflicts of Interest: The authors declare no conflict of interest.

References

1. Dyubankova, N.; Sochacka, E.; Kraszewska, K.; Nawrot, B.; Herdewijn, P.; Lescrinier, E. Contribution of dihydrouridine in folding of the D-arm in tRNA. *Organ. Biomol. Chem.* **2015**, *13*, 4960–4966. [CrossRef] [PubMed]

2. Sprinzl, M.; Horn, C.; Brown, M.; Ioudovitch, A.; Steinberg, S. Compilation of tRNA sequences and sequences of tRNA genes. *Nucleic Acids Res.* **1998**, *26*, 148–153. [CrossRef] [PubMed]

3. Yu, F.; Tanaka, Y.; Yamashita, K.; Suzuki, T.; Nakamura, A.; Hirano, N.; Suzuki, T.; Yao, M.; Tanaka, I. Molecular basis of dihydrouridine formation on tRNA. *Proc. Natl. Acad. Sci. USA* **2011**, *108*, 19593–19598. [CrossRef]

4. Jones, C.I.; Spencer, A.C.; Hsu, J.L.; Spremulli, L.L.; Martinis, S.A.; DeRider, M.; Agris, P.F. A counterintuitive Mg^{2+}-dependent and modification-assisted functional folding of mitochondrial tRNAs. *J. Mol. Biol.* **2006**, *362*, 771–786. [CrossRef] [PubMed]

5. Dalluge, J.J.; Hashizume, T.; Sopchik, A.E.; McCloskey, J.A.; Davis, D.R. Conformational flexibility in RNA: The role of dihydrouridine. *Nucleic Acids Res.* **1996**, *24*, 1073–1079. [CrossRef]

6. Kasprzak, J.M.; Czerwoniec, A.; Bujnicki, J.M. Molecular evolution of dihydrouridine synthases. *BMC Bioinform.* **2012**, *13*, 153. [CrossRef]

7. Whelan, F.; Jenkins, H.T.; Griffiths, S.C.; Byrne, R.T.; Dodson, E.J.; Antson, A.A. From bacterial to human dihydrouridine synthase: Automated structure determination. *Acta Crystallogr. Sect. D Biol. Crystallogr.* **2015**, *71*, 1564–1571. [CrossRef]
8. Alexandrov, A.; Chernyakov, I.; Gu, W.; Hiley, S.L.; Hughes, T.R.; Grayhack, E.J.; Phizicky, E.M. Rapid tRNA decay can result from lack of nonessential modifications. *Mol. Cell* **2006**, *21*, 87–96. [CrossRef] [PubMed]
9. Chen, W.; Xing, P.; Zou, Q. Detecting N6-methyladenosine sites from RNA transcriptomes using ensemble Support Vector Machines. *Sci. Rep.* **2017**, *7*, 40242. [CrossRef] [PubMed]
10. Jia, C.; Zuo, Y.; Zou, Q. O-GlcNAcPRED-II: An integrated classification algorithm for identifying O-GlcNAcylation sites based on fuzzy undersampling and a K-means PCA oversampling technique. *Bioinformatics* **2018**, *34*, 2029–2036. [CrossRef]
11. Zou, Q.; Guo, J.; Ju, Y.; Wu, M.; Zeng, X.; Hong, Z. Improving tRNAscan-SE Annotation Results via Ensemble Classifiers. *Mol. Inform.* **2015**, *34*, 761–770. [CrossRef] [PubMed]
12. Chen, W.; Feng, P.M.; Lin, H.; Chou, K.C. iRSpot-PseDNC: Identify recombination spots with pseudo dinucleotide composition. *Nucleic Acids Res.* **2013**, *41*, e68. [CrossRef] [PubMed]
13. Wan, S.; Duan, Y.; Zou, Q. HPSLPred: An Ensemble Multi-label Classifier for Human Protein Subcellular Location Prediction with Imbalanced Source. *Proteomics* **2017**, *17*, 1700262. [CrossRef] [PubMed]
14. Xuan, J.J.; Sun, W.J.; Lin, P.H.; Zhou, K.R.; Liu, S.; Zheng, L.L.; Qu, L.H.; Yang, J.H. RMBase v2.0: Deciphering the map of RNA modifications from epitranscriptome sequencing data. *Nucleic Acids Res.* **2018**, *46*, D327–D334. [CrossRef] [PubMed]
15. Fu, L.; Niu, B.; Zhu, Z.; Wu, S.; Li, W. CD-HIT: Accelerated for clustering the next-generation sequencing data. *Bioinformatics* **2012**, *28*, 3150–3152. [CrossRef] [PubMed]
16. Zhang, J.; Feng, P.; Lin, H.; Chen, W. Identifying RNA N(6)-Methyladenosine Sites in *Escherichia coli* Genome. *Front. Microbiol.* **2018**, *9*, 955. [CrossRef] [PubMed]
17. Feng, P.; Ding, H.; Yang, H.; Chen, W.; Lin, H.; Chou, K.-C. iRNA-PseColl: Identifying the Occurrence Sites of Different RNA Modifications by Incorporating Collective Effects of Nucleotides into PseKNC. *Mol. Ther.-Nucleic Acids* **2017**, *7*, 155–163. [CrossRef]
18. Chen, W.; Lei, T.-Y.; Jin, D.-C.; Lin, H.; Chou, K.-C. PseKNC: A flexible web server for generating pseudo K-tuple nucleotide composition. *Anal. Biochem.* **2014**, *456*, 53–60. [CrossRef]
19. Chen, W.; Zhang, X.; Brooker, J.; Lin, H.; Zhang, L.; Chou, K.-C. PseKNC-General: A cross-platform package for generating various modes of pseudo nucleotide compositions. *Bioinformatics* **2014**, *31*, 119–120. [CrossRef]
20. Chen, W.; Feng, P.-M.; Deng, E.-Z.; Lin, H.; Chou, K.-C. iTIS-PseTNC: A sequence-based predictor for identifying translation initiation site in human genes using pseudo trinucleotide composition. *Anal. Biochem.* **2014**, *462*, 76–83. [CrossRef]
21. Chen, W.; Feng, P.-M.; Lin, H.; Chou, K.-C. iSS-PseDNC: Identifying Splicing Sites Using Pseudo Dinucleotide Composition. *BioMed Res. Int.* **2014**. [CrossRef] [PubMed]
22. Lin, H.; Liang, Z.Y.; Tang, H.; Chen, W. Identifying sigma70 promoters with novel pseudo nucleotide composition. *IEEE/ACM Trans. Comput. Biol. Bioinform.* **2017**. [CrossRef]
23. Chen, W.; Lin, H.; Chou, K.-C. Pseudo nucleotide composition or PseKNC: An effective formulation for analyzing genomic sequences. *Mol. BioSyst.* **2015**, *11*, 2620–2634. [CrossRef]
24. Chen, W.; Feng, P.; Ding, H.; Lin, H.; Chou, K.-C. iRNA-Methyl: Identifying N-6-methyladenosine sites using pseudo nucleotide composition. *Anal. Biochem.* **2015**, *490*, 26–33. [CrossRef]
25. Freier, S.M.; Kierzek, R.; Jaeger, J.A.; Sugimoto, N.; Caruthers, M.H.; Neilson, T.; Turner, D.H. Improved free-energy parameters for predictions of RNA duplex stability. *Proc. Natl. Acad. Sci. USA* **1986**, *83*, 9373–9377. [CrossRef] [PubMed]
26. Xia, T.; SantaLucia, J., Jr.; Burkard, M.E.; Kierzek, R.; Schroeder, S.J.; Jiao, X.; Cox, C.; Turner, D.H. Thermodynamic parameters for an expanded nearest-neighbor model for formation of RNA duplexes with Watson-Crick base pairs. *Biochemistry* **1998**, *37*, 14719–14735. [CrossRef] [PubMed]
27. Lu, X.J.; Olson, W.K.; Bussemaker, H.J. The RNA backbone plays a crucial role in mediating the intrinsic stability of the GpU dinucleotide platform and the GpUpA/GpA miniduplex. *Nucleic Acids Res.* **2010**, *38*, 4868–4876. [CrossRef]
28. Lorenz, R.; Bernhart, S.H.; Honer Zu Siederdissen, C.; Tafer, H.; Flamm, C.; Stadler, P.F.; Hofacker, I.L. ViennaRNA Package 2.0. *Algorithms Mol. Biol. AMB* **2011**, *6*, 26. [CrossRef]

29. Feng, C.Q.; Zhang, Z.Y.; Zhu, X.J.; Lin, Y.; Chen, W.; Tang, H.; Lin, H. iTerm-PseKNC: A sequence-based tool for predicting bacterial transcriptional terminators. *Bioinformatics* **2018**. [CrossRef]

30. Su, Z.D.; Huang, Y.; Zhang, Z.Y.; Zhao, Y.W.; Wang, D.; Chen, W.; Chou, K.C.; Lin, H. iLoc-lncRNA: Predict the subcellular location of lncRNAs by incorporating octamer composition into general PseKNC. *Bioinformatics* **2018**. [CrossRef]

31. Chen, W.; Yang, H.; Feng, P.; Ding, H.; Lin, H. iDNA4mC: Identifying DNA N4-methylcytosine sites based on nucleotide chemical properties. *Bioinformatics* **2017**, *33*, 3518–3523. [CrossRef] [PubMed]

32. Li, D.; Ju, Y.; Zou, Q. Protein Folds Prediction with Hierarchical Structured SVM. *Curr. Proteomics* **2016**, *13*, 79–85. [CrossRef]

33. Wang, S.P.; Zhang, Q.; Lu, J.; Cai, Y.D. Analysis and Prediction of Nitrated Tyrosine Sites with the mRMR Method and Support Vector Machine Algorithm. *Curr. Bioinform.* **2018**, *13*, 3–13. [CrossRef]

34. Yang, H.; Lv, H.; Ding, H.; Chen, W.; Lin, H. iRNA-2OM: A Sequence-Based Predictor for Identifying 2'-O-Methylation Sites in Homo sapiens. *J. Comput. Biol. J. Comput. Mol. Cell Biol.* **2018**, *25*, 1266–1277. [CrossRef] [PubMed]

35. Dao, F.Y.; Lv, H.; Wang, F.; Feng, C.Q.; Ding, H.; Chen, W.; Lin, H. Identify origin of replication in Saccharomyces cerevisiae using two-step feature selection technique. *Bioinformatics* **2018**. [CrossRef] [PubMed]

36. Feng, P.-M.; Chen, W.; Lin, H.; Chou, K.-C. iHSP-PseRAAAC: Identifying the heat shock protein families using pseudo reduced amino acid alphabet composition. *Anal. Biochem.* **2013**, *442*, 118–125. [CrossRef] [PubMed]

37. Song, J.; Wang, Y.; Li, F.; Akutsu, T.; Rawlings, N.D.; Webb, G.I.; Chou, K.C. iProt-Sub: A comprehensive package for accurately mapping and predicting protease-specific substrates and cleavage sites. *Briefings Bioinform.* **2018**. [CrossRef]

38. Zhu, X.J.; Feng, C.Q.; Lai, H.Y.; Chen, W.; Lin, H. Predicting protein structural classes for low-similarity sequences by evaluating different features. *Knowl.-Based Syst.* **2018**. [CrossRef]

39. Yang, H.; Qiu, W.R.; Liu, G.Q.; Guo, F.B.; Chen, W.; Chou, K.C.; Lin, H. iRSpot-Pse6NC: Identifying recombination spots in Saccharomyces cerevisiae by incorporating hexamer composition into general PseKNC. *Int. J. Biol. Sci.* **2018**, *14*, 883–891. [CrossRef]

40. Tang, H.; Zhao, Y.W.; Zou, P.; Zhang, C.M.; Chen, R.; Huang, P.; Lin, H. HBPred: A tool to identify growth hormone-binding proteins. *Int. J. Biol. Sci.* **2018**, *14*, 957–964. [CrossRef]

41. Chou, K.C. Some remarks on protein attribute prediction and pseudo amino acid composition. *J. Theor. Biol.* **2011**, *273*, 236–247. [CrossRef] [PubMed]

42. Feng, P.M.; Lin, H.; Chen, W. Identification of antioxidants from sequence information using naive Bayes. *Comput. Math. Methods Med.* **2013**, *2013*, 567529. [CrossRef] [PubMed]

43. Feng, P.M.; Ding, H.; Chen, W.; Lin, H. Naive Bayes classifier with feature selection to identify phage virion proteins. *Comput. Math. Methods Med.* **2013**, *2013*, 530696. [CrossRef] [PubMed]

44. Lai, H.Y.; Chen, X.X.; Chen, W.; Tang, H.; Lin, H. Sequence-based predictive modeling to identify cancerlectins. *Oncotarget* **2017**, *8*, 28169–28175. [CrossRef] [PubMed]

45. Yang, H.; Tang, H.; Chen, X.X.; Zhang, C.J.; Zhu, P.P.; Ding, H.; Chen, W.; Lin, H. Identification of Secretory Proteins in Mycobacterium tuberculosis Using Pseudo Amino Acid Composition. *BioMed Res. Int.* **2016**, *2016*, 5413903. [CrossRef] [PubMed]

46. Chen, X.X.; Tang, H.; Li, W.C.; Wu, H.; Chen, W.; Ding, H.; Lin, H. Identification of Bacterial Cell Wall Lyases via Pseudo Amino Acid Composition. *BioMed Res. Int.* **2016**, *2016*, 1654623. [CrossRef] [PubMed]

Sample Availability: Samples of the compounds are not available from the authors.

molecules

MDPI

Article

Integrating Multiple Interaction Networks for Gene Function Inference

Jingpu Zhang [1] and Lei Deng [2,*

[1] School of Computer and Data Science, Henan University of Urban Construction, Pingdingshan 467000, China; zhangjp@csu.edu.cn
[2] School of Software, Central South University, Changsha 410075, China
* Correspondence: leideng@csu.edu.cn; Tel.: +86-731-8253-9736

Academic Editor: Xiangxiang Zeng
Received: 21 November 2018; Accepted: 20 December 2018; Published: 21 December 2018

Abstract: In the past few decades, the number and variety of genomic and proteomic data available have increased dramatically. Molecular or functional interaction networks are usually constructed according to high-throughput data and the topological structure of these interaction networks provide a wealth of information for inferring the function of genes or proteins. It is a widely used way to mine functional information of genes or proteins by analyzing the association networks. However, it remains still an urgent but unresolved challenge how to combine multiple heterogeneous networks to achieve more accurate predictions. In this paper, we present a method named ReprsentConcat to improve function inference by integrating multiple interaction networks. The low-dimensional representation of each node in each network is extracted, then these representations from multiple networks are concatenated and fed to gcForest, which augment feature vectors by cascading and automatically determines the number of cascade levels. We experimentally compare ReprsentConcat with a state-of-the-art method, showing that it achieves competitive results on the datasets of yeast and human. Moreover, it is robust to the hyperparameters including the number of dimensions.

Keywords: multiple interaction networks; function prediction; multinetwork integration; low-dimensional representation

---

1. Introduction

With the advent of high-throughput experimental techniques, genome-scale interaction networks have become an indispensable way to carry relevant information [1–5]. Researchers can extract functional information of genes and proteins by mining the networks [6,7]. These methods are based on the fact that proteins (or genes) that are colocated or have similar topological structures in the interaction network are more likely to be functionally related [8–18]. Thus, we are able to infer the unknown characteristics of proteins based on the knowledge of known genes and proteins.

An important challenge to the methods of network based prediction is how to integrate multiple interaction networks constructed according to heterogeneous information sources (for example, physical binding, gene interactions, co-expression, coevolution, etc.). The existing methods of integrating multiple networks for functional prediction mainly combine multiple networks into a representative network, and then perform prediction algorithms [19] (for example, label propagation algorithm [20] and graph clustering algorithm [21]) on the integrated network. There are two main methods for integrating the edges of different networks: one is the weighted averaging method of edge weights [12,22] with GeneMANIA [23] as a representative. In GeneMANIA, the weight of each network is obtained by optimizing according to the functional category. The other is a method based on Bayesian inference [24,25], which is used to combine multiple networks into the protein interaction network in database STRING [26]. A key drawback of these methods of projecting various data sets into a single network representation is that the projection process can result in a large loss of information. For example, a particular context interaction pattern that exists only in a particular data sets (e.g., tissue-specific gene modules) is likely to be obscured by the edges from other data sources in the integrated network. Recently, Cho et al. proposed a new integration method, Mashup [27], which integrates multiple networks by compressing representations of topological relationships between nodes. Vladimir and the coauthors [28] developed deepNF to derive functional labels of proteins using deep neural networks for calculating network embeddings. The method could explore underlying structure of networks and showed improved performance. However, tuning the hyperparameters requires efforts and expertise.

In this paper, we propose a multinetwork integration method, ReprsentConcat, based on gcForest [29], which builds a deep forest ensemble with a cascade structure. The cascade structure enables gcForest to learn representations. Moreover, by multigrained scanning of high-dimensional input data, gcForest can further enhance the learning ability of representation and learn the context or structure information of features. In gcForest, the number of cascade levels can be automatically determined, improving the effect of classification. In ReprsentConcat, first, a feature representation of each node in the network is obtained according to the topological structure of one network, and these features could represent the intrinsic topology of the network. Secondly, considering that the high-dimensional features contain noise, we compact these features to obtain the low dimensional representations which explain the connectivity patterns in the networks. Finally, the features of the nodes in each network are concatenated to train the classifier as the input of gcForest. A 5-fold cross-validation experiment is performed on the networks including six protein interaction networks, and the experimental results show that ReprsentConcat outperforms state-of-the-art Mashup.

2. Results

2.1. Experimental Data Set

In order to verify the effectiveness of our proposed multinetwork integration algorithm, the function prediction of proteins is performed on multiple networks consisting of six protein–protein interaction networks. The six protein interaction networks and the annotations of proteins are derived from the work of Cho et al [27]. The raw datasets are available online at http://denglab.org/ ReprsentConcat. In the dataset, protein interaction networks include species such as humans and yeast and so on, from the STRING database v9.1 [26]. Moreover, the networks constructed from text mining of the academic literature are excluded. As a result, the six yeast heterogeneous networks include a total of 6400 proteins, and the number of edges in these networks ranges from 1361 to 314,013 (as shown in Table 1). The six human heterogeneous networks include 18,362 proteins, and the number of edges in the networks ranged from 1880 to 788,166 (as shown in Table 1). The weights of edges in these networks are between 0 and 1, representing the confidence of the interaction.

Table 1. Interaction network and its corresponding number of edges.

Network	Human	Yeast
coexpression	788,166	314,014
co-occurrence	18,064	2664
database	159,502	33,486
experimental	309,287	219,995
fusion	1880	1361
neighborhood	52,479	45,610

The functional annotations for yeast proteins comes from Munich Information Center for Protein Sequences (MIPS) [30], and the annotations for human from the Gene Ontology (GO) database [31]. The functions in MIPS are organized in a tree structure and are divided into three levels, where Level 1 includes 17 most general functional categories, Level 2 includes 74 functional categories, and Level 3 includes 154 most specific functional categories. It is noted that each protein can have more than one function. The GO terms in the GO database are organized in a directed acyclic graph. The GO terms are divided into three categories including biological process (BP), molecular function (MF), and cellular component (CC), representing three different functional categories. In this dataset, these GO terms are divided into three groups where each consists of GO terms with 11–30, 31–100, and 101–300 annotated genes (see Table 2). In order to maintain the consistency of the predicted GO labels, the GO label is propagated in the GO hierarchy by applying the "is a" and "part of" relationships, i.e., if a gene is labeled as a GO term, then the gene is also annotated with all the ancestral terms of the term.

Table 2. Number of Gene Ontology (GO) terms by the number of annotated genes in human biological process (BP)/molecular function (MF)/cellular component (CC).

	11–30	31–100	101–300
BP	262	100	28
MF	153	72	18
CC	82	46	18

2.2. Evaluation Metrics

In our ReprsentConcat, the output for each class is a real number between 0 and 1, and we obtain the final predictions by applying an appropriate threshold, t, on the outputs. For a given sample, if the corresponding output for a class is equal to or greater than the threshold t, this class is assigned to the sample; otherwise it is not assigned to the sample. However, choosing the "optimal" threshold is a difficult task. Low thresholds will bring about more classes being assigned to the sample, resulting in high recall and low precision. On the contrary, a larger threshold allows fewer classes to be assigned to the sample, resulting in high precision and low recall. To tackle this problem, we use Precision–Recall (PR-curve) as an evaluation metric. In order to plot the PR-curve of a given classifier, different thresholds in [0, 1] are respectively applied to the output of the classifier, so as to obtain the corresponding precision and recall. The area under the PR-curve (AUPR) can also be calculated, and different methods can be compared based on their area under the PR-curve.

2.3. Impact of Feature Dimension on Performance

In this paper, the topology features of each node (entity) in one network are extracted by running random walk algorithm on the network, but the obtained features tend to have higher dimensions and contain noise. For this reason, the diffusion component analysis (DCA) method is used to reduce the dimension [32,33]. In this section, the sensitivity of the feature dimension is discussed. Specifically, we evaluate how the feature dimension of each network affects the performance. In this experiment, 5-fold cross-validation is used to evaluate the effect of feature dimensions on performance based on yeast six protein interaction networks and functional labels of Level 1. We preset the random walk restart probability $a = 0.5$ and vary the dimension of the feature, setting the dimensions to 50, 100, 200, 300, 400, 500, etc. The predictive performance of the gene function is tested through Macro-averaged F1, Micro-averaged F1, and AUPR (the micro-averaged area under the precision–recall curve) metrics. As shown in Figure 1, the abscissa stands for the feature dimension of each network and the ordinate for the score. The predicted scores is the average of five trials.

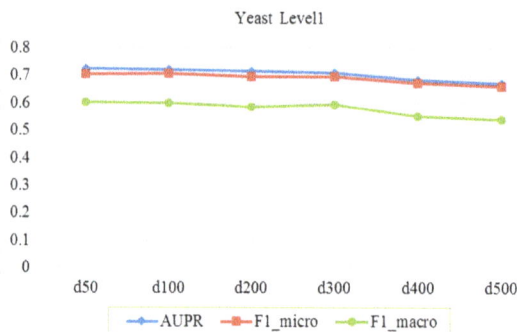

Figure 1. Performance comparison under different network feature dimensions.

As shown in the figure, when the dimension is increased from 50 to 500, the scores of metrics such as Macro-averaged F_1, Micro-averaged F_1, and AUPR do not change greatly. It is only when the dimension is greater than 300 that the corresponding score begins to slowly decline. In the experiments, the feature dimension of each network is set to 100.

2.4. Performance Evaluaton of Multinetwork Integration

An important factor that ReprsentConcat proposed in this paper can improve accuracy is the compactness of its feature representations, which not only helps to eliminate noise in the data, but also extracts functionally related topological patterns. In order to demonstrate the effectiveness of integrating multiple STRING networks, ReprsentConcat is applied to respectively single network in STRING, and the evaluation of function prediction for MIPS yeast annotations for Level 1 is performed. We compare the predictive performance on each individual network in STRING to using all networks simultaneously through 5-fold cross-validation. As shown in Figure 2, the cross-validation performance of ReprsentConcat is measured by metrics including Macro-averaged F1, Micro-averaged F1, and AUPR, as well as others. The results show that the prediction performance of all networks used at the same time (the bar with the horizontal axis of 'all' in the figure) is significantly better than the prediction performance of a single network (rank-sum test p value < 0.01). The results are summarized over five trials.

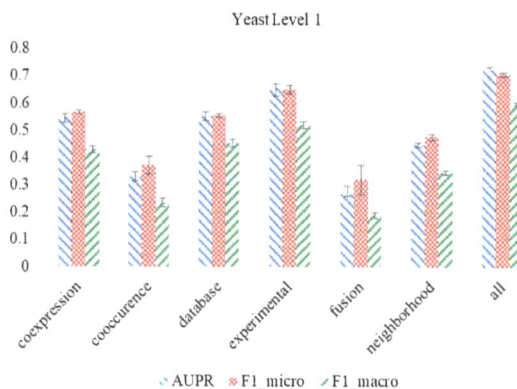

Figure 2. Comparison of predictive performance of multiple network integration with performance of single network.

2.5. Comparison of Different Integrative Methods

The results of gene function prediction on multiple networks in the STRING database using ReprsentConcat are shown in Figures 3–7. In the ReprsentConcat method, the restart probability, which is a parameter in random walk algorithm, is set to 0.5. We also experimentally confirm that the performance of ReprsentConcat is stable when the restart probability varies between 0.1 and 0.9. Due to the different protein interaction networks between yeast and humans, different dimensions are chosen when reducing the dimension of network topology features. For six yeast proteins interaction networks, the dimension is 100, and for human protein interaction networks, the dimension is 300. In the experiment, we employ gcForest for multinetwork integration and function prediction. Each level in the cascade uses eight random forest classifiers, and each forest contains 500 trees. In order to automatically determine the optimal number of cascade levels, it is especially important to select appropriate evaluation metric. Considering that gene function prediction belongs to multilabel classification problem, we use F_1 metric to determine the number of cascade levels. That is, if the prediction performance in the next four levels is not improved then, the current level is considered to be the optimal number of level, and the output of the current level is the final prediction result.

To evaluate the performance, ReprsentConcat is compared to the latest multinetwork integration methods: Mashup [27] and deepNF [28]. In the Mashup method, the high-dimension topological features of each node in the network were first obtained by random walk. When reducing the dimension of the high-dimension feature, it was assumed that the low-dimension features of the nodes in multiple networks were the same. Then the same low-dimension topology features of multiple networks were obtained by solving an optimization function. As shown in Figures 3–6, according to the PR-curve, the ReprsentConcat (denoted as RepCat) method is superior to the Mashup method in the cross-validation experiment of gene function prediction in the real data sets of yeast and human. We demonstrate that ReprsentConcat has significant performance improvements at the different annotation levels of the MIPS database and the GO database. For example, in the function annotation MIPS Level 1, the AUPR values of Mashup and ReprsentConcat are 0.70 and 0.728, respectively. Part of the reason for the improved performance of ReprsentConcat is that it obtains the topology pattern of each network and compacts the representation of topological features. The compressed low-dimension feature helps to eliminate noise in the network, while gcForest based on random forests does the feature selection.

deepNF integrated different heterogeneous networks of protein interactions and extracted the compact, low-dimensional feature representation for each node by using the stack denoising autoencoder, then fed the representations into SVM classifiers. The method was able to capture nonlinear information contained in large-scale biological networks and the experiments indicated that it had a good performance on human and yeast STRING networks. We compare ReprsentConcat and deepNF by running 5-fold cross-validation on yeast STRING networks. The results on different annotation levels of the MIPS hierarchy are summarized in Figure 7 (ReprsentConcat denoted as RepCat). We observe that the two methods share similar performance regarding the AUPR and F_1 at levels 1 and 2 of the MIPS hierarchy. At level 3, the AUPR value of deepNF is larger than that of ReprsentConcat while the F_1 value of ReprsentConcat is larger. Since deepNF is based on deep neural networks, there are a number of hyperparameters (e.g., hidden layers, nodes in the hidden layer, and learning rate) to tune and the procedure generally is difficult and needs tricks and expertise. Moreover, the computational cost is usually high. In DeepNF, there are more than three hundred million parameters in the yeast networks to be trained in total. The training consumes almost all of the memory of the GPU (two Geforce RTX 2080 GPUs with 22GB memory in our server). Relatively few hyperparameters (the number of forests and trees in each forest) need to be set in ReprsentConcat, and the training can be performed on CPU.

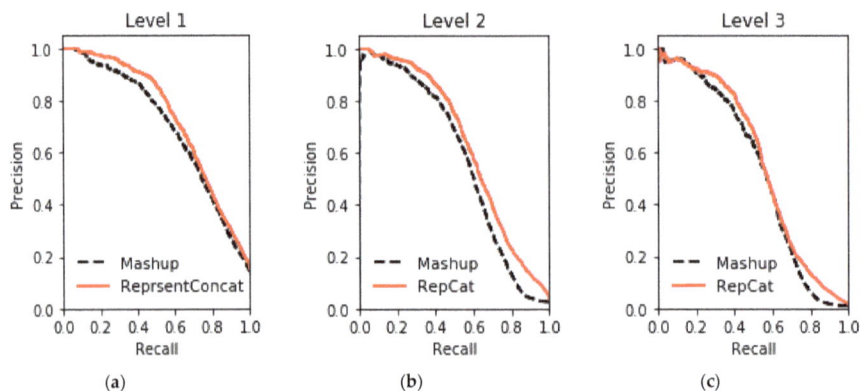

Figure 3. Comparison of performance on yeast datasets between ReprsentConcat and Mashup: (**a**) Level 1; (**b**) Level 2; and (**c**) Level 3.

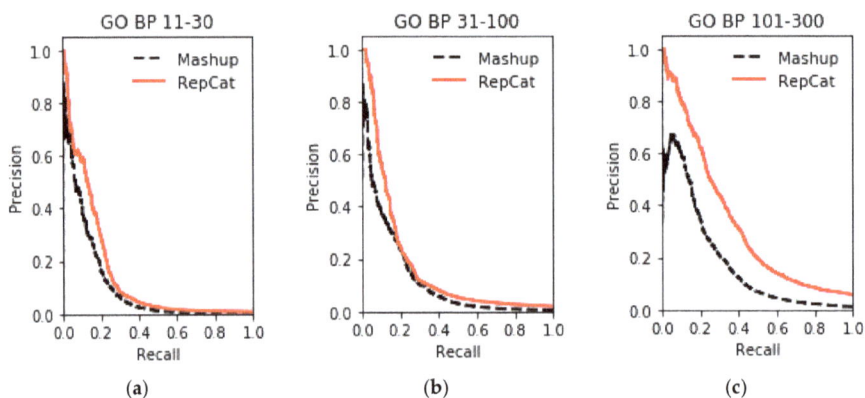

Figure 4. Performance Comparison of GO BP function prediction on human datasets between ReprsentConcat and Mashup. (**a**): GO BP 11-30; (**b**): GO BP 31-100; (**c**): GO BP 101-300.

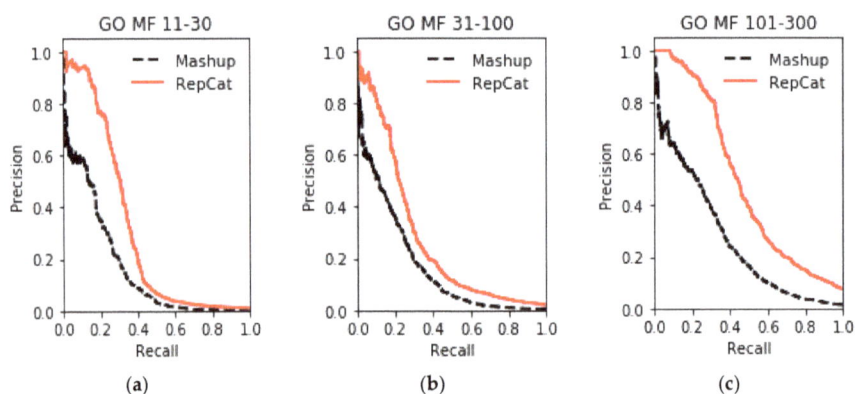

Figure 5. Performance comparison of GO MF function prediction on human datasets between ReprsentConcat and Mashup: (**a**) GO MF 11-30; (**b**) GO MF 31-100; and (**c**) GO MF 101-300.

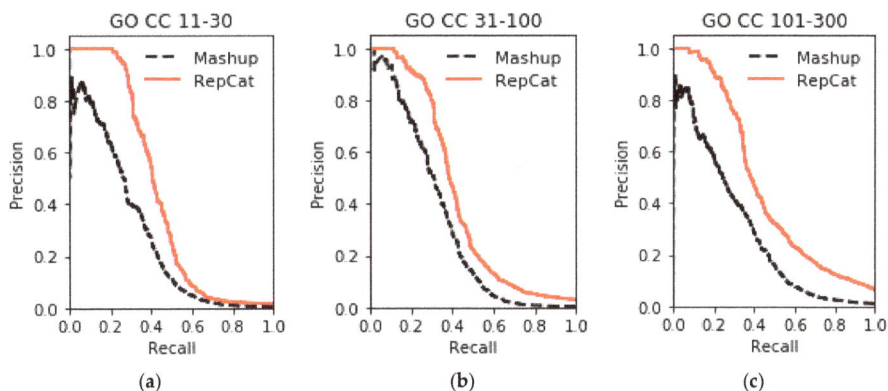

Figure 6. Performance comparison of GO CC function prediction on human datasets between ReprsentConcat and Mashup: (**a**) GO CC 11-30; (**b**) GO CC 31-100; and (**c**) GO CC 101-300.

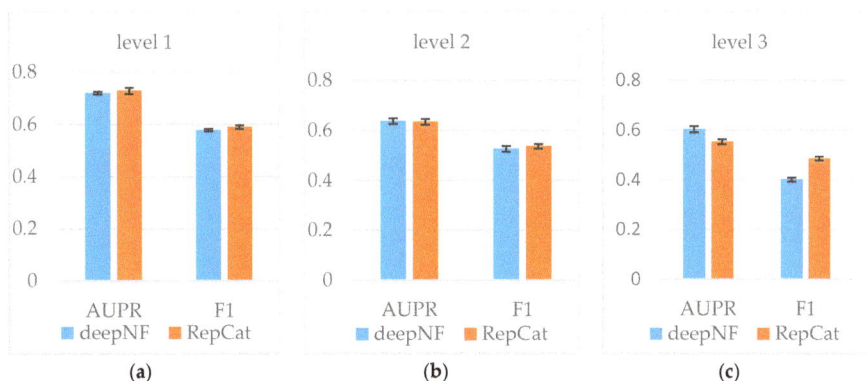

Figure 7. Performance comparison on yeast dataset between ReprsentConcat and deepNF: (**a**) Level 1; (**b**) Level 2; and (**c**) Level 3.

2.6. Case Study: ESR1

Estrogen signaling is mediated by binding to estrogen receptors (ERs), which are ligand-dependent transcription factors composed of several domains important for hormone binding, DNA binding, and activation of transcription. There exist two ER subtypes in humans, namely ERα and ERβ, coded by the *ESR1* and *ESR2* genes, respectively [34]. Gene *ESR1* is located on chromosome 6q25.1 and consists of eight exons spanning >140 kb. The protein coded by *ESR1* localizes to the nucleus where it may form a homodimer or a heterodimer with estrogen receptor 2. The researches have demonstrated that estrogen and its receptors are essential for sexual development and reproductive function, but are also involved in other tissues such as bone. Estrogen receptors are also involved in pathological processes including breast cancer, endometrial cancer, and osteoporosis [35,36]. There is strong evidence for a relationship between genetic variants on the ESR1 gene and cognitive outcomes. The relationships between ESR1 and cognitive impairment tend to be specific to or driven by women and restricted to risk for Alzheimer's disease rather than other dementia causes [37].

We employ ReprsentConcat to predict the functions of gene ESR1. As described above, the GO terms, which are divided into three categories (namely, BP, MF, and CC), which are further split into three groups for each category according to the number of annotated genes. In the category of BP, there are 28 GO terms with 101–300 annotated genes. In this experiment, we predict the functions of ESR1 by using the protein interaction networks and the 28 GO labels. The output of ReprsentConcat is a 28-dimensional probability vector in which each entry represents the probability of having the function. The vector is sorted and the result is listed in Table 3. The GO terms marked with the character '#', which have been confirmed in our annotation datasets, are ranked 2nd and 16th, respectively. The GO terms marked with character '*', which are new annotations and confirmed in 2017 from UniProt-GOA [38], ranked 1st, 4th, 9th, 10th, and 15th, respectively. The result shows ReprsentConcat generates relatively satisfactory predictions.

Table 3. The rank of GO terms according the predictions of ReprsentConcat. The GO terms marked with the character '#' indicate that they have been confirmed in the annotation datasets, and the GO terms marked with the character '*' represent they are new annotations for 2017 from UniProt-GOA.

Rank	GO Term	GO Name
1	GO:0000122 *	negative regulation of transcription by RNA polymerase II
2	GO:0071495 #	cellular response to endogenous stimulus
3	GO:0016265	obsolete death
4	GO:0048878*	chemical homeostasis
5	GO:0051241	negative regulation of multicellular organismal process
6	GO:0051098	regulation of binding
7	GO:0008284	positive regulation of cell population proliferation
8	GO:0007399	nervous system development
9	GO:0006259*	DNA metabolic process
10	GO:0009057*	macromolecule catabolic process
11	GO:0010564	regulation of cell cycle process
12	GO:0043900	regulation of multi-organism process
13	GO:0002520	immune system development
14	GO:0006928	movement of cell or subcellular component
15	GO:0006325*	chromatin organization
16	GO:0018130#	heterocycle biosynthetic process
17	GO:0016192	vesicle-mediated transport
18	GO:0031647	regulation of protein stability
19	GO:0003008	system process
20	GO:0008283	cell population proliferation
21	GO:0051259	protein complex oligomerization
22	GO:0030111	regulation of Wnt signaling pathway
23	GO:0006629	lipid metabolic process
24	GO:0034622	cellular protein-containing complex assembly
25	GO:0010608	posttranscriptional regulation of gene expression
26	GO:0055085	transmembrane transport
27	GO:0016311	dephosphorylation
28	GO:0007186	G protein-coupled receptor signaling pathway

3. Multinetwork Integration Based on gcForest

3.1. gcForest

Ensemble learning has been well studied and widely deployed in many applications [39–43]. As described in Section 1, gcForest is an ensemble method based on forest. Its structure mainly includes cascade forest and multigrained scanning.

3.1.1. Cascade Forest

gcForest's cascade structure adapts a level after level structure of deep network, that is, each level in the cascade structure receives the processed result of the preceding level, and passes the processed result of the level to the next level, as shown in Figure 8. Each level is composed of multiple random forests made up of decision trees. In Figure 8, there are two random forests, which are completely random forest (black) and random forest (blue), respectively.

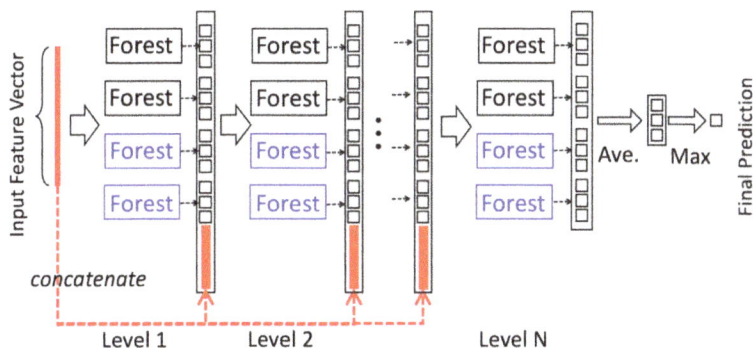

Figure 8. The cascade structure of gcForest.

Each forest will generate a probability vector of length C. If each level of gcForest is composed of N forests, then the output of each level is N C-dimensional vectors connected together, namely, $C*N$ dimensional vectors. The vector is then spliced with the original feature vector of the next level (the thick red line portion of each level in Figure 8) as the input to the next level. For example, in the three-classification problem in Figure 8, each level consists of four random forests, and each forest will generate a 3-dimensional vector. Hence, each level produces a 4*3=12-dimensional feature vector. This feature vector will be used as augmented feature of the original feature for the next level. To reduce the risk of overfitting, the class vector generated in each forest is produced by k-fold cross-validation. Specifically, after extending a new level, the performance of the entire cascade will be evaluated on the validation set, and the training process will terminate if there is no significant performance improvement. Therefore, the number of cascade levels in cascade is automatically determined.

3.1.2. Multigrained Scanning

Since there may be some relationships between the features of the data, for example, in image recognition, there is a strong spatial relationship between pixels close in position, and sequential relationships between sequence data. Cascade forest is enhanced through multigrained scanning, i.e., it samples by sliding windows with a variety of sizes to obtain more feature subsamples, so as to achieve the effect of multigrained scanning.

By employing multiple sizes of sliding windows, the final transformed feature vector will include more features, as shown in Figure 9. In Figure 9, it is assumed that the 100-dimensional, 200-dimensional, and 300-dimensional windows are used to slide on the raw 400-dimensional features.

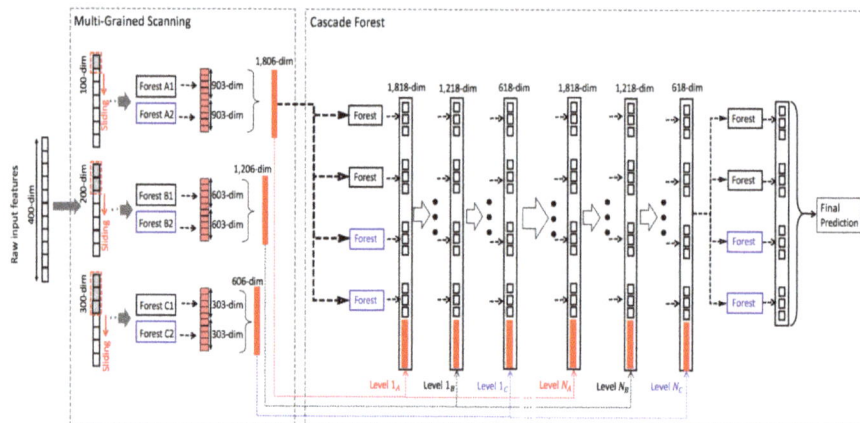

Figure 9. The overall structure of gcForest.

3.2. Network Feature Extraction

The method of random walk with restart (RWR) has been widely used in network structure analysis [44–48]. The RWR algorithm allows the restart of a random walk from the initial node at each step with a certain probability. It can capture local and global topology information to identify important nodes in the network. Assuming that a protein interaction network containing n nodes is represented by $G = (V, E)$, where V is the set of nodes, each node representing a protein, and E is the set of edges. A is the adjacency matrix of G. M represents the Markov possibility transition matrix of A, and each element M_{ij} denotes the probability walking from node j to node i, then,

$$M_{ij} = \frac{A_{ij}}{\sum_{i'} A_{i'j}} \tag{1}$$

The iterative equation for the random walk from node i is as follows,

$$s_i^{t+1} = (1 - \alpha)s_i^t M + \alpha s_i^0 \tag{2}$$

a is the restart probability, which determines the relative importance of local and global topology information. The larger its value, the greater the chances of restart, and the more important the local structure information. s_i is an distribution vector of n-dimension, where each entry represents the probability that a node is visited after t-walk; s_i^0 denotes the initial probability, and $s_i^0(i) = 1, s_i^0(j) = 0$. After several iterations, s_i can converge to a stable distribution, then this distribution represents the probability of a transition from node i to node j, including the topological information of the path from node i to node j. Then, if there are similar diffusion states between node i and node j, it means that they have similar positions in the network, which implies that they might have similar functions. Hence, when the RWR is stable, we obtain the diffusion state feature of each node.

The feature dimension obtained by random walk is high. We use diffusion component analysis (DCA) [27] to reduce the dimension. To extract a fewer dimensional vector representation of nodes, we employ the logistic model to approximate diffusion state s_i of each node. In detail, the probability of random walk from node i to node j is specified by

$$\hat{s}_{ij} = \frac{\exp\{x_i^T w_j\}}{\sum_{j'} \exp\{x_i^T w_{j'}\}} \tag{3}$$

Where x_i and w_j are d-dimension vectors and d is much smaller than n. x_i represents the node features, and w_i represents the context features, both of which capture the topology information of the network. The inner product is larger when the x_i and w_j are closer in direction, which implies that random walks starting from node i will frequently visit node j. In order to calculate w and x, we define the KL-divergence distance between the real distribution s_i and the transformed distribution $\hat{s}_i$ and minimize it, namely, the loss function for n nodes is

$$\min_{w,x} C(s, \hat{s}) = \frac{1}{n} \sum_{i=1}^{n} D_{KL}(s_i || \hat{s}_i). \tag{4}$$

We can obtain the low-dimensional feature by solving the minimum value of this loss function

3.3. Training and Prediction of ReprsentConcat

In ReprsentConcat, the d-dimension topology features of each network are first obtained according to the method described above, and then the topological features of multiple networks are concatenated to generate a one-dimension feature vector as the input features of gcForest. Considering that there is no spatial or sequential relationship between these features, we do not perform the multigrained process on these features. In the training, the prediction performance of each level is evaluated by k-fold cross-validation. We use Micro-averaged F_1 as the metric to determine the number of cascade levels. The outputs of the current level are considered to be the final predictions if there is no improvement in the next m levels in term of F_1. The pseudocode of ReprsentConcat is shown in Algorithm 1.

In order to obtain the predictions in a test set, the features of a test sample are fed to the cascade forest. The output of the optimal level which is determined by the training process is a multidimensional class vector. Each entry of the class vector is a probability indicating the possibility that the sample belongs to one class. Hence, a threshold t is applied to the class vector to obtain predictions for all classes. If the jth value of the class vector is equal to or larger than the given threshold, the sample is assigned to the class C^j where C represents the set of classes. The final classification result of ReprsentConcat is given by a binary vector V with the length of $|C|$. If the jth output is equal to or larger than the given threshold, V_j is set to 1. Otherwise, it is set to 0. Obviously, different thresholds may result in different predictions. Since the output of cascade forest is between 0 and 1, the thresholds also vary between 0 and 1. The larger the threshold used, the less the predicted classes. Conversely, the smaller the threshold used, the more the predicted classes.

Algorithm 1: ReprsentConcat Algorithm

Input: *network_files*: paths to adjacency list files, *n*: number of genes in input networks, *d*: number of output dimensions, *onttype*: which type of annotations to use, *early_stopping_rounds*: number of stopping the rounds

Output: *opt_pred_results*: prediction results

  for *i=1*: length(*network_files*)

    A=load_network(*network_files*(i), *n*)

    Q=rwr(*A*, 0.5)

    R=ln(*Q*+1/*n*)

    U, Σ, *V* =svd(*R*)

    $X\_cur = U_d \Sigma_d^{1/2}$

    X=hstack(*X*, *X_cur*)

  end for

  Y=load_annotation(*onttype*) //load annotations

  //split the data into train data and test data

  X_train, Y_train, X_test, Y_test=train_test_split(*X, Y*)

  layer_id=0

  while 1

    if *layer_id*==0

      X_cur_train=zeros(*X_train*)

      X_cur_test=zeros(*X_test*)

    else

      X_cur_train=*X_proba_train*.copy()

      X_cur_test= *X_proba_test*.copy()

    end if

    X_cur_train=hstack(*X_cur_train, X_train*)

    X_cur_ test =hstack(*X_cur_ test, X_ test*)

    for *estimator* **in** *n_randomForests*

      //train each forest through k-fold cross validation

      y_probas= estimator.fit_transform(*X_cur_train, Y_train*)

      y_train_proba_li+= *y_probas*

      y_test_probas= estimator.predict_proba(*X_cur_ test*)

      y_test_proba_li+= *y_test_probas*

    end for

    y_train_proba_li /=length(*n_randomForests*)

    y_test_proba_li /=length(*n_randomForests*)

    train_avg_F_1=calc_F1(*Y_train, y_train_proba_li*) // calculate the F_1 value

    test_avg_F_1=calc_F1(*Y_test, y_test_proba_li*)

    test_F_1_list.append(*test_avg_F_1*)

    opt_layer_id=get_opt_layer_id(*test_F_1_list*)

    if *opt_layer_id* = layer_id

      opt_pred_results=[*y_train_proba_li, y_test_proba_li*]

    end if

    if *layer_id - opt_layer_id* >= *early_stopping_rounds*

      return *opt_pred_results*

    end if

    layer_id+=1

  end while

---

4. Conclusions

In this paper, we propose ReprsentConcat, an integrative method, to combine multiple networks from heterogeneous data sources. In ReprsentConcat, the topological features are extracted by running random walks on each network, and the features are represented using low-dimensional vectors. Then the low-dimensional features are concatenated as the input of gcForests for prediction. To verify the performance of this method, we performed gene function prediction on multiple protein interaction networks of yeast and humans. The experimental results demonstrated that the prediction performance by integrating multiple networks is much better than that using a single network. Moreover, ReprsentConcat is not sensitive to multiple parameters such as the number of dimensions for function prediction. We also compare with the latest network integration method Mashup. According to the result of 5-fold cross-validation, ReprsentConcat outperforms Mashup in terms of precision–recall curves.

Besides the network data, other non-network information, such as sequence features, can be integrated into ReprsentConcat for function prediction by concatenating them. As a note, there are still further improvements in the predictions of protein function in our method. For example, the topological features of nodes are extracted through semisupervised learning by combining label information. As a result, the learned features might be more effective in this manner.

Author Contributions: L.D. conceived this work and designed the experiments. J.Z. conducted the experiments. L.D. and J.Z. collected the data and analyzed the results. J.Z. conducted the research of previous studies. L.D. and J.Z. wrote, revised, and approved the manuscript.

Funding: This work was funded by the National Natural Science Foundation of China (grant number 61672541) and the Natural Science Foundation of Hunan Province (grant number 2017JJ3287).

Acknowledgments: The authors would like to thank the Experimental Center of School of Software of Central South University for providing computing resources.

Conflicts of Interest: The authors declare no conflicts of interest.

References

1. Donghyeon, Y.; Minsoo, K.; Guanghua, X.; Tae Hyun, H. Review of biological network data and its applications. *Genom. Inform.* **2013**, *11*, 200–210.
2. Batushansky, A.; Toubiana, D.; Fait, A. Correlation-Based Network Generation, Visualization, and Analysis as a Powerful Tool in Biological Studies: A Case Study in Cancer Cell Metabolism. *BioMed Res. Int.* **2016**, *2016*, 8313272. [CrossRef]
3. Jiang, X.; Zhang, H.; Quan, X.W.; Yin, Y.B. A Heterogeneous Networks Fusion Algorithm Based on Local Topological Information for Neurodegenerative Disease. *Curr. Bioinform.* **2017**, *12*, 387–397. [CrossRef]
4. Luo, J.W.; Liu, C.C. An Effective Method for Identifying Functional Modules in Dynamic PPI Networks. *Curr. Bioinform.* **2017**, *12*, 66–79. [CrossRef]
5. Zeng, X.; Zhang, X.; Zou, Q. Integrative approaches for predicting microRNA function and prioritizing disease-related microRNA using biological interaction networks. *Brief. Bioinform.* **2016**, *17*, 193–203. [CrossRef] [PubMed]
6. Zeng, C.; Zhan, W.; Deng, L. Curation, SDADB: A functional annotation database of protein structural domains. *Database (Oxford)* **2018**, *2018*, 64. [CrossRef] [PubMed]
7. Zou, Q.; Li, J.; Wang, C.; Zeng, X. Approaches for Recognizing Disease Genes Based on Network. *Biomed Res. Int.* **2014**, *2014*, 416323. [CrossRef]
8. Chua, H.N.; Sung, W.; Wong, L. Exploiting indirect neighbours and topological weight to predict protein function from protein–protein interactions. *Bioinformatics* **2006**, *22*, 1623–1630. [CrossRef]
9. Milenković, T.; Pržulj, N. *Topological Characteristics of Molecular Networks*; Springer: New York, NY, USA, 2012; pp. 15–48.
10. Sharan, R.; Ulitsky, I.; Shamir, R. Network-based prediction of protein function. *Mol. Sys.Biol.* **2007**, *3*, 88–88. [CrossRef]

11. Wang, S.; Cho, H.; Zhai, C.; Berger, B.; Peng, J. Exploiting ontology graph for predicting sparsely annotated gene function. *Bioinformatics* **2015**, *31*, 357–364. [CrossRef]

12. Yu, G.; Zhu, H.; Domeniconi, C.; Guo, M. Integrating multiple networks for protein function prediction. *BMC Sys. Biol.* **2015**, *9*, 1–11. [CrossRef]

13. Zhang, J.; Zhang, Z.; Chen, Z.; Deng, L. Integrating Multiple Heterogeneous Networks for Novel LncRNA-Disease Association Inference. *IEEE/ACM Trans. Comput. Biol. Bioinform.* **2017**. [CrossRef] [PubMed]

14. Jiang, J.; Xing, F.; Zeng, X.; Zou, Q. RicyerDB: A Database For Collecting Rice Yield-related Genes with Biological Analysis. *Int. J. Biol. Sci.* **2018**, *14*, 965–970. [CrossRef] [PubMed]

15. Wang, L.; Ping, P.Y.; Kuang, L.N.; Ye, S.T.; Lqbal, F.M.B.; Pei, T.R. A Novel Approach Based on Bipartite Network to Predict Human Microbe-Disease Associations. *Curr. Bioinform.* **2018**, *13*, 141–148. [CrossRef]

16. Liu, Y.; Zeng, X.; He, Z.; Zou, Q. Inferring MicroRNA-Disease Associations by Random Walk on a Heterogeneous Network with Multiple Data Sources. *IEEE/ACM Trans. Comput. Biol. Bioinform.* **2017**, *14*, 905–915. [CrossRef] [PubMed]

17. Zhu, L.; Su, F.; Xu, Y.; Zou, Q. Network-based method for mining novel HPV infection related genes using random walk with restart algorithm. *Biochim. Biophys. Acta Mol. Basis Dis.* **2018**, *1864*, 2376–2383. [CrossRef] [PubMed]

18. Zeng, X.; Liu, L.; Lü, L.; Zou, Q. Prediction of potential disease-associated microRNAs using structural perturbation method. *Bioinformatics* **2018**, *34*, 2425–2432. [CrossRef]

19. Zhang, Z.; Zhang, J.; Fan, C.; Tang, Y.; Deng, L. KATZLGO: Large-scale Prediction of LncRNA Functions by Using the KATZ Measure Based on Multiple Networks. *IEEE/ACM Trans. Comput. Biol. Bioinform.* **2017**. [CrossRef]

20. Mostafavi, S.; Ray, D.; Wardefarley, D.; Grouios, C.; Morris, Q. GeneMANIA: A real-time multiple association network integration algorithm for predicting gene function. *Genome Biol.* **2008**, *9*, 1–15. [CrossRef]

21. Dutkowski, J.; Kramer, M.; Surma, M.A.; Balakrishnan, R.; Cherry, J.M.; Krogan, N.J.; Ideker, T. A gene ontology inferred from molecular networks. *Nat. Biotechnol.* **2013**, *31*, 38–45. [CrossRef]

22. Yu, G.; Fu, G.; Wang, J.; Zhu, H. Predicting protein function via semantic integration of multiple networks. *IEEE/ACM Trans. Comput. Biol. Bioinform.* **2016**, *13*, 220–232. [CrossRef] [PubMed]

23. Mostafavi, S.; Morris, Q. Fast integration of heterogeneous data sources for predicting gene function with limited annotation. *Bioinformatics* **2010**, *26*, 1759–1765. [CrossRef] [PubMed]

24. Lee, I.; Blom, U.M.; Wang, P.I.; Shim, J.E.; Marcotte, E.M. Prioritizing candidate disease genes by network-based boosting of genome-wide association data. *Genome Res.* **2011**, *21*, 1109–1121. [CrossRef] [PubMed]

25. Meng, J.; Zhang, X.; Luan, Y. Global Propagation Method for Predicting Protein Function by Integrating Multiple Data Sources. *Curr. Bioinform.* **2016**, *11*, 186–194. [CrossRef]

26. Franceschini, A.; Szklarczyk, D.; Frankild, S.; Kuhn, M.; Simonovic, M.; Roth, A.; Lin, J.; Minguez, P.; Bork, P.; Von Mering, C. STRING v9.1: Protein–protein interaction networks, with increased coverage and integration. *Nucleic Acids Res.* **2012**, *41*, 808–815. [CrossRef] [PubMed]

27. Cho, H.; Berger, B.; Peng, J. Compact integration of multi-network topology for functional analysis of genes. *Cell Syst.* **2016**, *3*, 540–548. [CrossRef]

28. Gligorijevic, V.; Barot, M.; Bonneau, R.J.B. deepNF: Deep network fusion for protein function prediction. *Bioinformatics* **2017**, *34*, 3873–3881. [CrossRef]

29. Zhou, Z.; Feng, J. Deep forest: Towards an alternative to deep neural networks. *Int. Joint Conf. Artif. Intell.* **2017**, 3553–3559.

30. Ruepp, A.; Zollner, A.; Maier, D.; Albermann, K.; Hani, J.; Mokrejs, M.; Tetko, I.V.; Guldener, U.; Mannhaupt, G.; Munsterkotter, M. The FunCat, a functional annotation scheme for systematic classification of proteins from whole genomes. *Nucleic Acids Res.* **2004**, *32*, 5539–5545. [CrossRef]

31. Consortium, G.O. The Gene Ontology (GO) project in 2006. *Nucleic Acids Res.* **2006**, *34*, 322–326. [CrossRef]

32. Cho, H.; Berger, B.; Peng, J. Diffusion component analysis: Unraveling functional topology in biological networks. *Res. Comput. Mol. Biol.* **2015**, *9029*, 62–64. [PubMed]

33. Zhang, B.; Li, L.; Lü, Q. Protein solvent-accessibility prediction by a stacked deep bidirectional recurrent neural network. *Biomolecules* **2018**, *8*, 33. [CrossRef]

34. Signe, A.E.; Kadri, H.; Maire, P.; Outi, H.; Anneli, S.E.; Helle, K.; Andres, M.; Andres, S. Allelic estrogen receptor 1 (ESR1) gene variants predict the outcome of ovarian stimulation in in vitro fertilization. *Mol. Hum. Reprod.* **2007**, *13*, 521–526.

35. Toy, W.; Yang, S.; Won, H.; Green, B.; Sakr, R.A.; Will, M.; Li, Z.; Gala, K.; Fanning, S.; King, T.A.; et al. ESR1 ligand-binding domain mutations in hormone-resistant breast cancer. *Nat. Genet.* **2013**, *45*, 1439–1445. [CrossRef] [PubMed]

36. Ioannidis, J.P.A.; Ralston, S.H.; Bennett, S.T.; Maria Luisa, B.; Daniel, G.; Karassa, F.B.; Bente, L.; Van Meurs, J.B.; Leif, M.; Serena, S. Differential genetic effects of ESR1 gene polymorphisms on osteoporosis outcomes. *Jama* **2004**, *292*, 2105–2114. [CrossRef] [PubMed]

37. Sundermann, E.E.; Maki, P.M.; Bishop, J.R. A review of estrogen receptor α gene (esr1) polymorphisms, mood, and cognition. *Menopause* **2010**, *17*, 874–886. [CrossRef] [PubMed]

38. Huntley, R.P.; Tony, S.; Prudence, M.M.; Aleksandra, S.; Carlos, B.; Martin, M.J.; Claire, O.D.J. The GOA database: Gene Ontology annotation updates for 2015. *Nucleic Acids Res.* **2015**, *43*, 1057–1063. [CrossRef]

39. Pan, Y.; Wang, Z.; Zhan, W.; Deng, L. Computational identification of binding energy hot spots in protein-RNA complexes using an ensemble approach. *Bioinformatics* **2018**, *34*, 1473–1480. [CrossRef]

40. Pan, Y.; Liu, D.; Deng, L. Accurate prediction of functional effects for variants by combining gradient tree boosting with optimal neighborhood properties. *PLoS ONE* **2017**, *12*, e0179314. [CrossRef]

41. Wang, H.; Liu, C.; Deng, L. Enhanced prediction of hot spots at protein–protein interfaces using extreme gradient boosting. *Sci. Rep.* **2018**, *8*, 14285. [CrossRef]

42. Kuang, L.; Yu, L.; Huang, L.; Wang, Y.; Ma, P.; Li, C.; Zhu, Y. A personalized qos prediction approach for cps service recommendation based on reputation and location–aware collaborative filtering. *Sensors* **2018**, *18*, 1556. [CrossRef] [PubMed]

43. Li, C.; Zheng, X.; Yang, Z.; Kuang, L. Predicting Short–Term Electricity Demand by Combining the Advantages of ARMA and XGBoost in Fog Computing Environment. *Wirel. Commun. Mob. Comput.* **2018**, *2018*, 5018053. [CrossRef]

44. Glaab, E.; Baudot, A.; Krasnogor, N.; Schneider, R.; Valencia, A. EnrichNet: Network–based gene set enrichment analysis. *Bioinformatics* **2012**, *28*, 451–457. [CrossRef] [PubMed]

45. Smedley, D.; Kohler, S.; Czeschik, J.C.; Amberger, J.S.; Bocchini, C.; Hamosh, A.; Veldboer, J.; Zemojtel, T.; Robinson, P.N. Walking the interactome for candidate prioritization in exome sequencing studies of Mendelian diseases. *Bioinformatics* **2014**, *30*, 3215–3222. [CrossRef] [PubMed]

46. Perozzi, B.; Alrfou, R.; Skiena, S. DeepWalk: Online learning of social representations. In Proceedings of the 20th Acm Sigkdd International Conference on Knowledge Discovery Data Mining, New York, NY, USA, 24–27 August 2014; pp. 701–710. [CrossRef]

47. Grover, A.; Leskovec, J. Node2vec: Scalable feature learning for networks. In Proceedings of the 22th Acm Sigkdd International Conference on Knowledge Discovery Data Mining, San Francisco, CA, USA, 13–17 August 2016; pp. 855–864. [CrossRef]

48. Deng, L.; Wu, H.; Liu, C.; Zhan, W.; Zhang, J. Probing the functions of long non-coding RNAs by exploiting the topology of global association and interaction network. *Comput. Biol. Chem.* **2018**, *74*, 360–367. [CrossRef] [PubMed]

molecules

MDPI

Article

Genome-Wide Identification and Comparative Analysis for OPT Family Genes in *Panax ginseng* and Eleven Flowering Plants

He Su [1,†], Yang Chu [2,†], Junqi Bai [1], Lu Gong [1], Juan Huang [1], Wen Xu [1], Jing Zhang [1], Xiaohui Qiu [1], Jiang Xu [2,*] and Zhihai Huang [1,*]

[1] The Second Clinical College of Guangzhou University of Chinese Medicine, Guangzhou 510006, China; suhe@cau.edu.cn (H.S.); baijunqi@126.com (J.B.); gonglu0904@126.com (L.G.); juanhuangzi@126.com (J.H.); freexuwen@163.com (W.X.); ginniezj@163.com (J.Z.); qiuxiaohui@gzucm.edu.cn (X.Q.)

[2] Institute of Chinese Materia Medica, China Academy of Chinese Medicinal Sciences, Bejing 100700, China; ychu@icmm.ac.cn

* Correspondence: jxu@icmm.ac.cn (J.X.); zhhuang7308@163.com (Z.H.); Tel.: +86-20-39318571 (Z.H.)

† These authors contributed equally to this work.

Received: 9 November 2018; Accepted: 17 December 2018; Published: 20 December 2018

Abstract: Herb genomics and comparative genomics provide a global platform to explore the genetics and biology of herbs at the genome level. *Panax ginseng* C.A. Meyer is an important medicinal plant for a variety of bioactive chemical compounds of which the biosynthesis may involve transport of a wide range of substrates mediated by oligopeptide transporters (OPT). However, information about the OPT family in the plant kingdom is still limited. Only 17 and 18 OPT genes have been characterized for *Oryza sativa* and *Arabidopsis thaliana*, respectively. Additionally, few comprehensive studies incorporating the phylogeny, gene structure, paralogs evolution, expression profiling, and co-expression network between transcription factors and OPT genes have been reported for ginseng and other species. In the present study, we performed those analyses comprehensively with both online tools and standalone tools. As a result, we identified a total of 268 non-redundant OPT genes from 12 flowering plants of which 37 were from ginseng. These OPT genes were clustered into two distinct clades in which clade-specific motif compositions were considerably conservative. The distribution of OPT paralogs was indicative of segmental duplication and subsequent structural variation. Expression patterns based on two sources of RNA-Sequence datasets suggested that some OPT genes were expressed in both an organ-specific and tissue-specific manner and might be involved in the functional development of plants. Further co-expression analysis of OPT genes and transcription factors indicated 141 positive and 11 negative links, which shows potent regulators for OPT genes. Overall, the data obtained from our study contribute to a better understanding of the complexity of the OPT gene family in ginseng and other flowering plants. This genetic resource will help improve the interpretation on mechanisms of metabolism transportation and signal transduction during plant development for *Panax ginseng*.

Keywords: *Panax ginseng*; oligopeptide transporter; flowering plant; phylogeny; transcription factor

1. Introduction

Peptide transportation is a widely observed phenomenon of translocating small peptides across a membrane in a carrier-mediated, energy-dependent manner [1]. Transported peptides are often hydrolyzed and the resulting amino acids are used as substrates for protein synthesis, sources of nitrogen and carbon [2,3], and signals for biological processes such as quorum sensing [4], yeast mating [5], and metal homeostasis regulation [6,7]. There are three distinct protein families related to peptide transportation. The ATP binding cassette (ABC, TC 3.A.1) transporter superfamily is the largest transporter gene family. The members are able to translocate a wide variety of substrates including amino acids, sugars, peptides, proteins, and a large number of hydrophobic compounds and metabolites across extra-cellular and intracellular membranes [8,9]. In contrast to the ABC family, the proton-dependent oligopeptide transporter (PTR, TC 2.A.17) family utilizes a proton gradient other than ATP hydrolysis for dipeptide and tripeptide translocation [10]. The members of PTR proteins have been found in all kingdoms of life except the Archaea [1,11]. PTR also participates in amino acid and nitrate transportation [12]. In addition to dipeptides and tripeptides that are translocated by PTR proteins, tetra-peptides, penta-peptides, and some longer oligopeptides are translocated by a novel protein family known as the oligopeptide transporter (OPT, TC 2.A.67) family [10].

The OPT family is a group of electrochemical potential-driven transporters that catalyze their solutes in an energy-dependent symport manner. CaOPT1 was first cloned from *Candida albicans* (Robin) Berkhout and functional verified in *Schizosaccharomyces pombe* (Lindner) and subsequently defined as OPT but not an ABC or PTR protein by Jeff Becker's laboratory [13–15]. OPTs are suggested to play diverse roles in long-distance sulfur distribution, metal homeostasis, nitrogen mobilization, heavy metal sequestration by transporting glutathione, peptides, and meta-chelates [16]. Phylogenetically, the OPTs can be divided into Oligopeptide Transporter (PT) and Stripe-like (YSL) clades [16,17]. Genes in the YSL clade have been found in Archaea, eubacteria, fungi, and plants but not in animals, which function as metal chelate transporters [6,18,19] consisting of mugeneic acids (MA) or nicotianamine (NA) while genes in the PT clade have only been identified in plants and fungi mediating long-distance metal distribution, nitrogen mobilization, glutathione translocation, and heavy metal sequestration [16,20–25].

In plants, the OPTs may play important roles in plant growth and abiotic and biotic stress responses [13,26]. The OPT member *ZmYS1*, which was first cloned by Curie et al. [6] but re-defined by Yen et al. [27], was proven to mediate the import of Fe-phytosiderophore complexes from soils and long-distance transport of iron-NA complexes [13]. Studies of two *AtOPT3* T-DNA mutants indicated that *AtOPT3* is of importance in both embryo development and iron deficiency signal transduction [1,7]. In addition, *AtOPT3* is found to be expressed in the phloem and functions in long-distance shoot-to-root signaling for Fe/Zn/Mn status. A lack of *AtOPT3* in *Arabidopsis thaliana* (*Arabidopsis*) led to the over-accumulation of cadmium in seeds [23]. Glutathione (GSH) is an essential sulfur-containing tripeptide that performs various important roles in plant processes, including detoxification of xenobiotics, heavy metal transport and resistance, controlling redox status, and long-distance transport of organic sulfur [22]. GSH is a precursor for plants to use to produce phytochelatins (PCs), which is the polymerized form of GSH, by which heavy metals can be transported to a central vacuole for detoxification [28]. AtOPT4 and AtOPT6 from *Arabidopsis* [21,22] and BjGT1 [25] from *Brassica juncea* (*B. juncea*) are all capable of translocating Cd-GSH conjuncts. Moreover, GSH also plays important roles in plant growth and development in response to abiotic and biotic stresses.

The majority of members of OPT proteins seem to contain 16 TMSs including a few of which appear to have 17 TMSs. A homology-based analysis for each TMSs in the OPT family indicated that the 16-TMS proteins might have been generated by three sequential duplications from 2-TMS protein precursors. Additionally, gene fusion might be responsible for the 17-TMS proteins [17]. However, although the OPT proteins have been studied for more than two decades, the majority of studies still focus on model plants such as yeast, *Arabidopsis*, and *Oryza sativa* (rice) [1,21,29,30]. Thanks to the rapid development of whole-genome sequencing techniques, an exponential increase in genome information has provided us with great opportunities to identify more OPT genes in non-model plants and make comparisons among multiple species simultaneously. However, to the best of our knowledge, genome-wide identification of OPT proteins has only been conducted in *Ganoderma lucidum* [31], *Populus trichocarpa*, and *Vitis vinifera* [32]. *Panax ginseng*, which is a Traditional Chinese Medicine, has been used for several millennia and has become more and more popular around the world. It is the most commonly used medicinally species in the *Panax* genera in contrast to the other four species: *Panax quinquefolius*, *Panax vietnamensis*, *Panax japonicus*, and *Panax notoginseng* [33]. Since we finished the genome assembly for *Panax ginseng* (*P. ginseng*) in our previous report [34], the interest in characterizing OPT genes in *P. ginseng* and comparing it with other genome-assembly-available species has increased.

Although more plant genomes have been mapped in the last decade, studies on the genome-wide identification and comparison of OPT genes among species are still limited. Information on the phylogeny, gene structure, expression patterns, and regulatory networks of OPT genes remains to be discovered. In the present study, we identified OPT genes from *P. ginseng* and 11 flowering plants with the purpose of uncovering the phylogenetic relationships and gene structures of OPT genes in flowering plants as well as investigating the expression profiles and regulators of OPT genes in *P. ginseng*. Our analysis, which combines these types of information, provides new insights into both the structural and functional roles of OPT genes in ginseng and serves as a valuable resource for further study of the roles OPT genes play in plant development and transport of secondary metabolism.

2. Results and Discussion

2.1. Identification of OPT Genes in P. ginseng and 11 Other Flowering Plants

We identified the OPT genes for *P. ginseng* and other species with TransportTP by setting as reference organisms Oryza sativa and *Arabidopsis thaliana* [35]. As a result, a total of 364 OPT candidates were identified in our study (Table 1). Seventeen of the 18 identified genes from *Arabidopsis* were in accordance with the reviewed records deposited in Swiss-Prot (release 2018_10), and, although the other gene *At5g45450*, was not recorded in Swiss-Prot, it was regarded as an OPT gene recorded in GenBank. However, only 12 OPT genes were identified from rice, of which the accession numbers were not in accordance with those records deposited in Swiss-Prot. This might be due to a different version of the rice genome being used. Considering our genome assembly did not scale to a chromosome level, we conducted a manual curation of the 39 OPTs identified from *P. ginseng*. Thereafter, 37 OPT genes were kept for further analysis. In addition, we identified 54 and 26 OPT genes from poplar and grape in our study, respectively, while only 20 and 18 genes were identified by Cao et al. [32]. These results suggested that our identification of OPT genes was accurate and comprehensive.

Table 1. Statistics of OPT genes predicted from *P. ginseng* and 11 flowering plants.

Species	Predicted	De-Redundant	Final
Ginseng	39	37	37
Arabidopsis	18	16	17
Rice	12	10	18
Sorghum	38	26	26
Carrot	25	16	16
Potato	42	29	29
Tomato	23	17	17
Cassava	29	21	21
Clover	31	25	25
Cacao	27	19	19
Poplar	54	28	28
Grape	26	24	23
Total	364	268	276

37 OPT genes were left after manual curation for OPT genes from Ginseng. OPT genes predicted for *Arabidopsis* and rice were replaced with 17 and 18 reviewed OPT genes retrieved from Swiss-Prot. De-redundant indicated only one gene could be kept if there were genes that had 100% similarity to it.

Since we found some OPTs were highly redundant (similar to each other with 100% similarity) within species, we removed the redundant OPTs in order to reduce the subsequent calculation consumption using CD-HIT software [36] by setting the sequence identity threshold to 100%. Lastly, a total of 268 OPT genes were kept. Because a candidate OPT gene named "GSVIVT01007176001" identified from grape contained too many "X"s, we excluded it from further analyses. In order to generate robust results from subsequent studies, we replaced those predicted OPT genes from *Arabidopsis* and rice with reviewed OPT genes retrieved from Swiss-Prot. Furthermore, we introduced two other experimentally verified OPT genes from *B. juncea* and *Zea mays* (*BjGT1* and *Maize_YS_1* respectively [6,25]) into our study. Lastly, 278 OPT genes were used for further analysis (Supplementary File 1).

2.2. Protein Properties of OPT Genes for OPT Genes Identified in P. ginseng and 11 Other Flowering Plants

By examining the properties of OPT genes for each plant species, we found that the number of amino acid residues varied among species. Generally, the number of amino acid residues for OPT genes in *Arabidopsis*, rice, sorghum, and cassava ranged from 552 to 766, which is higher than the rest of those studied species (ranged from 184 to 941, as for *P. ginseng*). The number of residues ranged from 348 to 919 (Table S1, Figure 1D). The distribution of molecular weight for OPT genes was similar to the distribution pattern of residue numbers (Figure 1B). The grand average of hydropathicity (GRAVY) value is a measure of protein hydrophobicity [37]. Our results suggested that GRAVY for those OPT genes mainly ranged from 0.30 to 0.60 (Figure 1A). As OPT genes with the lowest and the highest GRAVY values (0.029 for Potri.017G150620.2.p and 0.87 for PGSC0003DMP400037534) were filtered out for lacking OPT-specific information for further phylogenetic analysis (Figure 2), we expanded the confident range of GRAVY values from 0.329–0.628 to 0.114–0.659 compared with the previous study [32]. In addition, the isoelectric point (pI) of the majority of OPT genes was around 9.0, which suggests that the electrochemical properties of OPT genes might be less varied in the plant kingdom (Figure 1C).

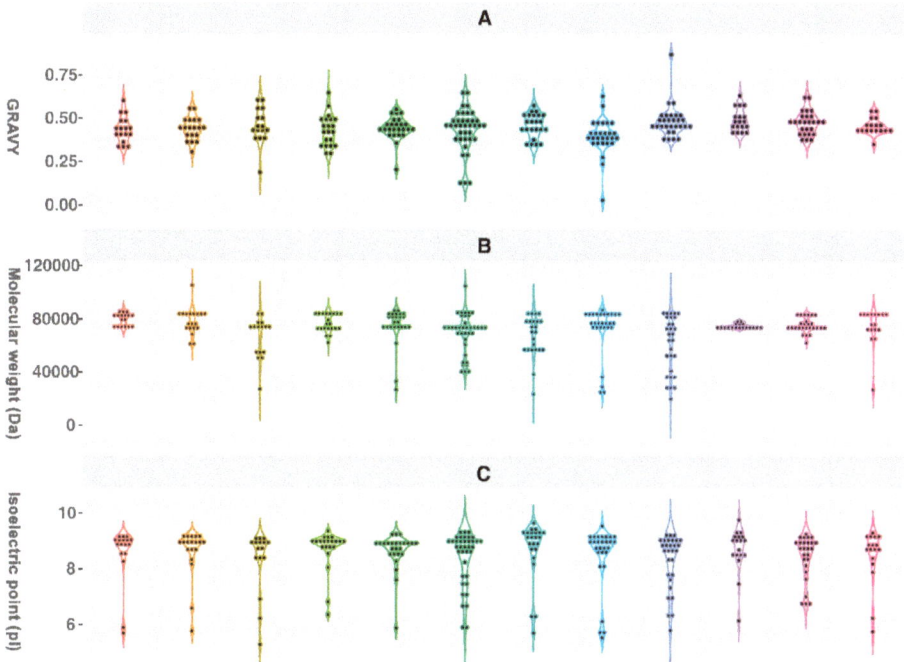

Figure 1. *Cont.*

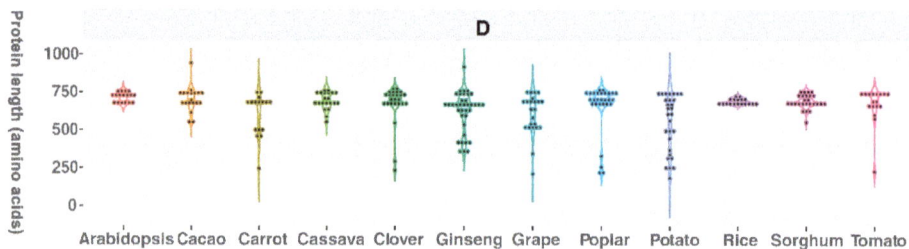

Figure 1. Protein properties for OPT genes identified from *P. ginseng* and 11 flowering plants. (**A**) Grand average of hydropathicity, GRAVY. (**B**) Molecular weight. (**C**), Isoelectric point, pI. (**D**) Number of amino acid residues.

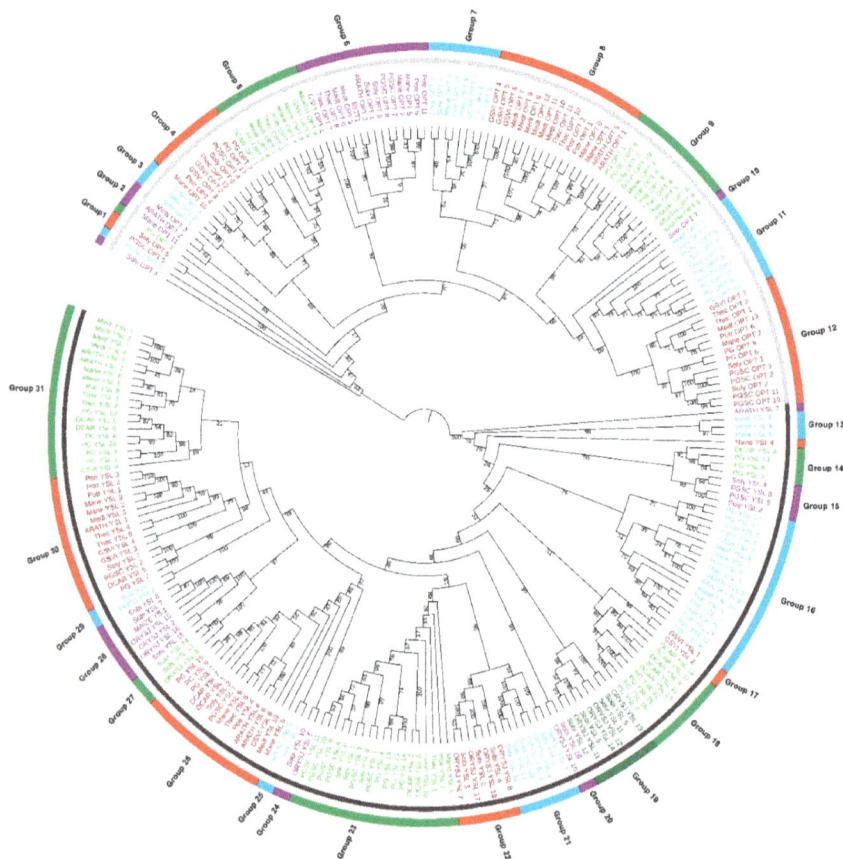

Figure 2. Phylogenetic relationships of OPT genes in *P. ginseng* and 11 other species. Tomato (Soly), Potato (PGSC), Cassava (Mane), *Arabidopsis* (ARATH), Clover (Medt), Poplar (Potr), Grape (GSVI), Cacao (Thec), Sorghum (Sobi), Carrot (DCAR), and Rice (ORYSJ). BjGT1 from *Brassica juncea* and Maize YS1 from *Zea Mays* (maize) are experimentally validated OPT proteins that were retrieved from GenBank database. Light gray in the inner circle indicates the PT clade. Dark gray refers to the YSL clade.

Further analysis conducted with WOLF PSORT (http://woltpsort.org) enabled us to predict the probable protein localization for each candidate OPT identified in our study. It was found that all candidate OPTs were most likely to be located in the plasma and vacuolar membranes. The results were in accordance with a previous study [32]. Furthermore, 17 OPT genes were predicted to only be located in the plasma membrane. The remaining OPT genes were predicted to be not only in the plasma but also in at least one of the following: vacuole, chloroplast, cytoplasm, nucleus, mitochondria, Golgi apparatus, or endoplasmic reticulum (Table S1).

2.3. Phylogenetic Analyses, Classification, and Functional Relatedness of the OPT Genes Identified in P. ginseng and 11 Other Flowering Plants

To unravel the phylogenetic relationships of OPT genes in flowering plants, we conducted a phylogenetic analysis for those genes from 12 flowering plants. All OPT genes were clustered into two major distinct clades known as PT and YSL clade for which the results were consistent with those of previous reports [13,16,32]. However, what was different from previous studies was that the rice OPT genes in the PT clade were not included because no rice OPT genes in this clade were available in the Swiss-Prot database (release 2018_10). Therefore, only OPTs from the YSL clade were used in this phylogenetic analysis. Based on the bootstrap permutation test and the relationships of each OPT gene, we further classified the PT clade into 12 subgroups (Groups 1-12) and the YSL clade into 19 subgroups (Groups 13-31). Groups 23 and 31 included the largest number of members in the YSL clade (each with 19 members). Groups 9-12 formed a highly confident larger group with a bootstrap value of 96% in the OPT clade and Groups 23 and 27-30 formed another group in the YSL clade with a supporting value of 99%, which suggested that those members were likely to have evolved by recent gene duplication from a common ancestor. However, Soly_OPT_4 (Tomato), PGSC_OPT_1 (Potato), and PG_OPT_2 (Ginseng) failed to be grouped with any other PT genes due to a lack of supporting information by maximum likelihood analyses. In addition, ARTH_YSL_7 (*Arabidopsis*) and Mane_YSL_4 (Cassava) also failed to be grouped with any other genes in the YSL clade. Although Sobi_OPT_7 (Sorghum) seemed likely to stand alone, it was in fact grouped with Groups 11 and 12 with a bootstrap value of 93% (Figure 2). Furthermore, the motif structures of the genes described below also supported the group classifications (Figure S1). Moreover, ARATH_OPT_3 and BjGT3 (*Brassica juncea*), ARATH_OPT_1 and ARATH_OPT_5, ARATH_OPT_6 and ARATH_OPT_8 and ARATH_OPT_9, ARATH_YSL_5 and ARATH_YSL_8, ARATH_YSL_4 and ARATH_YSL_6, ORYSJ_YSL_7 (rice) and ORYSJ_YSL_17 were grouped together, with the phylogenetic relationships in accordance with previous study reports [16,30,32]. The consistency of our findings with previous findings indicated that our phylogenetic study was properly conducted and the results were reliable. However, it was interesting to find out that ARATH_YSL_7, which has been reported to be sub-grouped with ARATH_YSL_5 and ARATH_YSL_8 [30], failed to be grouped with any other OPT members in the YSL clade in our study.

Genes with the same functions were often closely related, as found in both a previous study [32] and our study. *BjGT1*, which is the first cloned and characterized OPT gene from *Brassica juncea*, was experimentally validated to be a glutathione transporter mediating cadmium absorption [25]. ARATH_OPT_3, which is another OPT gene that was cloned and characterized in *Arabidopsis*, was reported to be involved in the sensing and translocation of Cd (as well as Fe and Zn) [1,7,23]. These two functionally similar genes were clustered together in our study. In addition, Mazie_YS_1, the first experimentally validated OPT gene responsible for transport of Fe(III)-phytosiderophore chelates, was clustered together with ORYSJ_YSL_15 and ORYSJ_YSL_2 (Figure 2). ORYSJ_YSL_15 has been suggested to be responsible for iron uptake from rhizosphere and for phloem transport of iron by transporting Fe(III)-phytosiderophore chelates while ORYSJ_YSL_2 has been suggested to be responsible for phloem transport of iron by transporting Fe(III)-nicotianamine chelates [38,39]. Furthermore, ARATH_YSL_2 and ARATH_YSL_3 clustered together in Group 31 were both reported to be involved in transport of nicotianamine-chelated metals in the vasculature [40,41]. These results supported the idea that genes with the same functions were closely related. Based on the hypothesis, it would be interesting to test if PG_YSL_1 is involved in iron-transportation since it was clustered together with ARATH_YSL_1 that was found to be involved in transport of iron-nicotianamine chelates [41,42]. Similarly, it would be interesting to test if PG_OPT_1 is akin to ARATH_OPT_6, which was reported to be involved in the transport of glutathione derivatives and metal complexes [21,43,44] and to test whether PG_OPT_9,10 and PG_OPT_11 are involved in increasing plant sensitivity to Cd like ARATH_OPT_7 functions [43].

We identified a total of 45 pairs of paralogs from the phylogenetic analyses (Table S2), which accounted for 11.1% to 70.6% of all OPT candidates in each studied species and shared similar structures within each group (Figure S1). We found that some OPT genes in ginseng were tandemly clustered on the same scaffold (Table S3) and those genes were location-related. For example, PG_OPT_10 and PG_OPT_11 were neighbor paralogs with 1122 bp in between. These genes might be formed by tandemly segmental duplication. PG_YSL_8 and PG_YSL_10 constitute a special tandemly clustered paralogs with a 3214 bp-long shared region. This paralogs pair might be generated by a crossover of chromosome after whole-genome duplication or by gene fusion. It would be interesting to test whether this OPT cluster was functional in further studies. In addition, PG_YSL_4-PG_YSL_5 and PG_YSL_14-PG_YSL_16 formed a special type of gene cluster block in which PG_YSL_4-PG_YSL_14 and PG_YSL_5-PG_YSL_16 were identified as paralogs oriented in the same direction. PG_YSL_18-PG_YSL_19 and PG_YSL_21-PG_YSL_22 constituted another special type of block, in which PG_YSL_18-PG_YSL_21 and PG_YSL_19-PG_YSL_22 were paralogs oriented in opposite directions (Figure 3). From this section of the study, we speculated that both types of cluster blocks were generated from segmental duplication or whole-genome duplication. Since *P. ginseng* is a tetraploid plant, we prefer to believe that genes from those blocks were more likely to be generated by whole-genome duplication. The paralogs blocks arranged in the opposite direction were likely to be generated by subsequent segmental inversion of the chromosome after segmental duplication.

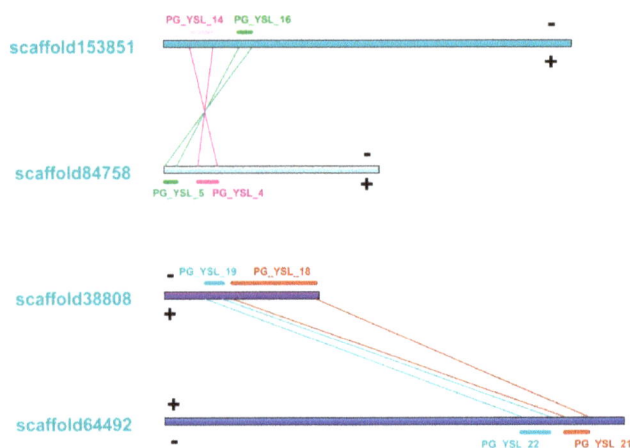

Figure 3. Chromosome locations for two special types of clusters of paralogs blocks. We used the ginseng genome version1 finished by Xu et al. [34] in this study. The scaffold refers to the DNA sequences in the ginseng genome that were generated by bridging non-gapped contigs (assembled with short gun sequencing reads) with mate-pair sequencing reads. A scaffold is equivalent to a chromosome segment.

Ks (synonymous substitution rate) is a widely accepted concept for gene duplication time estimation. In general, the lower *Ks* is, the more recently gene duplication occurred [32]. Since a codon-based alignment of PG_YSL_6 and PG_YSL_17 failed to generate, calculation of *Ks* was excluded from this study (Table S2). Aligned sequences were nearly identical after removing gaps from

Potr_YSL_2/Potr_YSL_3 (poplar), Thec_OPT_7/Thec_OPT_8 (cacao), and Thec_YSL_4/Thec_YSL_5. The estimation of *Ks* for these paralogs also failed. Additionally, *Ks* values for PG_OPT_9/PG_OPT_10 and Sobi_YSL_13/Sobi_YSL_14 were estimated as 0, suggested that they were generated by a very recent duplication event. It was interesting to find that gene duplication of OPT paralogs occurred more recently in the YSL clade than in the PT clade in *P. ginseng*. The phenomenon was similar to grape and clover but contrary to cacao, cassava, and *Arabidopsis*. The duplication event for paralogs occurred more recently in ginseng, potato, poplar, cacao, grape, clover, and sorghum (about 0 to 5 MYA) than in carrot, cassava, *Arabidopsis*, and rice, which indicates that ginseng and other species or their common ancestor might have suffered a high level of gene loss during evolution because of the lack of an older duplication event such as 94.2 MYA for ARATH_OPT_6/ARATH_OPT_9 [45].

2.4. Conserved Domains and Motif Analysis for OPT Genes Identified in P. ginseng and 11 Other Flowering Plants

By searching against the Conserved Domain Database (CDD) [46] with 278 OPT genes, all genes were annotated as OPT genes. However, only 267 were predicted to have specific domains, wherein all the 258 OPT genes used in the phylogeny analysis were covered. Because domain analysis could not provide information about smaller individual motifs and more divergent patterns, we conducted a study of motif analysis with MEME software (Supplementary File 2). As a result, 30 distinct motifs were identified in these genes. Detailed information of those motifs is presented in Supplementary File 3. It is interesting that the motif composition of OPT members in the PT clade is distinct from that in the YSL clade (Figure S1), which was in accordance with the conclusions generated by phylogenetic analysis. In addition, the number of motifs of OPT genes from the PT clade (ranging from 4 to 11, with a median value of 8) was distinct from that of the YSL clade (ranging from 5 to 12, with a median value of 11), which suggests the clade-specific structure of each OPT gene (Figure 4). Furthermore, we found nine motifs (Motif_1,3,6,13,14,19,23,15,29) unique to the PT clade and 10 motifs (Motif_7,8,12,16,17,21,22,24,26,30) unique to the YSL clade, respectively. Six motifs (Motif_10,14,19,23,102,106) were frequently shared by PT clade members (94.4%, 94.4%, 86.9%, 91.6%, 95.3%, and 99.1%, respectively) and 10 motifs (Motif_2,8,9,15,16,18,21,22,26,28) were frequently shared by the YSL clade members (92.0%, 88.7%, 71.3%, 38.7%, 92.0%, 92.0%, 92.7%, 81.3%, 84.7%, 97.3%, and 94.0%, respectively) (Table S4). Those findings might give us new insights into how OPT genes evolved since being separated from their common ancestor and how they functionally diverged during the subsequent evolution process.

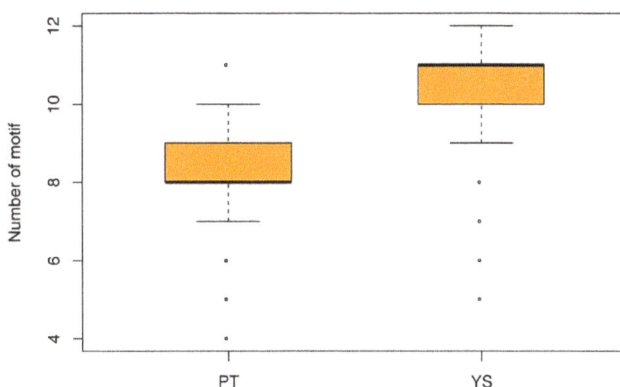

Figure 4. Number of motifs in OPT genes from PT and YS clades. These static results were calculated with xml output of MEME analysis (Supplementary File 5) by our custom R scripts. The boxplot was generated by the built-in function "boxplot" in R.

2.5. Profiling of Expression Patterns for OPT Genes Identified in P. ginseng

In order to examine the expression patterns of the OPT genes in *P. ginseng*, we performed a comprehensive expression analysis by using two sets of RNA-Seq datasets: one from our previous study about *P. ginseng* root [47] and one from a public study about 18 kinds of tissues. In general, genes in the YSL clade were more highly expressed than genes in the PT clade except in the periderm (Figure 5). PG_YSL_2,13 and PG_YSL_15 were expressed evenly in the root with little difference among tissues, which suggests that they might be constitutive OPTs. OPT genes exhibited distinct tissue-specific expression manners. For example, PG_OPT_4,5 and PG_YSL_12 were more likely to be expressed highly in periderm than in the stele or cortex. PG_YSL_11 and PG_YSL_7 had the highest expression in the stele and cortex, respectively, while they were still expressed at a considerably high level in other tissues. The different expression patterns for those OPT genes indicated that a wide range of substrates might be transported in different parts of the plant root.

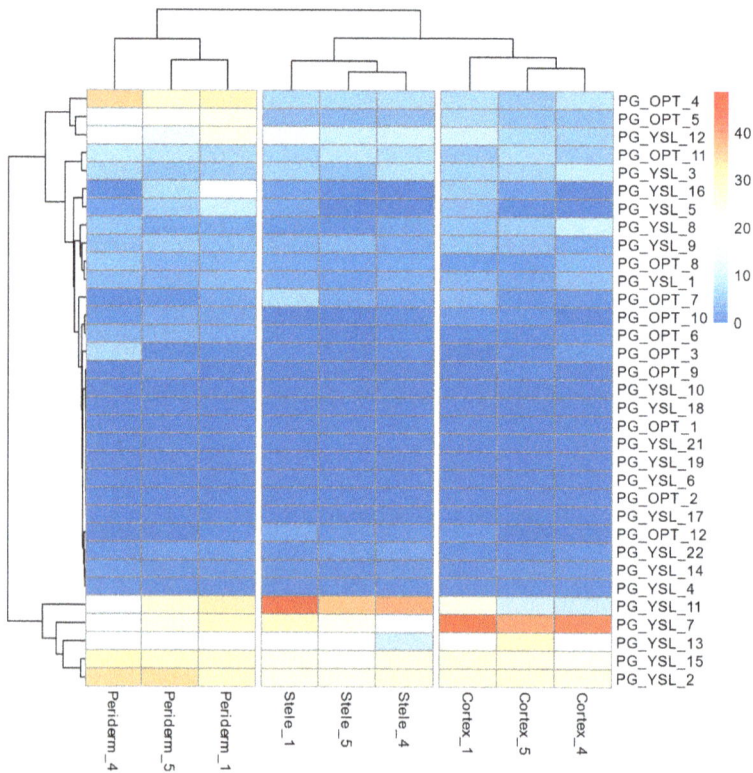

Figure 5. Expression of OPT genes of different tissues in ginseng root. Gene expression of OPT genes was calculated with RNA-Seq data generated by our previous study on *P. ginseng* root. The hierarchical clustered heat map was plotted with 'pheatmap' planted in R package named 'pheatmap'.

Due to the nature of sink tissue of fruit and seeds in plants, the expression characteristics of OPT genes of these tissues are expected to share more common traits than those of other tissues. Based on the expression data, fruit flesh, fruit pedicel, fruit peduncle, and seeds were clustered together, which suggests that the similar expression pattern of those OPT genes might contribute to methods of metabolism relocation. The expression of PG_YSL_1,3 and PG_YSL_7 both peaked in fruit flesh compared with the fruit pedicel, fruit peduncle, and seed, which indicates that lateral transportation might be the most active transportation process during seed development. Additionally, the leaf blade, leaf pedicel, leaf peduncle, and stem, which are physically connected organs forming a complex vascular transportation system in plants, were clustered together by their similar expression pattern, wherein PG_YSL_8 was expressed at the highest level (except for the stem). Moreover, the arm root, fiber root, and leg root were clustered together, and PG_OPT_13 and PG_YSL_16 were highly expressed. It was interesting that PG_YSL_7 was highly expressed in 12-year-old and 25-year-old roots but minimally expressed in five-year-old and 18-year-old roots that were clustered with the main root cortex, the main root epiderm, and the rhizome. Taken together, the expression patterns found in our study and Wang's [48] both suggested that OPT genes were expressed in tissue-specific and location-specific manners by which the transportation and distribution of oligopeptides and their conjugates with metals, signals, etc. were shaped in different ginseng tissues [41] (Figure 6).

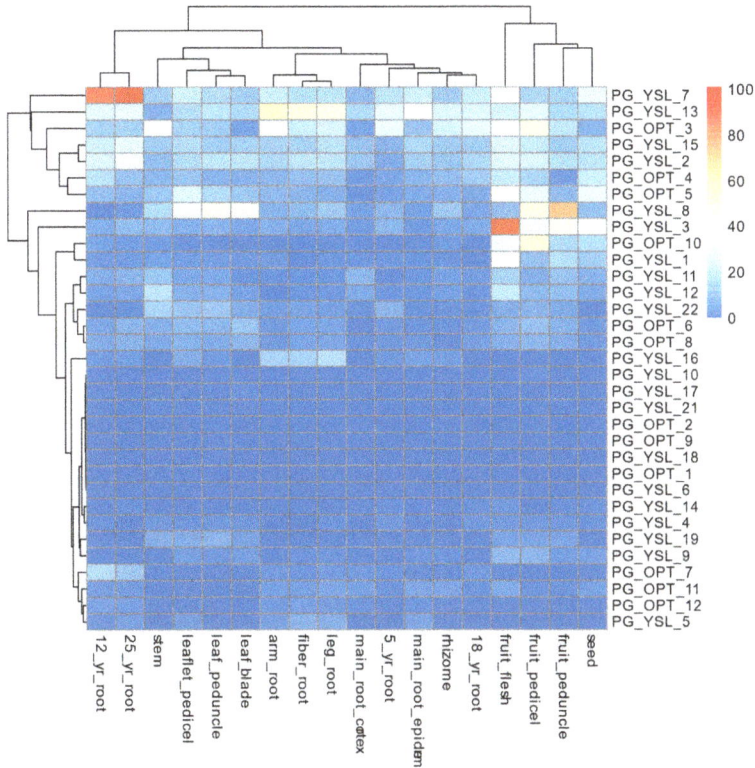

Figure 6. Expression of the OPT genes in 18 organs of ginseng. Gene expression of OPT genes was re-calculated with RNA-seq data from public research by taking our published genome as a reference.

Based on the phylogenetic analysis described above, PG_OPT_4 and PG_OPT_5 were grouped with ARATH_OPT_1 and ARATH_OPT_5, which were proven to be OPT transporters for penta-peptide (KLLLG) in an energy-dependent manner by yeast complementation assay [20,49]. High expression of those genes exclusively in the root suggested that the penta-peptide-related metabolism (metabolism substrates, signal molecules, etc.) transportation might be activated. PG_YSL_12 was identified as another periderm-specific expressed gene found in this study. It was expressed more highly in the stem and fruit flesh than in other tissues (Figure 6). PG_YSL_12 was clustered with ARATH_YSL_5 and ARATH_YSL_8 into one group, which indicates that it might be involved in the transport of nicotianamine-chelated metals (metals-NA) just as ARATH_YSL_2 was in the transport of Fe-NA across the plasma membrane in leaf cells, involving lateral movement of iron away from the xylem [40]. Furthermore, ARATH_YSL_8 might also be involved directly in iron uptake by leaf cells [13].

ARATH_YSL_1 and ARATH_YSL_3 were experimentally verified OPT proteins, which were found to be able to mediate Fe transportation to and from vascular tissues [41]. ARATH_YSL_3 was a sister branch to ARATH_YSL_2, which had been functionally confirmed to be expressed in both roots and shoots and to mediate transport of metal-NA complexes [40], which indicates their functional similarity. ARATH_YSL_1 was clustered into another sister group to a larger group including ARATH_YSL_2 and ARATH_YSL_3, which suggests that members of these two larger group might share some functional similarities. MAIZE_YS_1 (*ZmYS1*) known as a proton-coupled symporter transports iron complexed by plant-derived Fe(III) chelators (phytosiderophores, PS) by scavenging from soil, termed Strategy II [13]. It formed another cluster in the YSL clade with some YSL genes from sorghum and rice with a bootstrap value of 97% (Figure 2). Considering the functional similarity of ARATH_YSL_1,2,3 and *ZmYS1*, OPT genes from groups 27 and 28 and groups 29-31 were suggested to form two sister groups that might be involved in the transport of Fe.

The OPT paralogs were more likely to be generated by segmental tandem duplication rather than transposition [32]. The expression pattern of those duplicated genes may differ if they suffered evolutionary divergence such as neofunctionalization. [50]. No similar expression patterns of duplicated paralogs were identified in the study about poplar and grape [32]. However, we detected five similarly expressed paralogs pairs in this study wherein PG_OPT_4-PG_OPT_5 had similar expression patterns both in our previous study and in Wang's study. PG_YSL_4-PG_YSL_14 and PG_YSL_18-PG_YSL_21 were expressed similarly but with a very low expression level. PG_YSL_6-PG_YSL_8 and PG_YSL_11-PG_YSL_12 were reported to be expressed similarly in Wang's report but not in ours. However, PG_YSL_5/PG_YSL_16 was similarly expressed in our study but not in Wang's study. The similar expression patterns found in our study might be because of the relatively short time has been experienced in *Ginseng* paralogs compared with those paralogs in poplar and grape. On the other hand, the phenomenon that a majority of the identified paralogs in *Ginseng* did not have similar expression patterns, which indicates functional diversificationmight be a result of long-term evolution—adapting to changing environmental conditions after gene duplication.

2.6. Analysis of Co-Expression Network between OPT Genes and Potent Transcription factOr for P. ginseng

The regulation of gene expression in all living cells is dominated by transcriptional initiation, which is regulated by transcription factors, ancillary transcription regulators, and chromatin regulators. Therefore, we conducted an analysis focusing on the co-expression between all transcriptionally modulated genes in the ginseng genome and all transcription factors in order to reveal regulators for OPT genes in *Ginseng*. PlantTFcat is a useful tool for identifying proteins with signature domains specific to 108 major transcription regulators families [51]. We assessed the *Ginseng* genome for identifying those proteins. A total of 5073 distinct genes in the *P. ginseng* genome have been predicted to be transcription factors wherein there are 5457 members (Supplementary File 4). Genes annotated as different transcription factors by PlantTFcat (such as PG39956, annotated as Znf-B, LisH, WD40-like, or PLATZ) were removed from further analysis. The expression values of transcription factors that were mapped by many genes (such as MYB-HB-like, mapped by 334 genes) were determined by the median values of those genes. Lastly, a total of 59 transcription factors and 13 OPT genes were used for network analysis. We used non-parametric Spearman's rank-order correlation for our co-expression analysis due to its robustness for generating biologically relevant gene networks [52].

The matrix of all correlation values for expression values between each pair of transcription factor and OPT gene from a set of nine biological samples is shown in Table S5. At a conservative threshold of $\rho \geq |0.85|$, 141positive and 11 negative correlations involving 13 OPT genes were found (Table S6). The number of transcription factors correlated to an OPT gene ranged from 1 to 27, while it ranged only 1 to 5 for the number of OPT genes correlated with a transcription factor (Tables S7 and S8, Figure 7). For example, PG_OPT_5 and PG_OPT_6 positively correlated with 26 and 27 transcription factors, respectively, while *bHLH* and *WRKY* only correlated with five OPT genes. Our findings suggested that the initiation of transcription of OPT genes might be dominated by a complicated synergetic regulation system consisting of a number of transcription factors. Additionally, transcription factors might act as pleiotropic regulators participating in a variety of transcription regulations for OPT genes. On the other hand, because 19 out of 59 transcription factors were linked to only one OPT gene and two out of 13 OPT genes were linked to only one transcription factor, the results suggested that the transcription of some OPT genes was regulated by specific transcription factors and some transcription factors had specific target genes to regulate.

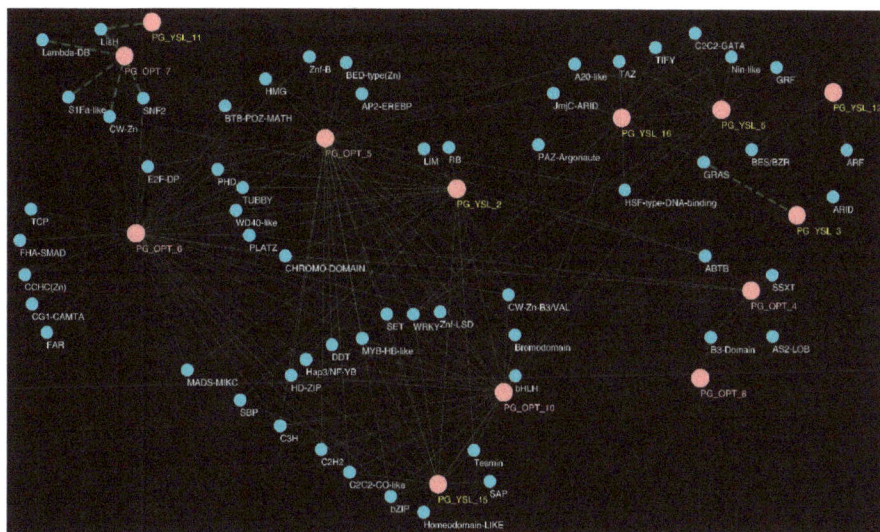

Figure 7. Regulatory gene networks involving transcription master genes and OPT genes. A stringent threshold ($\rho \geq |0.85|$) was used and the visualization was produced in Cytoscape-3.6.1. Nodes for OPT genes are represented by pink circles (yellow labels represent OPT genes in the YSL clade. Pink labels represent OPT genes in PT clade). Nodes for blue circles represent transcriptional regulators. Positive interactions are indicated by green lines and negative interactions are indicated by green dashed lines.

3. Materials and Methods

3.1. Sequence Retrieval and Identification of OPT Genes

We identified OPT genes from *P. ginseng* and 11 other flowering plants by using TransportTP (http://bioinfo3.noble.org/transporter/ [35]). Proteome sequences from 11 flowering plants (*Arabidopsis* TAIR10, rice v7, tomato iTAG2.4, potato v4.03, carrot v2.0, *Manihot esculenta* v6.1, *Medicago truncatula* Mt4.0v1, Poplar v3.1, Grape Genoscope 12X, Cacao v1.1, Sorghum v3.1.1) were retrieved from the Phytozome database (V12.1) [53], wherein the assembly version is followed by each species name. Ginseng proteome sequences were retrieved from http://ginseng.vicp.io:23488/index.php/index/download.html. The proteome sequences for each species were then used for the identification of OPT genes by searching against TransporterTP, setting the E-value threshold to 0.1, and setting *Arabidopsis thaliana* and *Oryza sativa* as the reference organisms.

Subcellular localization of those OPT proteins was predicted with WOLF-PSORT [54,55]. Isoelectric point (pI), molecular weight, and grand average hydropathicity (GRAVY) values were estimated with functions planted in the Peptides package (https://github.com/dosorio/Peptides/) for R.

3.2. Phylogenetic Analysis for OPT Genes

Phylogenetic analysis of these OPT genes was conducted on their conserved domains identified by CDD (Conserved Domain Database, [56,57]) and planted in NCBI with default parameters (50369 PSSMs, e-value of 0.01, maximum number of hits 500). Multiple sequence alignments of those conserved OPT protein were performed with MAFFT v7.158b [58] and followed by manual comparison and refinement. Aligned regions that contain over 50% gaps or ambiguous sites were removed and sequences that contained gaps for more than 50% of the remaining sequence were deleted. Lastly, 258 OPT genes were left for further phylogenetic analysis. In order to select the best evolutionary model for phylogeny reconstruction, we used the function 'modelTest' planted in the R package named 'phangorn' [59] with the parameter 'model' set to 'all' and found 'LG+G+F' was the best model. A maximum likelihood method of phylogenetic analysis based on RAxML v 8.2.9 [60] software was conducted, with the parameter "–bootstop-perms" set to 1000, "-m" set to PROTGAMMALG, and the "-e" set to 0.001. After finishing the reconstruction of the phylogeny of OPT genes for these species, the topology was plotted by the online tool iTOL [61].

3.3. Estimation of Duplication Time for OPT Paralogs

Pairwise alignment of protein sequences of the OPT paralogs was aligned with MAFFT software, and codon-based pairwise alignment of nucleotide sequences were generated by using PAL2NAL [62]. The Ka and Ks values for paralogous genes were estimated by the program yn00 planted in the PAML package with default parameters [63]. Assuming a molecular clock, the synonymous substitution rates (Ks) of the paralogous genes could be regarded as a proxy for time estimation of the segmental duplication events. The approximation of date for duplication events was estimated with the following formula: $T = Ks/2\lambda$, where λ denotes clock-like rates of synonymous substitution. In this study, 1.5×10^{-8} substitutions/synonymous site/year was used for *Arabidopsis*, 6.5×10^{-9} for rice, sorghum, cassava, grape and cacao, 9.1×10^{-9} for poplar [32], 1.08×10^{-8} for clover, 6.68×10^{-9} for *P. ginseng*, 2.69×10^{-9} for potato, and 2.91×10^{-9} for carrot. λ for each species was deduced or collected from previous studies [64,65].

3.4. Analysis of Motif Composition for OPT Genes

Conserved motif analysis for OPT genes in the *P. ginseng* genome was conducted with MEME (http://meme.sdsc.edu). The OPT candidates were run locally with MEME with the following parameters: number of repetitions = any, maximum number of motifs = 30. The other parameters were kept as default values.

3.5. Profiling Expression of OPT Genes for P. ginseng

The gene expression of *P. ginseng* was profiled by RNA-Seq datasets from our previous study [47] and public research [48]. Those datasets could be retrieved from the SRA database by searching BioProject id PRJNA369187 and PRJNA302556. These raw datasets from the SRA database were first converted into FASTQ files by sratoolkit.2.8.0 [66] and then quality controlled by Trimmomatic-0.36 [67]. Lastly, reference-based gene expression of those biological samples was estimated with the HISAT2+StringTie pipeline [68]. FPKM values for each gene were used as gene expression levels. A hierarchical clustered heatmap for OPT genes was plotted with the pheatmap package [69], wherein "manhattan" distance was used for both row-based and column-based clustering.

3.6. Identification of Regulatory Network between OPT Genes and Transcription Factors for P. ginseng

The *P. ginseng* proteome sequence dataset was submitted to the PlantTFcat analysis tool (http://plantgrn.noble.org/PlantTFcat/ [51]) for the identification and classification of transcription factors, chromatin modifiers, and other transcriptional regulators into protein families. Genes that mapped to more than one transcription factors were removed from further analysis. In addition, median values of those genes referring to the same transcription factors were regarded as the transcription factors' expression value. In this study, we used our previous RNA-Seq dataset for construction of the co-expression network between OPT genes and transcription factors. FPKMs of all genes including OPT genes and transcription factors were combined and used for the calculation of Spearman's rank correlation coefficient to predict potential gene regulatory networks. The correlation coefficient (ρ) for each gene pair was calculated by the built-in function "cor" in R, and a threshold of $\rho \geq |0.85|$ was regarded as significant co-expression. Visualization of the network was created in Cytoscape 3.6.1 [70].

4. Conclusions

This study is the first to investigate the chromosomal location, expression profiling, and transcriptional regulation networks of *P. ginseng* OPT genes and provide a comparative genome analysis addressing the phylogeny, gene structure, and paralogs duplication history of the OPT gene family in *P. ginseng* and 11 flowering plants. Chromosomal location analyses revealed that structural variation occurred after segmental duplication, expression profiling, and transcriptional co-expression networks analyses, which indicates that both specific and pleiotropic transcription regulators might be involved in the regulation of OPT genes' expression. Phylogenetic analyses suggested two well-supported clades in the OPT family, which can be further classified into 12 or 19 distinct groups. Motif compositions are conserved in each clade and clade-specific motifs were frequently occupied within each clade. Estimations for paralogs divergence history indicated that the majority of OPT paralogs in *P. ginseng* might have emerged from recent duplications, which was different from the history of *Arabidopsis* or cassava. The study of expression profiles in different organs and tissues of *P. ginseng* has provided insights into possible functional divergence among OPT members and important functional roles in the plant development of some OPT members. These data may provide valuable information for future functional investigations of this gene family.

Supplementary Materials: See the word file of "The list of supplementary materials." All supplementary materials are available online.

Author Contributions: Z.H. conceived and designed the research framework. H.C., J.X., and Y.C. prepared the sample and performed the experiments. J.X. and Y.C. provided many important suggestions for data analysis. H.S. analyzed the data. H.S. and J.X. wrote the manuscript. J.B., L.G., J.H., W.X., J.Z., X.Q, and Z.H make revisions to the final manuscript. All authors have read and approved the final manuscript.

Funding: This work was supported by grants from several founds supported by Guangdong Forestry Department, Guangdong food and Drug Administration and Guangdong Provincial Bureau of traditional Chinese Medicine (2017KT1835, 2018KT1050, 2018TDZ16, 2018KT1138, 2018KT1228, and 2018KT1230), National Nature Science Foundation of China (81803672), standardized research and application of precise powder decoction pieces in traditional Chinese Medicine, and Construction Project of TCM Hospital Preparation by Special Fund of Strong Province Construction in TCM, Guangdong, China (No. 6).

Conflicts of Interest: The authors declare no conflict of interest.

References

1. Stacey, M.G.; Koh, S.; Becker, J.; Stacey, G. Atopt3, a member of the oligopeptide transporter family, is essential for embryo development in *Arabidopsis*. *Plant Cell* **2002**, *14*, 2799–2811. [CrossRef] [PubMed]
2. Perry, J.R.; Basrai, M.A.; Steiner, H.Y.; Naider, F.; Becker, J.M. Isolation and characterization of a saccharomyces cerevisiae peptide transport gene. *Mol. Cell. Biol.* **1994**, *14*, 104–115. [CrossRef] [PubMed]
3. Steiner, H.Y.; Naider, F.; Becker, J.M. The ptr family: A new group of peptide transporters. *Mol. Microbiol.* **1995**, *16*, 825–834. [CrossRef] [PubMed]
4. Swift, S.; Throup, J.P.; Williams, P.; Salmond, G.P.; Stewart, G.S. Quorum sensing: A population-density component in the determination of bacterial phenotype. *Trends Biochem. Sci.* **1996**, *21*, 214–219. [CrossRef]
5. Kuchler, K.; Sterne, R.E.; Thorner, J. Saccharomyces cerevisiae ste6 gene product: A novel pathway for protein export in eukaryotic cells. *Embo J.* **1989**, *8*, 3973–3984. [CrossRef] [PubMed]
6. Curie, C.; Panaviene, Z.; Loulergue, C.; Dellaporta, S.L.; Briat, J.-F.; Walker, E.L. Maize yellow stripe1 encodes a membrane protein directly involved in Fe(III) uptake. *Nature* **2001**, *409*, 346–349. [CrossRef] [PubMed]

7. Stacey, M.G.; Patel, A.; Mcclain, W.E.; Mathieu, M.; Remley, M.; Rogers, E.E.; Gassmann, W.; Blevins, D.G.; Stacey, G. The Arabidopsis atopt3 protein functions in metal homeostasis and movement of iron to developing seeds. *Plant Physiol.* **2008**, *146*, 589–601. [CrossRef] [PubMed]

8. Dean, M.; Hamon, Y.; Chimini, G. The human atp-binding cassette (abc) transporter superfamily. *J. Lipid Res.* **2001**, *42*, 1007–1017. [CrossRef] [PubMed]

9. Higgins, C.F. Abc transporters: From microorganisms to man. *Ann. Rev. Cell Biol.* **1992**, *8*, 67–113. [CrossRef]

10. Hauser, M.; Narita, V.; Donhardt, A.M.; Naider, F.; Becker, J.M. Multiplicity and regulation of genes encoding peptide transporters in saccharomyces cerevisiae. *Mol. Membr. Biol.* **2001**, *18*, 105–112. [CrossRef]

11. Newstead, S. Recent advances in understanding proton coupled peptide transport via the pot family. *Curr. Opin. Struct. Biol.* **2017**, *45*, 17–24. [CrossRef] [PubMed]

12. Williams, L.; Miller, A. Transporters responsible for the uptake and partitioning of nitrogenous solutes. *Ann. Rev. Plant Biol.* **2001**, *52*, 659–688. [CrossRef] [PubMed]

13. Lubkowitz, M. The opt family functions in long-distance peptide and metal transport in plants. In *Genetic Engineering: Principles and Methods*; Setlow, J.K., Ed.; Springer: Boston, MA, USA, 2006; pp. 35–55.

14. Lubkowitz, M.A.; Hauser, L.; Breslav, M.; Naider, F.; Becker, J.M. An oligopeptide transport gene from candida albicans. *Microbiology* **1997**, *143*, 387–396. [CrossRef] [PubMed]

15. Lubkowitz, M.A.; Barnes, D.; Breslav, M.; Burchfield, A.; Naider, F.; Becker, J.M. Schizosaccharomyces pombe isp4 encodes a transporter representing a novel family of oligopeptide transporters. *Mol. Microbiol.* **1998**, *28*, 729–741. [CrossRef] [PubMed]

16. Lubkowitz, M. The oligopeptide transporters: A small gene family with a diverse group of substrates and functions? *Mol. Plant* **2011**, *4*, 407–415. [CrossRef] [PubMed]

17. Gomolplitinant, K.M.; Saier, M., Jr. Evolution of the oligopeptide transporter family. *J. Membr. Biol.* **2011**, *240*, 89. [CrossRef] [PubMed]

18. Feng, S.; Tan, J.; Zhang, Y.; Liang, S.; Xiang, S.; Wang, H.; Chai, T. Isolation and characterization of a novel cadmium-regulated yellow stripe-like transporter (snysl3) in solanum nigrum. *Plant Cell Rep.* **2017**, *36*, 281–296. [CrossRef] [PubMed]

19. Murata, Y.; Ma, J.F.; Yamaji, N.; Ueno, D.; Nomoto, K.; Iwashita, T. A specific transporter for iron(III)-phytosiderophore in barley roots. *Plant J.* **2006**, *46*, 563–572. [CrossRef] [PubMed]

20. Koh, S.; Wiles, A.M.; Sharp, J.S.; Naider, F.R.; Becker, J.M.; Stacey, G. An oligopeptide transporter gene family in arabidopsis. *Plant Physiol.* **2002**, *128*, 21–29. [CrossRef]

21. Wongkaew, A.; Asayama, K.; Kitaiwa, T.; Nakamura, S.-I.; Kojima, K.; Stacey, G.; Sekimoto, H.; Yokoyama, T.; Ohkama-Ohtsu, N. Atopt6 protein functions in long-distance transport of glutathione in arabidopsis thaliana. *Plant Cell Physiol.* **2018**. [CrossRef]

22. Zhang, Z.; Xie, Q.; Jobe, T.O.; Kau, A.R.; Wang, C.; Li, Y.; Qiu, B.; Wang, Q.; Mendoza-Cózatl, D.G.; Schroeder, J.I. Identification of atopt4 as a plant glutathione transporter. *Mol. Plant* **2016**, *9*, 481–484. [CrossRef] [PubMed]

23. Mendoza-Cózatl, D.G.; Xie, Q.; Akmakjian, G.Z.; Jobe, T.O.; Patel, A.; Stacey, M.G.; Song, L.; Demoin, D.W.; Jurisson, S.S.; Stacey, G. Opt3 is a component of the iron-signaling network between leaves and roots and misregulation of opt3 leads to an over-accumulation of cadmium in seeds. *Mol. Plant* **2014**, *7*, 1455–1469. [CrossRef] [PubMed]

24. Vasconcelos, M.W.; Li, G.W.; Lubkowitz, M.A.; Grusak, M.A. Characterization of the pt clade of oligopeptide transporters in rice. *Plant Genome* **2008**, *1*, 77–88. [CrossRef]

25. Bogs, J.; Bourbouloux, A.; Cagnac, O.; Wachter, A.; Rausch, T.; Delrot, S. Functional characterization and expression analysis of a glutathione transporter, bjgt1, from brassica juncea: Evidence for regulation by heavy metal exposure. *Plant Cell Environ.* **2003**, *26*, 1703–1711. [CrossRef]

26. Carole, D.M.; Beno, T.P. Role of glutathione in plant signaling under biotic stress. *Plant Signal. Behav.* **2012**, *7*, 210–212.

27. Yen, M.-R.; Tseng, Y.-H.; Saie, M., Jr. Maize yellow stripe1, an iron-phytosiderophore uptake transporter, is a member of the oligopeptide transporter (opt) family. *Microbiology* **2001**, *147*, 2881–2883. [CrossRef] [PubMed]

28. Cobbett, C.; Goldsbrough, P. Phytochelatins and metallothioneins: Roles in heavy metal detoxification and homeostasis. *Annu. Rev. Plant Biol.* **2003**, *53*, 159–182. [CrossRef]

29. Bourbouloux, A.; Shahi, P.; Chakladar, A.; Delrot, S.; Bachhawat, A.K. Hgt1p, a high affinity glutathione transporter from the yeast saccharomyces cerevisiae. *J. Biol. Chem.* **2000**, *275*, 13259–13265. [CrossRef]

30. Liu, T.; Zeng, J.; Xia, K.; Fan, T.; Li, Y.; Wang, Y.; Xu, X.; Zhang, M. Evolutionary expansion and functional diversification of oligopeptide transporter gene family in rice. *Rice* **2012**, *5*, 1–14. [CrossRef]
31. Xiang, Q.; Shen, K.; Yu, X.; Zhao, K.; Gu, Y.; Zhang, X.; Chen, X.; Chen, Q. Analysis of the oligopeptide transporter gene family in ganoderma lucidum: Structure, phylogeny, and expression patterns. *Genome* **2017**, *60*, 293–302. [CrossRef]
32. Cao, J.; Huang, J.; Yang, Y.; Hu, X. Analyses of the oligopeptide transporter gene family in poplar and grape. *BMC Genom.* **2011**, *12*, 465. [CrossRef] [PubMed]
33. Yun, T.K. Brief introduction of panax ginseng c.A. Meyer. *J. Korean Med. Sci.* **2001**, *16* (Suppl.), S3–S5.
34. Jiang, X.; Yang, C.; Baosheng, L.; Shuiming, X.; Qinggang, Y.; Rui, B.; He, S.; Linlin, D.; Xiwen, L.; Jun, Q. *Panax ginseng* genome examination for ginsenoside biosynthesis. *GigaScience* **2017**, *6*, 1–15.
35. Li, H.; Benedito, V.A.; Udvardi, M.K.; Zhao, P.X. Transporttp: A two-phase classification approach for membrane transporter prediction and characterization. *BMC Bioinform.* **2009**, *10*, 418. [CrossRef] [PubMed]
36. Li, W.; Godzik, A. Cd-hit: A fast program for clustering and comparing large sets of protein or nucleotide sequences. *Bioinformatics* **2006**, *22*, 1658–1659. [CrossRef] [PubMed]
37. Kyte, J.; Doolittle, R.F. A simple method for displaying the hydropathic character of a protein. *J. Mol. Biol.* **1982**, *157*, 105–132. [CrossRef]
38. Shintaro, K.; Haruhiko, I.; Daichi, M.; Michiko, T.; Hiromi, N.; Satoshi, M.; Nishizawa, N.K. Osysl2 is a rice metal-nicotianamine transporter that is regulated by iron and expressed in the phloem. *Plant J.* **2010**, *39*, 415–424.
39. Haruhiko, I.; Takanori, K.; Tomoko, N.; Michiko, T.; Yusuke, K.; Kazumasa, S.; Mikio, N.; Hiromi, N.; Satoshi, M.; Nishizawa, N.K. Rice osysl15 is an iron-regulated iron(III)-deoxymugineic acid transporter expressed in the roots and is essential for iron uptake in early growth of the seedlings. *J. Biol. Chem.* **2009**, *284*, 3470–3479.
40. DiDonido, D., Jr.; Roberts, L.A.; Sanderson, T.; Eisley, R.B.; Walker, E.L. Arabidopsis yellow stripe-like2 (ysl2): A metal-regulated gene encoding a plasma membrane transporter of nicotianamine–metal complexes. *Plant J.* **2004**, *39*, 403–414. [CrossRef]
41. Waters, B.M.; Chu, H.H.; Didonato, R.J.; Roberts, L.A.; Eisley, R.B.; Lahner, B.; Salt, D.E.; Walker, E.L. Mutations in arabidopsis yellow stripe-like1 and yellow stripe-like3 reveal their roles in metal ion homeostasis and loading of metal ions in seeds. *Plant Physiol.* **2006**, *141*, 1446–1458. [CrossRef]
42. Marie, L.J.; Adam, S.; Stéphane, M.; Jean-François, B.; Catherine, C. A loss-of-function mutation in atysl1 reveals its role in iron and nicotianamine seed loading. *Plant J.* **2010**, *44*, 769–782.
43. Cagnac, O.; Bourbouloux, A.; Chakrabarty, D.; Zhang, M.-Y.; Delrot, S. Atopt6 transports glutathione derivatives and is induced by primisulfuron. *Plant Physiol.* **2004**, *135*, 1378–1387. [CrossRef] [PubMed]
44. Pike, S.; Patel, A.; Stacey, G.; Gassmann, W. Arabidopsis opt6 is an oligopeptide transporter with exceptionally broad substrate specificity. *Plant Cell Physiol.* **2009**, *50*, 1923–1932. [CrossRef] [PubMed]
45. Jaillon, O.; Aury, J.M.; Noel, B.; Policriti, A.; Clepet, C.; Casagrande, A.; Choisne, N.; Aubourg, S.; Vitulo, N.; Jubin, C. The grapevine genome sequence suggests ancestral hexaploidization in major angiosperm phyla. *Nature* **2007**, *449*, 463–467. [PubMed]
46. Marchlerbauer, A.; Bo, Y.; Han, L.; He, J.; Lanczycki, C.J.; Lu, S.; Chitsaz, F.; Derbyshire, M.K.; Geer, R.C.; Gonzales, N.R. Cdd/sparcle: Functional classification of proteins via subfamily domain architectures. *Nucleic Acids Res.* **2017**, *45*, D200–D203. [CrossRef] [PubMed]
47. Zhang, J.J.; Su, H.; Zhang, L.; Liao, B.S.; Xiao, S.M.; Dong, L.L.; Hu, Z.G.; Wang, P.; Li, X.W.; Huang, Z.H. Comprehensive characterization for ginsenosides biosynthesis in ginseng root by integration analysis of chemical and transcriptome. *Molecules* **2017**, *22*, 889. [CrossRef]
48. Wang, K.; Jiang, S.; Sun, C.; Lin, Y.; Rui, Y.; Yi, W.; Zhang, M. The spatial and temporal transcriptomic landscapes of ginseng, panax ginseng c. A. Meyer. *Sci. Rep.* **2015**, *5*, 18283. [CrossRef]
49. Osawa, H.; Stacey, G.; Gassmann, W. Scopt1 and atopt4 function as proton-coupled oligopeptide transporters with broad but distinct substrate specificities. *Biochem. J.* **2006**, *393*, 267–275. [CrossRef]
50. Prince, V.E.; Pickett, F.B. Splitting pairs: The diverging fates of duplicated genes. *Nat. Rev. Genet.* **2002**, *3*, 827–837. [CrossRef]
51. Dai, X.; Sinharoy, S.; Udvardi, M.; Zhao, P.X. Planttfcat: An online plant transcription factor and transcriptional regulator categorization and analysis tool. *BMC Bioinform.* **2013**, *14*, 321. [CrossRef]

52. Sapna, K.; Nie, J.; Chen, H.S.; Hao, M.; Ron, S.; Xiang, L.; Lu, M.Z.; Taylor, W.M.; Wei, H. Evaluation of gene association methods for coexpression network construction and biological knowledge discovery. *PLoS ONE* **2012**, *7*, e50411.

53. Goodstein, D.M.; Shu, S.; Russell, H.; Rochak, N.; Hayes, R.D.; Joni, F.; Therese, M.; William, D.; Uffe, H.; Nicholas, P. Phytozome: A comparative platform for green plant genomics. *Nucleic Acids Res.* **2012**, *40*, D1178–D1186. [CrossRef] [PubMed]

54. Yu, N.Y.; Wagner, J.R.; Laird, M.R.; Melli, G.; Lo, R.; Dao, P.; Sahinalp, S.C.; Ester, M.; Foster, L.J.; Brinkman, F.S.L. Psortb 3.0. *Bioinformatics* **2010**, *26*, 1608–1615. [CrossRef] [PubMed]

55. Horton, P.; Park, K.J.; Obayashi, T.; Fujita, N.; Harada, H.; Adamscollier, C.J.; Nakai, K. Wolf psort: Protein localization predictor. *Nucleic Acids Res.* **2007**, *35*, 585–587. [CrossRef] [PubMed]

56. Marchlerbauer, A.; Anderson, J.B.; Chitsaz, F.; Derbyshire, M.K.; Deweesescott, C.; Fong, J.H.; Geer, L.Y.; Geer, R.C.; Gonzales, N.R.; Gwadz, M. Cdd: Specific functional annotation with the conserved domain database. *Nucleic Acids Res.* **2009**, *37*, D205–D210. [CrossRef] [PubMed]

57. Marchler-Bauer, A.; Lu, S.; Anderson, J.B.; Chitsaz, F.; Derbyshire, M.K.; DeWeese-Scott, C.; Fong, J.H.; Geer, L.Y.; Geer, R.C.; Gonzales, N.R.; et al. Cdd: A conserved domain database for the functional annotation of proteins. *Nucleic Acids Res.* **2011**, *39*, D225–D229. [CrossRef]

58. Katoh, K.; Kuma, K.; Toh, H.; Miyata, T. Mafft version 5: Improvement in accuracy of multiple sequence alignment. *Nucleic Acids Res.* **2005**, *33*, 511–518. [CrossRef]

59. Schliep, K.P. Phangorn: Phylogenetic analysis in r. *Bioinformatics* **2011**, *27*, 592–593. [CrossRef]

60. Stamatakis, A. Raxml version 8: A tool for phylogenetic analysis and post-analysis of large phylogenies. *Bioinformatics* **2014**, *30*, 1312–1313. [CrossRef]

61. Letunic, I.; Bork, P. *Interactive Tree of Life (Itol): An Online Tool for Phylogenetic Tree Display and Annotation*; Oxford University Press: Oxford, UK, 2007; pp. 78–82.

62. Suyama, M.; Torrents, D.; Bork, P. Pal2nal: Robust conversion of protein sequence alignments into the corresponding codon alignments. *Nucleic Acids Res.* **2006**, *34*, W609–W612. [CrossRef]

63. Yang, Z. Paml: A program package for phylogenetic analysis by maximum likelihood. *Comput. Appl. Biosci. Cabios* **1997**, *13*, 555–556. [CrossRef] [PubMed]

64. The Potato Genome Sequencing Consortium; Xu, X.; Pan, S.; Cheng, S.; Zhang, B.; Mu, D.; Ni, P.; Zhang, G.; Yang, S.; Li, R.; et al. Genome sequence and analysis of the tuber crop potato. *Nature* **2011**, *475*, 189–195. [CrossRef] [PubMed]

65. Young, N.D.; Debellé, F.; Oldroyd, G.E.D.; Geurts, R.; Cannon, S.B.; Udvardi, M.K.; Benedito, V.A.; Mayer, K.F.X.; Gouzy, J.; Schoof, H.; et al. The medicago genome provides insight into the evolution of rhizobial symbioses. *Nature* **2011**, *480*, 520–524. [CrossRef] [PubMed]

66. Sherry, S. Ncbi sra toolkit technology for next generation sequence data. *Pump Ind. Anal.* **2000**, *3*, 2230–2234.

67. Bolger, A.M.; Lohse, M.; Usadel, B. Trimmomatic: A flexible trimmer for illumina sequence data. *Bioinformatics* **2014**, *30*, 2114–2120. [CrossRef] [PubMed]

68. Pertea, M.; Kim, D.; Pertea, G.M.; Leek, J.T.; Salzberg, S.L. Transcript-level expression analysis of rna-seq experiments with hisat, stringtie and ballgown. *Nat. Protoc.* **2016**, *11*, 1650–1667. [CrossRef] [PubMed]

69. Kolde, R. Pheatmap: Pretty Heatmaps. R Package Version 1.0.8. Available online: https://CRAN.R-project.org/package=pheatmap (accessed on 18 December 2018).

70. Shannon, P.; Markiel, A.; Ozier, O.; Baliga, N.S.; Wang, J.T.; Ramage, D.; Amin, N.; Schwikowski, B.; Ideker, T. Cytoscape: A software environment for integrated models of biomolecular interaction networks. *Genome Res.* **2003**, *13*, 2498–2504. [CrossRef] [PubMed]

Sample Availability: Root samples of the *Panax ginseng* are available from the authors.

molecules

MDPI

Article

Adverse Drug Reaction Predictions Using Stacking Deep Heterogeneous Information Network Embedding Approach

Baofang Hu [1,2,3], Hong Wang [1,3,*], Lutong Wang [1,3] and Weihua Yuan [1,3]

[1] School of Information Science and Engineering, Shandong Normal University, Jinan 250014, China; hubaofang@sdwu.edu.cn (B.H.); wanglutong1002@163.com (L.W.); weihuayuan_qingdao@126.com (W.Y.)
[2] School of Data and Computer Science, Shandong Women's University, Jinan 250014, China
[3] Shandong Provincial Key Laboratory for Distributed Computer Software Novel Technology, Shandong Normal University, Jinan 250014, China
* Correspondence: wanghong106@163.com; Tel.: +86-531-8961-0769

Academic Editor: Xiangxiang Zeng
Received: 5 November 2018; Accepted: 30 November 2018; Published: 4 December 2018

Abstract: Inferring potential adverse drug reactions is an important and challenging task for the drug discovery and healthcare industry. Many previous studies in computational pharmacology have proposed utilizing multi-source drug information to predict drug side effects have and achieved initial success. However, most of the prediction methods mainly rely on direct similarities inferred from drug information and cannot fully utilize the drug information about the impact of protein–protein interactions (PPI) on potential drug targets. Moreover, most of the methods are designed for specific tasks. In this work, we propose a novel heterogeneous network embedding approach for learning drug representations called SDHINE, which integrates PPI information into drug embeddings and is generic for different adverse drug reaction (ADR) prediction tasks. To integrate heterogeneous drug information and learn drug representations, we first design different meta-path-based proximities to calculate drug similarities, especially target propagation meta-path-based proximity based on PPI network, and then construct a semi-supervised stacking deep neural network model that is jointly optimized by the defined meta-path proximities. Extensive experiments with three state-of-the-art network embedding methods on three ADR prediction tasks demonstrate the effectiveness of the SDHINE model. Furthermore, we compare the drug representations in terms of drug differentiation by mapping the representations into 2D space; the results show that the performance of our approach is superior to that of the comparison methods.

Keywords: adverse drug reaction prediction; heterogeneous information network embedding; stacking denoising auto-encoder; meta-path-based proximity

1. Introduction

Adverse drug reactions (ADRs) are side effects caused by the use of one or several drugs. Some ADRs may be part of the natural pharmacological action of a drug that cannot be avoided, but more often, they may be unpredictable at the development stage. ADRs have caused a global and substantial burden that accounts for considerable mortality and morbidity [1]. Before clinical application of a drug, it should go through two ADR detection stages, including preclinical in vitro safety profiling and clinical drug safety trials. However, since so many side effect types and drug combinations exist, many potential side effects cannot be detected during the early drug development stage [2].

Recently, with the development of data mining and computational prediction methods, researchers have collected extensive drug data from the literature, and reports and have utilized these data to predict unknown ADRs [3–6]. ADR predictions based on computational methods can point drug safety tests in the right direction and consequently shorten the time requirement and save financial costs during drug development. A large number of machine learning methods have been proposed to predict potential ADRs [7–9]. Vilar et al. [10] utilized known side effect information of drugs to construct an associated matrix of drugs and adverse effects and adopted the matrix completion method to predict unknown side effects. LaBrute et al. [11] processed multi-source drug target information to find association relationships between ADRs and drug targets. These prediction methods are based on single drug information, and these mined drug datasets usually contain much noise. For example, the SIDER dataset [5], which was extracted from the public ADR reports, may contain some fake or unconfirmed noise data. Researchers have established different drug databases that describe drug features from different aspects, including chemical, biological, phenotypic, and interaction relationships [4–6,12,13]. It is more logical to combine different drug information to reduce the prediction error. Integrating this useful complex drug information to obtain more accurate ADR predictions is more effective. Yamanishi et al. [7] used multi-source drug data from the SIDER, PubChem, DrugBank, and Matador databases to predict side effects. The prediction method they adopted was based on multiple kernel regression and canonical correlation analysis. Zhang et al. [14] integrated different drug information to calculate drug similarities and utilized the linear neighborhoods method to transform the similarities into the side effect space and predict side effects. These prediction methods are mainly for the side effects caused by a single drug. However, in real life, many patients, especially the elderly, are on multiple prescriptions to treat different diseases. Drug–drug interactions (DDIs) may change the effects of drugs and cause some potential ADRs. Therefore, predicting the potential side effects induced by DDIs is imperative. Segura-Bedmar et al. [15] utilized a text mining method to predict the occurrence of DDIs based on a shallow language learning model. Jin et al. [16] formulated the DDI type prediction problem as a multi-task dyadic regression problem and utilized the model to predict the side effect types induced by DDIs. Zhang et al. [17] collected a variety of drug data that might influence DDIs and adopted an ensemble learning method to predict the occurrence of DDIs. Motivated by the success of deep learning in many areas, Zitnik et al. [18] developed a new graph convolutional neural network for multi-relational link prediction in multimodal networks to predict the DDI types.

Although the above methods have achieved great success, the methods are mostly designed for specific ADR prediction tasks and lack generic abilities. With the development of the network embedding, learning combined characteristic embeddings of drugs has attracted great attention from researchers [19–21]. Every drug can be embedded into a low-dimensional feature vector, which integrates different drug information, including chemical, biological, phenotypic, and interaction relationships. The drug representations are more general and can be used for different ADR prediction tasks. Li et al. [22] proposed a matrix completion method to integrate multiple sources of drug data and predicted ADRs. Ma et al. [23] proposed a drug embedding method based on multi-view deep auto-encoders to predict ADRs. However, their works only considered immediately relevant information of side effects and neglected potential indirect information. There are some potential association relationships between different biological data. For example, drug targets propagate to another protein through the protein–protein interactions (PPI) network, because the biological function signal cascade propagates through different proteins via PPI [24,25]. When one drug acts on a known target protein, it may change another potential target protein through protein–protein interaction effects and consequently cause potential ADRs.

In this work, we propose a general drug embedding method to learn the representations of drugs and predict different types of ADRs. The flowchart for ADR prediction is shown in Figure 1. We firstly modeled different drug information in a drug heterogeneous information network (drug HIN) framework and then proposed a stacking deep heterogeneous information network embedding approach based on semantic meta-paths. The generated drug embedding integrates multi-source drug information and multi-relationship side effect information to improve the ADR prediction accuracy.

Especially, we utilized the target propagation strategy to recognize the potential drug targets and improve the prediction accuracy. At the target propagation stage, we need to search for the proteins that are more obviously affected by the known targets of the drug. Finding the nearest node of one node is challenging because tens of thousands of nodes exist in the PPI network. We propose using transition probability based on the random walk procedure [26,27], which is common in recommendation systems, to calculate the target propagation proximity and reconstruct the drug-target network based on the target propagation meta-path.

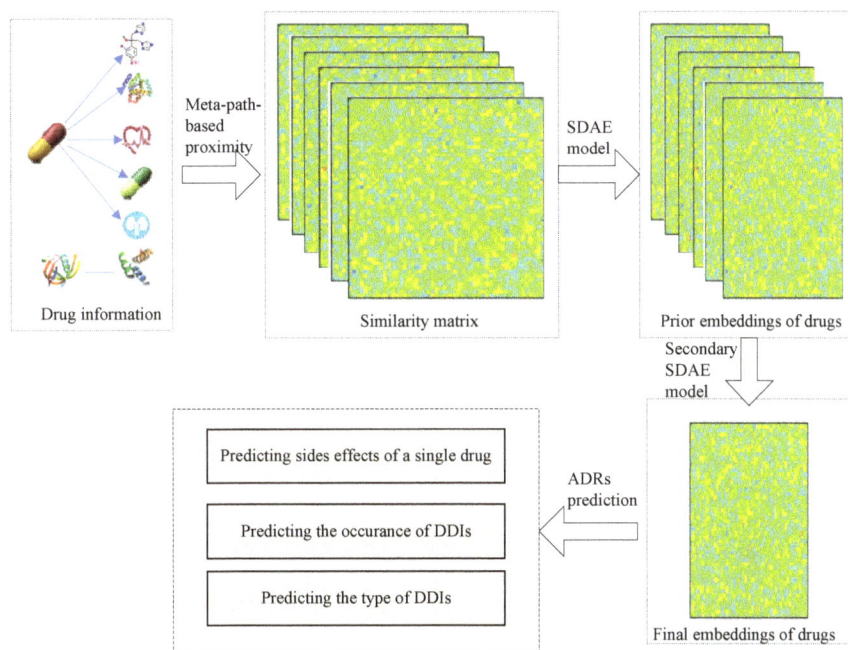

Figure 1. The flowchart for adverse drug reactions' (ADRs) prediction. SDAE, stacked denoising auto-encoder; DDIs, Drug-Drug interaction.

2. Datasets and Method

2.1. Datasets

We collected six types of drug data from seven public databases.

- Drug-Drug interaction information (DDI): Tatonetti [12] mined side effects induced by DDIs from the FDA Adverse Event Reporting System (FAERS, http://www.fda.gov/cder/aers/default.htm) and developed a database called "TWOSIDES". The database contains 645 drugs and ADRs caused by 63,473 combinations of different drugs.
- Protein-Protein interaction data (PPI): We downloaded the PPI network data from the Human Protein Reference Database (HPRD, http://www.hprd.org). The dataset contains 9519 proteins and 37,062 protein-protein interactions.
- Other drug information: We also obtained other drug information from four online drug information databases (DrugBank [4], the PubChem Compound database [6], the SIDER database [5], and the OFFSIDES database [12]). DrugBank is a widely-used public drug information database. From the DrugBank database, we collected drug target protein and disease treatment information. The PubChem system generates a binary substructure fingerprint

for chemical structures. From the PubChem database, we searched every drug's chemical substructure. We also extracted drug side effect information from the SIDER and OFFSIDES databases. These two databases include most associations between drugs and side effects, and we integrated the drug-side effect data obtained from the two databases.

We mapped drug ids in the TWOSIDES dataset to the other aforementioned datasets and finally constructed an integrated dataset that contained multi-source drug information, including DDIs, drug chemical substructures, drug targets, drug side effects, drug treatment, and PPI. The dataset used in this work is shown in Table 1. In this article, we did not consider the probabilities of ADR events. If a type of ADR occurred, the corresponding element in the DDI dataset or side effect dataset was labeled 1.

Table 1. Description of the drug data.PPI, protein–protein interaction.

Data Type	Data	Data Source	Dimension
Chemical	Substructures	PubChem	548×881
Biological	Target protein	DrugBank	548×695
Phenotypic	Treatment disease	DrugBank	548×718
Phenotypic	Side effect	SIDER, OFFSIDES	548×1318 (1318 ADR events)
Interaction	DDIs	TWOSIDES	$548 \times 548 \times 1318$ (1318 ADR events)
Interaction	PPI	HPRD	9519×9519 (37,062 interactions)

2.2. Drug HIN

Multi-source drug information describes different aspects of drugs and forms a typical heterogeneous information network (HIN). An HIN is a network that contains multiple types of objects or multiple types of relationships [28]. The drug HIN consists of five types of objects: drug (D), chemical substructure (C), protein (P), side effect (S_E), and disease (D_I). The five types of objects are connected through six types of links (as shown in Figure 2). A drug-drug link indicates a type of drug-drug interaction, whereas the link between a drug and its chemical substructure indicates that the drug consists of some type of chemical substructure. In Table 2, we present the semantics of the different link types in the drug HIN.

Table 2. Semantics of link types in the drug heterogeneous information network (HIN).

Link Types	Abbreviated Form	Semantics of Link Types
Drug-Drug	D-D	Drug-drug interactions
Drug-Chemical	D-C	The chemical substructure of a drug
Drug-Protein	D-P	The target protein of a drug
Protein-Protein	P-P	Protein-protein interactions
Drug-Disease	D-D_I	The therapeutic effect between a drug and a disease
Drug-Side Effect	D-S_E	The side effect between a drug and a disease

In HINs, two objects connect via different link types, which are called semantics meta-paths [29,30]. Given an HIN, a meta-path is a sequence of objects connected by different link types. Different types of meta-paths in the drug HIN are shown in Figure 3. Because our final goal is to learn drug representations, we only consider the meta-paths in which the starting objects are all drugs. The detailed meta-paths used in this study are summarized in Table 3.

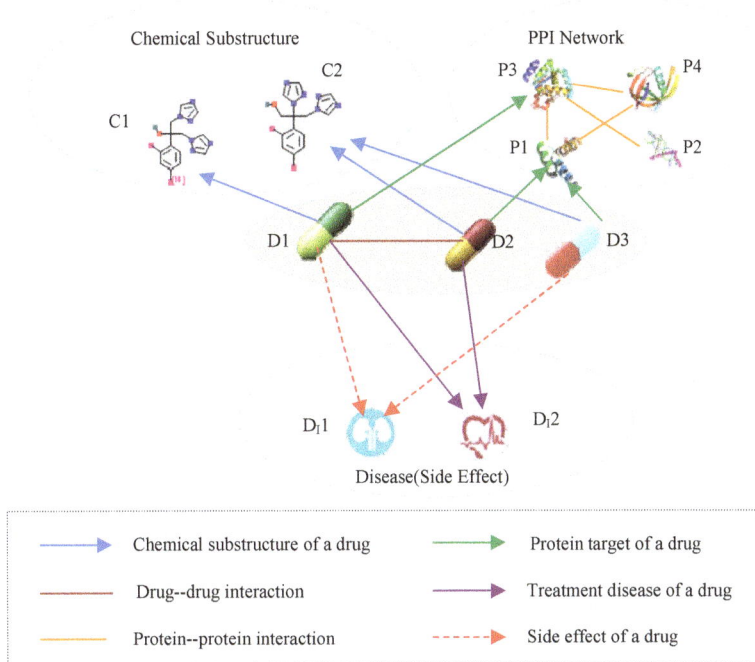

Figure 2. Heterogeneous drug information. PPI, protein-protein interaction.

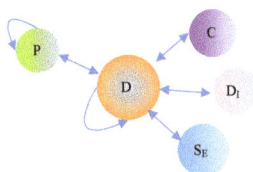

Figure 3. Meta-paths in drug HIN.

Table 3. Meta-paths in drug HIN.

Meta-Paths	Abbreviated Form	Semantics of Meta-Paths
Drug-Drug	DD	Drug-Drug interactions (at the drug embedding stage, interaction types are not considered).
Drug-Chemical-Drug	DCD	Two drugs have a similar chemical substructure.
Drug-Protein-Drug	DPD	Two drugs have the same target protein.
Drug-Protein-...-Protein-Drug	$DP^{(n)}D$ $(n \geq 2)$	There are protein-protein interactions between the targets of two drugs. For example, the path $D_1 P_1 P_2 D_2$ in Figure 4 indicates that the targets of D_1 and D_2 are P_1 and P_2, respectively. Meanwhile, there is an interaction between P_1 and P_2 (in meta-path $DP^{(n)}D$, there are $n-1$ protein-protein interactions).
Drug-Disease-Drug	$DD_I D$	Two drugs have the same therapeutic effect.
Drug-Side Effect-Drug	$DS_E D$	Two drugs have the same side effect.

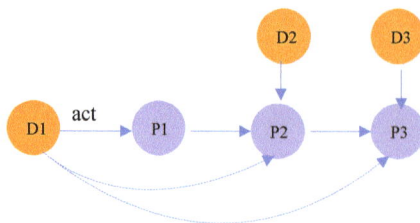

Figure 4. Illustration of meta-path $DP^{(n)}D$.

2.3. Stacking Deep HIN Embedding

Our goal is to learn the low-dimensional vector representations of drugs that highly summarize the drug information and preserve the original proximity of drugs in different drug relationships, and then to predict the different types of ADRs. In this work, we proposed a semi-supervised deep model SDHINE to perform HIN embedding; the framework of the model is shown in Figure 5. In detail, first, we defined the meta-path-based proximities and constructed several homogeneous sub-networks based on the defined proximities. Then, we adopted semi-supervised stacked denoising auto-encoders (SDAE) to encode each sub-network. The supervision information is the meta-path-based proximity in every sub-network. Next, we concatenated the drug embeddings together and further learned the final drug embeddings through the secondary encoding process.

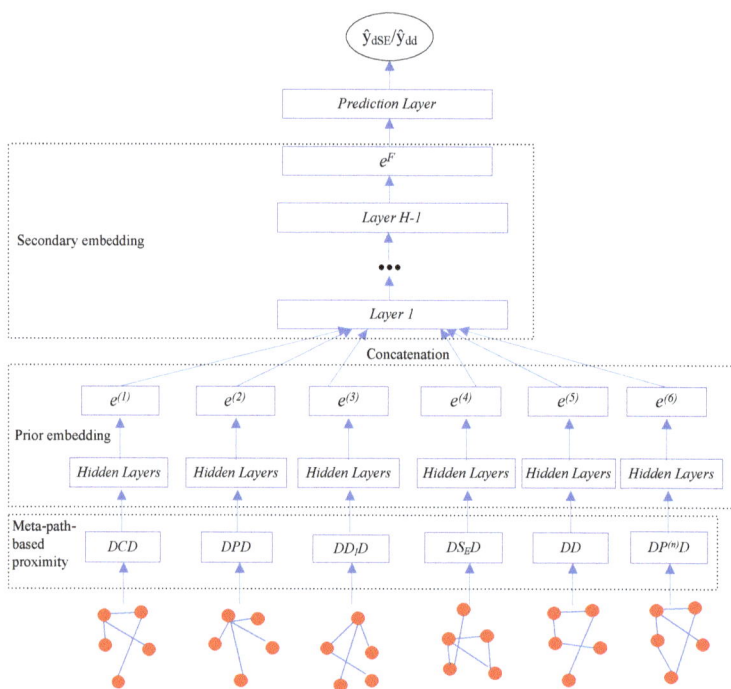

Figure 5. The framework of our proposed semi-supervised deep model SDHINE.

2.3.1. Meta-Path-Based Proximity

We defined three types of meta-path-based proximities and constructed corresponding sub-networks based on the defined proximities.

- Constructing drug-drug interaction sub-network.

For the Drug-Drug (DD) meta-path shown in Table 3, in which the nodes are all drug nodes, we utilized the Jaccard similarity as the edge weights to construct the Drug-Drug interaction sub-network. Notably, we do not consider the interaction types at the drug embedding stage. The proximity between drug i and drug j based on meta-path DD can be calculated as shown in Equation (1).

$$s(i,j) = \frac{|D_i \cap D_j|}{|D_i \cup D_j|} \tag{1}$$

where D_i is a column vector with 0 and 1 elements that represent the Drug-Drug interactions between the ith drug and other drugs.

- Constructing sub-networks using PathSim proximity.

The meta-paths Drug-Chemical-Drug (DCD), Drug-Protein-Drug (DPD), DD_ID, and DS_ED in drug HIN contain different types of nodes and path semantic information. For example, meta-path $D_1C_1D_2$ indicates that drug D_1 and drug D_2 have the same chemical substructure C_1 and there is a path from D_1 to D_2 via C_1. Therefore, we adopted PathSim [29] as the proximity measure in these meta-paths. The PathSim proximity $S(i,j)$ is defined in Equation (2).

$$s(i,j) = \frac{2 \times |\{p_{i \to j} : p_{i \to j} \in P\}|}{|\{p_{i \to i} : p_{i \to i} \in P\}| + |\{p_{j \to j} : p_{j \to j} \in P\}|} \tag{2}$$

where $p_{i \to j}$ is a path instance between i and j, $p_{i \to i}$ is a path instance between i and i, and $p_{j \to j}$ is a path instance between j and j.

The proximities of drugs under meta-paths DCD, DPD, DD_ID, and DS_ED (as shown in Table 3) are directly calculated using PathSim. Then, we constructed corresponding sub-networks using the proximities as the edge weights to form corresponding sub-networks.

- Reconstructing the drug-target sub-network using the target propagation method.

One innovation of our proposed approach is calculating target protein transition probabilities based on the PPI network and reconstructing the drug-target sub-network. As previously mentioned, potential association relationships exist between different biological data, especially for drug target information. When one target protein is activated by a drug, another potential protein may be activated by protein-protein interactions and consequently cause an unreported ADR. Therefore, we should reconstruct the drug-target sub-network using the target propagation strategy according to the meta-path $DP^{(n)}D(n \geq 2)$. The target propagation in the meta-path $DP^{(k)}D$ can be seen as a random walk procedure. A walker walks on the PPI network and achieves the destination protein via $k-1$ steps. Suppose node transition probabilities in the PPI network converge after n steps. The global proximity based on meta-paths $DP^{(n)}D(n \geq 2)$ is:

$$s(i,j) = \sum_{k=2}^{n} S^{(k)}(i,j) = \sum_{k=2}^{n} (DP^{(k)}D)_{ij} \tag{3}$$

In random walk theory, the k-step random walk transition probability is the kth power of the transition probability matrix P. For example, as shown in Figure 4, the probability that drug D_1 acts on protein P_3 is equal to the product of the transition probability from P_1 to P_2 and the probability from P_2 to P_3.

The proximity between drug i and drug j based on meta-path $DP^{(k)}D$ can be unfolded as follows:

$$s^{(k)}(i,j) = (DP^{(k)}D)_{ij} = (DP^kD)_{ij} = D_iPP^{k-2}(D_jP)^T = D_iP \times \underbrace{P \times P \times \cdots \times P}_{k-2} \times (D_jP)^T \tag{4}$$

where P is the transition probability matrix in the PPI network, D is the original drug-target matrix, and D_iP represents the target row vector of the ith drug.

$$s(i,j) = \sum_{k=2}^{n} s^{(k)}(i,j) = \sum_{k=2}^{n} (DP^kD)_{ij} = D_iP(\sum_{k=2}^{n} P^{k-2})(D_jP)^T \tag{5}$$

Given tens of thousands of nodes in the PPI network, calculating P^k is very difficult. However, P^k is very common in the random walk theory. During the random walk procedure, a walker starts from an initial node and moves to neighbors with probability μ and back to the initial node with probability $1 - \mu$. Based on the Katz model [31], which is a method of computing similarities between nodes in a graph, taking into account not only the direct edges but also the indirect edges, Equation (5) can be rewritten as follows:

$$s(i,j) \approx D_iP[\sum_{k=2}^{n} (\mu P)^{k-2}](D_jP)^T = D_iP[(I - \mu P)^{-1} - I](D_jP)^T \tag{6}$$

where I is the identity matrix and the damping factor μ usually is 0.98. The inverse of the matrix in Equation (6) can be calculated using the SVD-based matrix factorization method.

2.3.2. Prior Drug Embedding

A stacking auto-encoder is a multi-layer deep neural network based on layer-wise training in which different multi-granularity data features are learned layer by layer and higher complex features are learned in higher layers. To enhance the robustness of sub-network embedding, we adopted stacked denoising auto-encoders (SDAE) in which the input neurons in every layer were randomly discarded by assigning some of the input neurons to 0 with a certain probability.

Traditional SDAE is an unsupervised model, which is composed of the encoder stage and decoder stage. At the encoder stage, the input data x_i are mapped into representation vector space, whereas at the decoder stage, the output data $\hat{x}_i$ are the reconstructed data from x_i. The optimizer objective function of the SDAE is to minimize the reconstruction error of the output and input. The loss function is shown as follows:

$$L_1 = \sum_{i=1}^{n} \|\hat{x}_i - x_i\|_2^2 \tag{7}$$

Here, to protect the meta-path-based proximity of every sub-network, we adopted a semi-supervised SDAE framework [32]. The different meta-path-based proximities are the supervision information that preserves the proximity of the representation of two nodes. The optimizer objective function for this goal is defined as follows:

$$L_2 = \sum_{i,j=1}^{n} S^{(p)}{}_{ij} \|e^{(p)}{}_i - e^{(p)}{}_j\|_2^2 \tag{8}$$

where $S^{(p)}{}_{ij}$ is the proximity of drug i and drug j based on meta-path p and $e^{(p)}{}_i$ is the embedding of x_i based on the corresponding meta-path.

The objective function of the semi-supervised SDAE model, which combines Equations (7) and (8), is as follows:

$$L = L_1 + \alpha L_2 + \beta L_{reg} \tag{9}$$

where L_{reg} (as shown in Equation (12)) is an L2-norm regularizer term to prevent overfitting and α and β are hyperparameters. $W^{(k)}$ and $\hat{W}^{(k)}$ are the kth layer weight matrices at the encoder and decoder stages, respectively.

$$L_{reg} = \sum_{k=1}^{K} \left(\left\| W^{(k)} \right\|_F^2 + \left\| \hat{W}^{(k)} \right\|_F^2 \right) \tag{10}$$

2.3.3. Secondary Drug Embedding

After obtaining the sub-network embeddings, we concatenated the embeddings together and used the secondary semi-supervised stacking denoising auto-encoder to obtain the final drug embedding. Given the embedding node $e^{(p)}{}_i$ of drug i in a different meta-path p, we concatenated them to obtain a new representation vector e_i. Then, we utilized auto-encoder layers to learn the final embedding $e^F{}_i$ of drug i (as shown in Equation (11)).

$$e^{F(1)}{}_i = \sigma(W^{F(0)} e_i + b^{F(0)})$$
$$e^{F(2)}{}_i = \sigma(W^{F(1)} e^{F(1)}{}_i + b^{F(1)})$$
$$...$$
$$e^F{}_i = e^{F(h)}{}_i = \sigma(W^{F(h-1)} e^{F(h-1)}{}_i + b^{F(h-1)}) \tag{11}$$

Here, we continued to adopt a semi-supervised SDAE framework to protect the original proximities of drugs in every sub-network. The supervision information in the optimizer objective function is defined as follows:

$$L_2 = \sum_{i,j=1}^{n} \frac{\sum_{p=1}^{K} \alpha^{(p)} S^{(p)}{}_{ij}}{\sum_{p=1}^{K} \alpha^{(p)}} \left\| e^F{}_i - e^F{}_j \right\|_2^2 \tag{12}$$

where $S^{(p)}{}_{ij}$ is the proximity of drug i and drug j based on meta-path p and $\alpha^{(p)}$ is a hyperparameter, which is the weight coefficient of the meta-path p. At the experimental stage, the best hyperparameters $\alpha^{(p)}$ are learned using 10-fold cross-validation on 10% labeled data with a grid search over $\alpha^{(p)} \in \{0.1, 0.2, 0.3, 0.4, 0.5\}$.

The objective function of this part is shown in Equation (13).

$$L_G = \sum_{i=1}^{n} \| \hat{e}_i - e_i \|_2^2 + \alpha L_2 + \beta L_{reg} \tag{13}$$

2.4. Prediction Formulation

For prediction tasks, learning a classifier that can be generalized to unknown ADRs is desirable. We predict the labels on training data using a fully-connected layer $y = h(e^F{}_i) = \sigma(W^P e^F{}_i + b)$. The prediction loss is formulated by Equation (14).

$$L = \sum_{y \in Y_{tarin}} [-y \ln y' - (1-y) \ln(1 - y')] + \lambda \left\| W^P \right\|_F^2 \tag{14}$$

3. Experiment

3.1. Implementation and Evaluation Strategy

Proposed model: We implemented the proposed model with TensorFlow 1.2 and trained the model using the adaptive learning rate optimizer Adam [33]. All neurons were activated by the sigmoid function. We optimized the hyperparameters in the model using validation data and then fixed them for all denoising auto-encoder layers.

Baseline: In addition, we implemented the following three network embedding baselines for comparison:

- Concatenate drug features: This method is a simple original HIN embedding method [28]. The approach constructs a feature vector for each drug by concatenating the PCA representation of each correlation matrix, which represents one aspect of the drug character.
- GraphCNN [34]: GraphCNN is a recently-proposed network embedding method based on spectral convolutional operation and achieves state-of-the-art performance on important prediction problems in recommender systems. Here, first, we linearly integrated similarity matrices based on all meta-paths except the target propagation meta-path $DP^{(n)}D$ and then learned the drug embeddings using the same GraphCNN structure described in [35].
- metapath2vec++ [36]: metapath2vec++ is a heterogeneous information network embedding method based on a meta-path-guided random walk strategy.

For further validation of the impact of target propagation on improving the quality of ADR predictions, we designed a network embedding algorithm without regard for the impact of protein–protein interactions and discarded the PPI dataset; this algorithm was named SDHINE-no-target propagation.

The subsequent ADR prediction methods after the network embedding stage were all based on the same loss function in every prediction task.

Evaluation: We evaluated and compared these algorithms using a 10-fold cross-validation methodology. We randomly selected a fixed percentage (10%) of drugs as the test set and moved all ADRs associated with these drugs from the dataset. The side effects and DDIs of these drugs were all set to 0. The other 90% of the drugs were further divided into the training set and validation set. The training set was formed with 95% of the remaining drugs and was used to train the model. The validation set was formed with the other 5% of the drugs and was used to test the model performance. The independent validation experiments were repeated 30 times with different random divisions of the data for the three sets.

The metrics used to evaluate the model performance were two common ranking metrics: mean average precision at K (MAP@K) and area under the receiver operating characteristic curve (ROC-AUC).

Average precision at K (AP@K) reflects the accuracy of the top-ranked ADRs by a model and can be computed as the mean of Precision@k for each drug or drug pair in the test set. The formula for computing AP@K is given as follows:

$$AP@K = \sum_{k=1}^{K} \Pr ecision(k) / \min(L, K) \tag{15}$$

where Precision(k) is the precision at cut-off k in the return list. L is the total number of true ADRs for the test drug or drug pairs.

3.2. Experimental Results

3.2.1. Visualization Results

First, we compared the performances of all network embedding approaches for the visualization task, which aimed to layout the drug HIN in a 2-dimensional vector space. We mapped the representation vectors of drugs obtained from all comparison approaches to a 2D vector space using the t-SNE [37]. Once a drug is successfully developed, the chemical substructure is fixed. The targets and side effects of a drug are all affected by the chemical substructures of the drug. Therefore, to compare the dimensional reduction performance of different network embedding approaches, the drugs are firstly clustered into different clusters based on their chemical substructures.

The results are shown in Figure 6, in which drugs belonging to the same cluster are represented by the same color. The concatenated drug features method and metapath2vec++ could not separate drugs from different groups. GraphCNN and SDHINE-no-target propagation basically separated drugs from different groups, but some dark green points were mixed with the other groups. The results obtained with SDHINE were the best among these methods, because it separated most of the drugs from the different groups. This result was consistent with the fact that deep integration of different characteristics can effectively eliminate noise from data and recover the original signal.

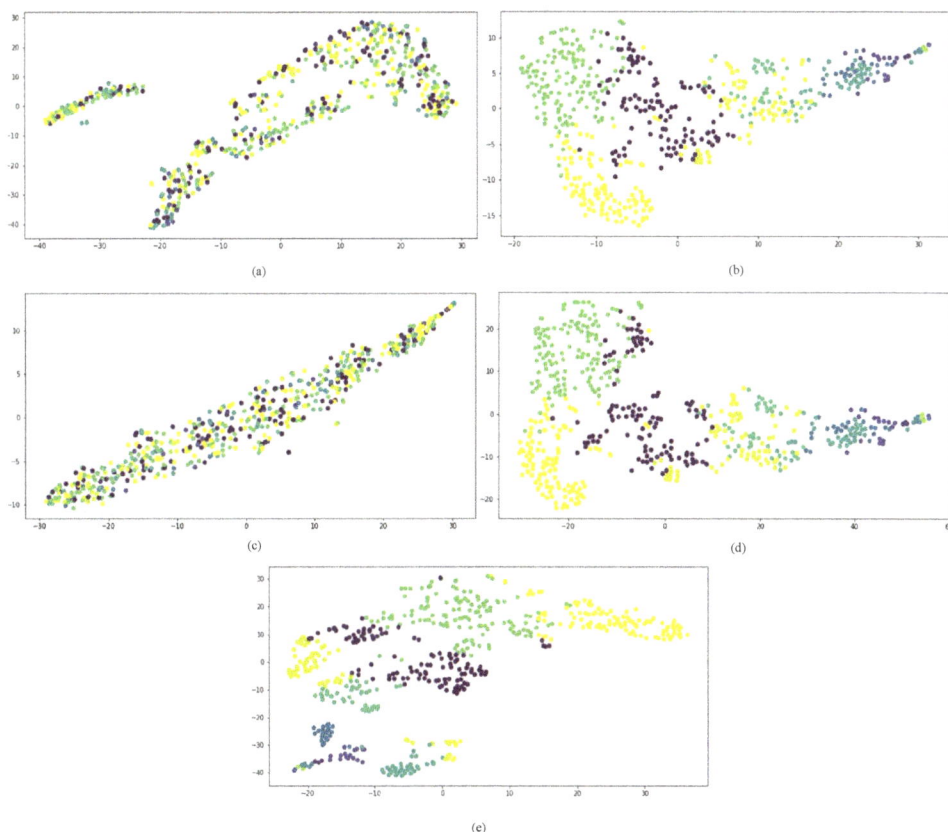

Figure 6. Visualization of the different representations: (**a**) concatenate drug features; (**b**) GraphCNN; (**c**) metapath2vec++ ; (**d**) SDHINE-no-target propagation; (**e**) SDHINE.

3.2.2. Prediction Results

Our experiments further evaluated the drug embeddings obtained through different network embedding methods on different tasks, including side effect predictions for a single drug, binary predictions of the occurrence of DDIs, and multi-label predictions of specific DDI types.

- Task 1: Predicting side effects of a single drug.

To demonstrate the side effect prediction performance based on our network embedding approach, we performed comparison experiments with the aforementioned three baselines and our two proposed models. Predicting the types of side effects caused by one single drug can be formulated into a multi-label classification problem. The output value y of the prediction formulation in Section 2.4 is a column vector with 1318 dimensions, and W^P is a weighted matrix. Each type of side effect was trained one by one. The negative sampling method was adopted to settle the sample unbalanced problem.

Detailed comparisons of the experimental results are shown in Table 4. Our model based on target propagation clearly performed best compared with other models without a target propagation process in terms of MAP@20 and MAP@100. It was also very close to the best result in terms of MAP@50. Analogously, our approach improved ROC-AUC by 5.87% (84.07% vs. 78.20%) compared to the worst result. From the perspective of the approaches based on deep architecture, GraphCNN and the two SDHINE models performed better than the other models. Meanwhile, the model based on target propagation clearly improved the performance by 3.86% (84.07% vs. 80.21%) compared with the similar model without a target propagation process in terms of ROC-AUC.

Table 4. Side effect identification performance comparison.

Models	MAP@20	MAP@50	MAP@100	ROC-AUC
Concatenate drug features	0.5590	0.5475	0.5310	0.7820
GraphCNN	0.6510	**0.6493**	0.6321	0.8190
metapath2vec++	0.5835	0.5760	0.5628	0.7845
SDHINE-no-target propagation	0.6508	0.6416	0.6356	0.8021
SDHINE	**0.6653**	0.6479	**0.6361**	**0.8407**

- Task 2: Binary prediction of the occurrence of DDIs.

When one drug is administered with another drug, the effect of the drug may be changed, and an unknown side effect may be caused by the DDI. Detecting the occurrence of DDIs is preparation for further research on the ADRs induced by DDIs. When predicting the occurrence of a DDI without regard for the type of DDI, the prediction task can be modeled as a binary classification problem. In this situation, a probability value can be the output layer of the prediction formulation in Section 2.4. W^P can be written as a weighted vector. The input layer is formed by the embedding vectors of the two drugs. Table 5 shows a detailed comparison of the experimental results obtained from the binary prediction task of the occurrence of DDIs. The model based on target propagation performed better compared with the similar model without a target propagation process in terms of the mean average precision at k and ROC-AUC. The target propagation strategy and deep architecture were still useful for improving the prediction of DDI occurrence.

Table 5. DDI occurrence identification performance comparison.

Models	MAP@20	MAP@50	MAP@100	ROC-AUC
Concatenate drug features	0.6122	0.5624	0.5432	0.7409
GraphCNN	0.6874	0.6715	0.6219	0.7918
metapath2vec++	0.6542	0.6326	0.5986	0.7332
SDHINE-no-target propagation	0.6813	0.6718	0.6211	0.7814
SDHINE	**0.7015**	**0.6854**	**0.6328**	**0.8124**

- Task 3: Multi-label prediction of specific adverse DDI types.

Compared with the prediction of DDI occurrence, most often, we need to address which types of side effects are caused by the DDI. This issue is a multi-label classification problem in which the output layer y in the prediction formulation is a column vector with 1318 DDI events. The input layer is concatenated by two drug representation vectors, and W^P is a weighted matrix.

Detailed comparison experimental results for specific adverse DDI type identification tasks are shown in Table 6. The model based on target propagation and deep architecture was superior to the models without a target propagation process or deep architecture in terms of not only the mean average precision at k, but also the ROC-AUC value.

Table 6. DDI type identification performance comparison.

Models	MAP@20	MAP@50	MAP@100	ROC-AUC
Concatenate drug features	0.6596	0.6144	0.5045	0.74322
GraphCNN	0.6823	0.6681	**0.6137**	0.7851
metapath2vec++	0.6766	0.6567	0.5118	0.7543
SDHINE-no-target propagation	0.6804	0.6622	0.6119	0.7996
SDHINE	**0.6881**	**0.6745**	0.6126	**0.8175**

Based on the results of the three prediction tasks, the network embedding approach with target propagation performance was superior to the approaches without target propagation processing. Moreover, approaches based on deep architecture performed better than the other linear network embedding methods and the combination methods. This result indicates the feasibility of predicting ADRs based on target propagation and proves that the deep learning process is effective at heterogeneous information network embedding.

3.3. Performance Comparison of Different Embedding Dimensions

To examine the impact of embedding size on prediction performance, we compared SDHINE and SDHINE-no-propagation with different dimensions of drug embeddings for three prediction tasks in terms of ROC-AUC. The results are shown in Figure 7. The prediction performances gradually increased with the increase of embedding dimension and reached the top when embedding dimensions were 64. The prediction performances at 256 dimensions were worse than that at 64 dimensions. This is because the higher dimensional embedding reduced the drug's differentiability. From the results, we also can find that SDHINE performed better than the same model without the target propagation process at the same embedding dimension on all three prediction tasks. It further verified our assumption that target propagation based on PPI can improve ADRs' prediction performance.

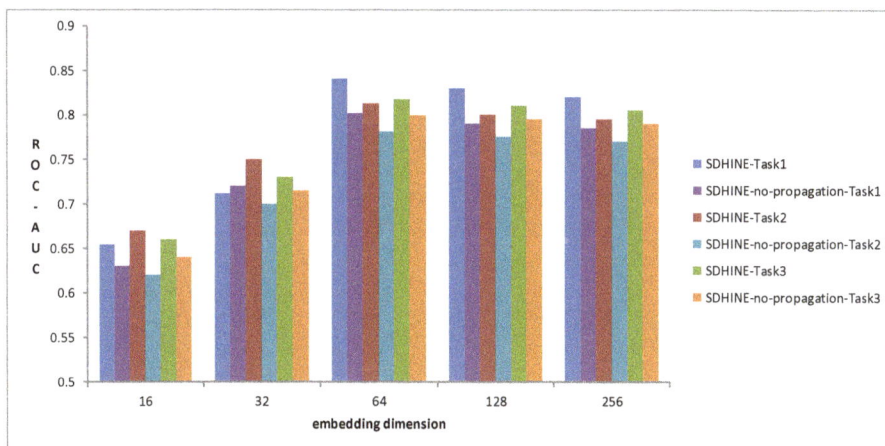

Figure 7. Performance comparison of different embedding dimensions.

3.4. Case Studies

We examined how the proposed network embedding method predicted potential unknown side effects based on learned drug embeddings. In this article, we only can query the drugs that are in the selected datasets by inputting their id. We took triamcinolone, which is an intermediate-acting synthetic glucocorticoid given orally, by muscular or intra-articular injection, or as a topical ointment or cream and is used to treat various medical conditions (e.g., eczema and ulcerative colitis), as an example. It has been confirmed that it may cause many kinds of side effects, such as cough, headache, influenza, and so on. In the SIDER database, triamcinolone may cause 147 kinds of side effects, and 37 of them have been confirmed in preclinical in vitro safety profiling and clinical drug safety trials.

Table 7 shows the top 10 side effects of triamcinolone based on our model. We found from the results that most of the returned side effects were confirmed and that only two side effects were not confirmed. We analyzed the reasons for this result by taking eye redness as an example. First, we analyzed the target protein of triamcinolone from the DrugBank database and found that triamcinolone activated the target protein NR3C1. Then, we calculated the similarity based on the target propagation meta-path and found that NR3C1 might walk to protein EGFR with a high transition probability in the PPI network. EGFR is the only target protein of the drug gefitinib. We searched the side effects of gefitinib in the SIDER and OFFSIDES databases and found that gefitinib had an eye redness side effect. Thus, we found that eye redness is a potential side effect of triamcinolone based on the above logical inference. The works in [38–40] reported that peribulbar injections of triamcinolone may cause intraocular pressure (IOP) elevation, keratitis, and cataract. It is reasonable to believe that eye redness could be one of the ADRs of triamcinolone.

Table 7. Prediction of the top 10 side effects for triamcinolone based on SDHINE.

Top K	Side Effect	Confirmation
K = 1	headache	yes
K = 2	cough	yes
K = 3	fever	yes
K = 4	eye redness	no
K = 5	sneezing	yes
K = 6	nausea	yes
K = 7	rash	yes
K = 8	fatigue	yes
K = 9	dry skin	no
K = 10	conjunctivitis	yes

4. Conclusions

In this work, we proposed to utilize the impact of protein–protein interactions on drug targets to improve the prediction performance of adverse drug reactions. We designed a meta-path-based heterogeneous information network embedding approach (SDHINE) to integrate multi-source drug information, especially the PPI network. Different meta-path-based proximity calculation methods are designed for different semantic meta-paths. We adopted a semi-supervised stacked denoising auto-encoder to learn drug embeddings in each type of meta-path and integrated them into a second auto-encoder to learn the final drug embeddings. Extensive experiments were performed to compare our algorithm with several state-of-the-art network embedding methods for three ADR prediction tasks, which demonstrated the effectiveness of SDHINE. We also verified the ability of SDHINE to distinguish side effect types and performed a case study by examining the impact of protein-protein interactions on side effects.

In this work, we only considered the meta-paths in which the start and end nodes are all drugs. In future work, we will investigate how to use meta-paths starting from other objects (e.g., side effect nodes) under the guarantee of rationality and interpretability. As a major issue in ADR prediction, we will also consider how to further enhance the interpretability of prediction methods and results based on the semantics of meta-paths.

Molecules **2018**, 23, 3193

Author Contributions: H.W. guided the research; B.H. designed the experiments and wrote the manuscript; W.Y. analyzed the data; L.W. performed the experiments.

Funding: The work is partially supported by the National Natural Science Foundation of China (Nos. 61672329, 61373149, 61472233, 61572300, 81273704), the Shandong Province Science and Technology Plan Supported Project (No. 2014GGX101026), the Project of Shandong Province Higher Educational Science and Technology Program (No.J18KA370), and the Taishan Scholar Fund of Shandong Province (No. TSHW201502038, 20110819). We also gratefully acknowledge the support of NVIDIA Corporation with the donation of the TITAN X GPU used for this research and the support of Shandong provincial key laboratory for distributed computer software novel technology.

Conflicts of Interest: The authors declare no conflict of interest.

References

1. Giacomini, K.M.; Krauss, R.M.; Dan, M.R.; Eichelbaum, M.; Hayden, M.R.; Nakamura, Y. When good drugs go bad. *Nature* **2007**, *446*, 975–977. [CrossRef] [PubMed]
2. Whitebread, S.; Hamon, J.; Bojanic, D.; Urban, L. Keynote review: In vitro safety pharmacology profiling: An essential tool for successful drug development. *Drug Discov. Today* **2005**, *10*, 1421–1433. [CrossRef]
3. Kanehisa, M.; Goto, S.; Furumichi, M.; Tanabe, M.; Hirakawa, M. KEGG for representation and analysis of molecular networks involving diseases and drugs. *Nucleic Acids Res.* **2010**, *38*, 355–360. [CrossRef] [PubMed]
4. Knox, C.; Law, V.; Jewison, T.; Liu, P.; Ly, S.; Frolkis, A.; Pon, A.; Banco, K.; Mak, C.; Neveu, V. DrugBank 3.0: A comprehensive resource for 'Omics' research on drugs. *Nucleic Acids Res.* **2011**, *39*, D1035. [CrossRef] [PubMed]
5. Kuhn, M.; Campillos, M.; Letunic, I.; Jensen, L.J.; Bork, P. A side effect resource to capture phenotypic effects of drugs. *Mol. Syst. Biol.* **2010**, *6*, 343. [CrossRef] [PubMed]
6. Li, Q.; Cheng, T.; Wang, Y.; Bryant, S.H. PubChem as a public resource for drug discovery. *Drug Discovery Today* **2010**, *15*, 1052–1057. [CrossRef] [PubMed]
7. Yamanishi, Y.; Pauwels, E.; Kotera, M. Drug side effect prediction based on the integration of chemical and biological spaces. *J. Chem. Inf. Model.* **2012**, *52*, 3284–3292. [CrossRef]
8. Li, J.; Zheng, S.; Chen, B.; Butte, A.J.; Swamidass, S.J.; Lu, Z. A survey of current trends in computational drug repositioning. *Brief. Bioinf.* **2016**, *17*, 2–12. [CrossRef]
9. Xu, B.; Shi, X.F.; Zhao, Z.H.; Zheng, W. Leveraging Biomedical Resources in Bi-LSTM for Drug Drug Interaction Extraction. *IEEE Access* **2018**, *17*, 33432–33439. [CrossRef]
10. Vilar, S.; Tatonetti, N.P.; Hripcsak, G. 3D Pharmacophoric Similarity improves Multi Adverse Drug Event Identification in Pharmacovigilance. *Sci. Rep.* **2015**, *5*, 8809. [CrossRef]
11. Labute, M.X.; Zhang, X.; Lenderman, J.; Bennion, B.J.; Wong, S.E.; Lightstone, F.C. Adverse drug reaction prediction using scores produced by large-scale drug-protein target docking on high-performance computing machines. *PLoS ONE* **2014**, *9*, e106298. [CrossRef] [PubMed]
12. Tatonetti, N.P.; Ye, P.P.; Daneshjou, R.; Altman, R.B. Data-Driven Prediction of Drug Effects and Interactions. *Sci. Transl. Med.* **2012**, *4*, 125–131. [CrossRef] [PubMed]
13. Ping, Z.; Fei, W.; Hu, J. Towards Drug Repositioning: A Unified Computational Framework for Integrating Multiple Aspects of Drug Similarity and Disease Similarity. *AMIA Annu. Symp. Proc.* **2014**, *2014*, 1258–1267.
14. Zhang, W.; Chen, Y.; Tu, S.; Liu, F.; Qu, Q. Drug side effect prediction through linear neighborhoods and multiple data source integration. In Proceedings of the IEEE International Conference on Bioinformatics and Biomedicine, Shenzhen, China, 15–18 December 2016; pp. 427–434.
15. Segura-Bedmar, I. Using a shallow linguistic kernel for drug–drug interaction extraction. *J. Biomed. Inf.* **2011**, *44*, 789–804. [CrossRef] [PubMed]
16. Jin, B.; Yang, H.; Xiao, C.; Zhang, P.; Wei, X.; Wang, F. Multitask Dyadic Prediction and Its Application in Prediction of Adverse Drug-Drug Interaction. In Proceedings of the Thirty-First AAAI Conference on Artificial Intelligence, San Francisco, CA, USA, 4–9 February 2017; pp. 1367–1373.
17. Zhang, W.; Chen, Y.; Liu, F.; Luo, F.; Tian, G.; Li, X. Predicting potential drug–drug interactions by integrating chemical, biological, phenotypic and network data. *BMC Bioinf.* **2017**, *18*, 18. [CrossRef] [PubMed]
18. Zitnik, M.; Agrawal, M.; Leskovec, J. Modeling polypharmacy side effects with graph convolutional networks. *Bioinformatics* **2018**, *34*, i457–i466. [CrossRef] [PubMed]

19. Yan, S.; Xu, D.; Zhang, B.; Zhang, H.J.; Yang, Q.; Lin, S. Graph Embedding and Extensions: A General Framework for Dimensionality Reduction. *IEEE Trans. Pattern Anal. Mach. Intell.* **2007**, *29*, 40–51. [CrossRef]

20. Cao, S.; Lu, W.; Xu, Q. Deep neural networks for learning graph representations. In Proceedings of the Thirtieth AAAI Conference on Artificial Intelligence, Phoenix, AZ, USA, 12–17 February 2016; pp. 1145–1152

21. Huang, Z.; Mamoulis, N. Heterogeneous Information Network Embedding for Meta Path based Proximity. Available online: https://arxiv.org/abs/1701.05291 (accessed on 19 Jan 2017).

22. Li, R.; Dong, Y.; Kuang, Q.; Wu, Y.; Li, Y.; Zhu, M.; Li, M. Inductive matrix completion for predicting adverse drug reactions (ADRs) integrating drug–target interactions. *Chemom. Intell. Lab. Syst.* **2015**, *144*, 71–79. [CrossRef]

23. Ma, T.; Xiao, C.; Zhou, J.; Wang, F. Drug Similarity Integration Through Attentive Multi-view Graph Auto-Encoders. Available online: https://arxiv.org/abs/1804.10850 (accessed on 28 Apr 2018).

24. Kelley, B.P.; Sharan, R.; Karp, R.M.; Sittler, T.; Root, D E.; Stockwell, B.R.; Ideker, T. Conserved pathways within bacteria and yeast as revealed by global protein network alignment. *Proc. Nat. Acad. Sci. USA* **2003**, *100*, 11394–11399. [CrossRef]

25. Yeh, C.Y.; Yeh, H.Y.; Arias, C.R.; Soo, V.W. Pathway Detection from Protein Interaction Networks and Gene Expression Data Using Color-Coding Methods and A* Search Algorithms. *Sci. World J.* **2012**, *2012*, 315797. [CrossRef]

26. Codling, E.A.; Plank, M.J.; Benhamou, S. Random walk models in biology. *J. R. Soc. Interface* **2008**, *5*, 813–834. [CrossRef] [PubMed]

27. Zou, Q.; Li, J.; Song, L.; Zeng, X.; Wang, G. Similarity computation strategies in the microRNA-disease network: A Survey. *Brief. Funct. Genom.* **2016**, *15*, 55–64. [CrossRef] [PubMed]

28. Shi, C.; Li, Y.; Zhang, J.; Sun, Y.; Yu, P.S. A Survey of Heterogeneous Information Network Analysis. *IEEE Trans. Knowl. Data. Eng.* **2016**, *29*, 17–37. [CrossRef]

29. Shakibian, H.; Charkari, N.M. Mutual information model for link prediction in heterogeneous complex networks. *Sci. Rep.* **2017**, *7*, 44981. [CrossRef] [PubMed]

30. Chang, S.; Han, W.; Tang, J.; Qi, G. J.; Aggarwal, C.C.; Huang, T.S. Heterogeneous Network Embedding via Deep Architectures. In Proceedings of the 21th ACM SIGKDD International Conference on Knowledge Discovery and Data Mining, Sydney, New South Wales, Australia, 10–13 August 2015; pp. 119–128.

31. Katz, L. A new status index derived from sociometric analysis. *Psychmetrika* **1953**, *18*, 39–43. [CrossRef]

32. Wang, D.; Cui, P.; Zhu, W. Structural Deep Network Embedding. In Proceedings of the 22nd ACM SIGKDD international conference on Knowledge discovery and data mining, San Francisco, CA, USA, 13–17 August 2016; pp. 1225–1234.

33. Kingma, D.P.; Ba, J. Adam: A Method for Stochastic Optimization. Available online: https://arxiv.org/abs/1412.6980 (accessed on 22 December 2014).

34. Kipf, T.N.; Welling, M. Semi-Supervised Classification with Graph Convolutional Networks. Availableonline: https://arxiv.org/abs/1609.02907 (accessed on 9 September 2016).

35. Kipf, T.N.; Welling, M. Variational Graph Auto-Encoders. Available online: https://arxiv.org/abs/1611.0730821 (accessed on 21 November 2016).

36. Dong, Y.; Chawla, N.V.; Swami, A. In metapath2vec: Scalable Representation Learning for Heterogeneous Networks. In Proceedings of the ACM SIGKDD International Conference on Knowledge Discovery and Data Mining, New York, NY, USA, 24–27 August 2017; pp. 135–144.

37. Maaten, L.V.D.; Hinton, G. Viualizing data using t-SNE. *J. Mach. Learn. Res.* **2008**, *9*, 2579–2605.

38. Hashizume, K.; Nabeshima, T.; Fujiwara, T.; Machida, S.; Kurosaka, D. A case of herpetic epithelial keratitis after triamcinolone acetonide subtenon injection. *Cornea* **2009**, *28*, 463–464. [CrossRef] [PubMed]

39. Suarez-Figueroa, M.; Contreras, I.; Noval, S. Side-effects of triamcinolone in young patients. *Arch. Soc. Esp. Oftalmol.* **2006**, *81*, 405–407. [PubMed]

40. Chew, E.Y.; Glassman, A.R.; Beck, R.W. Ocular side effects associated with peribulbar injections of triamcinolone acetonide for diabetic macular edema. *Retina* **2011**, *31*, 284. [CrossRef]

![molecules logo]

MDPI

Article

An Efficient Classifier for Alzheimer's Disease Genes Identification

Lei Xu [1], Guangmin Liang [1], Changrui Liao [2], Gin-Den Chen [3,*] and Chi-Chang Chang [4,5,*]

[1] School of Electronic and Communication Engineering, Shenzhen Polytechnic, Shenzhen 518055, China; csleixu@szpt.edu.cn (L.X.); gmliang@szpt.edu.cn (G.L.)
[2] Key Laboratory of Optoelectronic Devices and Systems of Ministry of Education and Guangdong Province, College of Optoelectronic Engineering, Shenzhen University, Shenzhen 518060, China; cliao@szu.edu.cn
[3] Department of Obstetrics and Gynecology, Chung Shan Medical University Hospital, Taichung 40201, Taiwan
[4] School of Medical Informatics, Chung Shan Medical University, Taichung 40201, Taiwan
[5] IT Office, Chung Shan Medical University Hospital, Taichung 40201, Taiwan
* Correspondences: gdchentw@hotmail.com (G.-D.C.); threec@csmu.edu.tw (C.-C.C.); Tel.: +86-64-2473-0022 (ext.12218) (C.-C.C.)

Academic Editor: Xiangxiang Zeng
Received: 24 October 2018; Accepted: 19 November 2018; Published: 29 November 2018

Abstract: Alzheimer's disease (AD) is considered to one of 10 key diseases leading to death in humans. AD is considered the main cause of brain degeneration, and will lead to dementia. It is beneficial for affected patients to be diagnosed with the disease at an early stage so that efforts to manage the patient can begin as soon as possible. Most existing protocols diagnose AD by way of magnetic resonance imaging (MRI). However, because the size of the images produced is large, existing techniques that employ MRI technology are expensive and time-consuming to perform. With this in mind, in the current study, AD is predicted instead by the use of a support vector machine (SVM) method based on gene-coding protein sequence information. In our proposed method, the frequency of two consecutive amino acids is used to describe the sequence information. The accuracy of the proposed method for identifying AD is 85.7%, which is demonstrated by the obtained experimental results. The experimental results also show that the sequence information of gene-coding proteins can be used to predict AD.

Keywords: Alzheimer's disease; gene coding protein; sequence information; support vector machine; classification

---

1. Introduction

Prior research has shown that there were more than 26.6 million people with AD worldwide in 2010 [1]. It has been predicted that there will soon be a further significant increase in prevalence: specifically, it is expected that there will be 70 million people with AD in 2030 and more than 115 million people with AD in 2050, respectively. In other words, in 2050, one in 85 people are expected to have AD. Unfortunately, to date, there is no treatment in existence that can cure AD. During disease progression, the neurons of AD patients are destroyed gradually, resulting in the loss of cognitive ability and ultimately death. Thus, it is important to identify AD, an age-related disease [2], as early as possible so as to manage the advancement of the condition.

Most existing diagnosis methods focus on identifying AD by way of magnetic resonance imaging (MRI). The MRI method is based on neuroimaging data, for the reason that the imaging data can reflect the structure of brain. Using this technique, the results of classification accuracy are encouraging. However, MRI scans are expensive and the time required for scanning is significant because of the large size of the images. A diffusion map is extended to identify AD in Mattsson [3] and principal component analysis (PCA) is used to reduce features before classification.

Many biomarkers have been discovered for AD identification, such as structural MRI for brain atrophy measurement [4–6] functional imaging for hypometabolism quantification [7–9], and cerebrospinal fluid for the quantification of specific proteins [6,10,11]. Multimodel data have been employed by multiple biomarkers for identifying AD. Zu et al. [12] predicted AD by using multimodality data to mine the hidden information between features. In Zu [12], the subjects with the same label on a different modal are closer in the selected feature space; as such, a multikernel support vector machine (SVM) can be used to classify the multimodal data, which are represented by the selected features.

As is known, machine learning methods can learn a model from a training sample and then subsequently predict the label of the testing samples. Some machine learning methods have been used to predict AD and mild cognitive impairment (MCI) [13–20]. The information obtained via structural MRI—for example, hippocampal volumes [21,22], cortical thickness [23,24], voxel-wise tissue [23,25,26], and so on—is extracted to classify AD and MCI. Functional imaging, such as fluorodeoxyglucose positron-emission tomography [14,27,28] can also be used for AD and MCI prediction.

Although most existing research has focused on classifying AD based on MRI methods, the cost is expensive. Furthermore, patients often have to have their brain scanned several times in order to inspect the changes in its structure during whole process, increasing the cost even more. Thus, it would be beneficial to find other options for AD identification. Several researches proved that coding genes/noncoding RNAs/proteins were related to diseases, including AD [29–36]. Other investigations [12] have shown that protein structure is related to AD. The gene coding is related to Alzheimer's disease [37–39]. Different from previous work, in the present study, AD is predicted based on protein information. The information of every sequence is represented by a 400-dimension vector, and each dimension represents the frequency of two consecutive amino acids.

The flow chart of AD identification is shown in Figure 1. First, the data are selected by using the CD-HIT method to remove the most similar sequences. In this step, the input are the proteins related with AD, and the output are selected proteins. Second, the features are extracted from the selected sequences. Each sequence is represented by a 400-dimension (400D) vector. In the third step, the data are classified by a support vector machine method. The input are the feature vectors, and the output are peptides with labels. To the best of our knowledge, this study represents the first effort to identify AD by protein sequence information without the use of MRI. Moreover, a dataset including AD and non-AD samples was created in this work. The experimental results show that the classification accuracy for AD prediction is 85.7%. The contributions of our work include:

(1) A method for predicting AD is proposed in this work. The experimental results demonstrate that the classification accuracy of the proposed method is 85.7%.
(2) Our method is based on protein sequence information. The frequencies of two consecutive amino acids are extracted from the sequence with a 400-dimension vector.
(3) A dataset with AD and non-AD samples is created. This dataset could also be used for additional AD prediction studies.

The rest of the paper is organized as follows: Section 2 introduces the experimental results of the proposed method. The dataset and the proposed method are introduced in Section 3. Finally, the conclusion is made in Section 4.

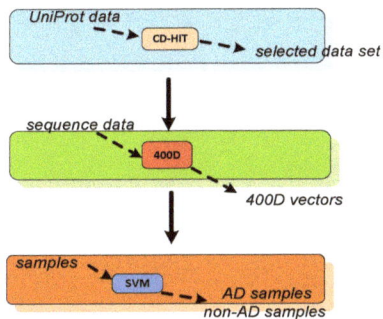

Figure 1. The flow chart of AD identification.

2. Results and Discussion

2.1. Results

Identifying AD by way of using protein sequence information has not been widely done yet. Moreover, most existing works use AD Neuroimaging Initiative (ADNI) database [40], which is based on MRI. Existing methods also use MRI information for classification, which is different from our method. Thus, it is difficult to compare the performance evaluation of our proposed method with the performance of existing methods. The performance of our method is shown in Table 1.

Table 1. The performance of our proposed method.

Performance Evaluation	Accuracy
ACC	0.8565
Precision	0.857
Recall	0.857
F-measure	0.856
MCC	0.714
AUC	0.857

As noted in the table, the method was evaluated according to accuracy, precision, recall, F-measure, Mathew coefficient (MCC), and receiver operating characteristic (ROC). The accuracy of the proposed method was 85.7%, which means that the more than 85% of AD and non-AD samples were able to be classified correctly using the method in question. F-measure is based on precision and recall. The recall of our method was 0.857, and the result shows that 85.7% of AD samples in the dataset could be identified in the experiment. Area under the curve (AUC) is related to the metrics of receiver operating characteristic (ROC). ROC is used to measure sensitivity and specificity, while AUC describes the area under the ROC curve. When the AUC is larger, the performance of the algorithm is better. The value of AUC for our method was 0.857 according to the UniProt dataset [41]. The experimental results show that the performance quality of our method in terms of accuracy, precision, and four other metrics as well as the results obtained are acceptable and encouraging.

2.2. The Comparison of Performance Evaluation on Feature Selection Methods

To demonstrate the efficiency of the feature extraction method we used, we compared the 400D features with information theory, which is another feature extraction method. Information theory is proposed in Wei [42], for exploring sequential information from multiple perspectives. Figure 2 shows that 400D performs better than information theory method on accuracy, precision, F-measure, AUC and MCC. The value of recall is higher by using information theory method than using 400D.

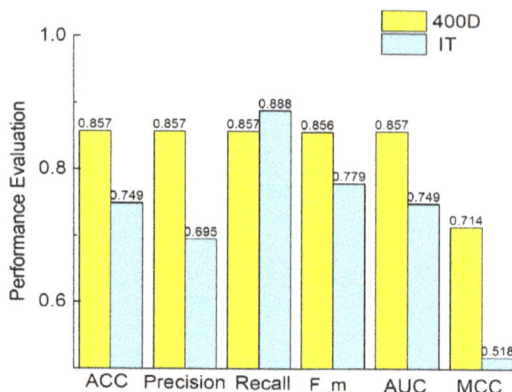

Figure 2. Comparison of 400D with information theory on SVM.

2.3. The Comparison of Performance Evaluation on Existing Classification Methods

Our method's performance is evaluated according to other classifiers, such as random forest, naïve Bayes, LibD3C, Adaptive Boosting (AdaBoost), and Bayes network. The classifiers are introduced briefly as follows:

- Random forest is an ensemble classifier, which learns more than one decision tree together. The decision will be made by voting process.
- Naïve Bayes assumes the features are independent of one other. The samples will be assigned to a class with the maximum posterior probability.
- LibD3C [43] is a hybrid ensemble model, which is based on k-means clustering and the framework of dynamic selection and circulating in combination with a sequential search method.
- AdaBoost can assemble classifiers together and, during the training process, the weights of the samples which are classified incorrectly will be increased. The weights of the samples classified correctly will be decreased.
- Bayes network is a probabilistic graph model. The variables and their relationships are represented by a directed acyclic graph.

Figure 3 shows the comparison of accuracy according to the six classifiers. The comparisons of precision, recall, F-measure, MCC, and AUC are shown in Figures 3–7. In Figure 3, we can see that accuracy performs better than the other classifiers. The value of accuracy of AdaBoost, Bayes network, and naïve Bayes is about 0.8, while the accuracy of SVM is 0.857. The accuracy of LibD3C is 0.84. The accuracy of random forest is 0.85, which is comparative with that of SVM. Thus, SVM improves the accuracy of other classifiers by nearly 1% to 7%.

Figures 4–7 show the comparisons of the classifiers on precision, recall, and F-measure. The results are similar to those of Figure 3. SVM performs better than the other methods. The performance is improved by SVM by approximately 1% to 7.5% as compared with in the case of the other methods. F-measure is calculated based on precision and recall, so the result here is consistent with that of precision and recall. AUC reflects the area under the ROC curve. AUC refers to the ratio of the specificity and sensitivity. The value of AUC on random forest is 0.93, which is better than the values achieved via other methods. The values of AUC for AdaBoost, Bayes network, SVM, and naïve Bayes are similar to one another. Figure 8 shows that the MCC of SVM is 0.714, which is better than the MCCs of the other mentioned methods. The values of MCC for random forest and SVM reach a level of 0.7. Moreover, the value of MCC is improved by 0.8% to 20% by using SVM. As a result, SVM performs better than other classifiers evaluated by the metrics.

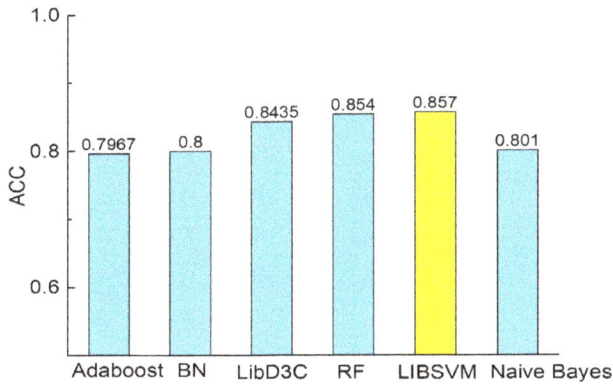

Figure 3. Comparison of ACC on different classifiers.

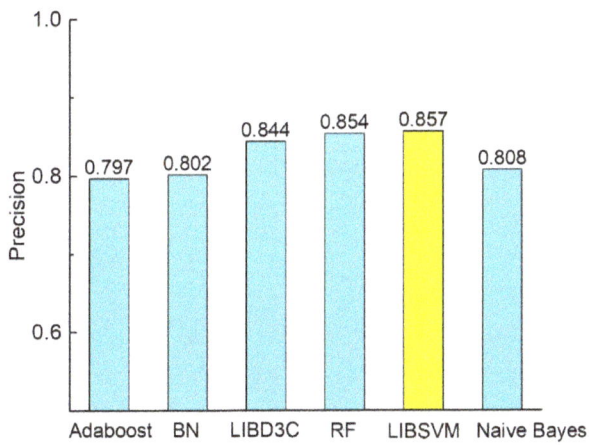

Figure 4. Comparison of precision on different classifiers.

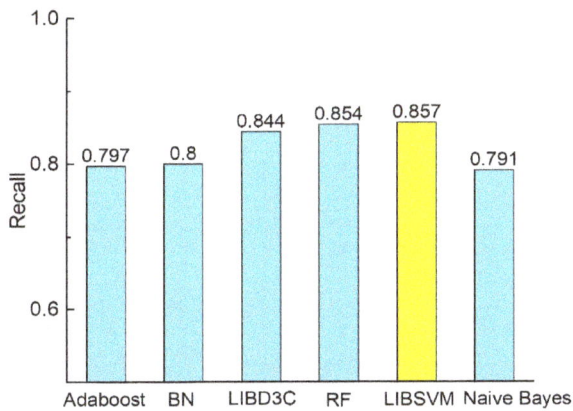

Figure 5. Comparison of recall on different classifiers.

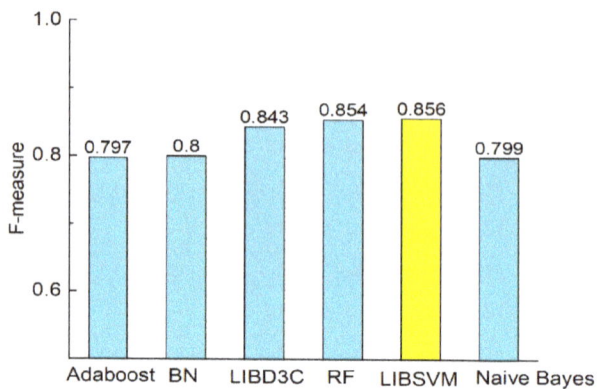

Figure 6. Comparison of F-measure on different classifiers.

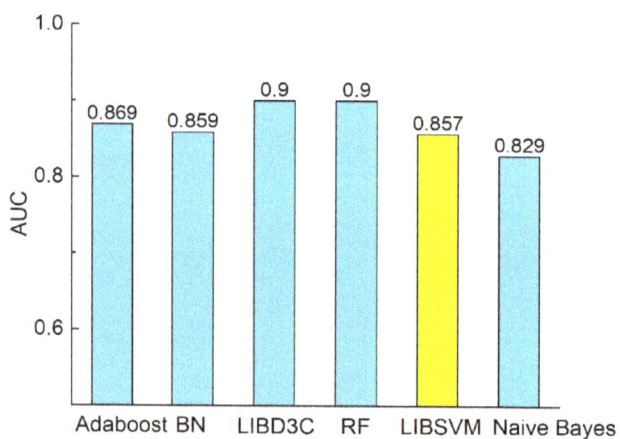

Figure 7. Comparison of AUC on different classifiers.

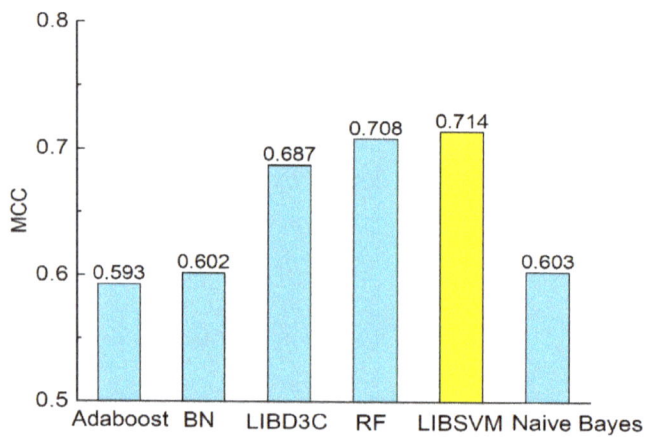

Figure 8. Comparison of MCC on different classifiers.

3. Materials and Methods

3.1. Benchmark Dataset

The data were selected from the UniProt database [41,44]. To guarantee the validity of the dataset, the proteins with ambiguous meanings (such as "B", "X", and so on) is removed, and only the proteins related to "Alzheimer's disease" are kept.

The benchmark dataset (D) is represented by a positive subset (D^+) and a negative subset (D^-), formulated as seen in Equation (1):

$$D = D^+ \cup D^- \tag{1}$$

where the symbol "$\cup$" represents the union of the sets in the set theory. After the selection process, there are 310 proteins related to AD and 312 non-AD proteins left in the benchmark dataset. Because some sequences are significantly similar, the redundancy of the sequences is considered. To avoid the overestimation of the performance of the methods, the homologous sequences with more than 60% similarity were removed from the dataset by using CD-HIT program [45]. As a result, a benchmark dataset with 279 proteins related to AD and 1,463 proteins not related to AD was used for the prediction model. In other words, the benchmark dataset contains 279 positive samples in the positive subset (D^+) and 1,463 negative samples in the negative subset (D^-), respectively.

3.2. Support Vector Machine

SVM is a supervised machine learning model. The labeled samples are trained based on the goal of maximizing the margin between the classes. Since SVM performs better than some the-state-of-art supervised learning methods, SVM is widely used in classification problems. Most works in bioinformatics [41,46–59] also use SVM for classification. SVM was used to identify AD in our work.

The principles of SVM were introduced in Chou and Cai [60,61], and more details are provided in Cristianini [62]. Above all, the key idea of SVM is that two groups are separated with a maximum margin by building a hyperplane. The objective function of SVM is described in Equation (2), as follows:

$$\underset{w,b}{argmax}\{\frac{1}{\|w\|}\underset{i=1,2,\dots,n}{\min}[y_i(w^T\varphi(x^{(i)})+b)]\} \tag{2}$$

In Equation (2), the input variable $x^{(i)}$ is mapped into a high dimensional feature space by the kernel function $\varphi(\cdot)$. Radial kernel function (RBF) is used in the experiment. RBF is used widely because of its effectiveness and efficiency. Equation (2) can be transferred to optimize Equation (3), as follows:

$$\max\frac{1}{\|w\|}, \; s.t. \; y_i(w^T\varphi(x^i)+b) \geq 1, i = 1,\dots,n \tag{3}$$

where n is the number of training samples. The condition $(y_i(w^T\varphi(x^i)+b) \geq 1)$ should be satisfied in Equation (3), which means that the samples must be classified correctly by the optimized hyperplane. However, the problem of overfitting will be caused. Soft SVM is proposed to tackle the problem. The objective function is refined into Equation (4), as follows:

$$\underset{w,b}{\min}(\frac{1}{2}\|w\|^2 + C\sum_{i=1}^{n}\delta_i)$$
$$s.t. \; y_i(w^T\varphi(x^i)+b) \geq 1-\delta_i, i=1,\dots,n \tag{4}$$
$$\delta_i \geq 0$$

where δ_i is the slack variable and C is the penalty parameter. The SVM used in our work is the package named LIBSVM written by Chang and Lin [63].

3.3. Sequence Representation

AD is classified based on protein sequence information, so, in this paper, we used the features extracted from the peptides. The sequence is represented by a 400-dimension vector, and each dimension describes the frequency of two consecutive amino acids. The feature extraction will be introduced later. To describe the information more clearly, the symbols used in the paper are summarized in Table 2.

Table 2. The symbols used in the present paper.

Symbol	Meaning
P_L	Peptide with L residual
R_i	The i-th residual
f_i	The frequency of the i-th amino acid
F_P	The feature vector of peptide P

P_L is a peptide with L residue, so P_L can be written into a sequence as $\{R_1 R_2 R_3 \ldots R_i \ldots R_L$. R_i represents the i-th residual of P_L in the sequence. The symbol f_i represents the normalized occurrence frequency of the i-th type of native amino acid in the peptide. There are, in total, 20 types of native amino acids. The peptide P can be represented by $F_P = [f_1, \ldots, f_i, \ldots, f_{20}]$, reflecting the occurrence frequency of every amino acid of P. It is obvious that the sequence information is lost in F_P. To overcome this limitation, we extracted the occurrence frequency of the combination of two consecutive amino acids, such as AR (A and R representing the amino acids). Since there are 20 native amino acids, the number of features of the combination of two consecutive amino acids is 400 (20^2). Thus, we call it a 400D sequence-based feature. The peptide P is straightly represented by $(f_{AA}, f_{AR}, \ldots, f_{VV})$.

3.4. Performance Evaluation

The classification quality is evaluated by accuracy, recall, precision, F-measure, MCC, and AUC. The metrics are used in evaluating the performance frequently [64–72]. In the experiments, n is the number of samples, so n^+ is the number of positive samples and n^- is the number of negative samples. TP (true positive) represents the number of samples that are labeled positive by the method correctly. FP (false positive) is the number of samples that are labeled positive but which are in fact negative. TN (true negative) means the number of sample which are classified correctly as negative sample. FN (false negative) is the number of samples that are positive but which are labeled as negative. The accuracy (ACC_G) represents the correct classification rate of a method G, which is shown in Equation (5). $Precision_G$, $recall_G$ and $F\text{-}measure_G$ are calculated in Equations (5) through (8). AUC is the area size of the ROC curve. The X-axis of ROC curve is the false positive rate, while the Y-axis is true positive rate. The MCC describes the rate of specificity and sensitivity, which is calculated by Equation (9). Specificity and sensitivity are used in evaluating the performance of protein prediction, such as in the case of Feng [47,48] and so on. Specificity (Sp, calculated by Equation (10)) is the rate of misclassification of AD proteins. Sensitivity (Sn, calculated by Equation (11)) is the rate of correctly classified AD proteins:

$$ACC_G = \frac{TP + TN}{n^+ + n^-} \tag{5}$$

$$Precision_G = \frac{TP}{TP + FP} \tag{6}$$

$$Recall_G = \frac{TP}{TP + FN} \tag{7}$$

$$F - measure_G = \frac{(1 + b^2) \times P \times R}{b^2 \times P + R} \tag{8}$$

$$MCC = \frac{Sp}{Sn} \tag{9}$$

$$Sp = \frac{TN}{TN + FP} \qquad (10)$$

$$Sn = \frac{TP}{TP + FN} \qquad (11)$$

4. Conclusions

In this paper, a computational method based on protein sequence information was introduced to predict the onset of AD. In our proposed method, the sequences are represented by the frequency of two consecutive amino acids, and then the data are classified by SVM. Our work is different from previous work that was completed using MRI, which is time-consuming and expensive. As demonstrated by the presented experimental results, the classification accuracy of our proposed method is 85.7%. Moreover, a dataset used for AD classification was created in our work. In future work, we will try to mine the relationships between the features to improve the classification performance of the predictions method. Furthermore, due to the wide use of webservers in bioinformatics, such as the work of RNA secondary structure comparison [73], we will also develop the a webserver for AD prediction.

Author Contributions: L.X. initially drafted the manuscript and did most of the codes work and the experiments. C.L. collected the features and analyzed the experiments. G.L., G.-D.C. and C.-C.C. revised to draft the manuscript. All authors read and approved the final manuscript.

Funding: This research was funded by the Natural Science Foundation of Guangdong Province (grant no. 2018A0303130084), the Science and Technology Innovation Commission of Shenzhen (grant nos. JCYJ20160523113602609, JCYJ20170818100431895), the Grant of Shenzhen Polytechnic (grant no. 601822K19011) and National Nature Science Foundation of China (grant no. 61575128), Chung Shan Medical University Hospital (grant no. CSH-2018-D-002), and Research projects of Shenzhen Institute of Information Technology (No. ZY201714).

Acknowledgments: We thank the reviewers for their great comments.

Conflicts of Interest: The authors declare no conflict of interest.

References

1. Brookmeyer, R.; Johnson, E.; Zieglergraham, K.; Arrighi, H.M. Forecasting the global burden of alzheimer's disease. *Alzheimers Dement.* **2007**, *3*, 186–191. [CrossRef] [PubMed]
2. Yang, J.; Huang, T.; Petralia, F.; Long, Q.; Zhang, B.; Argmann, C.; Zhao, Y.; Mobbs, C.V.; Schadt, E.E.; Zhu, J.; et al. Synchronized age-related gene expression changes across multiple tissues in human and the link to complex diseases. *Sci. Rep.* **2015**, *5*, 15145. [CrossRef] [PubMed]
3. Mattsson, N. Csf biomarkers and incipient alzheimer disease in patients with mild cognitive impairment. *JAMA* **2009**, *302*, 385. [CrossRef] [PubMed]
4. McEvoy, L.K.; Fennema-Notestine, C.; Roddey, J.C.; Hagler, D.J.; Holland, D.; Karow, D.S.; Pung, C.J.; Brewer, J.B.; Dale, A.M. Alzheimer disease: Quantitative structural neuroimaging for detection and prediction of clinical and structural changes in mild cognitive impairment. *Radiology* **2009**, *251*, 195–205. [CrossRef] [PubMed]
5. Du, A.T.; Schuff, N.; Kramer, J.H.; Rosen, H.J.; Gorno-Tempini, M.L.; Rankin, K.; Miller, B.L.; Weiner, M.W. Different regional patterns of cortical thinning in Alzheimer's disease and frontotemporal dementia. *Brain* **2007**, *130*, 1159–1166. [CrossRef] [PubMed]
6. Fjell, A.M.; Walhovd, K.B.; Fennema-Notestine, C.; McEvoy, L.K.; Hagler, D.J.; Holland, D.; Brewer, J.B.; Dale, A.M. Csf biomarkers in prediction of cerebral and clinical change in mild cognitive impairment and alzheimer's disease. *J. Neurosci.* **2010**, *30*, 2088–2101. [CrossRef] [PubMed]
7. de Leon, M.J.; Mosconi, L.; Li, J.; De Santi, S.; Yao, Y.; Tsui, W.H.; Pirraglia, E.; Rich, K.; Javier, E.; Brys, M.; et al. Longitudinal CSF isoprostane and MRI atrophy in the progression to AD. *J. Neurol.* **2007**, *254*, 1666–1675. [CrossRef] [PubMed]
8. Morris, J.C.; Storandt, M.; Miller, J.P.; McKeel, D.W.; Price, J.L.; Rubin, E.H.; Berg, L. Mild cognitive impairment represents early-stage Alzheimer disease. *Arch. Neurol.* **2001**, *58*, 397–405. [CrossRef] [PubMed]

9.   De, S.S.; de Leon, M.J.; Rusinek, H.; Convit, A.; Tarshish, C.Y.; Roche, A.; Tsui, W.H.; Kandil, E.; Boppana, M.; Daisley, K.; Wang, G.J.; et al. Hippocampal formation glucose metabolism and volume losses in MCI and AD. *Neurobiol. Aging* **2001**, *22*, 529–539.

10. Bouwman, F.H.; van der Flier, W.M.; Schoonenboom, N.S.; van Elk, E.J.; Kok, A.; Rijmen, F.; Blankenstein, M.A.; Scheltens, P. Longitudinal changes of CSF biomarkers in memory clinic patients. *Neurology* **2007**, *69*, 1006–1011. [CrossRef] [PubMed]

11. Shaw, L.M.; Vanderstichele, H.; Knapik-Czajka, M.; Clark, C.M.; Aisen, P.S.; Petersen, R.C.; Blennow, K.; Soares, H.; Simon, A.; Lewczuk, P.; Dean, R.; Siemers, E.; Potter, W.; Lee, V.M.-Y.; Trojanowski, J.Q. Cerebrospinal fluid biomarker signature in alzheimer's disease neuroimaging initiative subjects. *Ann. Neurol.* **2009**, *65*, 403–413. [CrossRef] [PubMed]

12. Zu, C.; Jie, B.; Liu, M.; Chen, S.; Shen, D.; Zhang, D. Label-aligned multi-task feature learning for multimodal classification of alzheimer's disease and mild cognitive impairment. *Brain Imaging Behav.* **2015**, *10*, 1148–1159. [CrossRef] [PubMed]

13. Xu, H.J.; Hu, S.X.; Cagle, P.T.; Moore, G.E.; Benedict, W.F. Absence of retinoblastoma protein expression in primary non-small cell lung carcinomas. *Cancer Res.* **1991**, *8*, 2735–2739.

14. Foster, N.L.; Heidebrink, J.L.; Clark, C.M.; Jagust, W.J.; Arnold, S.E.; Barbas, N.R.; DeCarli, C.S.; Turner, R.S.; Koeppe, R.A.; Higdon, R.; Minoshima, S. FDG-PET improves accuracy in distinguishing frontotemporal dementia and Alzheimer's disease. *Brain* **2007**, *130*, 2616–2635. [CrossRef] [PubMed]

15. Dai, Z.; Yan, C.; Wang, Z.; Wang, J.; Xia, M.; Li, K.; He, Y. Discriminative analysis of early alzheimer's disease using multi-modal imaging and multi-level characterization with multi-classifier (m3). *NeuroImage* **2012**, *59*, 2187–2195. [CrossRef] [PubMed]

16. Huang, S.; Li, J.; Ye, J.; Wu, T.; Chen, K.; Fleisher, A.; Reiman, E. Identifying Alzheimer's Disease-Related Brain Regions from Multi-Modality Neuroimaging Data using Sparse Composite Linear Discrimination Analysis. *Adv. Neural Inf. Process. Syst.* **2011**, 1431–1439.

17. Westman, E.; Muehlboeck, J.-S.; Simmons, A. Combining mri and csf measures for classification of alzheimer's disease and prediction of mild cognitive impairment conversion. *NeuroImage* **2012**, *62*, 229–238. [CrossRef] [PubMed]

18. Liu, F.; Shen, C. Learning Deep Convolutional Features for MRI Based Alzheimer's Disease Classification. *arXiv*, 2014; arXiv:1404.3366.

19. Herrera, L.J.; Rojas, I.; Pomares, H.; Guillén, A.; Valenzuela, O.; Baños, O. Classification of MRI Images for Alzheimer's Disease Detection. In Proceedings of the 2013 International Conference on Social Computing, Alexandria, VA, USA, 8–14 September 2013.

20. Liu, X.; Tosun, D.; Weiner, M.W.; Schuff, N. Locally linear embedding (LLE) for MRI based Alzheimer's disease classification. *Neuroimage* **2013**, *83*, 148–157. [CrossRef] [PubMed]

21. Gerardin, E.; Chételat, G.; Chupin, M.; Cuingnet, R.; Desgranges, B.; Kim, H.-S.; Niethammer, M.; Dubois, B.; Lehéricy, S.; Garnero, L.; Eustache, F.; Colliot, O. Multidimensional classification of hippocampal shape features discriminates alzheimer's disease and mild cognitive impairment from normal aging. *NeuroImage* **2009**, *47*, 1476–1486. [CrossRef] [PubMed]

22. West, M.J.; Kawas, C.H.; Stewart, W.F.; Rudow, G.L.; Troncoso, J.C. Hippocampal neurons in pre-clinical Alzheimer's disease. *Neurobiol. Aging* **2004**, *25*, 1205–1212. [CrossRef] [PubMed]

23. Desikan, R.S.; Cabral, H.J.; Hess, C.P.; Dillon, W.P.; Glastonbury, C.M.; Weiner, M.W.; Schmansky, N.J.; Greve, D.N.; Salat, D.H.; Buckner, R.L.; Fischl, B. Automated MRI measures identify individuals with mild cognitive impairment and Alzheimer's disease. *Brain* **2009**, *132*, 2048–2057. [CrossRef] [PubMed]

24. Oliveira, P.P., Jr.; Nitrini, R.; Busatto, G.; Buchpiguel, C.; Sato, J.R.; Amaro, E., Jr. Use of SVM methods with surface-based cortical and volumetric subcortical measurements to detect Alzheimer's disease. *J. Alzheimers Dis.* **2010**, *19*, 1263–1272. [CrossRef] [PubMed]

25. Fan, Y.; Shen, D.; Gur, R.C.; Gur, R.E.; Davatzikos, C. COMPARE: Classification of morphological patterns using adaptive regional elements. *IEEE Trans. Med. Imaging* **2007**, *26*, 93–105. [CrossRef] [PubMed]

26. Magnin, B.; Mesrob, L.; Kinkingnéhun, S.; Pélégrini-Issac, M.; Colliot, O.; Sarazin, M.; Dubois, B.; Lehéricy, S.; Benali, H. Support vector machine-based classification of alzheimer's disease from whole-brain anatomical mri. *Neuroradiology* **2008**, *51*, 73–83. [CrossRef] [PubMed]

27. Chetelat, G.; Desgranges, B.; de la Sayette, V.; Viader, F.; Eustache, F.; Baron, J.-C. Mild cognitive impairment: Can FDG-PET predict who is to rapidly convert to Alzheimer's disease? *Neurology* **2003**, *60*, 1374–1377. [CrossRef] [PubMed]

28. Higdon, R.; Foster, N.L.; Koeppe, R.A.; DeCarli, C.S.; Jagust, W.J.; Clark, C.M.; Barbas, N.R.; Arnold, S.E.; Turner, R.S.; Heidebrink, J.L.; Minoshima, S. A comparison of classification methods for differentiating fronto-temporal dementia from alzheimer's disease using fdg-pet imaging. *Stat. Med.* **2004**, *23*, 315–326. [CrossRef] [PubMed]

29. Zeng, X.; Liao, Y.; Liu, Y.; Zou, Q. Prediction and Validation of Disease Genes Using HeteSim Scores. *IEEE/ACM Trans. Comput. Biol. Bioinform.* **2017**, *14*, 687–695. [CrossRef] [PubMed]

30. Liu, Y.; Zeng, X.; He, Z.; Zou, Q. Inferring microrna-disease associations by random walk on a heterogeneous network with multiple data sources. *IEEE/ACM Trans. Comput. Biol. Bioinform.* **2017**, *14*, 905–915. [CrossRef] [PubMed]

31. Zeng, X.; Liu, L.; Lü, L.; Zou, Q. Prediction of potential disease-associated microRNAs using structural perturbation method. *Bioinformatics* **2018**, *34*, 2425–2432. [CrossRef] [PubMed]

32. Wang, L.; Ping, P.; Kuang, L.; Ye, S.; Pei, T. A Novel Approach Based on Bipartite Network to Predict Human Microbe-Disease Associations. *Curr. Bioinform.* **2018**, *13*, 141–148. [CrossRef]

33. Liao, Z.J.; Li, D.; Wang, X.; Li, L.; Zou, Q. Cancer Diagnosis Through IsomiR Expression with Machine Learning Method. *Curr. Bioinform.* **2018**, *13*, 57–63. [CrossRef]

34. Yang, J.; Huang, T.; Song, W.M.; Petralia, F.; Mobbs, C.V.; Zhang, B.; Zhao, Y.; Schadt, E.E.; Zhu, Y.; Tu, Z. Discover the network mechanisms underlying the connections between aging and age-related diseases. *Sci. Rep.* **2016**, *6*. [CrossRef] [PubMed]

35. Xiao, X.; Zhu, W.; Liao, B.; Xu, J.; Gu, C.; Ji, B.; Yao, Y.; Peng, L.; Yang, J. BPLLDA: Predicting lncRNA-Disease Associations Based on Simple Paths with Limited Lengths in a Heterogeneous Network. *Front. Genet.* **2018**, *9*. [CrossRef] [PubMed]

36. Lu, M.; Xu, X.; Xi, B.; Dai, Q.; Li, C.; Su, L.; Zhou, X.; Tang, M.; Yao, Y.; Yang, J. Molecular Network-Based Identification of Competing Endogenous RNAs in Thyroid Carcinoma. *Genes* **2018**, *9*, 44. [CrossRef] [PubMed]

37. Liu, G.; Wang, T.; Tian, R.; Hu, Y.; Han, Z.; Wang, P.; Zhou, W.; Ren, P.; Zong, J.; Jin, S.; Jiang, Q. Alzheimer's Disease Risk Variant rs2373115 Regulates GAB2 and NARS2 Expression in Human Brain Tissues. *J. Mol. Neurosci.* **2018**, *66*, 37–43. [CrossRef] [PubMed]

38. Jiang, Q.; Jin, S.; Jiang, Y.; Liao, M.; Feng, R.; Zhang, L.; Liu, G.; Hao, J. Alzheimer's disease variants with the genome-wide significance are significantly enriched in immune pathways and active in immune cells. *Mol. Neurobiol.* **2016**, *54*, 594–600. [CrossRef] [PubMed]

39. Liu, G.; Xu, Y.; Jiang, Y.; Zhang, L.; Feng, R.; Jiang, Q. Picalm rs3851179 variant confers susceptibility to alzheimer's disease in chinese population. *Mol. Neurobiol.* **2017**, *54*, 3131–3136. [CrossRef] [PubMed]

40. Wei, L.; Luan, S.; Nagai, L.A.E.; Su, R.; Zou, Q. Exploring sequence-based features for the improved prediction of DNA N4-methylcytosine sites in multiple species. *Bioinformatics* **2018**. [CrossRef] [PubMed]

41. Guo, S.-H.; Deng, E.-Z.; Xu, L.-Q.; Ding, H.; Lin, H.; Chen, W.; Chou, K.-C. Inuc-pseknc: A sequence-based predictor for predicting nucleosome positioning in genomes with pseudo k-tuple nucleotide composition. *Bioinformatics* **2014**, *30*, 1522–1529. [CrossRef] [PubMed]

42. Wei, L.; Xing, P.; Tang, J.; Zou, Q. Phospred-rf: A novel sequence-based predictor for phosphorylation sites using sequential information only. *IEEE Trans. NanoBiosci.* **2017**, *16*, 240–247. [CrossRef] [PubMed]

43. Lin, C.; Chen, W.; Qiu, C.; Wu, Y.; Krishnan, S.; Zou, Q. Libd3c: Ensemble classifiers with a clustering and dynamic selection strategy. *Neurocomputing* **2014**, *123*, 424–435. [CrossRef]

44. Available online: https://www.uniprot.org (accessed on 1 January 2007).

45. Fu, L.; Niu, B.; Zhu, Z.; Wu, S.; Li, W. Cd-hit: Accelerated for clustering the next-generation sequencing data. *Bioinformatics* **2012**, *28*, 3150–3152. [CrossRef] [PubMed]

46. Liu, B.; Fang, L.; Liu, F.; Wang, X.; Chen, J.; Chou, K.C. Identification of Real MicroRNA Precursors with a Pseudo Structure Status Composition Approach. *PLoS ONE* **2015**, *10*. [CrossRef] [PubMed]

47. Feng, P.; Chen, W.; Lin, H. Identifying antioxidant proteins by using optimal dipeptide compositions. *Interdiscip. Sci. Comput. Life Sci.* **2015**, *8*, 186–191. [CrossRef] [PubMed]

48. Feng, P.M.; Lin, H.; Chen, W. Identification of Antioxidants from Sequence Information Using Naïve Bayes. *Comput. Math. Methods Med.* **2013**, *2013*. [CrossRef] [PubMed]
49. Wei, L.; Xing, P.; Shi, G.; Ji, Z.L.; Zou, Q. Fast prediction of methylation sites using sequence-based feature selection technique. *IEEE/ACM Tran. Comput. Biol. Bioinform.* **2017**. [CrossRef] [PubMed]
50. Zhang, N.; Sa, Y.; Guo, Y.; Wang, L.; Wang, P.; Feng, Y. Discriminating ramos and jurkat cells with image textures from diffraction imaging flow cytometry based on a support vector machine. *Curr. Bioinform.* **2018**, *13*, 50–56. [CrossRef]
51. Chen, W.; Tang, H.; Ye, J.; Lin, H.; Chou, K.C. iRNA-PseU: Identifying RNA pseudouridine sites. *Mol. Ther. Nucleic Acids* **2016**, *5*. [CrossRef]
52. Lin, H.; Deng, E.-Z.; Ding, H.; Chen, W.; Chou, K.-C. Ipro54-pseknc: A sequence-based predictor for identifying sigma-54 promoters in prokaryote with pseudo k-tuple nucleotide composition. *Nucleic Acids Res.* **2014**, *42*, 12961–12972. [CrossRef] [PubMed]
53. Tseng, C.J.; Lu, C.J.; Chang, C.C.; Chen, G.D. Application of machine learning to predict the recurrence-proneness for cervical cancer. *Neural Comput. Appl.* **2014**, *24*, 1311–1316. [CrossRef]
54. Tseng, C.J.; Lu, C.J.; Chang, C.C.; Chen, G.D.; Cheewakriangkrai, C. Integration of data mining classification techniques and ensemble learning to identify risk factors and diagnose ovarian cancer recurrence. *Artif. Intell. Med.* **2017**, *78*, 47–54. [CrossRef] [PubMed]
55. Cheng, C.S.; Shueng, P.W.; Chang, C.C.; Kuo, C.W. Adapting an Evidence-based Diagnostic Model for Predicting Recurrence Risk Factors of Oral Cancer. *J. Univers. Comput. Sci.* **2018**, *24*, 742–752.
56. Zou, Q.C.; Chen, C.W.; Chang, H.C.; Chu, Y.W. Identifying Cleavage Sites of Gelatinases A and B by Integrating Feature Computing Models. *J. Univers. Comput. Sci.* **2018**, *24*, 711–724.
57. Ye, L.L.; Lee, T.S.; Chi, R. Hybrid Machine Learning Scheme to Analyze the Risk Factors of Breast Cancer Outcome in Patients with Diabetes Mellitus. *J. Univers. Comput. Sci.* **2018**, *24*, 665–681.
58. Das, A.K.; Pati, S.K.; Huang, H.H.; Chen, C.K. Cancer Classification by Gene Subset Selection from Microarray Dataset. *J. Univers. Comput. Sci.* **2018**, *24*, 682–710.
59. Xu, L.; Liang, G.; Wang, L.; Liao, C. A Novel Hybrid Sequence-Based Model for Identifying Anticancer Peptides. *Genes* **2018**, *9*, 158. [CrossRef] [PubMed]
60. Chou, K.-C. Using functional domain composition and support vector machines for prediction of protein subcellular location. *J. Biol. Chem.* **2002**, *277*, 45765–45769. [CrossRef] [PubMed]
61. Cai, Y.-D.; Zhou, G.-P.; Chou, K.-C. Support vector machines for predicting membrane protein types by using functional domain composition. *Biophys. J.* **2003**, *84*, 3257–3263. [CrossRef]
62. Cristianini, N.; Shawe-Taylor, J. *An Introduction to Support Vector Machines: And Other Kernel-Based Learning Methods*; Cambridge University Press: Cambridge, UK, 2000; pp. 1–28.
63. Chang, C.C.; Lin, C.J. LIBSVM: A library for support vector machines. *ACM TIST* **2011**, *2*, 1–27. [CrossRef]
64. Chou, K.-C. Some remarks on protein attribute prediction and pseudo amino acid composition. *J. Theor. Biol.* **2011**, *273*, 236–247. [CrossRef] [PubMed]
65. Chou, K.C. Using subsite coupling to predict signal peptides. *Protein Eng.* **2001**, *14*, 75–79. [CrossRef] [PubMed]
66. Lai, H.Y.; Chen, X.X.; Chen, W.; Tang, H.; Lin, H. Sequence-based predictive modeling to identify cancerlectins. *Oncotarget* **2017**, *8*, 28169–28175. [CrossRef] [PubMed]
67. Su, R.; Wu, H.; Xu, B.; Liu, X.; Wei, L. Developing a Multi-Dose Computational Model for Drug-induced Hepatotoxicity Prediction based on Toxicogenomics Data. *IEEE/ACM Trans. Comput. Biol. Bioinform.* **2018**. [CrossRef] [PubMed]
68. Wei, L.; Xing, P.; Zeng, J.; Chen, J.; Su, R.; Guo, F. Improved prediction of protein–protein interactions using novel negative samples, features, and an ensemble classifier. *Artif. Intell. Med.* **2017**, *83*, 67–74. [CrossRef] [PubMed]
69. Wei, L.; Ding, Y.; Su, R.; Tang, J.; Zou, Q. Prediction of human protein subcellular localization using deep learning. *J. Parallel Distrib. Comput.* **2018**, *117*, 212–217. [CrossRef]
70. Wei, L.; Chen, H.; Su, R. M6APred-EL: A Sequence-Based Predictor for Identifying N6-methyladenosine Sites Using Ensemble Learning. *Mol. Ther. Nucleic Acids* **2018**, *12*, 635–644. [CrossRef] [PubMed]
71. Liu, X.; Yang, J.; Zhang, Y.; Fang, Y.; Wang, F.; Wang, J.; Zheng, X.; Yang, J. A systematic study on drug-response associated genes using baseline gene expressions of the cancer cell line encyclopedia. *Sci. Rep.* **2016**, *6*, 22811. [CrossRef] [PubMed]

72. Xu, L.; Liang, G.; Shi, S.; Liao, C. SeqSVM: A Sequence-Based Support Vector Machine Method for Identifying Antioxidant Proteins. *Int. J. Mol. Sci.* **2018**, *19*, 1773. [CrossRef] [PubMed]
73. Li, Y.; Shi, X.; Liang, Y.; Xie, J.; Zhang, Y.; Ma, Q. RNA-TVcurve: A Web server for RNA secondary structure comparison based on a multi-scale similarity of its triple vector curve representation. *BMC Bioinform.* **2017**, *18*, 51. [CrossRef] [PubMed]

Sample Availability: Samples of the compounds are not available.

molecules

MDPI

Article

Inferring microRNA-Environmental Factor Interactions Based on Multiple Biological Information Fusion

Haiqiong Luo [1], Wei Lan [2,*], Qingfeng Chen [2,3,*], Zhiqiang Wang [3], Zhixian Liu [4], Xiaofeng Yue [5] and Lingzhi Zhu [6]

[1]   School of information and management, Guangxi Medical University, Nanning 530021, China; hqluo@163.com
[2]   School of Computer, Electronic and Information, Guangxi University, Nanning 530004, China
[3]   State Key Laboratory for Conservation and Utilization of Subtropical Agro-bioresources, Guangxi University, Nanning 530004, China; zhqwang@gxu.edu.cn
[4]   School of electronic and information engineering, Qinzhou University, Qingzhou 535011, China; qzxylzx@163.com
[5]   School of Automation, Huazhong University of Science and Technology, Wuhan 430074, China; xfyue@hust.edu.cn
[6]   Department of Computer and Information Science, Hunan Institute of Technology, Hengyang 421008, China; lz_zhu@csu.edu.cn
*   Correspondence: lanwei@gxu.edu.cn (W.L.); qingfeng@gxu.edu.cn (Q.C.); Tel.: +86-771-327-4658 (W.L.); +86-771-327-4658 (Q.C.)

Received: 15 August 2018; Accepted: 18 September 2018; Published: 24 September 2018

Abstract: Accumulated studies have shown that environmental factors (EFs) can regulate the expression of microRNA (miRNA) which is closely associated with several diseases. Therefore, identifying miRNA-EF associations can facilitate the study of diseases. Recently, several computational methods have been proposed to explore miRNA-EF interactions. In this paper, a novel computational method, MEI-BRWMLL, is proposed to uncover the relationship between miRNA and EF. The similarities of miRNA-miRNA are calculated by using miRNA sequence, miRNA-EF interaction, and the similarities of EF-EF are calculated based on the anatomical therapeutic chemical information, chemical structure and miRNA-EF interaction. The similarity network fusion is used to fuse the similarity between miRNA and the similarity between EF, respectively. Further, the multiple-label learning and bi-random walk are employed to identify the association between miRNA and EF. The experimental results show that our method outperforms the state-of-the-art algorithms.

Keywords: microRNA; environmental factor; structure information; similarity network

1. Introduction

There is increasing evidence demonstrating that phenotypes are associated with genetic factors (GFs) and environmental factors (EFs) [1,2]. Environmental factors, including stress, alcohol, pollution, radiation and drugs play important roles in many diseases [3]. The perturbation of GF-EF interactions may result in some diseases [4,5]. Thus, identifying the potential associations between GFs and EFs is useful for biologists to understand the molecular bases of diseases.

MiRNA is a kind of typical GF with the length from 18 nt to 25 nt. It has been proved that miRNA can regulate the expression of genes by binding to the 3′ untranslated region (UTR) or 5′ untranslated region of mRNA in organisms [6,7]. In addition, accumulated evidence has demonstrated that miRNA normally plays essential roles in many important biological processes, including cell growth, cell cycle control, cell differentiation, cell apoptosis, and so on [8]. Therefore, the functional abnormality of miRNA can cause a broad range of diseases. For example, miR-150 can regulate the expression of the genes GAB1 and FOXP1 and impact the B and T cell activity in chronic lymphocytic leukemia [9]. Recently, a growing number of studies have indicated that miRNAs interact with diverse EFs [10–12]. The perturbation of miRNA-EF interactions is also related to a number of human diseases. For example, gemcitabine can down-regulate the expression of hsa-let-7b in pancreatic cancer cells [13,14]. Therefore, identifying potential miRNA-EF interactions contributes to the study of diseases. In addition, with the development of biotechnology, several databases such as miRbase [15], miRecord [16], dbDEMC [17] and miREnvironment [18] have been developed to store miRNA and EF related data. Those databases provide reliable data resources for predicting miRNA-EF interactions.

In recent years, many computational methods have been proposed to predict miRNA-EF interactions [19]. Chen et al. [20] proposed a method called miREFScan based on Laplacian regularized least squares to predict the interactions between miRNAs and EFs. This method is based on the assumption that functionally similar miRNAs tend to be related with similar EFs [21]. Chen et al. [22] presented a computational approach (miREFRWR) to infer miRNA-EF interactions based on a random walk method. Jiang et al. [23] constructed a small molecule-miRNA interaction network in 23 cancers and then identified the miRNA-EF associations based on hypergeometric tests. Qiu et al. [24] revealed several important features of miRNA and EF by analyzing miRNA-EF interaction network and proposed a model based on Fisher tests to infer potential miRNA-EF interactions. Li et al. [25] presented a computational framework based on an EF structure and disease similarity method to predict the interaction. Although the above methods have achieve great successes, some of them use low quality datasets which may result in poor performance. For example, some approaches measure miRNA similarity and EF similarity by using network-based data only, which may result in a bias for ignoring the biological characteristics of miRNA and EF. Most cannot effectively integrate different biological data resources. Further, some methods are unsuitable for predicting interaction of new miRNA without any known related EFs or new EF without any known related miRNAs.

In this paper, we assume that functionally similar miRNAs tend to be related with similar EFs. Based on this assumption, a computational framework is developed to predict the interactions between miRNAs and EFs. Unlike traditional methods, we use different data sources to measure miRNA-miRNA similarity and EF-EF similarity. The former is calculated by using the miRNA sequences and miRNA-EF interaction information, and the EF-EF similarity is computed by the anatomical therapeutic chemical, chemical structure and miRNA-EF interaction information. In particular, the similarity network fusion is applied to integrate these two similarities. Further, the multiple-label learning and bi-random walk are employed to identify the association between miRNA and EF. The experimental results show that our method is effective in inferring miRNA-environmental factor interactions.

2. Datasets and Methods

2.1. Datasets

We downloaded the known miRNA-EF interaction data from the miREnvironment database (http://www.cuilab.cn/miren) [18], which includes 3857 entries from 24 species. Only the human-related data were used for the following experiments. We manually checked the data and removed the interactions which do not correspond to human diseases. After pruning the invalid information, 224 miRNAs, 124 EFs and 729 miRNA-EF interactions were extracted as the gold dataset. A matrix I is constructed to represent miRNA-EF interaction. The value 1 is assigned to $I (i, j)$ if the interaction between miRNA i and EF j can be found, otherwise 0.

miRNA sequence information is obtained from miRbase (version 22) [15], which contains more than 2400 human sequences. After mapping miRNA of the gold dataset to miRbase, 224 miRNA sequences were finally obtained.

We download the chemical structure and anatomical therapeutic chemical of drugs from KEGG database (in 2016) [26]. There are 81 drugs with chemical structure and 57 drugs with anatomical therapeutic chemical, respectively.

2.2. Measuring miRNA-miRNA Similarity and EF-EF Similarity

2.2.1. miRNA-miRNA Similarity

Based on assumption that miRNAs with similar function are tend to relate with similar EFs, the interaction profile similarity is utilized to measure the similarity of pairwise miRNAs [27]. The miRNA interaction profile similarity is defined as:

$$W_m^p(m_i, m_j) = e^{(-\gamma_m \|IP(m_i) - IP(m_j)\|^2)} \tag{1}$$

$$\gamma_m = \frac{1}{\frac{1}{n}\sum_{i=1}^{n} IP(m_i)} \tag{2}$$

where m_i and m_j represent miRNAs i and j. n represents the number of miRNAs. $IP(m_i)$ represents the interactions between miRNA i and all EFs in the known miRNA-EF interaction data, *i. e.* the i-th row of matrix I. The parameter γ_m is set to control the kernel bandwidth. The sequence information has been widely used to find miRNA-disease association and feature patterns of miRNA regulation inference [28]. The Emboss-needle tool is utilized to compute sequence similarity of pairwise miRNAs [29].

2.2.2. EF-EF Similarity

The chemical structure is an important piece of information for drug design and has been applied to measure drug similarity [20,30]. SIMCOMP [31] is used to calculate the similarity of pairwise drugs based on common substructures. In addition, the Anatomical Therapeutic Chemical (ATC) code obtained from the ATC Classification System [26] assists in calculating the pairwise similarity of drugs.

Based on the assumption that EFs with similar function are tend to relate with similar miRNA, the interaction profile similarity is employed to measure the similarity between EFs [27]. The EF interaction profile similarity is defined as:

$$W_e^p(e_i, e_j) = e^{(-\gamma_e \|IP(e_i) - IP(e_j)\|^2)} \tag{3}$$

$$\gamma_e = \frac{1}{\frac{1}{m}\sum_{i=1}^{m} IP(e_i)} \tag{4}$$

where e_i and e_j represent EFs i and j. m denotes the number of EFs. $IP(e_i)$ represents the interaction between EF i and all miRNAs in the known miRNA-EF interaction data, *i. e.* the i-th column of matrix I. The parameter γ_e is to control the kernel bandwidth.

2.3. Similarity Network Fusion

The similarity network fusion (SNF) is an approach for multiple omics fusion, which has been widely used for cancer data analysis [32,33]. It is able to capture the global and local features of different data. The SNF for miRNA is defined as follows:

$$F^m = \frac{F_m^s + F_m^p}{2} \tag{5}$$

$$F_m^p(t) = L_m^p \times G_m^s(t-1) \times \left(L_m^p\right)^T \tag{6}$$

$$F_m^s(t) = L_m^s \times G_m^p(t-1) \times (L_m^s)^T \tag{7}$$

$$L_m^s(i,j) = \begin{cases} \frac{W_m^s(i,j)}{\sum_{k \in N_i} W_m^s(i,k)}, & j \in N_i \\ 0, & otherwise \end{cases} \tag{8}$$

$$L_m^p(i,j) = \begin{cases} \frac{W_m^p(i,j)}{\sum_{k \in N_i} W_m^p(i,k)}, & j \in N_i \\ 0, & otherwise \end{cases} \tag{9}$$

$$G_m^s(i,j) = \begin{cases} \frac{W_m^s(i,j)}{2\sum_{k \neq i} W_m^s(i,k)}, & i \neq j \\ \frac{1}{2}, & i = j \end{cases} \tag{10}$$

$$G_m^p(i,j) = \begin{cases} \frac{W_m^p(i,j)}{2\sum_{k \neq i} W_m^p(i,k)}, & i \neq j \\ \frac{1}{2}, & i = j \end{cases} \tag{11}$$

where W_m^s and W_m^p denote the miRNA sequence similarity matrix and miRNA interaction profile similarity matrix, respectively. G_m^s, L_m^s, G_m^p and L_m^p denote the global matrix of miRNA sequence similarity, local matrix of miRNA sequence similarity, global matrix of miRNA interaction profile similarity, local matrix of miRNA interaction profile similarity, respectively. The N_i represents the K-nearest neighbors of miRNA i. F_m^s and F_m^p denote the fusional matrix of miRNA sequence similarity and the fusional matrix of miRNA interaction profile similarity, respectively. F^m denotes the final fusional matrix of miRNA. The final fusional matrix of EF F^e can be obtained in term of similar manner.

2.4. Inferring miRNA-EF Interaction by Using bi-Random Walk and Multi-Label Learning (MEI-BRWMLL)

Considering the features of bi-random walk and multi-label learning, we utilize a bi-random walk to infer interactions of known miRNA/EF and multi-label learning is used to infer interactions of new miRNA/EF. The reason for selecting these two methods is that the bi-random walk achieves good results in potential interaction prediction between known entities while multi-label learning is robust in predicting interactions between new entities.

2.4.1. Bi-Random Walk for Predicting Potential Interactions of Known miRNAs and EFs

Based on assumption that similar miRNAs tend to relate with similar EF, the bi-random walk is employed to predict potential miRNA-EF interaction.

Firstly, the miRNA similarity matrix and EF similarity matrix are normalized by using Laplace regularization, respectively. It is defined as:

$$N^m = D^{m-\frac{1}{2}} \times F^m \times D^{m-\frac{1}{2}} \tag{12}$$

$$N^e = D^{e-\frac{1}{2}} \times F^e \times D^{e-\frac{1}{2}} \tag{13}$$

where N^m and N^e represent normalized matrix of fusional miRNA similarity and EF similarity, respectively. D^m and D^e represent the diagonal matrix of F^m and F^e, respectively. In addition, the miRNA-EF interaction matrix I is normalized as follows:

$$N^I(i,j) = \frac{I(i,j)}{\sum_i \sum_j I(i,j)} \tag{14}$$

Then, we use bi-random walk to predict potential miRNA-EF interaction by walking on miRNA similarity network and EF similarity network. The iterative process of bi-random walk is defined as follows:

Left walk in miRNA similarity network:

$$R_L(t) = \alpha \times N^m \times R_L(t-1) + (1-\alpha) \times N^I \tag{15}$$

Right walk in EF similarity network:

$$R_R(t) = \alpha \times R_R(t-1) \times N^e + (1-\alpha) \times N^I \tag{16}$$

The final predicted score is defined as follows:

$$R(t) = \frac{R_L(t) + R_R(t)}{2} \tag{17}$$

where $R_L(t)$ and $R_R(t)$ denote the predicted score matrix of walk on miRNA similarity network and EF similarity network at step t, respectively. $R(t)$ denotes the final score matrix at step t. In addition, the miRNA similarity network and EF similarity network contain different topological and structural features, and the optimal iteration steps of the random walk on the two networks should be different. Therefore, we set two parameters l, r to control the maximal random walk steps on two networks, respectively. The iterative of bi-random walk will stop when the number of iteration t exceeds the maximum of parameters l and r. The parameters can accelerate the iteration termination. In here, the l and r are set as 4 and 2, respectively.

2.4.2. Multi-Label Learning for Predicting Interactions of New miRNAs and EFs

We employ multi-label learning to infer the interactions of new miRNA/EF, which predicts the label of unseen instances based on a maximum a posteriori rule [34,35]. For convenience, we define some notations. miRNAs and EFs are assigned two domains $D_M = \{m_1, m_2, \ldots m_x\}$ and $D_E = \{e_1, e_2, \ldots e_y\}$, respectively. x and y represent the numbers of miRNAs and EFs, respectively. The interactions between miRNAs and EFs are represented by matrix $I_{x \times y}$. P_{ij} denotes the interaction probability of miRNA m_i and EF e_j. P_{ij} is set to 1 if $I(i,j) = 1$; otherwise, 0. For a new miRNA m_c, the probability $P(m_c, e_j)$ between m_c and EF e_j demonstrates the confidence that miRNA m_c is linked to EF e_j. Based on the similarity of miRNA-miRNA, we select the k nearest neighbors of miRNA m_c. Then, the probability $P(m_c, e_j)$ is calculated as follows:

$$P(m_c, e_j) = \frac{P\left(L_1^j\right) P\left(E_s^j \middle| L_1^j\right)}{P\left(L_1^j\right) P\left(E_s^j \middle| L_1^j\right) + P\left(L_0^j\right) P\left(E_s^j \middle| L_0^j\right)} \tag{18}$$

$$P\left(L_1^j\right) = \frac{1 + \sum_{i=1}^{x} I(i,j)}{2 + x} \tag{19}$$

$$P\left(L_0^j\right) = 1 - P\left(L_1^j\right) \tag{20}$$

$$P\left(E_s^j \middle| L_1^j\right) = \frac{1 + e(s)}{k + 1 + \sum_{i=0}^{k} e(i)} \tag{21}$$

$$P\left(E_s^j \middle| L_0^j\right) = \frac{1 + e'(s)}{k + 1 + \sum_{i=0}^{k} e'(i)} \tag{22}$$

where k represents the number of nearest neighbors. $e(s)$ represents the number of miRNA related to EF e_j whose KNNs contain exactly s miRNAs related EF e_j. $e'(s)$ counts the number of miRNA unrelated to EF e_j whose KNNs contain exactly s miRNAs related EF e_j.

The flowchart for miRNA-EF interaction prediction is shown in Figure 1. Firstly, the similarities of miRNA and EF are calculated based on different similarity measures, respectively. Secondly, the similarity matrices of miRNA and EF are constructed in terms of similarity scores calculated previously. Further, the similarity network fusion is employed to integrating different similarity matrices of miRNA and EF, respectively. Finally, the bi-random walk and multi-label learning are used to infer potential miRNA-EF interactions.

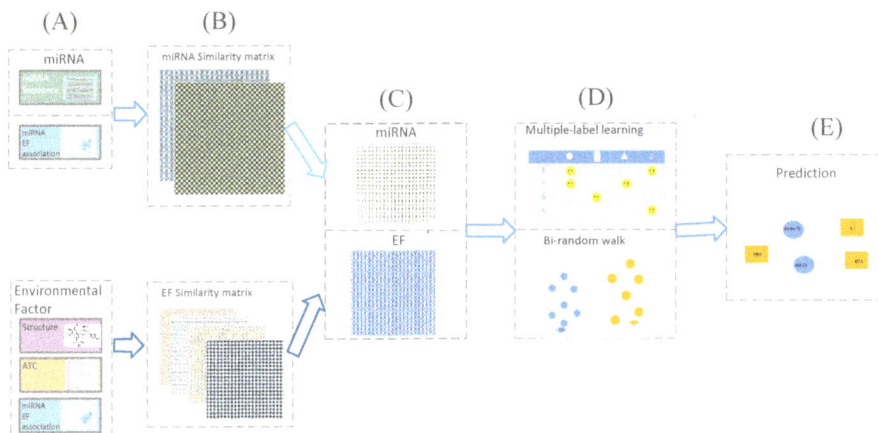

Figure 1. The flowchart of miRNA-EF interaction prediction. (**A**) Computing similarities of miRNA-miRNA and EF-EF, respectively. (**B**) Establishing similarity matrices of miRNA and EF, respectively. (**C**) Integrating similarity matrices of miRNA-miRNA and EF-EF by using similarity network fusion method, respectively. (**D**) Predicting miRNA-EF interactions by using multi-label learning and bi-random walk. (**E**) The final predicted results.

3. Experiments

3.1. Analyzing the miRNA-EF Interaction Network

There are 729 interactions between 224 miRNAs and 124 EFs in the whole miRNA-EF interaction network. The degree of EFs is shown in Figure 2. It is observed that the degree of most EFs is equal to 1. It means that most of EFs only have one related miRNA and a great amount of interactions are still unknown. The EF with the max degree is gemcitabine which has 56 related miRNAs.

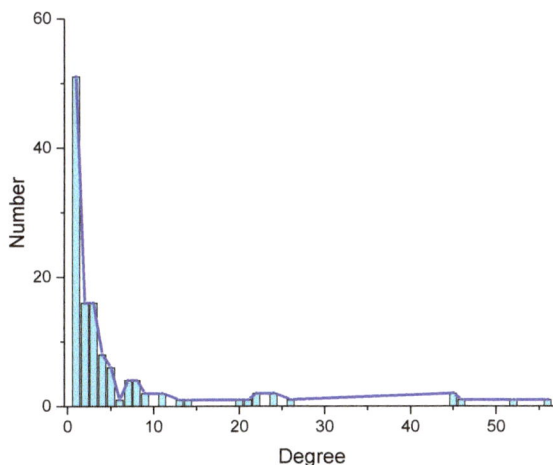

Figure 2. The degree of EFs.

In order to analyze the cluster feature of miRNA-EF interaction network, the ClusterViz [36] program is used to obtain clusters from the network. In Figure 3, three modules are obtained from the miRNA-EF interaction network. This demonstrates that EFs can regulate a group of functionally similar miRNAs rather than a single miRNA. Take the module (C) for example, it demonstrates that four EFs (DDT, E2, BPA and ionizing radiation) have associations with the let-7 family.

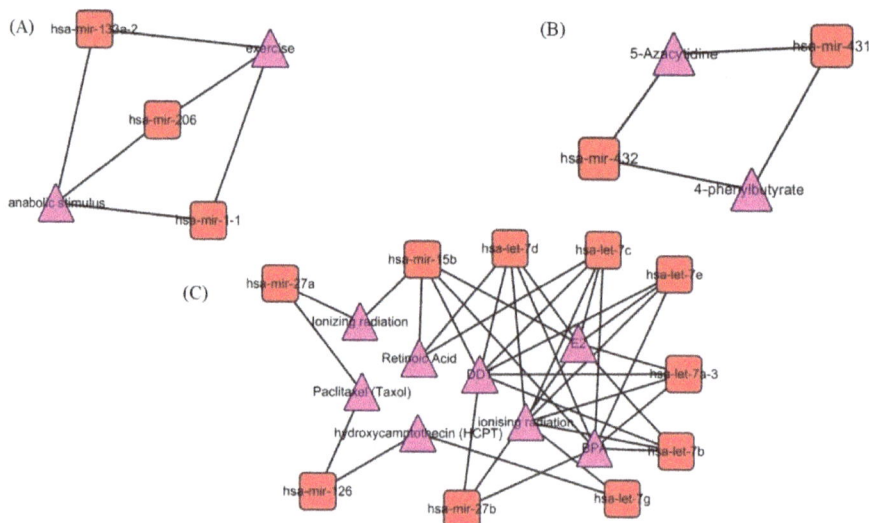

Figure 3. Three modules are obtained from miRNA-EF interaction network by utilizing ClusterViz. (**A**) The EFs (anabolic stimulus and exercise) are related with hsa-mir-133a-2, hsa-mir-206 and hsa-mir-1-1. (**B**) The EFs (5-Azacytidine and 4-phenylbutyrate) are associated with hsa-mir-431 and hsa-mir-432. (**C**) The EFs (DDT, E2, BPA and ionizing radiation) have associations with the let-7 family.

3.2. Experiment

To demonstrate the effectiveness of our method, a comparison between our method and three state-of-the-art methods (miREFScan [20], miREFRWR [22] and KBMF [6]) is conducted. The parameters of these methods are specified as the default value. The 10-fold cross validation is utilized to evaluate the performance of different methods. The known miRNA-EF interactions are divided into 10 subsets. One subset is used as test set and the remaining nine subsets are treated as training set. Then, the true positive rates (TPR) and false positive rates (FPR) are calculated by using different classification thresholds. The receiver operating characteristics (ROC) curve is drawn based on the value of TPR and FPR and the area under the ROC curve (AUC) is calculated to measure the performance. The higher of AUC value, the better performance is. The experimental result is shown in Figure 4. It can be found that our method achieves an AUC of 0.8208 which is better than other two methods (miREFRWR: 0.7905, miREFScan: 0.7963 and KBMF: 0.677).

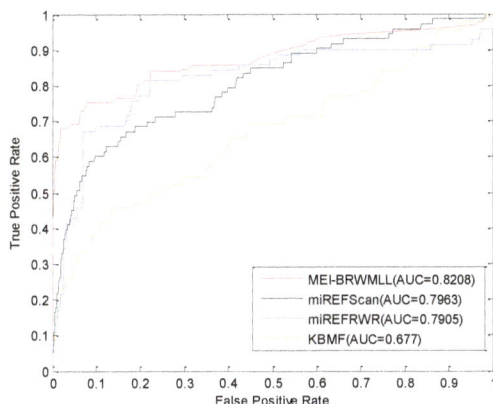

Figure 4. Comparison of different methods in miRNA-EF interaction prediction.

3.3. Case Study

3,3′-Diindolylmethane (DIM) is a kind of compound widely found in *Brassica* vegetables [37]. An increasing number of studies have shown that DIM has a close relationship with many cancers. For example, it has been proved that the expression of HDAC1 can be inhibited by DIM in colon cancer tissue [38]. Table 1 shows the top 15 potential miRNAs related with DIM which are identified by using MEI-BRWMLL nine miRNAs are confirmed to connect to DIM by the recent literature. It has been proved that the expression of hsa-mir-146a (ranked at first) is induced by DIM in pancreatic cancer cells [39]. In addition, the DIM has been certified to up-regulate miRNA-16 (ranked second) in CD4+ T cells [40]. The literature shows DIM has relationship with hsa-mir-181d, hsa-mir-125b and hsa-mir-34a (ranked at 6th, 8th and 12th), respectively [41,42]. DIM can inhibit the expression of these three miRNAs in SEB-mediated liver injury. The hsa-mir-200b (ranked at 9th) is upregulated by DIM in SKBR3 breast cancer cells [43]. It has been proved that the expression of hsa-mir-221 (ranked at 11th) can be downregulated in pancreatic cancer [44]. The DIM can inhibit the expression of EZH2 by up-regulating hsa-let-7e (ranked at 13th) in castration-resistant prostate cancer [45]. The literature [43] shows that the expression of hsa-mir-200c is up-regulated by DIM and herceptin in breast cancer. In addition, it can be found that several miRNAs are identified to be related with DIM. However, the functions of these miRNAs are still unknown. This requires biologists to validate them by using biological experiments.

Table 1. The top 15 potential miRNAs related to 3,3′-diindolylmethane predicted by MEI-BRWMLL.

Rank	miRNA	Evidence
1	hsa-mir-146 a	PMID: 20124483
2	hsa-mir-16	PMID: 24899890
3	hsa-mir-24	Unknown
4	hsa-mir-155	Unknown
5	hsa-mir-223	Unknown
6	hsa-mir-181 d	PMID: 25706292
7	hsa-mir-181 b	Unknown
8	hsa-mir-125 b	PMID: 25706292
9	hsa-mir-200 b	PMID: 23372748
10	hsa-mir-126	Unknown
11	hsa-mir-221	PMID: 24224124
12	hsa-mir-34 a	PMID: 25706292
13	hsa-let-7 e	PMID: 22442719
14	hsa-mir-200 c	PMID:23372748
15	hsa-mir-222	Unknown

4. Conclusions

Understanding the complex pathogenesis of diseases is still a significant challenge in disease research [46,47]. Increasing studies have demonstrated that diseases have close relationship with GFs and EFs [48,49]. miRNAs are a group of important GFs which have been proved to play critical roles in many diseases [50,51]. Therefore, identifying miRNA-EF interactions is helpful for elucidating the pathogenesis of diseases. In this paper, a computational framework to predict interactions between miRNAs and EFs is proposed. Multiple biological data are used to measure the pairwise similarity of miRNA-miRNA and EF-EF, respectively. Then, the similarities of miRNA-miRNA and EF-EF are fused by using SNF, respectively. Further, the bi-random walk and multiple label learning are utilized to infer miRNA-EF interactions. The experimental results show that this method is effective for miRNA-EF interaction identification.

Author Contributions: Conceptualization, H.L. and W.L.; Methodology, H.Q.L., W.L., Z.Q.W. and Q.F.C.; Software, Z.X.L., X.F.Y. and L.Z.Z.; Writing-Original Draft Preparation, H.Q.L., W.L. and Q.F.C.

Funding: This research was funded by the National Natural Science Foundation of China under Grant No. 61702122, No.61751314 and No. 61802442; Key project of Natural Science Foundation of Guangxi 2017GXNSFDA198033; Key research and development plan of Guangxi AB17195055 and Director Open Fund of Qinzhou City Key Laboratory of Advanced Technology of Internet of Things IOT2017A04.

Conflicts of Interest: The authors declare no conflict of interest.

References

1. Barabási, A.L.; Gulbahce, N.; Loscalzo, J. Network medicine: A network-based approach to human disease. *Nat. Rev. Genet.* **2011**, *12*, 56–68. [CrossRef] [PubMed]
2. Moreau, Y.; Tranchevent, L.C. Computational tools for prioritizing candidate genes: Boosting disease gene discovery. *Nat. Rev. Genet.* **2012**, *13*, 523–536. [CrossRef] [PubMed]
3. Clayton, D.; McKeigue, P.M. Epidemiological methods for studying genes and environmental factors in complex diseases. *Lancet* **2001**, *358*, 1356–1360. [CrossRef]
4. Lan, W.; Wang, J.X.; Li, M.; Peng, W.; Wu, F.X. Computational approaches for prioritizing candidate disease genes based on PPI networks. *Tsinghua Sci. Technol.* **2015**, *20*, 500–512. [CrossRef]
5. Li, M.; Zheng, R.; Li, Y.; Wu, F.X.; Wang, J.X. MGT-SM: A Method for Constructing Cellular Signal Transduction Networks. *IEEE/ACM Trans. Comput. Biol. Bioinform.* **2017**. [CrossRef] [PubMed]
6. Zeng, X.; Liu, L.; Lu, L.; Zou, Q. Prediction of potential disease-associated microRNAs using structural perturbation method. *Bioinformatcis* **2018**, *1*, 8. [CrossRef]
7. Ha, M.; Ki, V.N. Regulation of microRNA biogenesis. *Nat. Rev. Mol. Cell. Biol.* **2014**, *15*, 509–524. [CrossRef] [PubMed]
8. Zou, Q.; Li, J.; Song, L.; Zeng, X.; Wang, G. Similarity computation strategies in the microRNA-disease network: A survey. *Brief. Func. Genom.* **2015**, *15*, 55–64. [CrossRef] [PubMed]
9. Zhou, B.; Wang, S.; Mayr, C.; Bartel, D.P.; Lodish, H.F. miR-150, a microRNA expressed in mature B. and T. cells, blocks early B cell development when expressed prematurely. *Proc. Natl. Acad. Sci. USA* **2007**, *104*, 7080–7085. [CrossRef] [PubMed]
10. Lan, W.; Huang, L.Y.; Lai, D.H.; Chen, Q.F. Identifying Interactions Between Long Noncoding RNAs and Diseases Based on Computational Methods. *Methods Mol. Biol.* **2018**, *1754*, 205–221. [PubMed]
11. Peng, W.; Lan, W.; Zhong, J.C.; Wang, J.X.; Pan, Y. A novel method of predicting microRNA-disease associations based on microRNA, disease, gene and environment factor networks. *Method* **2017**, *124*, 69–77. [CrossRef] [PubMed]
12. Mathers, J.C.; Strathdee, G.; Relton, C.L. Induction of epigenetic alterations by dietary and other environmental factors. *Adv. Genet.* **2009**, *71*, 3–39.
13. Wen, X.Y.; Wu, S.Y.; Liu, Z.Q.; Zhang, J.J.; Wang, G.F.; Jiang, Z.H.; Wu, S.G. Ellagitannin (BJA3121), an anti-proliferative natural polyphenol compound, can regulate the expression of MiRNAs in HepG2 cancer cells. *Phytother. Res.* **2009**, *23*, 778–784. [CrossRef] [PubMed]

14. Chiyomaru, T.; Yamamura, S.; Fukuhara, S.; Yoshino, H.; Kinoshita, T.; Majid, S.; Saini, S.; Chang, I.; Tanaka, Y.; Enokida, H.; et al. Genistein inhibits prostate cancer cell growth by targeting miR-34a and oncogenic HOTAIR. *PLoS ONE* **2013**, *8*, e70372. [CrossRef] [PubMed]

15. Kozomara, A.; Griffiths-Jones, S. miRBase: Annotating high confidence microRNAs using deep sequencing data. *Nucleic Acids Res.* **2014**, *42*, D68–D73. [CrossRef] [PubMed]

16. Wang, D.; Gu, J.; Wang, T.; Ding, Z. OncomiRDB: A database for the experimentally verified oncogenic and tumor-suppressive microRNAs. *Bioinformatics* **2014**, *30*, 2237–2238. [CrossRef] [PubMed]

17. Yang, Z.; Ren, F.; Liu, C.; He, S.; Sun, G.; Gao, Q.; Yao, L.; Zhang, Y.; Miao, R.; Cao, Y. dbDEMC: A database of differentially expressed miRNAs in human cancers. *BMC Genom.* **2010**, *11*, S5. [CrossRef] [PubMed]

18. Yang, Q.; Qiu, C.; Yang, J.; Wu, Q.; Cui, Q. miREnvironment database: Providing a bridge for microRNAs, environmental factors and phenotypes. *Bioinformatics* **2011**, *27*, 3329–3330. [CrossRef] [PubMed]

19. Baccarelli, A.; Bollati, V. Epigenetics and environmental chemicals. *Curr. Opin. Pediatr.* **2009**, *21*, 243–251. [CrossRef] [PubMed]

20. Luo, H.M.; Wang, J.X.; Li, M.; Luo, J.W.; Peng, X.Q.; Wu, F.X.; Pan, Y. Drug repositioning based on comprehensive similarity measures and Bi-Random Walk algorithm. *Bioinformatics* **2016**, *32*, 2664–2671. [CrossRef] [PubMed]

21. Chen, X.; Liu, M.X.; Cui, Q.H.; Yan, G.Y. Prediction of disease-related interactions between microRNAs and environmental factors based on a semi-supervised classifier. *PLoS ONE* **2012**, *7*, e43425. [CrossRef] [PubMed]

22. Chen, X. miREFRWR: A novel disease-related microRNA-environmental factor interactions prediction method. *Mol. Biosyst.* **2016**, *12*, 624–633. [CrossRef] [PubMed]

23. Jiang, W.; Chen, X.; Liao, M.; Li, W.; Lian, B.; Wang, L.; Meng, F.; Liu, X.; Chen, X.; Jin, Y.; et al. Identification of links between small molecules and miRNAs in human cancers based on transcriptional responses. *Sci. Rep.* **2012**, *2*, 282. [CrossRef] [PubMed]

24. Qiu, C.; Chen, G.; Cui, Q.H. Towards the understanding of microRNA and environmental factor interactions and their relationships to human diseases. *Sci. Rep.* **2012**, *2*, 318. [CrossRef] [PubMed]

25. Li, J.; Wu, Z.; Cheng, F.; Li, W.; Liu, G.; Tang, Y. Computational prediction of microRNA networks incorporating environmental toxicity and disease etiology. *Sci. Rep.* **2014**, *4*, 5576. [CrossRef] [PubMed]

26. Kanehisa, M.; Araki, M.; Goto, S.; Hattori, M.; Hirakawa, M.; Itoh, M.; Katayama, T.; Kawashima, S.; Okuda, S.; Tokimatsu, T.; et al. KEGG for linking genomes to life and the environment. *Nucleic Acids Res.* **2008**, *36*, D480–D484. [CrossRef] [PubMed]

27. Laarhoven, T.V.; Nabuurs, S.B.; Marchiori, E. Gaussian interaction profile kernels for predicting drug–target interaction. *Bioinformatics* **2011**, *27*, 3036–3043. [CrossRef] [PubMed]

28. Chen, Q.F.; Lan, W.; Wang, J.X. Mining featured patterns of MiRNA interaction based on sequence and structure similarity. *IEEE/ACM Trans. Comput. Biol. Bioinform.* **2013**, *10*, 415–422. [CrossRef] [PubMed]

29. McWilliam, H.; Li, W.; Uludag, M.; Squizzato, S.; Park, Y.M.; Buso, N.; Cowley, A.P.; Lopez, R. Analysis Tool Web Services from the EMBL-EBI. *Nucleic Acids Res.* **2013**, *12*, W597–W600. [CrossRef] [PubMed]

30. Hattori, M.; Tanaka, N.; Kanehisa, M. SIMCOMP/SUBCOMP: Chemical structure search servers for network analyses. *Nucleic Acids Res.* **2010**, *38*, W652–W656. [CrossRef] [PubMed]

31. Lan, W.; Wang, J.X.; Li, M.; Liu, J.; Li, Y.H.; Wu, F.X.; Pan, Y. Predicting drug–target interaction using positive-unlabeled learning. *Neurocomputing* **2016**, *206*, 50–57. [CrossRef]

32. Wang, B.; Mezlini, A.M.; Demir, F.; Fiume, M.; Tu, Z.; Brudno, M.; Haibe-Kains, B.; Goldenberg, A. Similarity network fusion for aggregating data types on a genomic scale. *Nat. Meth.* **2014**, *11*, 333–337. [CrossRef] [PubMed]

33. Liu, J.; Wang, X.; Zhang, X.; Pan, Y.; Wang, X.; Wang, J.X. MMM: Classification of schizophrenia using multi-modality multi-atlas feature representation and multi-kernel learning. *Multimed. Tool Appl.* **2017**, 1–17. [CrossRef]

34. Zhang, M.L.; Zhou, Z.H. ML-KNN: A lazy learning approach to multi-label learning. *Pattern Recognit.* **2007**, *40*, 2038–2048. [CrossRef]

35. Liu, J.; Li, M.; Lan, W.; Wu, F.X.; Pan, Y.; Wang, J.X. Classification of Alzheimer's disease using whole brain hierarchical network. *IEEE/ACM Trans Comput. Biol. Bioinform.* **2018**, *15*, 624–632. [CrossRef] [PubMed]

36. Wang, J.X.; Zhong, J.C.; Chen, G.; Li, M.; Wu, F.X.; Pan, Y. ClusterViz: A cytoscape APP for cluster analysis of biological network. *IEEE/ACM Trans. Comput. Biol. Bioinform.* **2015**, *12*, 815–822. [CrossRef] [PubMed]

37. Ge, X.; Yannai, S.; Rennert, G.; Gruener, N.; Fares, F.A. 3,3′-Diindolylmethane induces apoptosis in human cancer cells. *Biochem. Biophys. Res. Commun.* **1996**, *228*, 153–158. [CrossRef] [PubMed]

38. Li, Y.; Li, X.; Guo, B. Chemopreventive agent 3,3′-diindolylmethane selectively induces proteasomal degradation of class I histone deacetylases. *Cancer Res.* **2010**, *70*, 646–654. [CrossRef] [PubMed]

39. Li, Y.; VandenBoom, T.G.; Wang, Z.; Kong, D.; Ali, S.; Philip, P.A.; Sarkar, F.H. miR-146a suppresses invasion of pancreatic cancer cells. *Cancer Res.* **2010**, *70*, 1486–1495. [CrossRef] [PubMed]

40. Rouse, M.; Rao, R.; Nagarkatti, M.; Nagarkatti, P.S. 3,3′-diindolylmethane ameliorates experimental autoimmune encephalomyelitis by promoting cell cycle arrest and apoptosis in activated T cells through microRNA signaling pathways. *J. Pharmacol. Exp. Ther.* **2014**, *350*, 341–352. [CrossRef] [PubMed]

41. Busbee, P.; Nagarkatti, M.; Nagarkatti, P. Natural indoles, indole-3-carbinol (I3C) and 3,3′-diindolylmethane (DIM), attenuate staphylococcal enterotoxin B-mediated liver injury by downregulating miR-31 expression and promoting caspase-2-mediated apoptosis (IRC4P. 605). *J. Immunol.* **2015**, *194*, 57. [CrossRef] [PubMed]

42. Busbee, P.B.; Nagarkatti, M.; Nagarkatti, P.S. Natural Indoles, Indole-3-Carbinol (I3C) and 3,3′-Diindolylmethane (DIM), Attenuate Staphylococcal Enterotoxin B-Mediated Liver Injury by Downregulating miR-31 Expression and Promoting Caspase-2-Mediated Apoptosis. *PLoS ONE* **2015**, *10*, e0118506. [CrossRef] [PubMed]

43. Ahmad, A.; Ali, S.; Ahmed, A.; Ali, A.S.; Raz, A.; Sakr, W.A.; Rahman, K.M. 3,3′-Diindolylmethane enhances the effectiveness of herceptin against HER-2/neu-expressing breast cancer cells. *PLoS ONE* **2013**, *8*, e54657. [CrossRef] [PubMed]

44. Sarkar, S.; Dubaybo, H.; Ali, S.; Goncalves, P.; Kollepara, S.L.; Sethi, S.; Philip, P.A.; Li, Y. Down-regulation of miR-221 inhibits proliferation of pancreatic cancer cells through up-regulation of PTEN, p27 (kip1), p57 (kip2), and PUMA. *Am. J. Cancer Res.* **2013**, *3*, 465–477. [PubMed]

45. Kong, D.; Heath, E.; Chen, W.; Cher, M.L.; Powell, I.; Heilbrun, L.; Li, Y.; Ali, S.; Sethi, S.; Hassan, O.; et al. Loss of let-7 up-regulates EZH2 in prostate cancer consistent with the acquisition of cancer stem cell signatures that are attenuated by BR-DIM. *PLoS ONE* **2012**, *7*, e33729. [CrossRef] [PubMed]

46. Hinks, T.S.; Zhou, X.; Staples, K.J.; Dimitrov, B.D.; Manta, A.; Petrossian, T.; Lum, P.Y.; Smith, C.G.; Ward, J.A.; Howarth, P.H.; et al. Innate and adaptive T cells in asthmatic patients: Relationship to severity and disease mechanisms. *J. Allergy Clin. Immunol.* **2015**, *136*, 323–333. [CrossRef] [PubMed]

47. Lan, W.; Li, M.; Zhao, K.J.; Liu, J.; Wu, F.X.; Pan, Y.; Wang, J.X. LDAP: A web server for lncRNA-disease association prediction. *Bioinformatics* **2017**, *33*, 458–460. [CrossRef] [PubMed]

48. Lan, W.; Chen, Q.F.; Li, T.S.; Yuan, C.G.; Mann, S.; Chen, B.S. Identification of important positions within miRNAs by integrating sequential and structural features. *Curr. Protein Pept. Sci.* **2014**, *15*, 591–597. [CrossRef] [PubMed]

49. Liu, J.; Wang, J.X.; Tang, Z.Z.; Hu, B.; Wu, F.X.; Pan, Y. Improving Alzheimer's Disease Classification by Combining Multiple Measures. *IEEE/ACM Trans. Comput. Biol. Bioinform.* **2017**. [CrossRef]

50. Lan, W.; Wang, J.X.; Li, M.; Liu, J.; Wu, F.X.; Pan, Y. Predicting microRNA-disease associations based on improved microRNA and disease similarities. *IEEE/ACM Trans. Comput. Biol. Bioinform.* **2016**. [CrossRef] [PubMed]

51. Zeng, X.; Zhang, X.; Zou, Q. Integrative approaches for predicting microRNA function and prioritizing disease-related microRNA using biological interaction networks. *Brief. Bioinform.* **2016**, *17*, 193–203. [CrossRef] [PubMed]

Sample Availability: Samples of the compounds are not available from the authors.

molecules

MDPI

Article

Synstable Fusion: A Network-Based Algorithm for Estimating Driver Genes in Fusion Structures

Mingzhe Xu [1,2,3], Zhongmeng Zhao [1,3], Xuanping Zhang [1,3,*], Aiqing Gao [1,3], Shuyan Wu [4] and Jiayin Wang [1,3,*]

[1]   Department of Computer Science and Technology, School of Electronic and Information Engineering, Xi'an Jiaotong University, Xi'an 710049, China; mingzhe.xu@hnuahe.edu.cn (M.X.); zmzhao@mail.xjtu.edu.cn (Z.Z.); algoxjtu@163.com (A.G.)
[2]   Department of Automation, College of Intelligent Manufacturing and Automation, Henan University of Animal Husbandry and Economy, Zhengzhou 450011, China
[3]   Shaanxi Engineering Research Center of Medical and Health Big Data, School of Electronic and Information Engineering, Xi'an Jiaotong University, Xi'an 710049, China
[4]   Department of Network Technology, College of Intelligent Manufacturing and Automation, Henan University of Animal Husbandry and Economy, Zhengzhou 450011, China; xxxwljys@126.com
*   Correspondence: zxp@mail.xjtu.edu.cn (X.Z.); wangjiayin@mail.xjtu.edu.cn (J.W.); Tel.: +86-29-8266-8971 (J.W.)

Academic Editors: Xiangxiang Zeng, Alfonso Rodríguez-Patón and Quan Zou
Received: 25 June 2018; Accepted: 7 August 2018; Published: 16 August 2018

Abstract: Gene fusion structure is a class of common somatic mutational events in cancer genomes, which are often formed by chromosomal mutations. Identifying the driver gene(s) in a fusion structure is important for many downstream analyses and it contributes to clinical practices. Existing computational approaches have prioritized the importance of oncogenes by incorporating prior knowledge from gene networks. However, different methods sometimes suffer different weaknesses when handling gene fusion data due to multiple issues such as fusion gene representation, network integration, and the effectiveness of the evaluation algorithms. In this paper, Synstable Fusion (SYN), an algorithm for computationally evaluating the fusion genes, is proposed. This algorithm uses network-based strategy by incorporating gene networks as prior information, but estimates the driver genes according to the destructiveness hypothesis. This hypothesis balances the two popular evaluation strategies in the existing studies, thereby providing more comprehensive results. A machine learning framework is introduced to integrate multiple networks and further solve the conflicting results from different networks. In addition, a synchronous stability model is established to reduce the computational complexity of the evaluation algorithm. To evaluate the proposed algorithm, we conduct a series of experiments on both artificial and real datasets. The results demonstrate that the proposed algorithm performs well on different configurations and is robust when altering the internal parameter settings.

Keywords: gene fusion data; gene susceptibility prioritization; evaluating driver partner; gene networks

1. Introduction

Gene fusion is an important class of somatic mutational events in cancers [1]. A series of studies have shown that gene fusion structures, as well as the related genomic structural variations, are significantly associated with cancer susceptibilities across multiple cancer types [1–6]. With the development of sequencing technology, detecting gene fusion structures has become routine work in a number of computational pipelines for cancer sequencing data [7–9].

A fusion gene is typically formed by the interaction of two or more genes that are usually called partner genes. Normally, a fusion gene has a driver partner and one or more passenger partners, according to their roles in the evolution of tumor tissue [10]. The driver partner has a vital function in the carcinogenesis processes. Thus, identifying the driver partner is important for many downstream analyses and presents clinical implications. However, the throughput for validating driver genes is limited by current technology, which is both time consuming and expensive. A small number of the driver partners have demonstrated associations to cancer susceptibilities. Thus, computational approaches have been introduced to filter and prioritize the driver partner candidates, which facilitate and may further guide functional validations. To evaluate the importance of each partner in a gene fusion structure, gene networks are used in almost every existing approach, although different approaches vary in their use and application. A gene network is usually represented as a weighted graph, where each node in the graph denotes a gene, whereas each edge denotes a specific type of interaction between the two genes. Different types of approaches and interactions exist, including co-expression and co-localization networks [11–13], genetic interaction networks [14,15], pathway networks [16,17], physical interaction networks [18,19], shared protein domain networks [20], and predicted networks [21,22].

Along with the accumulation of gene network data, network-based approaches are faced with two major computational challenges. The first is determining how to incorporate knowledge from various types of networks. Multiple heterogeneous gene networks reflect different relationships. A common strategy involves establishing a virtual network by weighting the prior information from different networks. This is similar to the collapsing, or burden-test, strategy used in association studies [23] or the multi-source data-integration and decision-making process [24]. Benefiting from the amplification of the data signals via the newly collapsed network, the evaluation algorithms may be better for discovering potential associations and be more accurate in prioritizing the susceptibility genes [25–27]. Here, edge weight and graph structure are the two major evaluation strategies used to sort the important nodes (genes) through the collapsed network. The importance of edge weight is obvious, whereas the graph structure is considered by calculating the impact of each node based on the network, such as node degree [28] and node betweenness [29]. Node degree is a local topology strategy that only computes the weights on the edges that directly connect to the node. Node betweenness provides a global view by presenting the connectivity influence of nodes on the entire network. The existing approaches, however, are usually sensitive to the incorporation of the networks. When a neural network is collapsed into a disease-associated network, it may excessively dilute the data signal [23]. For example, in the multi-layer design of neural networks [24,30], multiple disease-associated network data are merged into a single output signal. Most of the data being processed within neural networks are eliminated by the weights of the input layer and the activation function of neurons in hidden layers [30].

The second major computational challenge is addressing the conflicting results from different networks. Different from the point mutation or indel calls, a gene fusion structure consists of two or more partner genes. Gene networks do not contain any "combined" nodes corresponding to a fusion gene. Thus, in many cases, the evaluation algorithms may provide conflicting results on the same virtual network. To solve the conflicts, after extensive experimental verifications [31–34], Wu et al. [34] provided the hypothesis that "if a fusion gene plays an important role in tumor formation, then the partner genes should be an important node in the gene network". This hypothesis, called "network fusion centrality", is based on many previous research works, which concluded that all partner genes of the carcinogenic fusion gene usually have higher network centrality, and suggested that oncogenes prefer hub nodes in the network. The network fusion centrality hypothesis allows the algorithms to merge the nodes that correspond to the partner genes into a burden node representing the fusion gene [34]. In this case, the importance of a gene fusion structure is the accumulation of the importance of the previous partner nodes. However, some of the information between the nodes, which may be lost due to the overlapped edges of merged partner gene nodes, is often ignored.

Multiple approaches are available for prioritizing partner genes, among which network fusion centrality (FC) strategy is popular [28,29,34], as it is able to process gene fusion data better than other existing approaches. In this strategy, the gene networks are obtained as prior knowledge, each of which contains a set of genes. Note that, each node of these networks represents a single gene, and each edge denotes a specific type of interaction between the two genes. As none of the nodes represent a fusion gene or a gene fusion structure, the fusion gene nodes are constructed by merging the corresponding partner genes. To achieve this, the merging step first maps the partner genes to the entire gene network, and then each partner inherits the functions of the original gene on the network to evaluate the potential influence.

Two evaluation strategies for measuring the importance of a node are widely used: node degree [28] criterion and node betweenness [29] criterion. In the node degree algorithm, the degree of node i is calculated with $K(i) = \frac{\sum_{j \in G} a_{ij}}{N-1}$, where N represents the number of nodes and G represents the set of nodes. For unweighted networks, $a_{ij} \in (0, 1)$, where 0 indicates that no edge exists between node i and node j, and 1 indicates that an edge exists. For weighted networks, a_{ij} denotes the edge weight between nodes, where $K(i)$ represents the weighted degree of a node. The degree of the node represents the direct connection state between the node and other nodes. The importance of the node is expressed by the number of directly connected nodes. This method evaluates the significance of a node based on how well the node is directly connected to other nodes in the network topology. The advantage of this method is that the calculation is simple and the algorithm's time complexity is $O(N^2)$. The disadvantage is that only the neighbors of the node are considered, and only the local importance of the nodes in the network is calculated. For nodes in different positions in a complex network, the node importance caused by various topologies is not considered.

Node betweenness is a parameter used by Freeman [29] to measure social status of individuals in their research on social networks. The betweenness of node k is defined as the number of shortest paths between any two nodes passing through node k. The betweenness centrality $B(k)$ of node k is defined as $B(k) = \frac{g(k)}{g}$, where g is the number of shortest paths between each pair of nodes, and $g(k)$ is the number of the shortest paths via node k. The larger the value of node betweenness, the greater the role played by the node in the connectivity between other nodes in the network. That is, the greater the influence of the node on the network connectivity, the more important the node to the entire network. The node betweenness mainly considers the impact of nodes on the connectivity between other nodes in the network. The advantage is that the global importance of a node is explained by the impact of the node on the shortest paths between nodes in the entire network. The disadvantage is that the interaction between directly connected nodes is ignored, and the method is highly complex because it is time consuming to find the shortest path between all nodes.

The algorithm based on fusion centrality degree (DEG) [28,34] uses the degree of fusion node as the evaluation measurement, whereas the algorithm based on fusion centrality betweenness (BET) [29,34] evaluates the fusion nodes based on the betweenness. However, these criteria have been further argued to have their own preferences; thus, more comprehensive strategies are suggested. Other than the degree and betweenness, graph stability is another important measurement in graph theory to describe destructiveness of a network. The graph stability state is gradually approximated if all of weights of the edges satisfy a necessary condition [35]. The necessary condition is determined by the size of the network, average connectivity among the nodes, and a coupling coefficient that relies on graph topology. Existing studies have proposed multiple synchronous stability criteria for various graph topologies [35]. For example, many networks have a semi-ring $2K$ adjacent sub-structure, which enables existing conclusions on synchronous stability criteria, widely extensible to more complicated gene network topologies. Specifically, when $k = 1$, the graph degenerates to a ring structure, whereas if $k = n/2$, the graph is a fully connected graph. Once the synchronous stability criteria are locked in the evaluation algorithm, the calculation complexity for the edge weight condition considerably decreases compared to the betweenness calculation.

To overcome the disadvantages of the current methods, and to evaluate the cancer susceptibility created by a fusion gene based on the synchronous stability method, an algorithm named Synstable Fusion is proposed in this paper. Synchronous stability means that the coupled network is synchronously stable if the internal coupling matrix and the network coupling matrix satisfy certain conditions [35]. The proposed algorithm calculates the importance of genes in the gene network according to the "destructiveness equals to importance" hypothesis [28,34,36], which states that the importance of a node in a connected graph is identical to the destructiveness of deleting the node, and evaluates the corresponding fusion genes through the importance of partner gene nodes. The Synstable Fusion algorithm, which is based on synchronous stability, evaluates the importance of the fusion gene according to the whether or not the gene network achieves a synchronously stable state. When a weighted network falls into a synchronously stable state, the network ignores the noise and insignificant information while retaining the important node edges and network structure as much as possible, thereby reducing the computational complexity when evaluating the overall impact of the node on the network. The destructiveness of deleting the node is measured by using the network difference criterion, which reflects the importance of the gene nodes. This approach not only considers the local importance of the node, but also measures the influence of the node on the overall network structure, so the gene node's importance can be accurately calculated. The performance of our algorithm is tested and compared to the DEG and BET algorithms in a series of experiments. The experimental results demonstrate that the Synstable Fusion algorithm is able to effectively evaluate cancer fusion genes and performs better than the existing method.

2. Results

In order to test and verify the effectiveness of the proposed algorithm, named Synstable Fusion, we applied the algorithm to a widely used whole-gene network [36] obtained by 17 heterogeneous data to evaluate the importance of the fusion genes represented by the nodes. This gene network was obtained from Wu et al. [36], and the edge weights in the network indicate the tendency of the two genes to be joined to work together in one pathway. This network not only represents direct interactions between genes, but also includes functional interactions in a broader sense and has been used in many pathological and therapeutic studies related to cancer genes [37–41]. The 40,230 genes included in the entire gene network are provided in the Supplementary Materials. In order to reflect as much key and useful information as possible, the network has to be further processed to retain reliable inter-gene interactions. In the experiments, we used the "network fusion centrality" hypothesis, which was also used in many subsequent studies [42–46].

2.1. Experimental Data

The experimental data were selected based on the above studies [34,36]. A gene whose mutation is associate to a disease is called a susceptible gene. We followed the hypothesis that fusion genes formed by the interaction of susceptible cancer genes have relatively high significance, since susceptible cancer genes are important for the production of cancer [31–34]. We extracted 699 professionally curated human oncogenes from the Cancer Gene Census (CGC) [47] project as the susceptible cancer fusion genes, from which cancer may result due to their mutations. The CGC project collects and validates all published cancer-related genetic mutation studies by professionals in the field, collating them into a database with filtering criteria, and updates and maintains the data. Oncogenic mutations include both single-gene mutations (amplification, insertion, deletion, etc.) and translocations (fusions). Thus, this oncogene list also contains all possible partner genes of known oncogenic fusion genes until the date (December 2017) we obtained the list (Supplementary Table S1).

In the test data, it is assumed that N_f represents the number of total fusion genes, N_i is the number of susceptible fusion genes, and N_o is non-susceptible fusion genes. So, $N_f = N_i + N_o$. Two partner genes form a fusion gene, thus the number of partner genes in the dataset is $2N_f = 2N_i + 2N_o$. To generate the dataset, we randomly selected $2N_i$ susceptible partner genes from the known susceptible cancer genes [47], then paired them to create N_i susceptible fusion genes. For non-susceptible fusion genes, we randomly picked $2N_o$ common partner genes from the whole-gene network [36] and selected pairs to create N_o ordinary fusion genes. Here, we simply used the random function in the programming language's built-in library to implement random sampling without replacement process, and reset the random seed before each random process to ensure irregularity. N_i important fusion genes were assembled from paired samples of $2N_i$ oncogenes by random sampling without replacement. The same random sampling method was applied to $2N_o$ common genes to extract pairs of genes into N_o common fusion genes. The possibility of repeating samples inside the N_i and N_o datasets was avoided because the non-return sampling method was adopted. Since the whole gene network also contained 699 oncogenes for formation of susceptible fusion genes, and the selection process of N_i and N_o was independent of each other, overlaps between the important fusion genes (N_i) and the ordinary fusion genes (N_o) of one dataset occurred. Once this happened, we re-selected $2N_o$ common genes and randomly generated N_o fusion genes until no duplication was present between susceptible fusions and ordinary fusions. We prepared two N_i configurations and three N_f ($N_f = N_i + N_o$) configurations for the experiment, and 20 sets of random data were selected for each $N_i + N_o$ configuration, so there were $2 \times 3 \times 20 = 120$ sets of data in total. Every set of experimental data was created accordingly. Real known oncogenic fusion genes can be created using this procedure. Three expert-curated carcinogenic fusion genes, *EWSR1-FEV*, *HMGA2-LPP*, and *EWSR1-ETV4*, were identified from the datasets. All were assessed at high importance rankings by our evaluation algorithm. The results of respective datasets are provided in Supplementary Table S2.

2.2. Experimental Results

The effects of the SYN algorithm are illustrated using three criteria: (1) distribution curve of susceptible fusion gene; (2) recognition rate; and (3) receiver operating characteristic curve. The test applied various values of N_f and N_i. The effectiveness of the SYN algorithm was validated by the comparison with the DEG and BET algorithms. Different experiment scenarios were created based on various N_f and N_i values. For $N_f \in \{150, 200, 250\}$ and $N_i \in \{15, 25\}$, a total of six parameter configurations were generated. For each configuration, 20 sets of data were randomly generated. Our algorithm and the other two algorithms were applied to each set to separately calculate the importance and then sort the fusion genes according to these values. The results of the different configurations and algorithms are statistically summarized and the average data calculated from the 20 results of each case are demonstrated in the following subsections.

2.2.1. Distribution Curve of Susceptible Fusion Gene

The susceptible fusion genes were divided into 10 intervals I_i ($i = 1, 2, \ldots, 10$), where $I_i = \left(\frac{i-1}{10} N_f, \frac{i}{10} N_f \right]$. For each dataset, all calculated fusion gene significance was sorted in descending order, and then separated into 10 ranking intervals. The number of susceptible fusion genes that fell under each interval were counted. The results showed the effect of SYN, DEG, and BET algorithms in six cases of $N_f \in \{150, 200, 250\}$ and $N_i \in \{15, 25\}$. Figure 1 shows the average distribution curves of the susceptible fusion genes identified by the three algorithms.

From the distribution curves of the susceptible fusion genes, SYN was able to find most of the significant fusion genes from the first two intervals. In $N_i = 15$ cases, the mean number of susceptible fusion genes in the top two intervals was 13.367, and this number was 21.4 in $N_i = 25$ situations. There were approximately zero susceptible fusion genes in the lowest five ranges. Therefore, we summarize the average number of susceptible fusions in the top 20% ranked fusion genes in various cases in Figure 2.

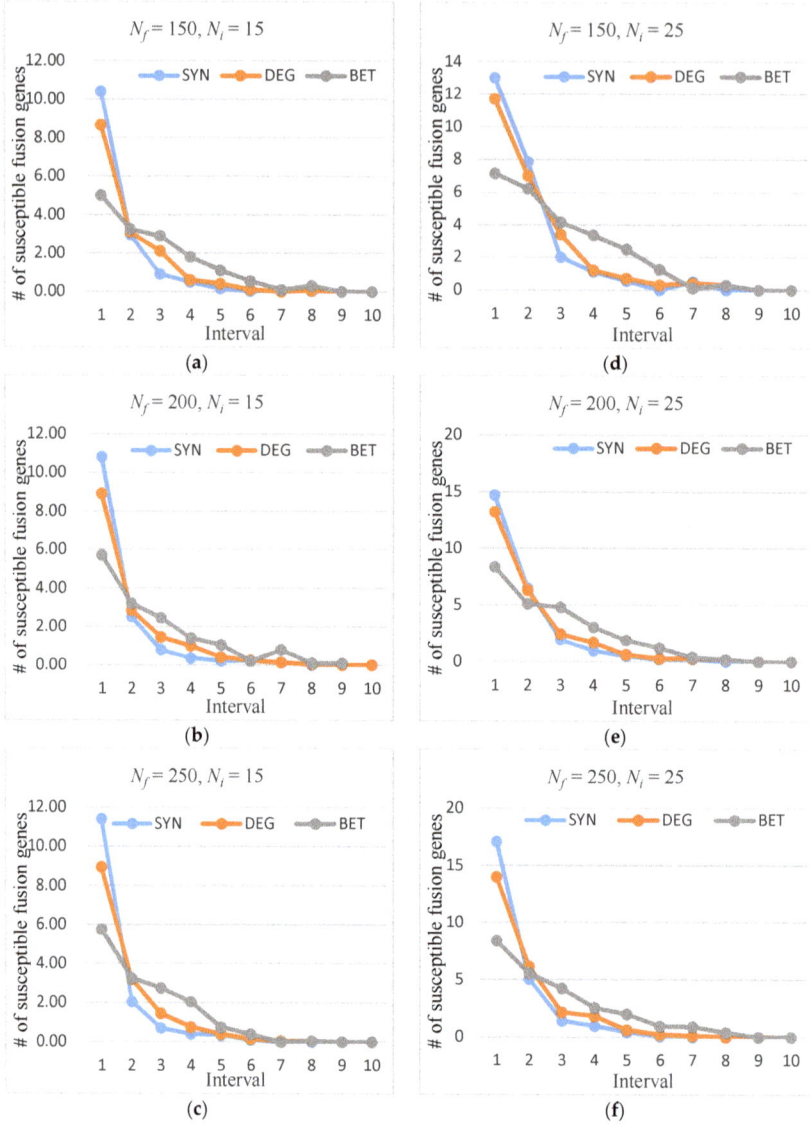

Figure 1. Average distribution curve of susceptible fusion genes: (**a**) $N_f = 150$, $N_i = 15$; (**b**) $N_f = 200$, $N_i = 15$; (**c**) $N_f = 250$, $N_i = 15$; (**d**) $N_f = 150$, $N_i = 25$; (**e**) $N_f = 200$, $N_i = 25$; and (**f**) $N_f = 250$, $N_i = 25$.

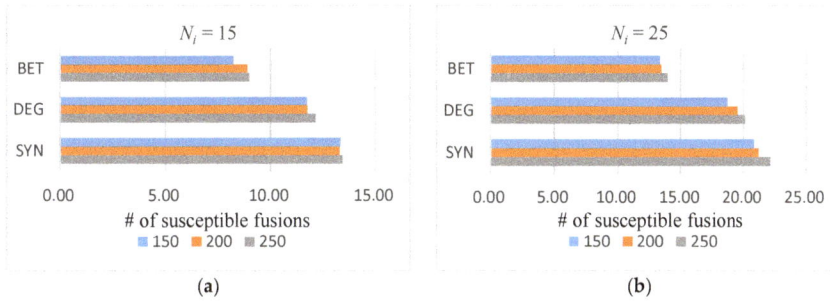

Figure 2. Average number of susceptible fusions in the top 20%. Different colors indicate different cases of total fusion gene amounts: (**a**) $N_i = 15$ and (**b**) $N_i = 25$.

From Figure 2, the average results of the SYN algorithm always outperform the results obtained with the DEG and BET methods in all situations. The largest difference occurred when $N_i = 25$ and $N_f = 250$: the average number of susceptible fusions found by the SYN algorithm was 58.8% more than the BET algorithm. The smallest gap occurred in comparing with the DEG algorithm when $N_i = 15$ and $N_f = 250$, as the difference percentage was 10.7%. From these results in Figure 2, we found that as the total number of fusion genes increased notably (by one-third or one-quarter), the amount of susceptible fusions within the top 20% area did not increase considerably, and the ratios were lower than the increasing rates of total fusion genes. We will discuss possible reasons for this result later in the Discussion section.

2.2.2. Recognition Rate

In order to illustrate the effectiveness of the proposed algorithm, the recognition rate of the susceptible fusion gene was adopted. The recognition rate represents the ratio of susceptible fusion genes located in a statistical interval to the total susceptible fusion genes. The recognition rate P is presented as $P(i) = \frac{f(R(i))}{N_i}$, where $R(i)$ denotes the ith statistical interval, $R(i) = \left[1, \frac{i}{10} N_f\right]$, $i \in \{1, 2, \ldots, 10\}$, and $f(R(i))$ indicates the number of susceptible fusion genes being found in the ith interval. We randomly selected 120 sets, and generated mixed experimental data in six cases ($N_f \in \{150, 200, 250\}$, $N_i \in \{15, 25\}$). As an illustrative case, Figure 3 demonstrates the $p(2)$ value of every experimental result.

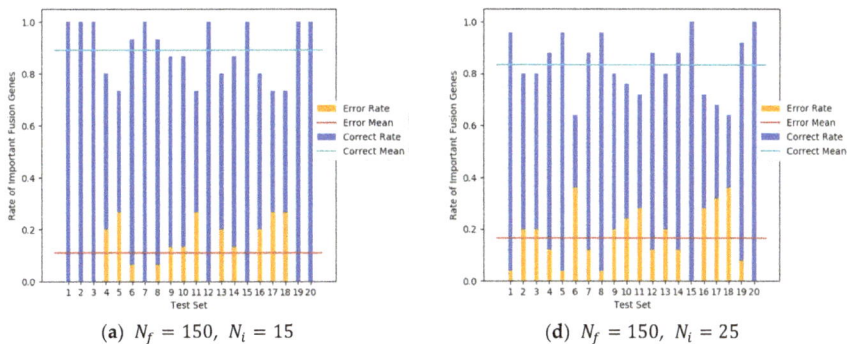

Figure 3. *Cont.*

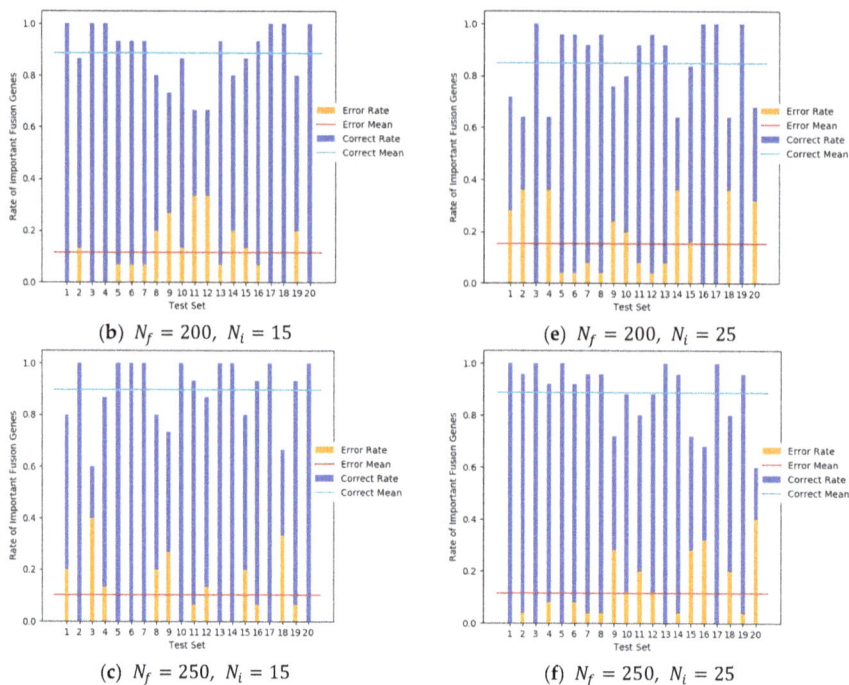

Figure 3. Recognition rates (correct rate) $p(2)$ (blue bars) of susceptible fusion genes in the top 20% of ranked fusion genes in each experimental result. The orange bar represents the rate of important fusion genes which is not included in this interval (error rate). (**a**) $N_f = 150$, $N_i = 15$; (**b**) $N_f = 200$, $N_i = 15$; (**c**) $N_f = 250$, $N_i = 15$; (**d**) $N_f = 150$, $N_i = 25$; (**e**) $N_f = 200$, $N_i = 25$; and (**f**) $N_f = 250$, $N_i = 25$.

The summarized average results are exhibited using radar panels, where vertices indicate the statistical intervals of the susceptible fusions, and axes indicate that recognition rate, which gradually increased outward. Because almost all susceptible fusion genes were included in the top 50% of ranked result, only the first five intervals are shown in the figures.

Figure 4 shows the $N_i = 15$. $p(1)$ results of the SYN algorithm. The recognition rate was about 70%, whereas those for the same interval obtained by the other two algorithms were less than 60%. The $p(2)$ value of the SYN algorithm was around 90%, which means approximately 90% of the susceptible fusion genes can be found using the SYN algorithm from its top 20% sorted results. As the range continuously increased, the $p(i)$ value increased as well. The differences among algorithms continually decreased and the recognition rates of all algorithms gradually approached 100%.

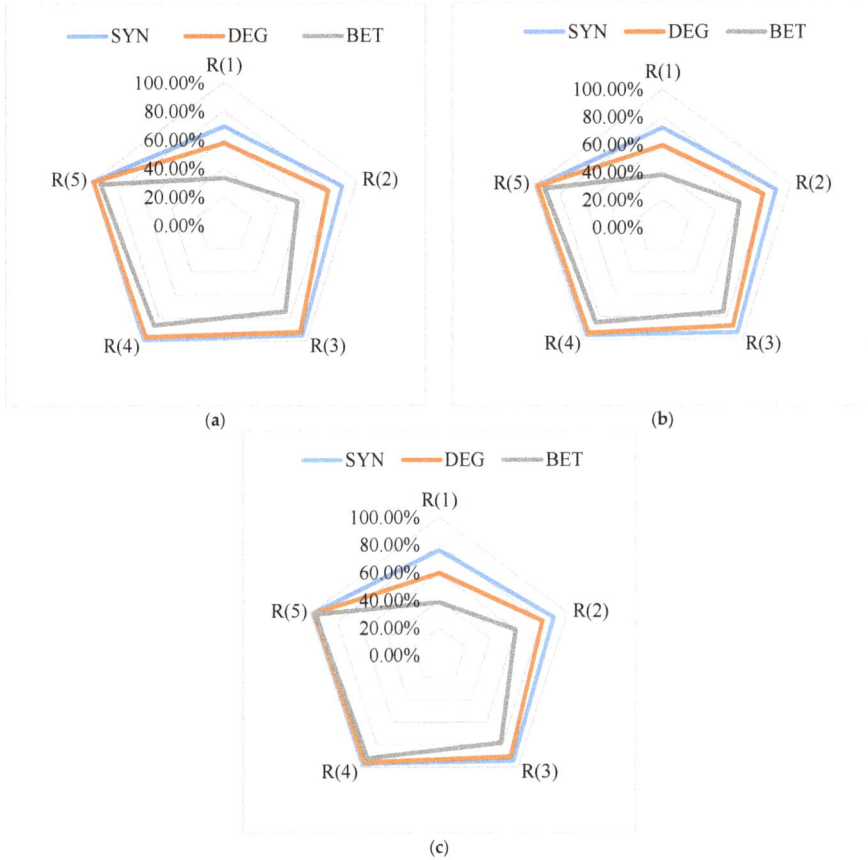

Figure 4. Average recognition rates of susceptible fusion gene in each interval when $N_i = 15$: (a) $N_f = 150$; (b) $N_f = 200$; and (c) $N_f = 250$.

Figure 5 highlights the statistical average results when $N_i = 25$. Compared with the case when $N_i = 15$, the overall recognition rate of the interval $R(1)$ decreased significantly, which was mainly because the total number of fusion genes in $R(1)$ was less than or equal to the number of pathogenic fusion genes in test samples ($N_i = 25$). The corresponding proportion of pathogenic fusion genes was relatively lower. Other factors also affected the experimental results. We discuss the possible causes in the Discussion section. The $p(2)$ value of the SYN algorithm was around 85%, whereas the recognition rates for the same interval obtained by the two other control algorithms were both less than 80%. The $p(3)$ value of the SYN algorithm was higher than 90%, whereas the values obtained by the two control algorithms were less than 90%. Figures 4 and 5 clearly illustrate that the recognition rate of the SYN algorithm in all situations was higher than those of the DEG and BET methods.

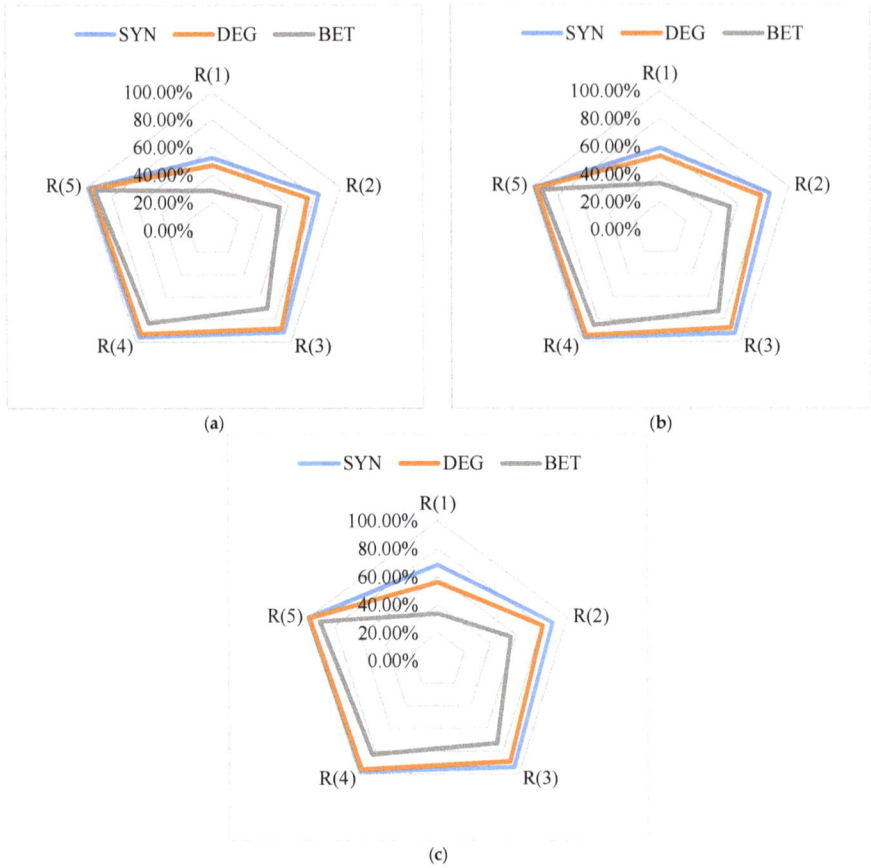

Figure 5. Average recognition rates of susceptible fusion gene in each interval when $N_i = 25$: (a) $N_f = 150$; (b) $N_f = 200$; and (c) $N_f = 250$.

2.2.3. Receiver Operating Characteristic Curve

By adding a classification boundary to the results of the algorithms, the original algorithm can be changed into a binary classification algorithm. Fusion genes above the classification limit can be classified as cancer pathogen fusion genes, and vice versa as normal fusion genes. As such, we calculated the algorithm's receiver operating characteristic curve (ROC). Figure 6 shows the ROC curves of the three algorithms in all six cases and the area under curve (AUC) values for each curve, where the Y-axis is the true positive (TP) rate and the X-axis is the false positive (FP) rate.

From the ROC results, the best classification performance occurred at $N_f = 150$ and $N_i = 15$, where the AUC value was around 0.945. The situation with the smallest AUC score was $N_f = 200$ and $N_i = 15$, which had a value around 0.93. The overall performance of the proposed algorithm remained high.

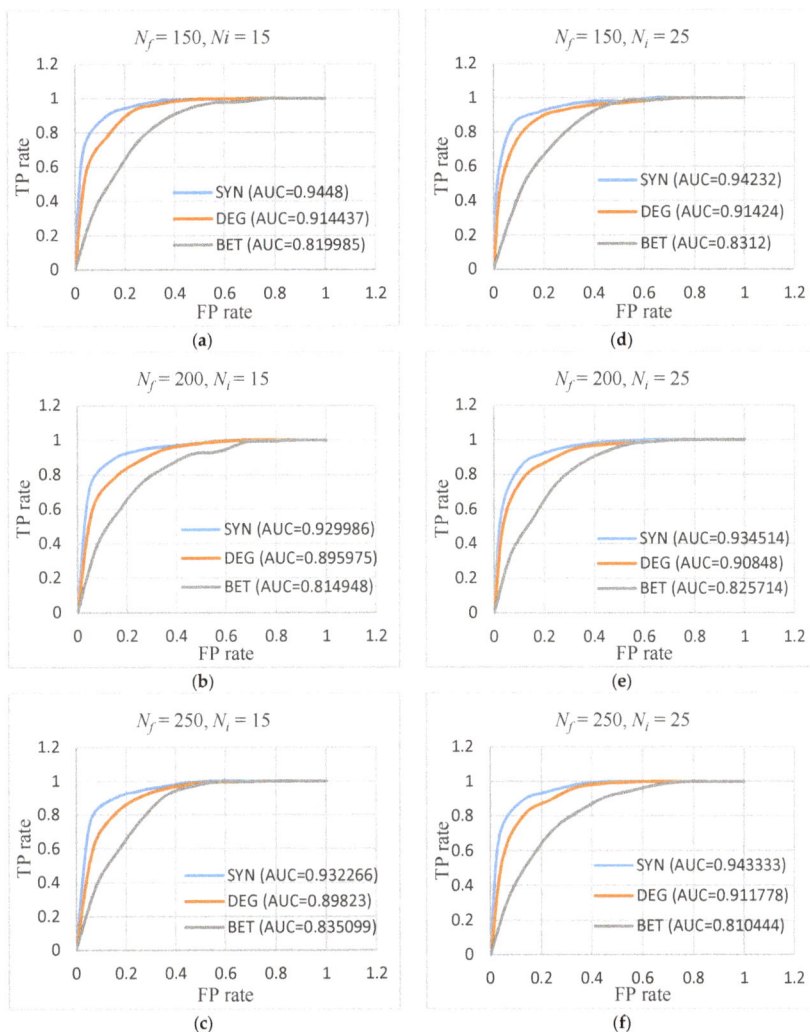

Figure 6. Receiver operating characteristic (ROC) curves of the three algorithms: (**a**) $N_f = 150$, $N_i = 15$; (**b**) $N_f = 200$, $N_i = 15$; (**c**) $N_f = 250$, $N_i = 15$; (**d**) $N_f = 150$, $N_i = 25$; (**e**) $N_f = 200$, $N_i = 25$; and (**f**) $N_f = 250$, $N_i = 25$. FP—false positive; TP—true positive.

3. Discussion

The experimental results clearly demonstrate that the proposed Synstable Fusion (SYN) algorithm performs better when calculating the importance of cancer-causing fusion genes in gene networks. More susceptible fusion genes were included in the top portion of the descending-sorted result, which means that possible oncogenic fusion genes have a greater tendency to be evaluated with higher importance values when using the SYN algorithm. As an example, three known oncogenic fusion genes found in the experimental results obtained by our algorithm received high importance rankings. The three fusion genes, EWSR1-FEV, HMGA2-LPP, and EWSR1-ETV4, were ranked first, tenth, and second in their respective datasets. Specific experimental datasets, importance calculation scores, and potential carcinogenic rankings are provided in Supplementary Table S2.

We found two phenomena worth noting. The first is that the number of cancer pathogenic fusion genes identified in the first 20% of the ranking results did not increase significantly as the total number of samples included in the test dataset increased. At $N_i = 15$, the discrepancy among the maximum and minimum numbers of pathogenic fusion genes in the first 20% of the results of SYN algorithm was only 0.15 in three cases, and this difference only increased to 1.3 at $N_i = 25$. The second phenomena is that the overall recognition effect slightly decreased when the total number of pathogenic fusion genes in the sample was high. For example, when $N_f = 250$, the five-interval average recognition rate of the cancer-causing fusion gene of the SYN algorithm was 91.27%; when $N_i = 15$, this value was 89.8%, and when $N_i = 25$, a decrease of about 1.5% was observed. In the following, we discuss the possible causes of these two phenomena and explain why the results of the proposed algorithm are better than those of the other two algorithms.

The first case phenomenon occurred when the number of identified pathogenic fusion genes did not increase with the total fusion gene number in the sample. This may have occurred because the calculation scores of most disease-causing fusion genes were high but the scores of a fixed fraction of the proportion of susceptible fusion genes were lower. This is because the algorithms' results are based on the genomics inference network derived by the classification algorithm of machine learning, and the result generated by classification algorithms must be partially consistent with the expected errors.

In the experimental dataset, the number of susceptible fusion genes was high whereas the recognition effect was slightly lower, possibly because the increase in the number of pathogenic fusion genes in the samples led to an increase in the occurrence probability of susceptible cancer fusion genes with low importance scores. The distribution of the number of pathogenic fusion genes in the first 20% results can provide support for this explanation. When $N_i = 15$, the recognition distribution of each algorithm (Figure 2a) was almost unchanged, and the number of high-importance pathogenic fusion genes remained unchanged at a high rate. The number of identifications at $N_i = 25$ (Figure 2b) slightly increased because some of the disease-causing fusion genes with slightly lower scores appeared in the test dataset. These genes were gradually identified as the range of recognition intervals increased.

From the experimental results, the performance of the proposed algorithm is better than that of the DEG and BET algorithms under various parameter settings with experimental data. The proposed algorithm uses more comprehensive information contained in the gene network to calculate the importance of nodes. When evaluating the importance of nodes, BET algorithm only considers the influence of the nodes on the network topology, whereas the DEG algorithm only considers the influence between the node and its directly related parts of the network. However, in our algorithm, the "destructiveness equals to importance" hypothesis is applied, which not only considers the degree of the node, but also the impact of deleting the node on the network topology. This is equivalent to a certain degree of the incorporation of the first two algorithms. Therefore, SYN outperforms the DEG and BET algorithms.

4. Materials and Methods

The Synstable Fusion algorithm is based on the synchronous stability method, which evaluates the node importance according to the influence on stability of gene network when a node is removed. Wu et al. [36] used a Relevance Vector Machine (RVM)-based [48] ensemble-learning model to construct a whole gene network. This model integrates 17 heterogeneous genomic data and proteomics data [36]. We used this model mainly because it incorporates many different kinds of data, and simultaneously better handles the problem of missing attribute values among heterogeneous data [49] and outputs probabilistic results. Weighted edges existed between paired nodes in the entire gene network of the human genome. The weight represents the probability of interactive works between two genes, not only reflecting the direct interactions between genes, such as activation, inhibition, binding, and dissociation, but also other broader relationships among genes, for example, the likelihood of genes working on the same or similar biological pathways. In order to evaluate the influence of a node on stability, the synchronously stable networks were identified from the original gene network and the network of deleting a node, and then the difference between these two synchronously stable networks was calculated. Based on the node influence on network stability, the cancer fusion gene was evaluated. In this section, the design of the proposed algorithm is described in detail.

4.1. Synchronous Stability Method

In order to identify the synchronously stable network from a gene network, the synchronous stability method was required to ensure the relative stability of the gene network. A network is considered to be in synchronously stable state when it satisfies a certain condition.

4.1.1. Synchronous Stability Condition

For the connected graph with ring of $2K$ adjacent nodes, the condition of synchronous stability is presented [35] as:

$$w > \varepsilon = \frac{a}{n}\left(\frac{n}{2K}\right)^3\left(1 + \frac{65}{4}\frac{K}{n}\right) \tag{1}$$

where w denotes the edge weight, ε represents the lowest limit of the w, a is an important parameter indicating the coupling state of network, n denotes the node amount of the network, and K indicates the number of half neighbors. Parameter a is called the coupling parameter that describes the coupling characteristic of the network. The value of a is determined by analyzing the adjacency matrix of graph. A previous study [50] indicated that λ_2 is the algebraic connectivity of the connected graph and a $< \lambda_2$. By inducing the Laplace matrix, we obtained a series of eigenvalues that satisfy $0 = \lambda_1 \leq \lambda_2 \leq \ldots \leq \lambda_n$. The algebraic connectivity λ_2 is one of the eigenvalues and it was the minimum nonzero-eigenvalue. λ_2 denotes the synchronous ability of the connected graph. Thus, $0 < a < \lambda_2$ and then we sequentially chose the fittest a value from this range based on some system analysis.

In order to find the most suitable value a, let $a = s\lambda_2$ and $s = [0.01, 0.02, ..., 0.99]$. For each s value, we calculated the proportion of the lost information filtered by the synchronously steady state of a given gene network. In the experimental gene data, 20 gene networks were randomly selected and generated. Figure 7 shows the result of one set of data, the average result of 20 sets of data, and the gradient of the average result.

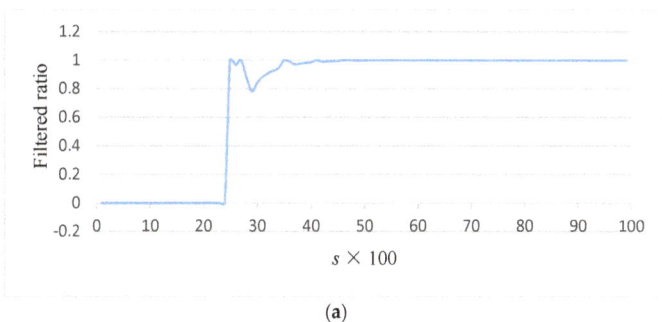

(a)

Figure 7. *Cont.*

(b)

(c)

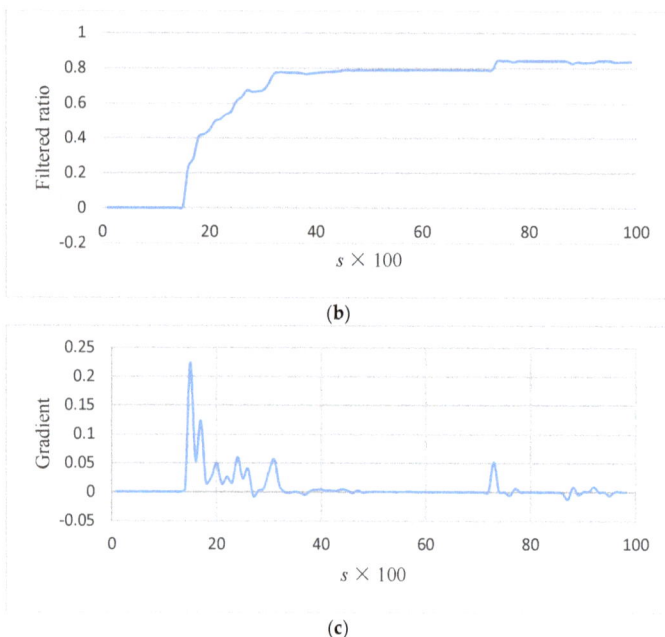

Figure 7. The results by various *s* (the product factor of coupling state parameter) values: (**a**) result of one set of data; (**b**) average result of 20 sets of data; and (**c**) Gradient of average result.

We tried to find a suitable *s* value for most experimental data to filter out most noise and insignificant information while retaining key information in gene network. From Figure 6a,b, we found that depression points always occurred in the proportion of filtered data in all 20 experimental results. Through analyzing the gradient of the average data, we found the depression point where the gradient was first close to zero. From Figure 6c, point 0.28 satisfies the requirement. So $a = 0.28\lambda_2$, which is inserted into Equation (1):

$$w > \varepsilon = \frac{0.28\lambda_2}{n}\left(\frac{n}{2K}\right)^3\left(1 + \frac{65}{4}\frac{K}{n}\right).\tag{2}$$

Our research considered two situations of the connected graphs: the gene networks have a fully connected topology, and the gene networks do not have a fully connected topology. The synchronous stability condition of fully connected networks can be obtained by letting $n = 2K$. Therefore, Equation (2) becomes:

$$w > \varepsilon = 2.555\frac{\lambda_2}{n}.\tag{3}$$

If the topology of graph is not fully connected, its maximum fully connected subgraph can be found. Let *m* denote half of the number of nodes in this subgraph. For the connected graph with a maximum ring of 2K adjacent nodes, the fully connected subgraph is a ring of 2m adjacent nodes, so we obtain $2m < 2K$. The value of ε is increased by replacing *k* with *m*:

$$w > \varepsilon = \frac{0.28\lambda_2}{n}\left(\frac{n}{2m}\right)^3\left(1 + \frac{65}{4}\frac{m}{n}\right).\tag{4}$$

The edge weight limit ε calculated by Equation (4) is greater than the lowest limit of the synchronously stable condition. Therefore, the new limit can also be used as the judging condition for synchronous stability.

4.1.2. Identification of Synchronously Stable Network

From Equations (3) and (4), we obtained the lower edge weight limit for every gene network of various topologies. The network is in a synchronously steady state if $\forall w_{ij} > \varepsilon$, where w_{ij} denotes the edge weight between node i and j. Otherwise, if $\exists w_{ij} < \varepsilon$, it is in a non-synchronously steady state and needs further processing to achieve synchronous stability. Different procedures were assigned based on whether or not the gene network was fully connected. All edges with a weight less than the lower limit were deleted if the network was fully connected. If the connected subgraph was a non-fully connected graph, the hanging nodes were detected. If there were hanging nodes, the hanging nodes were deleted. If no hanging nodes existed in the connected subgraph, then the edges with the weight less than the low limit were deleted. This procedure was iterated until all connected subgraphs were in the synchronously steady state. Figure 8 shows the flowchart of the identification of the synchronously stable network.

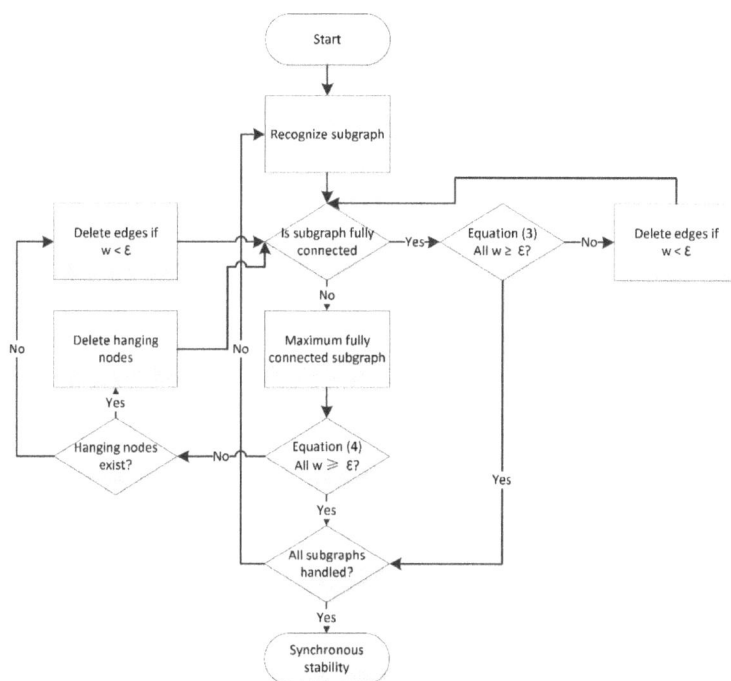

Figure 8. Flowchart followed for synchronously stable network identification.

In the identification process for non-fully connected graphs, the maximum fully connected subgraph had to be obtained. This is called the maximum clique problem and is *NP*-hard that no known algorithms can achieve optimized solution. To solve this problem, some widely used algorithms include the greedy search algorithm, intelligent search algorithm, and heuristic search algorithm. In this work, we chose the greedy search algorithm.

4.2. Evaluation Susceptible Fusion Gene

Here, we describe the algorithm for prioritizing the susceptible cancer fusion gene using graph theory and gene network. First, we estimated the importance of gene nodes in the gene network. Then, the cancer susceptibility of the fusion genes was evaluated based on the importance of the partner gene nodes. The algorithm estimates the importance of gene nodes by evaluating the destructiveness to the network of deleting the node, where the destructiveness is evaluated according to the difference between the synchronously stable networks before and after the removal of the node. To ensure the gene network stayed synchronously stable, an identification method for synchronously stable networks was used. Figure 8 shows the process followed for achieving a synchronously stable network.

4.2.1. Network Difference Evaluation

The impacts of deleting a node and its associated edges include two aspects: the impact on the degree of remained nodes, and the influence on connectivity. Considering these two aspects, we defined the metric for a gene network, $M(G)$, as the ratio of total edge weights of network to the number of subgraphs:

$$M(G) = \frac{\sum_{i \in G} \sum_{j \neq i \wedge j \in G} w_{ij}}{m_G} \tag{5}$$

where G indicates the gene network, m_G denotes the number of subgraphs of G, and w_{ij} denotes the weight between nodes i and j, $i, j \in G$. Once a node is deleted, the total edge weights decrease and the network connectivity decreases as well, which can be reflected by the increasing of number of subgraphs. All these influences can decrease the value of $M(G)$. The network difference $D(G,v)$ in deleting node v can be represented as:

$$
\begin{aligned}
D(G,v) &= M(G) - M(G - v) \\
&= \frac{\sum_{i \in G} \sum_{j \neq i \wedge j \in G} w_{ij}}{m_G} - \frac{\sum_{i \in (G-v)} \sum_{j \neq i \wedge j \in (G-v)} w_{ij}}{m_{G-v}}
\end{aligned}
\tag{6}
$$

where $G - v$ is the network obtained by removing node v and its corresponding edges from network G. Based on the difference $D(G,v)$, the importance $H(G,v)$ of node v in network G is defined as:

$$H(G,v) = \frac{D(v)}{M(G_s)} = \frac{M(G_s) - M((G_s - v)_s)}{M(G_s)} \tag{7}$$

where G_s and $(G_s - v)_s$ represent the network G in synchronously stable state depending on whether or not node v is deleted.

4.2.2. Calculation of Gene Node Importance

The algorithm for evaluating a gene node uses the synchronous stability method. Let $G = (V, W)$ represent the gene network. n is the number of nodes in the network, $V = \{v_1, v_2, \ldots, v_n\}$ indicates the set of nodes, and $W = \{w_{01}, w_{02}, \ldots, w_{ij}, \ldots, w_{nk}\}$ represents the set of edge weights, where w_{ij} is the edge weight between nodes i and j, $i, j \in V$. The algorithm processes are as follows:

Step 1: Utilize the susceptible cancer gene test data to generate gene network G from the human gene network.

Step 2: Process G by the synchronously stable network identification procedures described in Section 4.1.2., marked as G_s.

Step 3: Delete a node v_i and its associated edges in G_s, $(G_s - v_i)$.

Step 4: Use the synchronously stable network identification to process $(G_s - v_i)$ to obtain the synchronously stable state, $(G_s - v_i)_s$.

Step 5: Use Equation (6) to calculate the $D(G,v)$ value, and subsequently calculate the $H(G,v)$ value using Equation (7).

Step 6: Evaluate the importance of every node in the gene network by repeating Steps 3–5.

4.2.3. Evaluation of Susceptible Cancer Fusion Genes

By using the importance of the partner gene nodes, the significance of the fusion gene could be evaluated. The significance of a fusion gene is calculated by adding partner genes' importance together then multiplying the weight of the edge between two partners. The significance $S(f)$ of fusion gene f is:

$$S(f) = (1 + w_{ij})(H(i) + H(j)) \tag{8}$$

where i and j denote the partner gene node associated with f, w_{ij} is the edge weight between i and j, and $H(i)$ and $H(j)$ are the significance values of i and j, respectively. Fusion genes are formed by the interaction of partner genes. Edge weight between partner gene nodes reflects the interactive relationship between partners genes. Therefore, we consider this probabilistic value when evaluating the fusion gene's significance.

5. Conclusions

This study proposed a method called Synstable Fusion for prioritizing the importance of fusion nodes in a weighted graph, based on the synchronous stability of gene network. This method, when applied to a gene network, effectively evaluates important fusion genes and identifies possible cancer pathogenicity fusion genes. The experimental results showed that the effectiveness of the proposed algorithm is superior to the other two algorithms based on network fusion centrality. In the experiment, we also found some issues that need attention, which could be the focus of future research and development. First, a more accurate gene network generation method should be explored to increase the reliability of the evaluation calculations. Second, other relevant theories can be applied instead of the synchronous stability method to achieve a more efficient and accurate interference information filtering method. In addition, we will try to introduce other algorithms that consider the node's effect on network topology, so that we can more accurately evaluate the value of a node in the network.

Supplementary Materials: The following are available online, Table S1: Lists of genes used in experiment, Table S2: Experimental results of datasets including known oncogenic fusion genes.

Author Contributions: J.W. and X.Z. conducted this study; J.W., A.G., and X.Z. conceived and designed the algorithms; A.G., M.X., and S.W. designed and performed the experiments; M.X., X.Z., Z.Z., and J.W. wrote the manuscript. All authors read and approved the final version of this manuscript.

Funding: This work is supported by the National Science Foundation of China (Grant No: 31701150) and the Fundamental Research Funds for the Central Universities (CXTD2017003). The research funds cover the costs to publish in open access.

Conflicts of Interest: The authors declare no conflict of interest.

References

1. Mertens, F.; Johansson, B.; Fioretos, T.; Mitelman, F. The emerging complexity of gene fusions in cancer. *Nat. Rev. Cancer* **2015**, *15*, 371–381. [CrossRef] [PubMed]
2. Kumar-Sinha, C.; Kalyana-Sundaram, S.; Chinnaiyan, A.M. Landscape of gene fusions in epithelial cancers: Seq and ye shall find. *Genome Med.* **2015**, *7*, 129. [CrossRef] [PubMed]
3. Latysheva, N.S.; Babu, M.M. Discovering and understanding oncogenic gene fusions through data intensive computational approaches. *Nucleic Acids Res.* **2016**, *44*, 4487–4503. [CrossRef] [PubMed]
4. Persson, H.; Søkilde, R.; Häkkinen, J.; Pirona, A.C.; Vallon-Christersson, J.; Kvist, A.; Mertens, F.; Borg, Å.; Mitelman, F.; Höglund, M.; et al. Frequent miRNA-convergent fusion gene events in breast cancer. *Nat. Commun.* **2017**, *8*, 788. [CrossRef] [PubMed]
5. Lu, C.; Xie, M.; Wendl, M.C.; Wang, J.; McLellan, M.D.; Leiserson, M.D.; Huang, K.L.; Wyczalkowski, M.A.; Jayasinghe, R.; Banerjee, T.; et al. Patterns and functional implications of rare germline variants across 12 cancer types. *Nat. Commun.* **2015**, *6*, 10086. [CrossRef] [PubMed]

6. Huang, K.L.; Mashl, R.J.; Wu, Y.; Ritter, D.I.; Wang, J.; Oh, C.; Paczkowska, M.; Reynolds, S.; Wyczalkowski, M.A.; Oak, N.; et al. Pathogenic Germline Variants in 10,389 Adult Cancers. *Cell* **2018**, *173*, 355–370.e14. [CrossRef] [PubMed]

7. Kim, D.; Salzberg, S.L. TopHat-Fusion: An algorithm for discovery of novel fusion transcripts. *Genome Boil.* **2011**, *12*, R72. [CrossRef] [PubMed]

8. McPherson, A.; Hormozdiari, F.; Zayed, A. deFuse: An Algorithm for Gene Fusion Discovery in Tumor RNA-Seq Data. *PLoS Comput. Boil.* **2011**, *7*, e1001138. [CrossRef] [PubMed]

9. Zhang, J.; White, N.M.; Schmidt, H.K.; Fulton, R.S.; Tomlinson, C.; Warren, W.C.; Wilson, R.K.; Maher, C.A. INTEGRATE: Gene fusion discovery using whole genome and transcriptome data. *Genome Res.* **2016**, *26*, 108–118. [CrossRef] [PubMed]

10. Haber, D.A.; Settleman, J. Cancer: Drivers and passengers. *Nature* **2007**, *446*, 145–146. [CrossRef] [PubMed]

11. Grigoryev, Y.A.; Kurian, S.M.; Avnur, Z.; Borie, D.; Deng, J.; Campbell, D.; Sung, J.; Nikolcheva, T.; Quinn, A.; Schulman, H.; et al. Deconvoluting post-transplant immunity: Cell subset-specific mapping reveals pathways for activation and expansion of memory T, monocytes and B cells. *PLoS ONE* **2010**, *5*, e13358. [CrossRef] [PubMed]

12. Johnson, J.M.; Castle, J.; Garrett-Engele, P.; Kan, Z.; Loerch, P.M.; Armour, C.D.; Santos, R.; Schadt, E.E.; Stoughton, R.; Shoemaker, D.D. Genome-wide survey of human alternative pre-mRNA splicing with exon junction microarrays. *Science* **2003**, *302*, 2141–2144. [CrossRef] [PubMed]

13. Schadt, E.E.; Edwards, S.W.; GuhaThakurta, D.; Holder, D.; Ying, L.; Svetnik, V.; Leonardson, A.; Hart, K.W.; Russell, A.; Li, G.; et al. A comprehensive transcript index of the human genome generated using microarrays and computational approaches. *Genome Biol.* **2004**, *5*, R73. [CrossRef] [PubMed]

14. Wang, J.; Zhao, Z.; Cao, Z.; Yang, A.; Zhang, J. A probabilistic method for identifying rare variants underlying complex traits. *BMC Genomics* **2013**, *14*, S11. [CrossRef] [PubMed]

15. Blomen, V.A.; Májek, P.; Jae, L.T.; Bigenzahn, J.W.; Nieuwenhuis, J.; Staring, J.; Sacco, R.; Diemen, F.R.; Olk, N.; Stukalov, A.; et al. Gene essentiality and synthetic lethality in haploid human cells. *Science* **2015**, *350*, 1092–1096. [CrossRef] [PubMed]

16. Papin, J.A.; Hunter, T.; Palsson, B.O.; Subramaniam, S. Reconstruction of cellular signalling networks and analysis of their properties. *Nat. Rev. Mol. Cell Biol.* **2005**, 99–111. [CrossRef] [PubMed]

17. Wu, G.; Feng, X.; Stein, L. A human functional protein interaction network and its application to cancer data analysis. *Genome Biol.* **2010**, *11*, R53. [CrossRef] [PubMed]

18. Zhou, L.; Lyons-Rimmer, J.; Ammoun, S.; Müller, J.; Lasonder, E.; Sharma, V.; Ercolano, E.; Hilton, D.; Taiwo, I.; Barczyk, M.; et al. The scaffold protein KSR1, a novel therapeutic target for the treatment of Merlin-deficient tumors. *Oncogene* **2016**, *35*, 3443–3453. [CrossRef] [PubMed]

19. Soler-López, M.; Zanzoni, A.; Lluís, R.; Stelzl, U.; Aloy, P. Interactome mapping suggests new mechanistic details underlying Alzheimer's disease. *Genome Res.* **2011**, *21*, 364–376. [CrossRef] [PubMed]

20. Rodgers-Melnick, E.; Culp, M.; DiFazio, S.P. Predicting whole genome protein interaction networks from primary sequence data in model and non-model organisms using ENTS. *BMC Genom.* **2013**, *14*, 608. [CrossRef] [PubMed]

21. Stark, C.; Breitkreutz, B.J.; Reguly, T.; Boucher, L.; Breitkreutz, A.; Tyers, M. BioGRID: A general repository for interaction datasets. *Nucleic Acids Res.* **2006**, *34*, D535–D539. [CrossRef] [PubMed]

22. Zeng, X.; Liu, L.; Lü, L.; Zou, Q. Prediction of potential disease-associated microRNAs using structural perturbation method. *Bioinformatics* **2018**, *34*, 2425–2432. [CrossRef] [PubMed]

23. Geng, Y.; Zhao, Z.; Zhang, X.; Wang, W.; Cui, X.; Ye, K.; Xiao, X.; Wang, J. An improved burden-test pipeline for identifying associations from rare germline and somatic variants. *BMC Genomics* **2017**, *18*, 55–62. [CrossRef] [PubMed]

24. Wang, H.; Ding, S.; Wu, D.; Zhang, Y.; Yang, S. Smart connected electronic gastroscope system for gastric cancer screening using multi-column convolutional neural networks. *Int. J. Prod. Res.*. [CrossRef]

25. Warde-Farley, D.; Donaldson, S.L.; Comes, O.; Zuberi, K.; Badrawi, R.; Chao, P.; Franz, M.; Grouios, C.; Kazi, F.; Lopes, C.T.; et al. The GeneMANIA prediction server: Biological network integration for gene prioritization and predicting gene function. *Nucleic Acids Res.* **2010**, *38*, W214–W220. [CrossRef] [PubMed]

26. Cantini, L.; Medico, E.; Fortunato, S.; Caselle, M. Detection of gene communities in multi-networks reveals cancer drivers. *Sci. Rep.* **2015**, *5*, 17386. [CrossRef] [PubMed]

27. Cava, C.; Bertoli, G.; Colaprico, A.; Olsen, C.; Bontempi, G.; Castiglioni, I. Integration of multiple networks and pathways identifies cancer driver genes in pan-cancer analysis. *BMC Genom.* **2018**, *19*. [CrossRef] [PubMed]

28. Freeman, L.C. Centrality in Social Networks Conceptual Clarification. *Soc. Netw.* **1978**, *1*, 215–239. [CrossRef]

29. Freeman, L.C. A Set of Measures of Centrality Based on Betweenness. *Sociometry* **1977**, *40*, 35–41. [CrossRef]

30. Schmidhuber, J. Deep learning in neural networks: An overview. *Neural Netw.* **2015**, *61*, 85–117. [CrossRef] [PubMed]

31. Palanisamy, N.; Ateeq, B.; Kalyana-Sundaram, S.; Pflueger, D.; Ramnarayanan, K.; Shankar, S.; Han, B.; Cao, Q.; Cao, X.; Suleman, K.; et al. Rearrangements of the RAF kinase pathway in prostate cancer, gastric cancer and melanoma. *Nat. Med.* **2010**, *16*, 793–798. [CrossRef] [PubMed]

32. Robinson, D.R.; Kalyana-Sundaram, S.; Wu, Y.M.; Shankar, S.; Cao, X.; Ateeq, B.; Asangani, I.A.; Iyer, M.; Maher, C.A.; Grasso, C.S.; et al. Functionally recurrent rearrangements of the MAST kinase and Notch gene families in breast cancer. *Nat. Med.* **2011**, *17*, 1646–1651. [CrossRef] [PubMed]

33. Wang, X.S.; Prensner, J.R.; Chen, G.; Cao, Q.; Han, B.; Dhanasekaran, S.M.; Ponnala, R.; Cao, X.; Varambally, S.; Thomas, D.G.; et al. An integrative approach to reveal driver gene fusions from paired-end sequencing data in cancer. *Nat. Biotechnol.* **2009**, *27*, 1005–1011. [CrossRef] [PubMed]

34. Wu, C.C.; Kannan, K.; Lin, S.; Yen, L.; Milosavljevic, A. Identification of cancer fusion drivers using network fusion centrality. *Bioinformatics* **2013**, *29*, 1174–1181. [CrossRef] [PubMed]

35. Belykh, V.N.; Belykh, I.V.; Hasler, M. Connection graph stability method for synchronized coupled chaotic systems. *Phys. D Nonlinear Phenom.* **2004**, *195*, 159–187. [CrossRef]

36. Wu, C.C.; Asgharzadeh, S.; Triche, T.J.; D'Argenio, D.Z. Prediction of human functional genetic networks from heterogeneous data using RVM-based ensemble learning. *Bioinformatics* **2010**, *26*, 807–813. [CrossRef] [PubMed]

37. He, L.; Wang, Y.; Yang, Y.; Huang, L.; Wen, Z. Identifying the gene signatures from gene-pathway bipartite network guarantees the robust model performance on predicting the cancer prognosis. *Biomed. Res. Int.* **2014**, *2014*, 424509. [CrossRef] [PubMed]

38. Wang, H.; Huang, L.; Jing, R.; Yang, Y.; Liu, K.; Li, M.; Wen, Z. Identifying oncogenes as features for clinical cancer prognosis by Bayesian nonparametric variable selection algorithm. *Chemom. Intell. Lab. Syst.* **2015**, *146*, 464–471. [CrossRef]

39. Grover, M.P.; Ballouz, S.; Mohanasundaram, K.A.; George, R.A.; Sherman, C.D.; Crowley, T.M.; Wouters, M.A. Identification of novel theracassociation data. *BMC Med. Genom.* **2014**, *7* (Suppl. S1), S8. [CrossRef] [PubMed]

40. Schneider, L.; Stöckel, D.; Kehl, T.; Gerasch, A.; Ludwig, N.; Leidinger, P.; Huwer, H.; Tenzer, S.; Kohlbacher, O.; Hildebrandt, A.; et al. DrugTargetInspector: An assistance tool for patient treatment stratification. *Int. J. Cancer* **2016**, *138*, 1765–1776. [CrossRef] [PubMed]

41. Makhijani, R.K.; Raut, S.A.; Purohit, H.J. Identification of common key genes in breast, lung and prostate cancer and exploration of their heterogeneous expression. *Oncol. Lett.* **2018**, *15*, 1680–1690. [CrossRef] [PubMed]

42. Abate, F.; Zairis, S.; Ficarra, E.; Acquaviva, A.; Wiggins, C.H.; Frattini, V.; Lasorella, A.; Iavarone, A.; Inghirami, G.; Rabadan, R. Pegasus: A comprehensive annotation and prediction tool for detection of driver gene fusions in cancer. *BMC Syst. Boil.* **2014**, *8*, 97. [CrossRef] [PubMed]

43. Zhao, J.; Li, X.; Yao, Q.; Li, M.; Zhang, J.; Ai, B.; Liu, W.; Wang, Q.; Feng, C.; Liu, Y.; et al. RWCFusion: Identifying phenotype-specific cancer driver gene fusions based on fusion pair random walk scoring method. *Oncotarget* **2016**, *7*, 61054–61068. [CrossRef] [PubMed]

44. Gu, J.; Chukhman, M.; Lu, Y.; Liu, C.; Liu, S.; Lu, H. RNA-seq Based Transcription Characterization of Fusion Breakpoints as a Potential Estimator for Its Oncogenic Potential. *BioMed Res. Int.* **2017**, *2017*, 9829175. [CrossRef] [PubMed]

45. Frenkel-Morgenstern, M.; Gorohovski, A.; Tagore, S.; Sekar, V.; Vazquez, M.; Valencia, A. ChiPPI: A novel method for mapping chimeric protein–protein interactions uncovers selection principles of protein fusion events in cancer. *Nucleic Acids Res.* **2017**, *45*, 7094–7105. [CrossRef] [PubMed]

46. Hu, X.; Wang, Q.; Tang, M.; Barthel, F.; Amin, S.; Yoshihara, K.; Lang, F.M.; Martinez-Ledesma, E.; Lee, S.H.; Zheng, S.; et al. TumorFusions: An integrative resource for cancer-associated transcript fusions. *Nucleic Acids Res.* **2018**, *46*, D1144–D1149. [CrossRef] [PubMed]

47. Futreal, P.A.; Coin, L.; Marshall, M.; Down, T.; Hubbard, T.; Wooster, R.; Rahman, N.; Stratton, M.R. A census of human cancer genes. *Nat. Rev. Cancer* **2004**, *4*, 177–183. [CrossRef] [PubMed]
48. Tipping, M.E.; Smola, A. Sparse Bayesian Learning and the Relevance Vector Machine. *J. Mach. Learn. Res.* **2001**, *1*, 211–244. [CrossRef]
49. Tsechansky, M.S.; Provost, F. Handling Missing Values when Applying Classification Models. *J. Mach. Learn. Res.* **2007**, *8*, 1625–1657.
50. Liu, H.; Cao, M.; Wu, C.W. Graph comparison and its application in network synchronization. In Proceedings of the 12th European Control Conference, Zurich, Switzerland, 17–19 July 2013; IEEE: Piscataway, NJ, USA, 2013; pp. 3809–3814.

Sample Availability: Not available.

molecules

MDPI

Article

Putative Iron Acquisition Systems in *Stenotrophomonas maltophilia*

V. Kalidasan, Adleen Azman, Narcisse Joseph, Suresh Kumar, Rukman Awang Hamat and Vasantha Kumari Neela *

Department of Medical Microbiology and Parasitology, Faculty of Medicine and Health Sciences, Universiti Putra Malaysia, Serdang 43400 UPM, Selangor Darul Ehsan, Malaysia; petejuz@gmail.com (V.K.); adleen_azman@hotmail.com (A.A.); narcissemsjoseph@gmail.com (N.J.); sureshkudsc@gmail.com (S.K.); rukman@upm.edu.my (R.A.H.)
* Correspondence: neela2000@hotmail.com or vasantha@upm.edu.my; Tel.: +60-3-8947-2507

Received: 6 June 2018; Accepted: 23 July 2018; Published: 16 August 2018

Abstract: Iron has been shown to regulate biofilm formation, oxidative stress response and several pathogenic mechanisms in *Stenotrophomonas maltophilia*. Thus, the present study is aimed at identifying various iron acquisition systems and iron sources utilized during iron starvation in *S. maltophilia*. The annotations of the complete genome of strains K279a, R551-3, D457 and JV3 through Rapid Annotations using Subsystems Technology (RAST) revealed two putative subsystems to be involved in iron acquisition: the iron siderophore sensor and receptor system and the heme, hemin uptake and utilization systems/hemin transport system. Screening for these acquisition systems in *S. maltophilia* showed the presence of all tested functional genes in clinical isolates, but only a few in environmental isolates. NanoString nCounter Elements technology, applied to determine the expression pattern of the genes under iron-depleted condition, showed significant expression for FeSR (6.15-fold), HmuT (12.21-fold), Hup (5.46-fold), ETFb (2.28-fold), TonB (2.03-fold) and Fur (3.30-fold). The isolates, when further screened for the production and chemical nature of siderophores using CAS agar diffusion (CASAD) and Arnows's colorimetric assay, revealed *S. maltophilia* to produce catechol-type siderophore. Siderophore production was also tested through liquid CAS assay and was found to be greater in the clinical isolate (30.8%) compared to environmental isolates (4%). Both clinical and environmental isolates utilized hemoglobin, hemin, transferrin and lactoferrin as iron sources. All data put together indicates that *S. maltophilia* utilizes siderophore-mediated and heme-mediated systems for iron acquisition during iron starvation. These data need to be further confirmed through several knockout studies.

Keywords: *Stenotrophomonas maltophilia*; iron acquisition systems; iron-depleted; RAST server; NanoString Technologies; siderophores

1. Introduction

Iron is a vital nutritional component for all living organisms, including pathogens, and is crucial for the preservation of cellular morphology, DNA and RNA biosynthesis, cellular growth and proliferation, catalysis of tricarboxylic acid cycle (TCA), electron transport chain (ETC), oxidative phosphorylation, nitrogen fixation and many more [1]. In order to successfully sustain an infective state in the human host, bacterial cells require a continuous supply of iron [2]. However, the mammalian host captures the freely available iron through high-affinity proteins such as transferrin (Tf), lactoferrin (Lf), ferritin (Fn) and heme (Hm) or hemoproteins and thereby protects itself from cellular damage by reactive oxygen species (ROS). As a result, the amount of iron is considerably reduced and the pathogens encounter a period of iron starvation upon invading their hosts [3]. In these circumstances, most pathogens sense the nutritional immunity imposed by the host, thereby expressing proteins associated with iron uptake for continual survival. In Gram negative bacteria, iron uptake occurs via (1) a siderophore-mediated system, (2) a hemophore-mediated system, (3) Tf/Lf receptors, and (4) ferrous-iron transport (Feo) [4].

Siderophore is a low molecular weight iron-complexing molecule characterized by high affinity and specificity for ferric iron (Fe^{3+}) [5]. It is secreted by most bacteria including *Escherichia coli* [6] and fungi such as *Ustilago sphaerogena* [7] and *Aspergillus fumigatus* [8] under iron limited conditions to acquire iron from the host. Siderophores are considered to be pathogenic determinants, as these compounds chelate iron for proliferation in the host [9]. On the other hand, hemophores capture free heme or bind with heme from hemoglobin (Hb), or hemoglobin-haptoglobin (Hb-Hpt) complex or heme-hemopexin (Hm-Hpx) complex, and mediate further uptake into periplasm [4]. Tf/Lf receptors such as transferrin-binding proteins (TbpA and TbpB) and lacoferrin-binding proteins (LbpA and LbpB) are largely found in pathogenic *Neisseria* with the ability to directly interact with mammalian transferrin and lactoferrin for iron [10].

S. maltophilia is an emerging nosocomial, multiple-drug-resistant (MDR) and opportunistic pathogen, primarily from an environmental origin [11,12]. Infections are commonly seen among immunocompromised hosts, such as patients with invasive devices, prolonged hospitalization, and who are on broad spectrum antibiotics. Several infectious complications are associated with *S. maltophilia* ranging from bacteremia, bone and joint infections, wound infections, catheter related infections, meningitis, respiratory tract infections, endocarditis, typhlitis, biliary sepsis and peritonitis. *S. maltophilia* is also commonly isolated from the airways of cystic fibrosis patients [13], with an increased incidence seen in patients with hematological malignancy and among recipients of hematopoietic stem cell transplantation [14]. Furthermore, *S. maltophilia*'s complete genome revealed that the bacterium possesses a huge number of virulence factors and an antibiotics-resistance profile [15,16].

In *S. maltophilia*, iron has been shown to play an important role for biofilm formation, oxidative stress response, outer membrane proteins (OMPs) expression and other virulence factors via the regulation of ferric uptake regulation protein (FUR) [17]. Under iron-limited conditions, improved biofilm organization and formation, increased production of extracellular polymeric substances (EPS) and enhanced superoxide dismutase (SOD) activities were observed. Despite its clinical relevance and the role of iron in various pathogenic events, very little is known about the iron acquisition systems in *S. maltophilia* [18]. It was found that *S. maltophilia* uptakes ferrous iron through the Feo [19], and synthesizes siderophores under iron starvation to scavenge free ferric iron [20,21]. Nevertheless, genetic factors that possibly contribute to the siderophore-mediated iron uptake system in *S. maltophilia* are not well established [15]. When compared to most Gram-negative pathogens, the potential of heme, hemoproteins and other iron high-affinity ligands as nutrients during iron starvation has not been extensively studied in *S. maltophilia*. Therefore, this study is aimed at investigating the putative iron acquisition systems in *S. maltophilia* through various genotypic and phenotypic approaches with special focus on siderophore- and heme-mediated systems.

2. Results

2.1. Putative Iron Acquisition Systems in S. maltophilia

Targeted in-silico analysis of the four complete genomes of *S. maltophilia* strains (see Table 1), revealed the presence of shared iron acquisition and metabolism subsystems, with an additional system in some *S. maltophilia* strains. The subsystem information, generated through the Rapid Annotations using Subsystems Technology (RAST) server, is included in Supplementary Table S1. These subsystems include targets encoding iron siderophore sensor and receptor systems, heme, hemin uptake and utilization systems and the hemin transport system. In addition, encapsulating protein DyP-type peroxidase and ferritin-like protein oligomers were only detected in K279a.

Table 1. *S. maltophiia* strains for in-silico, clinical and environmental isolates used in this study.

Strain I.D.	Biological Source	Geographical Source	GenBank Accession Number	References
In-silico				
K279a	Clinical (Blood infection)	Bristol, UK	AE016879	[16]
R551-3	Environmental (Poplar tree endophyte)	Washington, USA	CP001111	[22]
D457	Clinical	Mostoles, Spain	HE798556	[23]
JV3	Environmental (Rhizosphere)	Brazil	CP002986	[24]
Clinical				
SM1 to SM 101	Various specimens	Malaysia	—	[25,26]
CS17	Blood	Malaysia	—	[26]
CS24	Wound swab	Malaysia	—	[26]
ATCC13637	Pleural fluid of a patient with oral carcinoma	Stafford, England	CP008838	[27]
Environmental				
LMG959	Rice paddy	Japan	—	[28]
LMG10871	Soil	Japan	—	[28]
LMG10879	Rice paddy	Japan	—	[28]
LMG11104	Roots	Unknown	—	[29]
LMG11108	Roots	Unknown	—	[29]

Table 2. Functional roles (RAST server), their abbreviations, locus tags respective to *S. maltophilia* K279a genome and BLAST identities respective to *S. maltophilia* K279a, D457 and 13637 genomes.

Targets [a]	Functional Role [a]	Locus Tag [b]	BLAST Identity [c]
	Subsystem: Iron siderophore sensor & receptor systems		
FeSreg	Sigma factor ECF subfamily	SMLT_RS12950	98% (13637)
FeSR	Iron siderophore receptor protein	SMLT_RS18575	99% (D457)
FeSS	Iron siderophore sensor protein	SMLT_RS18580	99% (K279a)
	Subsystem: Heme, hemin uptake and utilization systems in Gram-positives		
HemO/HO	Heme oxygenase, associated with heme uptake	SMLT_RS18565	100% (13637)
HmuV	Heme ABC transporter, ATPase component	SMLT_RS11325	99% (K279a)
Hyp1	Hypothetical protein related to heme utilization	SMLT_RS19415	98% (1337)
HmuU	Heme ABC transporter, permease protein	SMLT_RS11320	92% (K279a)
HmuT	Heme ABC transporter, cell surface heme and hemoprotein receptor	SMLT_RS11315	97% (D457)
	Subsystem: Heme, hemin uptake and utilization systems in Gram-negatives		
Rp2	Outer membrane receptor proteins, mostly Fe transport	SMLT_RS18050	95% (D457)
Hup	Hemin uptake protein	SMLT_RS03780	100% (D457)
ETFb	Electron transfer flavoprotein, beta subunit	SMLT_RS03080	100% (D457)
TonB	Ferric siderophore transport system, periplasmic binding protein	SMLT_RS21345	99% (D457)
ExbB	Ferric siderophore transport system, biopolymer transport protein	SMLT_RS07890	99% (K279a)
Htp	Hemin transport protein	SMLT_RS03790	99% (K279a)
FCR	TonB-dependent hemin, ferrichrome receptor	SMLT_RS03785	99% (K279a)
	Subsystem: Encapsulating protein DyP-type peroxidase and ferritin-like protein oligomers		
DyP	Predicted dye-decolorizing peroxidase, encapsulated subgroup	SMLT_RS00875	95% (K279a)
	Subsystem: Oxidative stress		
Fur	Ferric uptake regulation protein (FUR)	SMLT_RS09600	96% (K279a)

[a] The abbreviations of the targets and name of the functional roles are derived from RAST server; [b] Corresponding locus tag respective to *S. maltophilia* K279a from GenBank, NCBI; [c] Percentage of BLAST identities of sequenced PCR products respective to the *S. maltophilia* genome is indicated in the bracket.

Ferric uptake regulation protein (FUR), which functions as a pleiotropic transcriptional regulator involved in the control of diverse cellular processes, such as iron homeostasis, oxidative stress responses, and the production of virulence factors, was observed across all the strains analysed. The targets and functional roles obtained from RAST server and locus tag (respective to *S. maltophilia* K279a) are listed in Table 2.

2.2. Distribution of Iron Acquisition Genes and Systems in S. maltophilia

In order to determine the distribution of iron acquisition genes and systems identified by the in-silico analysis in *S. maltophilia*, a total of 109 isolates (refer Table 1) obtained from different clinical and environmental sources were screened by PCR. All clinical isolates accounted for the 100% amplification for Hyp1, Hup, ETFb, TonB, DyP and FUR targets, while the remaining were as follows: FeSR (99%), HemO/HO (98.1%), FeSreg (96.2%), Rp2 (94.2%), HmuT (87.5%), HmuU (81.7%), FCR (53.8%), Htp (35.6%), ExbB (33.7%), HmuV (26.9%) and FeSS (25%). On the other hand, in environmental isolates, of the 17 targets tested, only eight showed amplification, which include Hyp1 (100%), Hup (100%), ETFb (100%), TonB (100%), DyP (100%), Fur (100%), FeSR (80%) and Rp2 (60%). The results from BLAST identities are shown in Table 2 and the information on the sequence homologue to the available genomes in the database are included in Supplementary Table S2.

2.3. Expression Profile of the Iron Acquisition System in S. maltophilia

The differential gene expression investigation by NanoString nCounter Elements showed significant upregulation for the following targets among the clinical isolates tested: FeSR (6.15-fold, p = 0.023), HmuT (12.21-fold, p = 0.005), Hup (5.46-fold, p = 0.014), ETFb (2.28-fold, p = 0.010), TonB (2.03-fold, p < 0.01) and Fur (3.30-fold, p = 0.003). The remaining functional targets exhibited no or slight fold changes, which is statistically insignificant: FeSreg (2.40-fold), HemO/HO (3.34-fold), HmuU (8.14-fold), HmuV (2.34-fold), Hyp1 (3.16-fold), Rp2 (1.14-fold), ExbB (3.64-fold), Htp (1.28-fold), FCR (3.73-fold) and DyP (2.35-fold). One siderophore-mediated target, FeSS (−1.36-fold) was found to be down-regulated; however, this was not statistically significant. In the case of environmental isolates, none of the functional targets showed statistically significant changes in the gene expression for both iron conditions tested. The average normalized grouped data and the fold changes of gene expression under iron-depleted and iron-repleted conditions for each target, together with p-values, are listed in Supplementary Table S3. The differentially expressed targets during iron-depleted and iron-repleted conditions were generated using nSolver software. The agglomerative cluster of the heat map with a dendrogram tree showed an obvious clustering of up-regulated genes for clinical isolates (SM72, SM77, and SM79) that ranged from 1.00 to 3.00 under the iron-depleted condition, as illustrated in Figure 1. Red indicates an increase in gene expression, and green indicates a decrease in gene expression. This figure, which represents data from three clinical isolates and three environmental isolates tested against 17 targets, showed at least a two-fold differential expression based on normalized grouped counts under the different iron conditions tested.

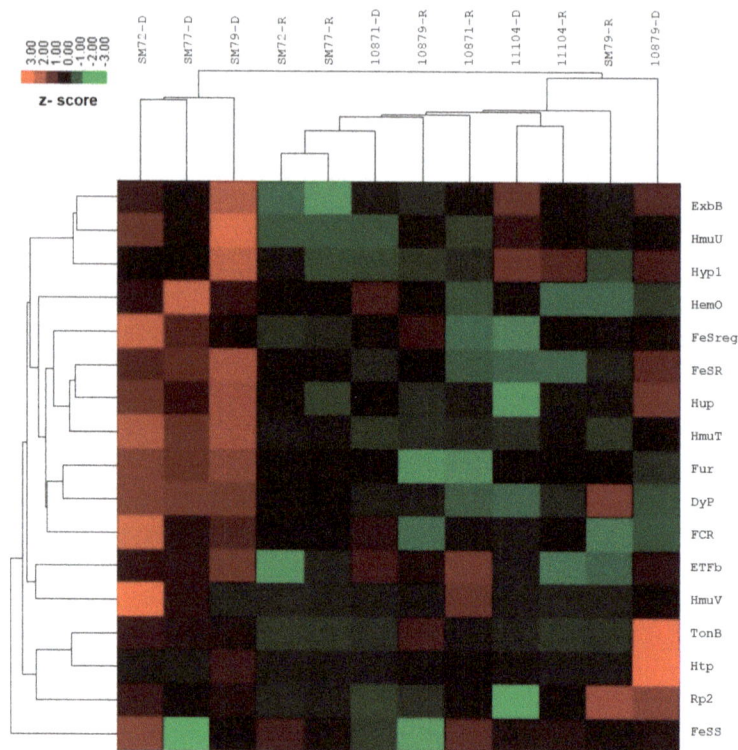

Figure 1. Heat map of nCounter (NanoString Technologies, Inc.) results comparing iron acquisition gene expressions of clinical (SM72, SM77, SM79) and environmental isolates (LMG10871, LMG10979, LMG11104) under iron-depleted and iron-repleted conditions. For clinical isolates, the up-regulated genes such as FeSR, Hup, HmuT, Fur, ETFb and TonB ranged from 1.00 to 3.00 under iron-depleted condition. While environmental isolates under both conditions, the targets remained neutral (no changes) or down-regulated but statistically insignificant. Red represents up-regulated targets under iron-depletion and green represents down-regulated targets.

2.4. Siderophore Production and Its Chemical Nature

Cultures that were grown to stationary phase for 48 h were subjected to an optical density (OD) measurement at 600 nm. The OD readings (mean, SD) of *S. maltophilia* grown in iron-depleted BHI broth were lower (1.007, 0.276), than those obtained in iron-repleted BHI broth (1.329, 0.485), showing sufficient iron starvation. Siderophore activity was observed in the cell-free culture supernatants of ten clinical and five environmental isolates tested. All isolates exhibited a prominent zone of an orange halo surrounding the well which was inoculated with the supernatants of cultures grown under iron-depletion, compared to the slight/lesser zone under iron-repleted conditions. The zone size and intensity of the orange halo was lower for environmental isolates when compared to clinical isolates, as seen in Figure 2. Arnow's assay, performed to identify the chemical nature of the siderophores, revealed that *S. maltophilia* secreted catechol-type as it formed a yellow color in nitrous acid, which then turned to pink-red when excess sodium hydroxide was added (data not shown). On another note, through liquid CAS assay, CS17 was found to produce greater percentage of siderophores (30.8%) when grown under iron-depleted compared to iron-repleted conditions (<5%) ($p < 0.05$). However, LMG10879 showed only 4% siderophore production, but this was not statistically significant.

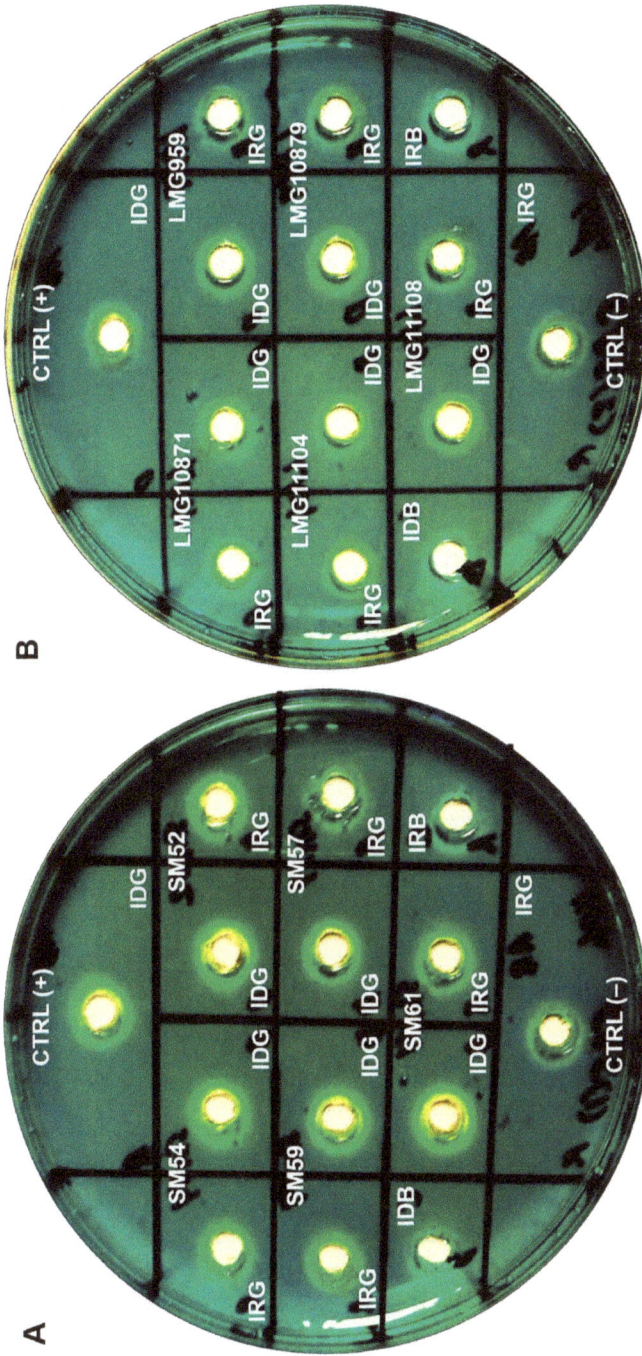

Figure 2. Representation of CAS agar diffusion (CASAD) agar plates showing zone of orange halo after 72 h incubation at room temperature. (**A**) Clinical isolates: SM52, SM54, SM57, SM59 and SM61; (**B**) Environmental isolates: LMG959, LMG10871, LMG10879, LMG11104 and LMG11108. Key: CTRL (+): Positive control; CTRL (−): negative control; IDG: iron-depleted growth; IRG: iron-repleted growth; IDB: iron-depleted uninoculated broth; IRB: iron-repleted uninoculated broth.

2.5. Iron Source Utilization during Iron Starvation

To identify the iron sources utilized by *S. maltophilia* during iron starvation, one clinical (SM77) isolate and one environmental (LMG10879) isolate grown in the presence of different iron sources were investigated. As seen in Figure 3, although growth was observed for all iron sources, the maximum and fast replication was seen in the presence of transferrin ($p < 0.001$), followed by hemoglobin ($p < 0.001$) for both clinical and environmental isolates. Both strains utilized similar iron sources for growth during iron starvation; however, the clinical isolate exhibited a higher growth rate compared to environmental isolates. The mean OD readings for the positive control, *N. meningitidis*, SM77, and LMG10879 under the iron-depleted and iron-repleted conditions, are shown in Supplementary Table S4. On the other hand, the growth under iron-depletion remained low, underscoring the importance of iron for the growth and replication of bacterial cells. From 54 h onwards, the growth under iron-depletion slowed and was similar to iron-repletion, indicating that the utilization of supplemented iron and the reach of the stationary phase. For the isolates from both sources, a decline in growth was observed from 72 h onwards.

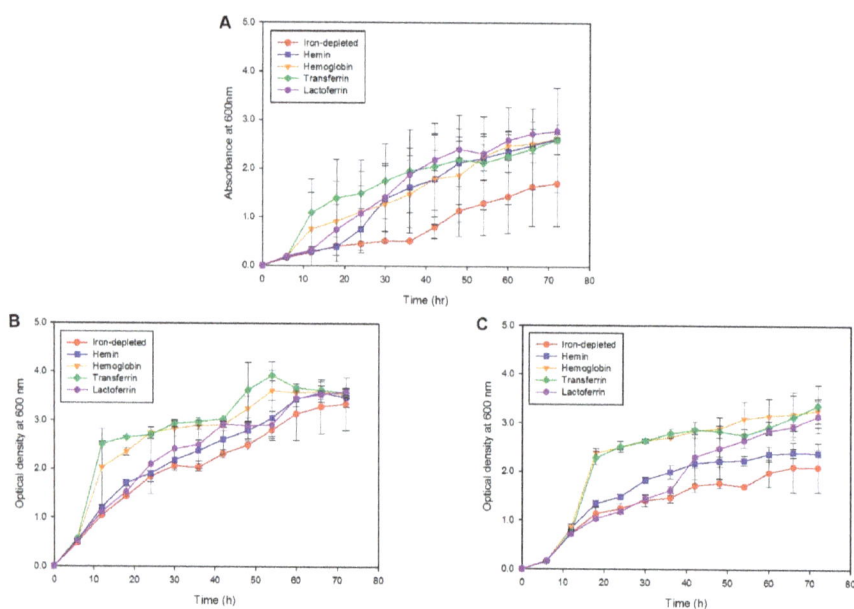

Figure 3. Growth kinetics curve (**A**) Positive control: *N. meningitidis*, (**B**) Clinical isolate: SM77 and (**C**) Environmental isolate: LMG10879. The symbol shape represents the mean reading and the error bars represent the standard deviation. Iron-depleted versus respective iron-repleted sources, post-hoc test (Duncan's Method) showed $p < 0.05$. The growth in the iron-depleted condition remains low in comparison to the iron-replete condition.

3. Discussion

The in-silico approach using the RAST server revealed two prominent subsystems for iron acquisition in *S. maltophilia* strains K279a, R551-3, D457 and JV3, which suggests that the bacteria may acquire the iron source through siderophore- and/or heme-mediated iron acquisition systems. These systems are regulated by FUR, which is expressed during oxidative stress response, such as for iron starvation, as described in an earlier study [17]. Among the 17 putative functional targets tested which are involved in iron acquisition, only eight showed a positive signal in PCR for environmental strains. The environmental isolates contained fewer of the sequences predicted to be involved in iron transport and homeostasis compared to clinical isolates as observed by the molecular study. The genome of clinical *S. maltophilia* strain K279a [16] and environmental *S. maltophilia* strain R551-3 [30] showed a large variation in the sequences, which could suggest the differences in the mechanisms for pathogenicity in the human host [12]. On the contrary, a recent study showed that the coding DNA sequence (CDS) composition and the distribution of virulence genes among clinical and environmental *S. maltophilia* is not distinguishable [31]. Therefore, it is hypothesized that the presence and absence of the functional targets among clinical and environmental isolates could also be influenced by the molecular diversity of *S. maltophilia* strains and the nature of the availability of the source of iron in a specific environment [32].

In this study, iron removal was achieved by using 2,2′-dipyridyl as metal chelating ligand [33]. It forms charged complexes with metal cations, and this property is useful in the synthesis of iron-dipyridyl. The bacterial cells treated with 2,2′-dipyridyl become pink in color because of the uptake of the chelator and formation of Fe(II) complex inside the cell [34]. 2,2′-dipyridyl can cause iron depletion and the depression of iron-regulated proteins and siderophores, even in the presence of normally repressive levels of iron in the medium. However, it is important to note that, 2,2′-dipyridyl can cause the loss of cell viability, thus chelators to cause iron limitations must be monitored carefully. The suggested final concentration of 2,2′-dipyridyl to be added into a media is 100 to 400 μM [35]. During optimization, the chelator did not render iron removal in a concentration of 50 μM, while at 200 μM bacterial growth was inhibited. Furthermore, *S. maltophilia* was grown for 48 h, as different levels of regulation were noted at the onset of stationary phase in Gram-negative bacteria such as sigma factors [36] and expression of iron-regulated outer membrane protein (IROMP) [17].

Among the 17 functional targets analysed for expression using the NanoString nCounter Elements, seven targets showed significant fold changes in clinical isolates, indicating the derepression of these targets; although other targets showed some degree of fold changes, no significant up or down regulation was observed. The environmental isolates did not show any significant differential expression for the targets tested, when grown under both iron-depleted and iron-repleted conditions. As observed in molecular screening, the targets that participated in iron acquisition and metabolism are not well manifested among environmental isolates. Moreover, the existence of iron uptake mechanisms is certainly advantageous to the growth of pathogenic bacteria under limited iron availability [32]. Specifically, a considerable difference might be seen in the type of iron transporters and iron sources utilization among different bacteria. It is worthwhile to note that the expression of various iron acquisition system studied herein, under laboratory conditions, is not well-established among environmental isolates. Further validation of the assay using real-time quantitative PCR (RT-qPCR) was not necessary, as NanoString nCounter Elements results were found to be as accurate as RT-qPCR in bacterial gene expression study [37].

The discrepancy in the gene expression profile among clinical and environmental isolates may also be attributed to the biological origin of the strains. As the environment contains a high amount of nutrients such as iron, particularly in the soil [38], the necessity of expressing special systems to acquire iron may not be vital. Thus, a low concentration of siderophores is usually detected in soil extracts [39]. Under an extreme iron demand in pure culture, pseudomonads distribute most of their carbon and ATP to synthesize siderophore. The degree of iron stress experienced by the environmental bacteria in the rhizosphere is much lower than what occurs in pure cell cultures. Moreover, the necessity for siderophore production in the rhizosphere depends largely on the effectiveness of plant iron stress responses, which is important to raise the iron availability to both plants and rhizosphere bacteria.

In the present study, siderophore detection using CASAD assay showed a prominent zone of a halo under iron-depletion in comparison with the iron-replete condition. This suggests that extracellular siderophore is secreted during iron starvation to scavenge the free iron available in the blue-green agar. However, a notable variation of intensity among the clinical and environmental isolates was observed. In support of this, the percentage of siderophore production was investigated through liquid CAS, and the clinical isolate produced a greater amount of siderophore compared to environmental isolate when grown under iron-depleted condition. Our results are in agreement with earlier studies, which reported that the environmental strain did not produce siderophores or produced very minimal amounts compared to clinical isolates [20,40]. Arnow's assay concluded that the *S. maltophilia* isolates tested in this study are catechol-type siderophore producers, as reported previously [21,41,42]. *S. maltophilia* was reported to produce hydroxamate-type ornibactin siderophore [43]. The data from the molecular and phenotypic studies support the notion that *S. maltophilia* uses siderophore-mediated iron acquisition system for obtaining iron.

In order to identify the iron source utilized by *S. maltophilia*, an iron assimilation assay was performed. *N. meningitidis* was chosen as a positive control because it is a well-established model for iron uptake from heme, lactoferrin, and transferrin [44]. The growth in iron-depleted media remained low throughout the kinetic study in comparison to the iron-repleted conditions. Hemoglobin and transferrin stimulated the growth of both strains tested under iron-depleted conditions, with hemin and lactoferrin having less effect in enhancing the growth of SM77 and LMG10879. SM77 showed sufficient growth under iron-replete conditions compared to iron-depleted ones, whereas LMG10879 showed clear differences in iron utilization patterns. This suggests that the clinical isolate utilizes of even minor traces of iron which could be present in the medium. The comparatively lower growth of LMG10879 than SM77 in the presence of different iron sources explains the low number of amplified targets in PCR and non-expression of iron uptake genes in the environmental isolates during iron depletion. Overall, *S. maltophilia* utilizes iron sources such as ferric iron or other iron-containing proteins such as hemoglobin, lactoferrin, and transferrin for cellular growth and proliferation.

The utilization of hemin and hemoproteins by *S. maltophilia* may be contributed by the heme uptake locus (*hmu*) detected in-silico and also through PCR in this study, as these similar genes *hmuRSTUV* were found in *Yersinia pestis* [45–47]. The hemoprotein-receptor-based system encoded by *hmuRSTUV* operon is used for the utilization of both hemin and other hemoproteins in *Y. pestis*. All of the other three targets were observed in *S. maltophilia*, except for HmuR and HmuS. This indicates that both *S. maltophilia* and *Y. pestis* may use a similar mechanism in acquiring iron from hemoproteins. Hemin uptake system found in *Yersinia enterocolitica* is found to pose similarities with other TonB-dependent systems in Gram-negative bacteria [48]. Thereby, the similarity between the heme-mediated systems of *S. maltophilia* with *Yersinia* spp. in this study is affirmed.

The expression of siderophore- and heme-mediated system under the iron-depleted condition in term of genotypic and phenotypic profiles reveals that it is possible to elucidate how *S. maltophilia* could establish its pathogenicity upon invasion. FeSR acts as a receptor protein, which allows the internalization of an iron-bounded siderophore complex, which must pass the outer membrane (OM) and cytoplasmic membrane (CM) before reaching the cytoplasm [49]. The siderophore detected in CASAD assay could potentially scavenge not only the ferric iron, but is also capable of delivering iron-saturated Tf, Lf, or hemin and hemoproteins, investigated through iron assimilation assay [50]. For a heme-mediated system, the intake of hemin and hemoproteins from the extracellular space into the cytoplasm occurs via HmuTUV systems [51]. The hemin and hemoglobin utilization revealed how *S. maltophilia* could potentially utilize other iron sources apart from ferric iron. This would give an indication, upon bloodstream infection with *S. maltophilia*, that bacterial multiplication within the blood stream is possible, as a unit of packed erythrocyte contains approximately 200 mg of iron, which serves an alternative source of iron [52]. In support of this, blood transfusion for an anemic patient admitted to the medical-surgical-trauma intensive care unit (ICU) was found to be associated with nosocomial infections such as pneumonia, bacteremia, sepsis, and cystitis [53].

4. Materials and Methods

4.1. Bacterial Strains, Identification and Culture Conditions

A total of 103 clinical isolates (referred to as SM in Table 1) were isolated from blood, swab, urine, tracheal aspirates, cerebrospinal fluid (CSF), pus swab, nasopharyngeal aspirates (NPA) and sputum including CS17 (clinical invasive) and CS24 (clinical non-invasive) as reference strains obtained from the laboratory culture collections (Department of Medical Microbiology and Parasitology, Universiti Putra Malaysia, Serdang, Selangor, Malaysia) were used in this study. *S. maltophilia* ATCC13637 (clinical) purchased from the American Type Culture Collection (ATCC, Manassas, VA, USA) and five environmental isolates, LMG959 (rice paddy), LMG10871 (soil), LMG10879 (rice paddy), LMG11104 (*Cichorium intybus*, rhizosphere tuberous roots) and LMG11108 (Triticum, roots); purchased from Belgian Coordinated Collections of Microorganisms (BCCM) (Laboratorium voor Microbiologie, Universiteit Gent, Belgium) were also studied. The isolates were incubated aerobically for 24 h at 37 °C for clinical and 30 °C for environmental isolates.

All isolates were previously identified as *S. maltophilia* using standard biochemical assays, API 20 NE (bioMerieux, Marcy-l'Étoile, France) and confirmed by the VITEK® Mass Spectrometry System [25,26]. Besides this, the isolates were morphologically identified by culture characteristics on Columbia agar with 5% sheep blood (Isolac, Selangor, Malaysia) and Gram morphology. The isolates were re-confirmed genotypically by species-specific polymerase chain reaction (SS-PCR) as previously described [54].

4.2. In-Silico Analysis of Putative Iron Acquisition Systems

The complete genome sequences of all four *S. maltophilia* strains, K279a, R551-3, D457 and JV3 (refer Table 1) were downloaded from the National Centre for Biotechnology Information (NCBI, Bethesda, MD, USA) Genbank (www.ncbi.nlm.nih.gov/genome/browse/). The genomes were annotated by Rapid Annotations using Subsystem Technology (RAST) server (http://rast.nmpdr.org/) [55]. RAST is a fully automated annotation service that produces gene functions and an initial metabolic reconstruction. Iron acquisition genes and gene clusters were identified by intrinsic RAST subsystem profiling for each genome as well as through, gene homologs search by Basic Local Alignment Search Tool (BLAST). Comparative genomics in SEED viewer (Genome Viewer) were used to confirm the identification and conservation of putative iron acquisition genes within *S. maltophilia* genome sequences [56].

4.3. Screening of Iron Acquisition Systems by Polymerase Chain Reaction (PCR)

PCR primers targeting the different putative iron acquisition genes and gene clusters were designed. All primers were derived from consensus sequences of four complete genome of *S. maltophilia* strains aligned through multiple sequence alignment (CLUSTALW) program using Molecular Evolutionary Genetics Analysis (MEGA 6) software [57]. Primers were designed using PrimerQuest Tool in order to select the optimal primers for PCR assay and further comprehensive oligonucleotide analysis was conducted using OligoAnalyzer 3.1 [58]. The sequences of the primers, targeted functional roles and the amplification parameters used for each set of primers are listed in Table 3.

The genomic DNA was extracted from both clinical and environmental isolates of *S. maltophilia* using Wizard® Genomic DNA Purification Kit as per the manufacturer's protocol (Promega, Madison, WI, USA). All PCR mixtures were prepared using 25 μL per tube containing 12.5 μL of EconoTaq® PLUS GREEN Master Mix, 2X (Lucigen Corporation, Middleton, WI, USA); 10 μM of forward and reverse primers; 10 ng of DNA template and 10.5 μL nuclease-free water. Target amplification was carried out in a thermal cycler (MyCycler Personal Thermal Cycler, BioRad, Hercules, CA, USA) programmed for one step of initial denaturation at 95 °C for 2 min followed by 30 cycles comprised of denaturation at 95 °C for 30 s, primer annealing at 50–51 °C for 30 s (refer Table 3), primer extension at 72 °C for 1 min, and a final extension at 72 °C for 5 min. The purified PCR products with corresponding primer pairs were sequenced through commercial company (1st Base Sdn. Bhd., Selangor, Malaysia) and analyzed using Biology Workbench 3.2 (SDSC Biology Workbench: http://workbench.sdsc.edu/). The sequences were then subjected to Standard Nucleotide BLAST (https://blast.ncbi.nlm.nih.gov/Blast.cgi) to determine the percentage of query cover and identities as well as features against *S. maltophilia's* complete genome.

Table 3. Functional roles (RAST server), the sequences of the PCR primers and amplification parameters.

Targets	Abbrev. (RAST)	Primer Pairs	Target Sequence (5′–3′)	Amplicon Size (bp)	T_a (°C)
Iron siderophore sensor & receptor system					
Sigma factor ECF subfamily	FeSreg	FeSreg-F / FeSreg-R	TTCATCGCGCGCTATCTC / ACGGATGCTCCGGCTGAT	275	50
Iron siderophore receptor protein	FeSR	FeSR-F / FeSR-R	CAATCGCAGCGTACCTACC / CGGCCACGTTGAAGAACT	271	51
Iron siderophore sensor protein	FeSS	FeSS-F / FeSS-R	ACGTCGTGCAGAACGTAAC / GGGTTTCCACCAGGTCATC	237	51
Heme, hemin uptake and utilization systems in Gram-positives					
Heme oxygenase, associated with heme uptake	HemO/HO	HemO-F / HemO-R	CAGCAATTCGCCCGTTTC / GCTTGGCAGCCATCTTGTA	281	51
Heme ABC transporter, ATPase component	HmuV	HmuV-F / HmuV-R	GAAGCTGCATGAGGTGGT / TCTACCGTGAAGGCGAAAC	273	51
Hypothetical protein related to heme utilization	Hyp1	Hyp1-F / Hyp1-R	GGCATCGTCGGCATCTT / ACTTCACCCAGGCAATCG	247	51
Heme ABC transporter, permease protein	HmuU	HmuU-F / HmuU-R	ACGCCATTGGACATGCT / AACAAGCCCAGCGGAAT	299	50
Heme ABC transporter, cell surface heme and hemoprotein receptor	HmuT	HmuT-F / HmuT-R	CATGCCCCACGACTGAT / CATCACCCAGACCCGATTG	300	51
Heme, hemin uptake and utilization systems in Gram-negatives					
Outer membrane receptor proteins, mostly Fe transport	Rp2	Rp2-F / Rp2-R	AACGCATGCCCGACTAC / CTGGCTCATGCCCATCAT	221	51
Hemin uptake protein	Hup	Hup-F / Hup-R	ATGCTCATGAAATGCTCAACC / TACTTGGTCAGGATCAGCTTG	200	51
Electron transfer flavoprotein, beta subunit	ETFb	ETFb-F / ETFb-R	CCTGGAAAACGCTGGAAGT / CCTTGACCATCACACCCTT	215	51
Ferric siderophore transport system, periplasmic binding protein	TonB	TonB-F / TonB-R	CGCGAGAACCGCATGTAT / TCCTCGGCGTCCTTCTT	314	51
Ferric siderophore transport system, biopolymer transport protein	ExbB	ExbB-F / ExbB-R	GAGCGTTTCTGGTCCCTTC / CCCAGTCGCGTTCAGGAAT	251	51
Hemin transport protein	Htp	Htp-F / Htp-R	CACCGTGTTGTGCGTGTA / GCCTCGCTATCGTGTTCC	234	51
TonB-dependent hemin, ferrichrome receptor	FCR	FCR-F / FCR-R	CGGAAATGAAGGCCGGTATC / CCATTCGATGTAGCGCTTGT	380	51
Encapsulating protein DyP-type peroxidase and ferritin-like protein oligomers					
Predicted dye-decolorizing peroxidase, encapsulated subgroup	DyP	DyP-F / DyP-R	GTGCTGAAGGTGAAGGATGA / TGCACGGATGTGGTACAG	258	50
Oxidative stress related to iron uptake					
Ferric uptake regulation protein FUR	Fur	Fur-F / Fur-R	TGACCGCCGAAGACATCTA / GCGAGTCGTCTTCCAGTTC	279	51

Abbrev.: Abbreviation; **F**: forward; **R**: reverse; **bp**: base pair; T_a: annealing temperature.

4.4. Bacterial Culture under Iron-Depleted and Iron-Repleted Conditions

A few single colonies were picked from the agar plates and inoculated into brain–heart infusion (BHI) broth. Followed by overnight incubation, cultures were centrifuged for 5 min at 10,000 rpm. The cell pellets were resuspended in phosphate buffered saline (PBS), centrifuged and washed twice. The cell suspensions in PBS were adjusted to 0.2 with an Eppendorf BioPhotometer Plus (Hamburg, Germany) at an optical density (OD) of 600 nm. One milliliter of a standardized suspension was used to inoculate the media prepared under two conditions. An iron-depleted condition was achieved by adding an iron chelator, 100 µM 2,2′-dipyridyl (DIP) (Sigma Aldrich, Darmstadt, Germany) to BHI broth (BHI-DIP), while the iron-repleted condition was further defined by the addition of 100 µM ferric chloride (Sigma Aldrich) to the BHI-DIP. The tubes were incubated aerobically (37 °C for clinical and 30 °C for environmental isolates) for 48 h on an incubator shaker (Model IKA® KS 4000 i control, IKA® Works (Asia) Sdn Bhd, Selangor, Malaysia) at 200 rpm to ensure stationary phase bacterial growth [21]. All glassware were treated with 3M HCl followed by extensive washing with deionized water to remove any iron from the labware before proceeding with the experiments [59].

4.5. Gene Expression of Iron Acquisition Systems by NanoString Technologies

It was hypothesized that the expression of *S. maltophilia* iron acquisition genes will be enhanced under iron-depleted conditions to encounter iron starvation. Based on in-silico analysis and positive molecular screening, three clinical isolates (SM72, SM77 and SM79) and three environmental isolates (LMG10871, LMG10879 and LMG11104) that amplified most functional targets were selected for gene expression study. Total RNA was extracted from each culture conditions using Agilent Total RNA Isolation Mini Kit-Bacteria (Agilent Technologies, Santa Clara, CA, USA) as per the manufacturer's protocol. The extracted total RNA was used for expression study using nCounter® Elements technology (NanoString Technologies, Inc., Seattle, WA, USA). The technology is based on molecular barcoding and digital quantification of target RNA sequences through the use of nCounter Elements TagSet and target-specific oligonucleotide probe pairs (Probe A and B). The nCounter Elements probes consist of the targets' name, GenBank accession numbers, a position of the targets, target sequences and melting temperature (T_m) for both Probe A and Probe B as shown in Supplementary Table S4.

The nCounter assay comprising of three steps including hybridization, sample processing, and digital data acquisition were performed as per the manufacturer's instructions. The components including hybridization buffer, code set and RNA samples were added into a strip tube [60]. The hybridization was performed in a thermal cycler (Turbocycler2, Blue-Ray Biotech, Taipei City, Taiwan) programmed for 16 cycles at 67 °C for 60 min and holding at 4 °C for infinity. The samples were then processed by placing the strip tubes into the automated nCounter Prep Station with reagents and consumables from the nCounter Master Kit. After the purification and immobilization were completed, the cartridge was taken out from the Prep Station and placed into the Digital Analyzer for digital data counting. The cartridges were scanned at a maximum resolution of 555 Field of View (FOV) [61]. The fold changes in expression under iron-depleted and iron-repleted conditions were analyzed based on "all pairwise ratios" of the normalized grouped data using nSolver™ Analysis software version 3.0, considering iron-depleted as the baseline condition for comparison. Four parameters were used as quality control (QC) which include imaging QC, binding density QC, positive control linearity QC and positive control limit of detection QC [62]. Both positive control and housekeeping normalization were used to normalize all platform associated sources of variations.

4.6. Siderophore Detection Using CASAD and Colorimetric Assays

Extracellular siderophore production was examined using the CASAD method by modifying the classical CAS plate as described previously [63]. Based on in-silico analysis and positive molecular screening for most of the functional targets, ten randomly selected clinical isolates (CS17, CS24, ATCC13637, SM49, SM50, SM52, SM54, SM57, SM59, and SM61) and five environmental isolates

(LMG959, LMG10871, LMG10879, LMG11104, and LMG11108) were used for the assay. Briefly, 5 mm diameter holes were punched on the CASAD agar plate and each hole was filled with 70 μL (35 μL twice) of cell-free-supernatant containing secreted extracellular siderophore [32]. The plates were incubated at room temperature for 72 h. Formation of orange halo zone around the holes inoculated with cultures indicates positive reaction. *Acinetobacter baumannii* ATCC 19606 was used as positive control, while uninoculated BHI broth (both iron-depleted and iron-repleted) served as the negative control.

The liquid CAS assay was performed as described previously [64], with *A. baumannii* ATCC 19606 positive control. To estimate the quantity of siderophores produced, 500 μL of culture supernatant was added to 500 μL CAS solutions. Upon 30 min of incubation at room temperature, the absorbance was read at 630 nm [64,65]. The BHI-DIP was used as a blank, BHI-DIP plus CAS assay solution plus shuttle as a reference (r) and culture supernatant as a sample for testing (s). The percentage of iron-binding compounds of the siderophore type was calculated by subtracting the sample absorbance (A_s) values from the reference (A_r). Siderophore units are defined as $\frac{A_r - A_s}{A_r} \times 100 =$ percent siderophore units [35,66]. Percentages of siderophores units less than 10 were considered as negative, which is indicated by no change in the blue color of CAS solution.

The chemical nature of siderophores (phenolic-type and/or hydroxamate-type) produced in the cell-free supernatant was detected using the colorimetric assays described by Atkin [67] and Arnow [68] respectively. *Escherichia coli* ATCC 8739 (aerobactin hydroxamate-type siderophore producer) [69] was used as positive control for Atkin's method, while *A. baumannii* ATCC 19606 (catecholate-type siderophore producer) [70] served as a positive control for Arnow's method.

4.7. Iron Utilization Kinetics Using Liquid Assimilation Assay

To determine the iron source utilized by *S. maltophilia* during iron starvation, one clinical isolate (SM77) and one environmental isolate (LMG10879) were investigated, based on positive molecular screening and gene expression study for most of the functional targets. The BHI-DIP was supplemented with other iron sources to make the BHI broth iron-repleted. Iron sources used herein include 10 μM hemin chloride (bovine) (MP Biomedicals, Santa Ana, CA, USA) (dissolved in 1.4 M ammonia hydroxide), 2.5 μM hemoglobin (human) (Sigma Aldrich) (dissolved in deionized water), 1 μM iron-saturated lactoferrin (human) (Sigma Aldrich) (dissolved in PBS) and 5 μM iron-saturated transferrin (human) (MP Biomedicals) (dissolved in deionized water) [71,72]. The iron utilization kinetics was measured by observing the turbidity of the culture. 100 μL of culture was filled into the UV-Vis cuvette and turbidity was measured with an Eppendorf BioPhotometer Plus (Hamburg, Germany) at the optical density (OD) of 600 nm.

The bacterial growth was tested every 6 h up to 72 h for growth kinetics measurement. *Neisseria meningitidis* obtained from Department of Medical Microbiology and Immunology, Universiti Kebangsaan Malaysia Medical Centre (UKMMC) grown anaerobically on chocolate agar II (Isolac, Selangor, Malaysia) in a candle jar served as the positive control [73,74]. Bacterial growth in BHI iron-depleted broth was used as the negative control to show the comparison of growth kinetics between two conditions tested.

4.8. Data and Statistical Analysis

For gene expression study, the distribution of the *t*-statistic was calculated by the Welch-Satterthwaite equation using nSolver™ Analysis software version 3.0. The *p*-values were set at ($p < 0.05$) to be considered statistically significant, as the lower *p*-value, the stronger the evidence that the two different groups have different expression levels. On the other hand, the data obtained from iron assimilation assay were analyzed through SigmaPlot version 12.5 and statistically significant data was determined by using One Way Repeated Measures Analysis of Variance (One Way RM ANOVA) ($p < 0.05$). To determine the significance in the difference of means among the iron sources supplemented, all pairwise multiple comparison procedures (Duncan's Method) were performed.

Molecules **2018**, 23, 2048

5. Conclusions

In conclusion, it is revealed that *S. maltophilia* expresses two putative iron acquisition systems—the iron siderophore sensor and receptor system and the heme, hemin uptake and utilization systems/hemin transport system—during iron starvation. It is indicated that siderophore-based iron uptake is mediated through FeSR and TonB, while the heme-mediated uptake may involve targets such as HmuT, Hup, and ETFb. Both clinical and environmental isolates produced catechol-type siderophores and utilized all iron sources tested, such as hemin, hemoglobin, transferrin, and lactoferrin. This study is the first step towards understanding iron acquisition systems in *S. maltophilia* focusing on siderophore- and heme-mediated systems.

A major limitation of the study was that the role of each gene in iron acquisition during starvation could not be established. However, in future studies, a mutant construction for each gene will be considered to understand the roles of differentially expressed genes during iron starvation. Further investigation on the heme acquisition system would be able to provide firmer evidence on how *S. maltophilia* utilizes heme or hemoproteins.

Supplementary Materials: The following are available online, Table S1: RAST subsystem information for *S. maltophilia* strain K279a, R551-3, D457 and JV3. Table S2: BLAST output for 17 functional targets. Table S3: Fold changes in expression under iron-depleted and iron-repleted conditions screened through nCounter Elements for clinical and environmental isolates. Table S4: Mean OD reading for *N. meningitidis*, SM77 and LMG10879 under the iron-depleted and iron-repleted conditions. Table S5: nCounter Elements design details for iron acquisition gene expression in *S. maltophilia*.

Author Contributions: Conceptualization, V.K. and V.K.N.; Formal analysis, V.K.; Funding acquisition, V.K. and V.K.N.; Investigation, V.K., A.A. and N.J.; Methodology, V.K., A.A. and N.J.; Project administration, V.K., N.J. and V.K.N.; Resources, S.K. and V.K.N.; Software, V.K.; Supervision, S.K., R.A.H. and V.K.N.; Validation, V.K., N.J., S.K. and V.K.N.; Visualization, V.K.; Writing—original draft, V.K.; Writing—review & editing, V.K., N.J., S.K. and V.K.N.

Funding: This research was funded by Ministry of Higher Education, Malaysia through Fundamental Research Grant Scheme [04-01-14-53FR] and Universiti Putra Malaysia through Geran Putra—Inisiatif Putra Siswazah (IPS) [GP-IPS/2016/9478200].

Acknowledgments: We would like to show our gratitude to Wan Mohd Aizat, Research Fellow, Institute of Systems Biology, Universiti Kebangsaan Malaysia (UKM) who provided insight and expertise on analysis of the gene expression study.

Conflicts of Interest: The authors declare that they have no competing interests.

References

1. Symeonidis, A.; Marangos, M. Iron and Microbial Growth. In *Insight and Control of Infectious Disease in Global Scenario;* InTech: London, UK, 2012; pp. 289–330.
2. Braun, V.; Hantke, K. Recent insights into iron import by bacteria. *Curr. Opin. Chem. Biol.* **2011**, *15*, 328–334. [CrossRef] [PubMed]
3. Skaar, E.P. The battle for iron between bacterial pathogens and their vertebrate hosts. *PLoS Pathog.* **2010**, *6*, 1–2. [CrossRef] [PubMed]
4. Marx, J.J.M. Iron and infection: Competition between host and microbes for a precious element. *Best Pract. Res. Clin. Haematol.* **2002**, *15*, 411–426. [CrossRef] [PubMed]
5. Chu, B.C.; Garcia-Herrero, A.; Johanson, T.H.; Krewulak, K.D.; Lau, C.K.; Peacock, R.S.; Slavinskaya, Z.; Vogel, H.J. Siderophore uptake in bacteria and the battle for iron with the host; a bird's eye view. *BioMetals* **2010**, *23*, 601–611. [CrossRef] [PubMed]
6. Braun, V.; Braun, M. Iron transport and signaling in *Escherichia coli*. *FEBS Lett.* **2002**, *529*, 78–85. [CrossRef]
7. Ecker, D.J.; Passavant, C.W.; Emery, T. Role of two siderophores in *Ustilago sphaerogena*. Regulation of biosynthesis and uptake mechanisms. *BBA Mol. Cell Res.* **1982**, *720*, 242–249. [CrossRef]
8. Schrettl, M.; Ibrahim-Granet, O.; Droin, S.; Huerre, M.; Latgé, J.P.; Haas, H. The crucial role of the *Aspergillus fumigatus* siderophore system in interaction with alveolar macrophages. *Microbes Infect.* **2010**, *12*, 1035–1041. [CrossRef] [PubMed]

9. Schmidt, H.; Hensel, M. Pathogenicity islands in bacterial pathogenesis. *Clin. Microbiol. Rev.* **2004**, *17*, 14–56. [CrossRef] [PubMed]

10. Schryvers, A.B.; Stojiljkovic, I. Iron acquisition systems in the pathogenic *Neisseria*. *Mol. Microbiol.* **1999**, *32*, 1117–1123. [CrossRef] [PubMed]

11. Looney, W.J.; Narita, M.; Muhlemann, K. *Stenotrophomonas maltophilia*: An emerging opportunist human pathogen. *Lancet Infect. Dis.* **2009**, *9*, 312–323. [CrossRef]

12. Brooke, J.S. *Stenotrophomonas maltophilia*: An emerging global opportunistic pathogen. *Clin. Microbiol. Rev.* **2012**, *25*, 2–41. [CrossRef] [PubMed]

13. Pompilio, A.; Pomponio, S.; Crocetta, V.; Gherardi, G.; Verginelli, F.; Fiscarelli, E.; Dicuonzo, G.; Savini, V.; D'Antonio, D.; Di Bonaventura, G. Phenotypic and genotypic characterization of *Stenotrophomonas maltophilia* isolates from patients with cystic fibrosis: Genome diversity, biofilm formation, and virulence. *BMC Microbiol.* **2011**, *11*, 159. [CrossRef] [PubMed]

14. Al-Anazi, K.A.; Al-Jasser, A.M.; Alsaleh, K. Infections Caused by *Stenotrophomonas maltophilia* in Recipients of Hematopoietic Stem Cell Transplantation. *Front. Oncol.* **2014**, *4*, 231. [CrossRef] [PubMed]

15. Adamek, M.; Linke, B.; Schwartz, T. Virulence genes in clinical and environmental *Stenotrophomas maltophilia* isolates: A genome sequencing and gene expression approach. *Microb. Pathog.* **2014**, *67–68*, 20–30. [CrossRef] [PubMed]

16. Crossman, L.C.; Gould, V.C.; Dow, J.M.; Vernikos, G.S.; Okazaki, A.; Sebaihia, M.; Saunders, D.; Arrowsmith, C.; Carver, T.; Peters, N.; et al. The complete genome, comparative and functional analysis of *Stenotrophomonas maltophilia* reveals an organism heavily shielded by drug resistance determinants. *Genome Biol.* **2008**, *9*, R74. [CrossRef] [PubMed]

17. García, C.A.; Alcaraz, E.S.; Franco, M.A.; Rossi, B.N.P. De Iron is a signal for *Stenotrophomonas maltophilia* biofilm formation, oxidative stress response, OMPs expression, and virulence. *Front. Microbiol.* **2015**, *6*, 1–14. [CrossRef] [PubMed]

18. Huang, T.P.; Lee Wong, A.C. A cyclic AMP receptor protein-regulated cell-cell communication system mediates expression of a FecA homologue in *Stenotrophomonas maltophilia*. *Appl. Environ. Microbiol.* **2007**, *73*, 5034–5040. [CrossRef] [PubMed]

19. Su, Y.C.; Chin, K.H.; Hung, H.C.; Shen, G.H.; Wang, A.H.J.; Chou, S.H. Structure of *Stenotrophomonas maltophilia* FeoA complexed with zinc: A unique prokaryotic SH3—Domain protein that possibly acts as a bacterial ferrous iron-transport activating factor. *Acta Crystallogr. Sect. F Struct. Biol. Cryst. Commun.* **2010**, *66*, 636–642. [CrossRef] [PubMed]

20. Minkwitz, A.; Berg, G. Comparison of Antifungal Activities and 16S Ribosomal DNA Sequences of Clinical and Environmental Isolates of *Stenotrophomonas maltophilia*. *J. Clin. Microbiol.* **2001**, *39*, 139–145. [CrossRef] [PubMed]

21. García, C.A.; Passerini De Rossi, B.; Alcaraz, E.; Vay, C.; Franco, M. Siderophores of *Stenotrophomonas maltophilia*: Detection and determination of their chemical nature. *Rev. Argent Microbiol.* **2012**, *44*, 150–154. [PubMed]

22. Taghavi, S.; Garafola, C.; Monchy, S.; Newman, L.; Hoffman, A.; Weyens, N.; Barac, T.; Vangronsveld, J.; Van Der Lelie, D.D. Genome survey and characterization of endophytic bacteria exhibiting a beneficial effect on growth and development of poplar trees. *Appl. Environ. Microbiol.* **2009**, *75*, 748–757. [CrossRef] [PubMed]

23. Lira, F.; Hernandez, A.; Belda, E.; Sanchez, M.B.; Moya, A.; Silva, F.J.; Martinez, J.L. Whole-genome sequence of *Stenotrophomonas maltophilia* D457, A clinical isolate and a model strain. *J. Bacteriol.* **2012**, *194*, 3563–3564. [CrossRef] [PubMed]

24. Lucas, S.; Han, J.; Lapidus, A.; Cheng, J.-F.; Goodwin, L.; Pitluck, S.; Peters, L.; Ovchinnikova, G.; Teshima, H.; Detter, J.C.; et al. Complete Sequence of Stenotrophomonas Maltophilia Jv3. *EMBL/GenBank/DDBJ Databases* **2011**, submitted.

25. Neela, V.; Rankouhi, S.Z.R.; van Belkum, A.; Goering, R.V.; Awang, R. *Stenotrophomonas maltophilia* in Malaysia: Molecular epidemiology and trimethoprim-sulfamethoxazole resistance. *Int. J. Infect. Dis.* **2012**, *16*, e603–e607. [CrossRef] [PubMed]

26. Thomas, R.; Hamat, R.A.; Neela, V. Extracellular enzyme profiling of *Stenotrophomonas maltophilia* clinical isolates. *Virulence* **2014**, *5*, 326–330. [CrossRef] [PubMed]

27. Davenport, K.W.; Daligault, H.E.; Minogue, T.D.; Broomall, S.M.; Bruce, D.C.; Chain, P.S.; Coyne, S.R.; Gibbons, H.S.; Jaissle, J.; Rosenzweig, C.N.; et al. Complete Genome Sequence of *Stenotrophomonas maltophilia* Type Strain 810-2 (ATCC 13637). *Genome Announc.* **2014**, *2*, 7–8. [CrossRef] [PubMed]

28. Roscetto, E.; Rocco, F.; Carlomagno, M.S.; Casalino, M.; Colonna, B.; Zarrilli, R.; Di Nocera, P.P. PCR-based rapid genotyping of *Stenotrophomonas maltophilia* isolates. *BMC Microbiol.* **2008**, *8*, 202. [CrossRef] [PubMed]

29. Nicoletti, M.; Iacobino, A.; Prosseda, G.; Fiscarelli, E.; Zarrilli, R.; De Carolis, E.; Petrucca, A.; Nencioni, L.; Colonna, B.; Casalino, M. *Stenotrophomonas maltophilia* strains from cystic fibrosis patients: Genomic variability and molecular characterization of some virulence determinants. *Int. J. Med. Microbiol.* **2011**, *301*, 34–43. [CrossRef] [PubMed]

30. Rocco, F.; De Gregorio, E.; Colonna, B.; Di Nocera, P.P. *Stenotrophomonas maltophilia* genomes: A start-up comparison. *Int. J. Med. Microbiol.* **2009**, *299*, 535–546. [CrossRef] [PubMed]

31. Lira, F.; Berg, G.; Martínez, J.L. Double-face meets the bacterial world: The opportunistic pathogen *Stenotrophomonas maltophilia*. *Front. Microbiol.* **2017**, *8*, 1–15. [CrossRef] [PubMed]

32. Grim, C.J.; Kothary, M.H.; Gopinath, G.; Jarvis, K.G.; Jean-Gilles Beaubrun, J.; McClelland, M.; Tall, B.D.; Franco, A.A. Identification and characterization of *Cronobacter* iron acquisition systems. *Appl. Environ. Microbiol.* **2012**, *78*, 6035–6050. [CrossRef] [PubMed]

33. Kaes, C.; Katz, A.; Hosseini, M.W. Bipyridine: The most widely used ligand. A review of molecules comprising at least two 2,2′-bipyridine units. *Chem. Rev.* **2000**, *100*, 3553–3590. [CrossRef] [PubMed]

34. Barton, L.L.; Hemming, B.C. *Iron Chelation in Plants and Soil Microorganisms*; Academic Press: Cambridge, MA, USA, 1993.

35. Payne, S.M. Detection, Isolation and Characterization of Siderophores. *Methods Enzymol.* **1994**, *235*, 229–344. [CrossRef]

36. Navarro Llorens, J.M.; Tormo, A.; Martínez-García, E. Stationary phase in gram-negative bacteria. *FEMS Microbiol. Rev.* **2010**, *34*, 476–495. [CrossRef] [PubMed]

37. Beaume, M.; Hernandez, D.; Docquier, M.; Delucinge-Vivier, C.; Descombes, P.; François, P. Orientation and expression of methicillin-resistant *Staphylococcus aureus* small RNAs by direct multiplexed measurements using the nCounter of NanoString technology. *J. Microbiol. Methods* **2011**, *84*, 327–334. [CrossRef] [PubMed]

38. Berg, G.; Roskot, N.; Smalla, K. Genotypic and Phenotypic Relationships between Clinical and Environmental Isolates of *Stenotrophomonas maltophilia*. *J. Clin. Microbiol.* **1999**, *37*, 3594–3600. [PubMed]

39. Crowley, D.E. Microbial Siderophores in the plant rhizosphere. In *Iron Nutrition in Plants and Rhizospheric Microorganisms*; Springer Science & Business Media: Berlin, Germany, 2006; pp. 169–198. ISBN 1402047428.

40. Dunne, C.; Crowley, J.J.; Moenne-Loccoz, Y.; Dowling, D.N.; De Bruijn, F.J.; O'Gara, F. Biological control of Pythium ultimum by *Stenotrophomonas maltophilia* W81 is mediated by an extracellular proteolytic activity. *Microbiology* **1997**, *143*, 3921–3931. [CrossRef]

41. Mokracka, J.; Cichoszewska, E.; Kaznowski, A. Siderophore production by Gram-negative rods isolated from human polymicrobial infections. *Biol. Lett.* **2011**, *48*, 147–157. [CrossRef]

42. Ryan, R.P.; Monchy, S.; Cardinale, M.; Taghavi, S.; Crossman, L.; Avison, M.B.; Berg, G.; van der Lelie, D.; Dow, J.M. The versatility and adaptation of bacteria from the genus *Stenotrophomonas*. *Nat. Rev. Microbiol.* **2009**, *7*, 514–525. [CrossRef] [PubMed]

43. Chhibber, S.; Gupta, A.; Sharan, R.; Gautam, V.; Ray, P. Putative virulence characteristics of *Stenotrophomonas maltophilia*: A study on clinical isolates. *World J. Microbiol. Biotechnol.* **2008**, *24*, 2819–2825. [CrossRef]

44. Perkins-Balding, D.; Ratliff-Griffin, M.; Stojiljkovic, I. Iron transport systems in *Neisseria meningitidis*. *Microbiol. Mol. Biol. Rev.* **2004**, *68*, 154–171. [CrossRef] [PubMed]

45. Thompson, J.A.N.M.; Jones, H.A.; Jones, H.A.; Perry, R.D.; Perry, R.D. Molecular Characterization of the Hemin Uptake Locus (hmu) from Yersinia pestis and Analysis ofhmu Mutants for Hemin and Hemoprotein Utilization. *Infect. Immun.* **1999**, *67*, 3879–3892. [PubMed]

46. Hornung, J.M.; Jones, H.A.; Perry, R.D. The HMU locus of *Yersinia pestis* is essential for utilization of free haemin and haem-protein complexes as iron sources. *Mol. Microbiol.* **1996**, *20*, 725–739. [CrossRef] [PubMed]

47. Zhou, D.; Qin, L.; Han, Y.; Qiu, J.; Chen, Z.; Li, B.; Song, Y.; Wang, J.; Guo, Z.; Zhai, J.; et al. Global analysis of iron assimilation and fur regulation in *Yersinia pestis*. *FEMS Microbiol. Lett.* **2006**, *258*, 9–17. [CrossRef] [PubMed]

48. Stojiljkovic, I.; Hantke, K. Hemin uptake system of *Yersinia enterocolitica*: Similarities with other TonB-dependent systems in gram-negative bacteria. *EMBO J.* **1992**, *11*, 4359–4367. [PubMed]

49. Braun, V. Bacterial Iron Transport Related to Virulence. *Signal. Gene Regul.* **2005**, *12*, 210–233. [CrossRef]
50. Braun, V.; Hantke, K. Acquisition of Iron by Bacteria. *Mol. Microbiol. Heavy Met.* **2007**, *43*, 189–219. [CrossRef]
51. Kikuchi, G.; Yoshida, T.; Noguchi, M. Heme oxygenase and heme degradation. *Biochem. Biophys. Res. Commun.* **2005**, *338*, 558–567. [CrossRef] [PubMed]
52. Pieracci, F.M.; Barie, P.S. Iron and the Risk of Infection. *Surg. Infect.* **2005**, *6*, 41–46. [CrossRef] [PubMed]
53. Taylor, R.W.; Manganaro, L.; O'Brien, J.; Trottier, S.J.; Parkar, N.; Veremakis, C. Impact of allogenic packed red blood cell transfusion on nosocomial infection rates in the critically ill patient. *Crit. Care Med.* **2002**, *30*, 2249–2254. [CrossRef] [PubMed]
54. Gallo, S.W.; Ramos, P.L.; Ferreira, C.A.S.; de Oliveira, S.D. A specific polymerase chain reaction method to identify *Stenotrophomonas maltophilia*. *Mem. Inst. Oswaldo Cruz* **2013**, *108*, 390–391. [CrossRef] [PubMed]
55. Aziz, R.K.; Bartels, D.; Best, A.A.; DeJongh, M.; Disz, T.; Edwards, R.A.; Formsma, K.; Gerdes, S.; Glass, E.M.; Kubal, M.; et al. The RAST Server: Rapid annotations using subsystems technology. *BMC Genom.* **2008**, *9*. [CrossRef] [PubMed]
56. Overbeek, R.; Olson, R.; Pusch, G.D.; Olsen, G.J.; Davis, J.J.; Disz, T.; Edwards, R.A.; Gerdes, S.; Parrello, B.; Shukla, M.; et al. The SEED and the Rapid Annotation of microbial genomes using Subsystems Technology (RAST). *Nucleic Acids Res.* **2014**, *42*, 1–9. [CrossRef] [PubMed]
57. Tamura, K.; Stecher, G.; Peterson, D.; Filipski, A.; Kumar, S. MEGA 6: Molecular evolutionary genetics analysis version 6.0. *Mol. Biol. Evol.* **2013**, *30*, 2725–2729. [CrossRef] [PubMed]
58. Owczarzy, R.; Tataurov, A.V.; Wu, Y.; Manthey, J.A.; McQuisten, K.A.; Almabrazi, H.G.; Pedersen, K.F.; Lin, Y.; Garretson, J.; McEntaggart, N.O.; et al. IDT SciTools: A suite for analysis and design of nucleic acid oligomers. *Nucleic Acids Res.* **2008**, *36*, 163–169. [CrossRef] [PubMed]
59. Louden, B.C.; Haarmann, D.; Lynne, A. Use of Blue Agar CAS Assay for Siderophore Detection. *J. Microbiol. Biol. Educ.* **2011**, *12*, 51–53. [CrossRef] [PubMed]
60. Kulkarni, M.M. Digital multiplexed gene expression analysis using the NanoString nCounter system. *Curr. Protoc. Mol. Biol.* **2011**, 1–17. [CrossRef]
61. Geiss, G.K.; Bumgarner, R.E.; Birditt, B.; Dahl, T.; Dowidar, N.; Dunaway, D.L.; Fell, H.P.; Ferree, S.; George, R.D.; Grogan, T.; et al. Direct multiplexed measurement of gene expression with color-coded probe pairs. *Nat. Biotechnol.* **2008**, *26*, 317–325. [CrossRef] [PubMed]
62. Xu, W.; Solis, N.V.; Filler, S.G.; Mitchell, A.P. Gene Expression Profiling of Infecting Microbes Using a Digital Bar-coding Platform. *J. Vis. Exp.* **2016**, *107*, 1–5. [CrossRef] [PubMed]
63. Shin, S.H.; Lim, Y.; Lee, S.E.; Yang, N.W.; Rhee, J.H. CAS agar diffusion assay for the measurement of siderophores in biological fluids. *J. Microbiol. Methods* **2001**, *44*, 89–95. [CrossRef]
64. Schwyn, B.; Neilands, J.B. Universal chemical assay for the detection and determination of siderophores. *Anal. Biochem.* **1987**, *160*, 47–56. [CrossRef]
65. Pal, R.B.; Gokarn, K. Siderophores and pathogenecity of microorganisms. *J. Biosci. Technol.* **2010**, *1*, 127–134.
66. Machuca, A.; Milagres, A.M.F. Use of CAS-agar plate modified to study the effect of different variables on the siderophore production by *Aspergillus*. *Lett. Appl. Microbiol.* **2003**, *36*, 177–181. [CrossRef] [PubMed]
67. Atkin, C.L.; Neilands, J.B.; Phaff, H.J. Rhodotorulic Acid from Species of *Leucosporidium, Rhodosporidium, Rhodotorula, Sporidiobolus*, and *Sporobolomyces*, and a New Alanine-Containing Ferrichrome from *Cryptococcus melibiosum*. *J. Bacteriol.* **1970**, *103*, 722–733. [PubMed]
68. Arnow, L.E. Colorimetric determination of the components of 3,4-dihydroxyphenylalaninetyrosine mixtures. *J. Biol. Chem.* **1937**, 531–537.
69. Watts, R.E.; Totsika, M.; Challinor, V.L.; Mabbett, A.N.; Ulett, G.C.; De Voss, J.J.; Schembri, M.A. Contribution of siderophore systems to growth and urinary tract colonization of asymptomatic bacteriuria *Escherichia coli*. *Infect. Immun.* **2012**, *80*, 333–344. [CrossRef] [PubMed]
70. Dorsey, C.W.; Tomaras, A.P.; Connerly, P.L.; Tolmasky, M.E.; Crosa, J.H.; Actis, L.A. The siderophore-mediated iron acquisition systems of *Acinetobacter baumannii* ATCC 19606 and *Vibrio anguillarum* 775 are structurally and functionally related. *Microbiology* **2004**, *150*, 3657–3667. [CrossRef] [PubMed]
71. Kvitko, B.H.; Goodyear, A.; Propst, K.L.; Dow, S.W.; Schweizer, H.P. *Burkholderia pseudomallei* known siderophores and hemin uptake are dispensable for lethal murine melioidosis. *PLoS Negl. Trop. Dis.* **2012**, *6*. [CrossRef] [PubMed]

72. Genco, C.A.; Chen, C.Y.; Arko, R.J.; Kapczynski, D.R.; Morse, S.A. Isolation and characterization of a mutant of *Neisseria gonorrhoeae* that is defective in the uptake of iron from transferrin and haemoglobin and is avirulent in mouse subcutaneous chambers. *J. Gen. Microbiol.* **1991**, *137*, 1313–1321. [CrossRef] [PubMed]

73. Zhu, W.; Hunt, D.J.; Richardson, A.R.; Stojiljkovic, I. Use of Heme Compounds as Iron Sources by Pathogenic *Neisseriae* Requires the Product of the hemO Gene. *J. Bacteriol.* **2000**, *182*, 439–447. [CrossRef] [PubMed]

74. Stojiljkovic, I.; Srinivasan, N. *Neisseria meningitidis* tonB, exbB, and exbD genes: Ton-dependent utilization of protein-bound iron in *Neisseriae*. *J. Bacteriol.* **1997**, *179*, 805–812. [CrossRef] [PubMed]

Sample Availability: Samples of the compounds are not available from the authors.

![molecules logo]

Article

Theoretical Prediction of the Complex P-Glycoprotein Substrate Efflux Based on the Novel Hierarchical Support Vector Regression Scheme

Chun Chen [1], Ming-Han Lee [1], Ching-Feng Weng [2] and Max K. Leong [1,2,*]

[1] Department of Chemistry, National Dong Hwa University, Shoufeng, Hualien 97401, Taiwan; 610212011@ems.ndhu.edu.tw (C.C.); 610512018@gms.ndhu.edu.tw (M.-H.L.)
[2] Department of Life Science and Institute of Biotechnology, National Dong Hwa University, Shoufeng, Hualien 97401, Taiwan; cfweng@gms.ndhu.edu.tw
* Correspondence: leong@gms.ndhu.edu.tw; Tel: +886-3-890-3609

Academic Editors: Xiangxiang Zeng, Alfonso Rodríguez-Patón and Quan Zou
Received: 7 July 2018; Accepted: 20 July 2018; Published: 22 July 2018

Abstract: P-glycoprotein (P-gp), a membrane-bound transporter, can eliminate xenobiotics by transporting them out of the cells or blood–brain barrier (BBB) at the expense of ATP hydrolysis. Thus, P-gp mediated efflux plays a pivotal role in altering the absorption and disposition of a wide range of substrates. Nevertheless, the mechanism of P-gp substrate efflux is rather complex since it can take place through active transport and passive permeability in addition to multiple P-gp substrate binding sites. A nonlinear quantitative structure–activity relationship (QSAR) model was developed in this study using the novel machine learning-based hierarchical support vector regression (HSVR) scheme to explore the perplexing relationships between descriptors and efflux ratio. The predictions by HSVR were found to be in good agreement with the observed values for the molecules in the training set ($n = 50$, $r^2 = 0.96$, $q_{CV}^2 = 0.94$, RMSE = 0.10, $s = 0.10$) and test set ($n = 13$, $q^2 = 0.80$–0.87, RMSE = 0.21, $s = 0.22$). When subjected to a variety of statistical validations, the developed HSVR model consistently met the most stringent criteria. A mock test also asserted the predictivity of HSVR. Consequently, this HSVR model can be adopted to facilitate drug discovery and development.

Keywords: P-glycoprotein; efflux ratio; in silico; machine learning; hierarchical support vector regression; absorption; distribution; metabolism; excretion; toxicity

---

1. Introduction

Permeability glycoprotein also known as P-glycoprotein (P-gp), which belongs to the ATP-binding cassette (ABC) superfamily of transporters, can actively transport a wide range of structurally and mechanistically diverse endogenous and xenobiotic chemical agents across the cell membrane at the energy expense of ATP hydrolysis [1]. P-gp, a 170-kDa plasma membrane protein encoded by the multidrug resistance gene (*MDR1/ABCB1*), is expressed at high levels in various tissues such as blood–brain-barriers (BBB), gastrointestinal tract (GIT), liver, kidney, and placenta [2–6]. In addition, P-gp plays significant roles in cell and tissue detoxification and elimination of harmful substances per se [1]. For example, the accumulation of neurotoxic amyloid-β (Aβ) peptides in the brain represents a pathogenic hallmark of Alzheimer's disease (AD), which is the most common form of dementia in aging populations [7]. It has been found that the decreased clearance rather than production of Aβ is the primary formation of the deleterious Aβ plaques in the brain [8]. The decreased elimination of Aβ from the brain into the blood can be partially attributed to the dysfunction of P-gp function, leading to the progression of AD [9–11]. Furthermore, it has been shown that Aβ can downregulate the P-gp expression along with other transporters and consequently lead to further accelerated neurodegeneration [12]. Hence, it has been suggested to increase Aβ clearance from the brain by restoring P-gp function of BBB to reduce Aβ brain accumulation as a new strategy in the medical treatment of the early stages of AD [13,14].

Additionally, P-gp efflux can profoundly implicate the role of drug absorption, distribution, metabolism, excretion, and toxicity (ADME/Tox) [15] that can clinically alter the administrated drug efficacy or even lead to various adverse side-effects due to drug–drug interaction (DDI) in the case of polypharmacy [16]. For instance, rifampin can interact with the P-gp substrate digoxin, leading to a lower accumulation of digoxin, as demonstrated by a clinical study [17]. Moreover, it is of particular interest to observe the subtle role played by P-gp in the central nervous system (CNS) since P-gp can affect the BBB penetration and pharmacological activities of administrated drugs [18]. The CNS-related side-effects of non-CNS drugs can be eliminated by P-gp because of their limited BBB penetration [19,20]. For instance, the P-gp substrate loperamide, which is a long-acting anti-diarrheal agent by agonizing the μ-opioid receptor, does not cause any CNS side-effects when administrated alone due to the blockage of the BBB penetration by P-gp [21]. When co-administrated with the P-gp inhibitor quinidine, loperamide produces adverse respiratory depression without significant alteration of the plasma accumulation due to its central opioid effect [22]. Conversely, P-gp can restrict or even eliminate the entry of CNS-targeted drugs into the brain, resulting in the reduction of the clinical efficacy [23].

In addition to normal tissues and organs, various types of tumor can over-express P-gp, producing multidrug resistance (MDR) [24], in which a single drug causes a non-drug resistant cell or cell line to become cross-resistant to other pharmacologically unrelated drugs due to the increase of administrated drug efflux and the decrease of intracellular drug accumulation [25]. As a result, P-gp efflux remains a major obstacle in the success of various kinds of cancer treatment [26] as well as infectious diseases [3,27]. For instance, brain tumor is one of the leading forms of malignancy and one of highest causes of cancer-related mortality among young adults aged less than 40 years and children [28] and glioma is the most common type of primary brain cancer with limited survival time and rate [29]. The CNS penetration of cediranib, which is a tyrosine kinase inhibitor for the treatment of glioma, is severely limited by the P-gp active efflux [30]. Co-administration of P-gp inhibitors is conceptually plausible and yet infeasible to circumvent MDR because of ineffective P-gp inhibitors in practical clinical applications [31,32]. Alternatively, P-gp can be considered as an anti-target in pharmaceutical research [33] especially in the field of CNS-targeted therapeutics [34,35]. Nevertheless, not all of marketed drugs have to be P-gp non-substrates provided that their therapeutic index is large with respect to the P-gp efflux ratio (ER) [36,37]. For instance, risperidone and 9-hydroxyl risperidone are clinically approved therapeutic agents for the treatment of schizophrenia even though they are P-gp substrates [38]. Accordingly, it is conceivable to expect that quantitative measure, viz. P-gp substrate efflux ratio, is more clinically relevant than qualitative classification, viz. substrate/non-substrate classification.

Of various in vitro assays to measure the efflux ratio [39–42], the monolayer efflux assay is the most relevant to drug distribution and the most commonly used in practice [20], in which the polarized epithelial cells, such as Madin–Darby canine kidney (MDCK) cells, are transfected with the *MDR1* gene, followed by measuring the ratios between basolateral-to-apical (B→A) apparent permeability (P_{app}) and apical-to-basolateral (A→B) P_{app} [43].

$$ER = \frac{P_{app}(B \to A)}{P_{app}(A \to B)},$$ (1)

$$P_{app} = \frac{1}{AC_0} \cdot \frac{dQ}{dt},$$ (2)

where P_{app} is evaluated using the membrane surface area (A), initial dosing concentration of the test molecule (C_0) in the donor compartment, and the amount of molecule transported per time (dQ/dt) in the receiver compartment [44]. Normally, molecules with ER > 2 are classified as P-gp substrates [39].

In contrast to in vitro and in vivo assays, in silico approaches are usually swift, inexpensive, less labor intensive, and less time-consuming for drug discovery and ADME/Tox profiling [45,46]. In fact, numerous P-gp classification structure–activity relationship (CSAR) models have been published elsewhere [47–68], whereas in silico quantitative studies of efflux ratio are scant [69–71]. Nevertheless, it is highly challenging to accurately model P-gp–substrate interactions [72] since P-gp is highly promiscuous per se as the result of the fact that P-gp can undergo substantial conformational changes upon binding with various ligands as illustrated by Figure 1 of Leong et al. [73]. In addition, P-gp has multiple substrate binding sites, as reported [72,74–77]. The mechanism of P-gp substrate efflux is far more complicated than P-gp–substrate interactions since P-gp substrate efflux can take place through various routes in that substrates can be actively transported by P-gp from the cytoplasm into the extracellular environment in an energy-dependent manner or through a protein channel positioned between the inner and outer leaflets of the lipid membrane, as illustrated by Figure 2 of Edwards [78]. In addition to active transport, P-gp substrates can also passively diffuse from the cytoplasm into the extracellular environment through transcellular diffusion and/or paracellular route, as illustrated by Figure 1 of Balimane et al. [79]. Notably, the P-gp substrate vinblastine, for instance, can be both passively diffused and actively transported [80]. As such, those modeling schemes employed by previously published investigations can only render the direct protein-ligand interactions and they are not suitable to model the efflux ratio. Conversely, quantitative structure–activity relationship (QSAR) schemes, which are a mathematic means to establish the relationship between biological activity and chemical characteristics, provide better approaches to model the efflux ratio since they can take into account any mechanisms that can occur through complex routes [81].

The complexity of P-gp mediated efflux can be problematic once the delicate roles played by those associated chemical features, viz. descriptors in QSAR models, are considered. For instance, inhibitors, modulators, and substrates can interact with P-gp using the hydrophobicity, hydrogen-bond acceptor (HBA), and hydrogen-bond donor (HBD) features [47,73,82]. Accordingly, hydrophobicity, HBA, and HBD can simultaneously enhance and reduce the P-gp efflux, and it is plausible to expect extremely nonlinear relationships between those chemical features and efflux ratio, suggesting that those linear models can yield significant prediction errors once applied to the test samples that are very different from their training patterns.

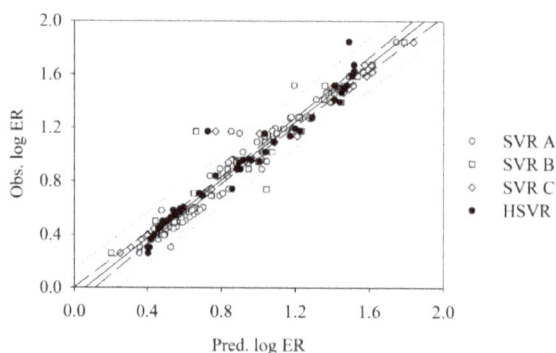

Figure 1. Observed log ER vs. the log ER predicted by SVR A (open circle), SVR B (open square), SVR C (open diamond), and HSVR (solid circle) for the molecules in the training set. The solid line, dashed line, and dotted lines correspond to the HSVR regression of the data, 95% confidence.

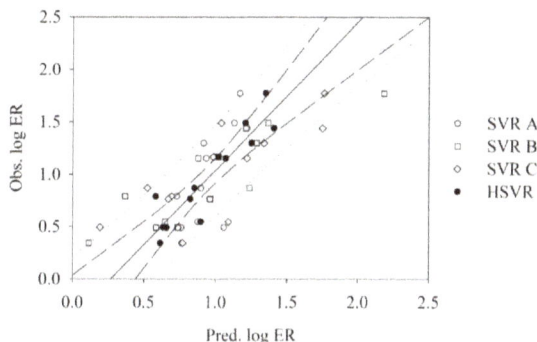

Figure 2. Observed log ER vs. the log ER predicted by SVR A (open circle), SVR B (open square), SVR C (open diamond), and HSVR (solid circle) for the molecules in the test set. The solid line, dashed line, and dotted lines correspond to the HSVR regression of the data, 95% confidence interval for the HSVR regression, and 95% confidence interval for the prediction, respectively.

Thus, it seems extremely difficult, if not completely impossible, to develop a sound in silico model to predict the P-gp substrate efflux ratio to compressively take into account those critical factors mentioned above. A solution to such challenge, however, can be obtained by the novel hierarchical support vector regression (HSVR) scheme proposed by Leong et al. [83] because HSVR can render the complex and varied dependencies of descriptors. As such, HSVR can simultaneously possess the advantageous characteristics of a local model and a global model, viz. broader coverage of applicability domain and higher level of predictivity, respectively. Furthermore, HSVR is designated to circumvent the "mesa effect" [84] in that the performance of a developed model deteriorates dramatically when applied to extrapolated predictions as demonstrated elsewhere [85,86]. In other words, HSVR is insensitive to outliers as compared with the other predictive models that is of critical importance to a predictive model [87]. Herein, the objective of this investigation was to develop an accurate, fast, and predictive in silico model based on the HSVR scheme to predict the P-gp substrate efflux ratio to facilitate drug discovery to design molecules with a preferable ADME/Tox profile.

2. Results

2.1. Data Compilation

More than 550 compounds were collected after comprehensive literature search. data curation was carefully carried out by eliminating those compounds: (i) with only qualitative array results (i.e., substrate or non-substrate); (ii) without specific ER values; or (iii) chemical structures. In addition, cells used to express P-gp protein also play a significant role in determining ER values. For instance, the measured ER values of astemizole were 2.16 and 0.6 assayed in MDCK and human colon adenocarcinoma (Caco-2) cells, respectively [51,88]. Of various assayed cells, 63 molecules tested in MDCK cells were selected from various sources [23,39,88–101] since it constituted the largest amount of data. The data size is seemingly small since several CSAR models have been derived based on rather large amounts of data. For instance, Li et al. [66] built various predictive models based on 423 P-gp substrates and 399 non-substrates compiled from numerous sources. Nevertheless, their data were generated from different assay conditions (e.g., different cell lines), leading to high levels of data heterogeneity. QSAR models, conversely, are vulnerable to data inhomogeneity [102]. Additionally, some molecules such as selenium-containing ones [103] were excluded because their topological descriptors, for instance, cannot be enumerated. Those ER values were discarded when they were not consistent with their measured P_{app} (B→A) and P_{app} (A→B) values [104]. Recently, the efflux ratios of more than 4000 Amgen in-house compounds were measured [105]. It is plausible to expect that the great sample amount and data consistency can furnish a good ER pool. Unfortunately, those chemical structures are proprietary, leading to the fact that there are only limited quantitative data with chemical structures available in the public domain to date. Those factors partially contribute to the fact that there is no genuine QSAR model has been published.

As such, only very limited data samples with available chemical structures and consistent assay conditions were recruited in this study to maximize the structural diversity and to maintain data homogeneity after purging inappropriate data based on above-mentioned criteria. Table S1 lists the SMILES strings, CAS registry numbers, efflux ratio values, and literature references of all molecules collected in this study.

2.2. Data Partition

Of all molecules adopted in this study, 50 and 13 molecules were randomly assigned to the training set and test set, respectively, with a ca. 4:1 ratio as suggested [106]. Figure S1 displays the projection of all molecules enrolled in this investigation in chemical space, spanned by the first three principal components (PCs), explaining 94.6% of the variance in the original data. As illustrated, both datasets exhibited high levels of similarity in the chemical space. Furthermore, the high levels of biological and chemical similarity between both datasets can also be validated by Figure S2, which shows the histograms of log ER, molecular weight (MW), polar surface area (PSA), number of HBA, and number of HBD in density form for all molecules in the training set and test set. Thus, it can be asserted that there was no substantial bias in datasets.

2.3. SVRE

Of all generated SVR models using various combinations of descriptors and runtime parameters, three SVR models, denoted by SVR A, SVR B, and SVR C, were assembled to construct the SVR ensemble, which was further subjected to regression by another SVR to generate the HSVR model. Table S2 summarizes the optimal runtime parameters of SVR A, SVR B, and SVR C. These three SVR models, which adopted 4, 6, and 3 descriptors (Table 1), respectively, were selected based on their individual performances on all molecules and statistical analyses in the training set and test set. Table S1 lists the predicted log ER values. Tables 2 and 3 summarize the associated statistical analyses of these three SVR models in the training set and test set, respectively. Figures 1 and 2 display the scatter plots of observed versus the predicted log ER values by SVR A, SVR B, and SVR C for the molecules in the training set and test set, respectively.

Table 1. Descriptor selected as the input of SVR models in the ensemble and their description.

Descriptor	SVR A	SVR B	SVR C	Description
SA			x[†]	Total surface area
n_{N+O}	x	x		Number of nitrogen and oxygen atoms
V_m	x	x	x	Molecule volume
PSA	x	x		Polar surface area
HBD	x	x		Number of hydrogen bond donating groups
n_{Rot}		x	x	Number of rotatable bonds
n_{Ar}		x		Number of aromatic rings

[†] Selected.

Figure 1 shows that the predictions by SVR A, SVR B, and SVR C are in good agreement with the observed values for most of the molecules in the training set as further manifested by their small RMSDs, average deviations, standard deviations (*s*), and larger r^2 parameters (Table 2). Of 50 training samples, SVR A, SVR B, and SVR C gave rise to 28, 3, and 2 predictions, which deviated from the experimental values by more than 0.10, respectively. It can be further observed in Figure 1 that most of the points predicted by SVR C generally lie on or are closer to the regression line when compared with

SVR A and SVR B. As a result, SVR C produced the lowest MAE (0.02), *s* (0.06), and RMSE (0.06) and the highest r^2 parameter (0.98), suggesting that SVR C performed better than SVR A and SVR B for the molecules in the training set. Nevertheless, the predictions of quinidine (**48**) by SVR A, SVR B and SVR C unanimously yielded the maximum residuals of 0.32, 0.51 and 0.40, respectively, denoting that SVR A executed better than SVR B and SVR C.

Table 2. Statistic evaluations, namely correlation coefficient (r^2), maximum residual (Δ_{Max}), mean absolute error (MAE), standard deviation (*s*), RMSE, and 10-fold cross-validation correlation coefficient (q^2_{CV}) evaluated by SVR A, SVR B, SVR C, and HSVR in the training set.

	SVR A	SVR B	SVR C	HSVR
r^2	0.95	0.95	0.98	0.96
Δ_{Max}	0.32	0.51	0.40	0.45
MAE	0.11	0.07	0.02	0.06
s	0.12	0.10	0.06	0.10
RMSE	0.12	0.10	0.06	0.10
q^2_{CV}	0.01	0.01	0.07	0.94

The predictions by SVR A, SVR B, and SVR C in the test set are also in good agreement with the experimental values (Figure 2). Nevertheless, most of the residuals obtained by the three SVR models in the test set are more than 0.15 (11, 11, and 8, respectively). It can be further observed in Table 3 that the mean absolute errors computed by SVR A, SVR B, and SVR C unequivocally increase from 0.11, 0.07, and 0.02 in the training set to 0.29, 0.22, and 0.24 in test set, respectively. The other statistical parameters also suggest that the performances of these three models in the SVRE slightly decline from the training set to the test set (Tables 2 and 3). The maximum residual computed by SVR C in the test set was yielded from the prediction of cimetidine (**13**) with an absolute residual of 0.55, which were only 0.34 and 0.10 by SVR A and SVR B, respectively. Similarly, vinblastine (**58**) was best predicted by SVR C with an absolute residual of 0.01, whereas SVR A and SVR B gave rise to absolute errors of 0.60 and 0.41, respectively.

Table 3. Statistic evaluations, correlation coefficients q^2, q^2_{F1}, q^2_{F2}, and q^2_{F3}, concordance correlation coefficient (CCC), maximal absolute residual (Δ_{Max}), mean absolute error (MAE), standard deviation (*s*), and RMSE evaluated by SVR A, SVR B, SVR C, and HSVR in the test set.

	SVR A	SVR B	SVR C	HSVR
q^2	0.54	0.75	0.60	0.83
q^2_{F1}	0.39	0.67	0.55	0.80
q^2_{F2}	0.39	0.67	0.54	0.80
q^2_{F3}	0.38	0.66	0.54	0.80
CCC	0.45	0.86	0.78	0.87
Δ_{Max}	0.60	0.42	0.55	0.42
MAE	0.29	0.22	0.24	0.17
s	0.35	0.26	0.30	0.22
RMSE	0.34	0.25	0.29	0.21

Furthermore, SVR A, SVR B, and SVR C yielded the q^2 values of 0.54, 0.75, and 0.60 in the test and the cross-validation correlation coefficients q^2_{CV} of 0.01, 0.01, and 0.07 in the training set, respectively (Tables 2 and 3). When subjected to the *Y*-scrambling test, SVR A, SVR B, and SVR C gave rise to the $\langle r^2_s \rangle$ values of 0.02, 0.03, and 0.03, respectively (Table 1). The almost zero values of $\langle r^2_s \rangle$ as well as substantial differences between corresponding r^2 and $\langle r^2_s \rangle$ signify that those three SVR models in the ensemble are not the result of chance correlation [107]. Conversely, the substantial differences between r^2 and q^2 and between r^2 and q^2_{CV} imply the over-fitting characteristics of these three models that actually can be further manifested by their small q^2_{F1}, q^2_{F2}, q^2_{F3}, and CCC values (Table 3). As a result, it is plausible to expect that these models are local models per se, which have limited coverage of applicability domain (vide infra) [108].

2.4. HSVR

The HSVR model was produced by the regression of the SVR ensemble based on the predictions of all molecules and statistical evaluations in the training set (Table S1 and Table 2). Table S2 lists the optimal runtime conditions for the final SVR model. It can be observed in Figure 1 that the HSVR model showed better prediction accuracy than SVR A, SVR B, and SVR C for the molecules in the training set because the distances between the predictions by HSVR and regression line are generally between the largest ones and smallest ones produced by its SVR counterparts in the ensemble. However, HSVR executed better than any of SVR models in the ensemble in some cases. The predictions of desloratadine (**19**) by SVR A, SVR B, SVR C, and HSVR, for instance, yielded absolute residuals of 0.10, 0.06, 0.01, and 0.00, respectively. Statistically, HSVR performed better than SVR A and SVR B, whereas SVR C, in turn, functioned negligibly better than HSVR, as manifested by those parameters listed in Table 2. For example, SVR A, SVR B, SVR C, and HSVR yielded the r^2 values of 0.95, 0.95, 0.98, and 0.96, respectively.

When applied to the test samples, HSVR only showed insignificant performance decreases from the training set to the test set. For instance, RMSE increased from 0.10 in the training set to 0.21 in the test set (Tables 2 and 3). However, the maximum residual declined from 0.45 in the training set to 0.42 in the test set. Figure 2 displays that HSVR showed better performance than SVR A, SVR B, and SVR C in the test set. The performance predominance of HSVR can be further manifested by those statistical parameters listed in Table 3. For instance, SVR A, SVR B, SVR C, and HSVR gave rise to MAE values of 0.29, 0.22, 0.24, and 0.17, respectively. Similar observation that HSVR generated smaller absolute residuals than its counterparts in the ensemble can also be found in the test set. The absolute prediction error of paliperidone (**41**), for instance, was 0.14 given rise by HSVR, whereas SVR A, SVR B, and SVR C produced residuals of 0.57, 0.25, and 0.30, respectively. When compared with its counterparts in the ensemble, HSVR generally produced consist and small errors in both training set and test set as manifested by those parameters associated with error listed in Tables 2 and 3, suggesting that HSVR has broader coverage of applicability domain. Additionally, HSVR yielded the smallest differences between r^2 and q^2_{CV} (0.02) and between r^2 and q^2 (0.13), indicating that HSVR was a well-trained model or no over-fitting effect was observed because it will otherwise produce at least one significant difference among those parameters. Similarly, the possibility of chance correlation of HSVR can be eliminated by Y-scrambling since it also produced an almost zero $\langle r^2_s \rangle$ (0.03) and marked difference between r^2 and $\langle r^2_s \rangle$ (Table 2) [107].

2.5. Predictive Evaluations

Figure 3 displays the scatter plots of the residual vs. the log ER values predicted by HSVR for the molecules in the training set and test set. It can be observed that the residuals are approximately evenly distributed on both sides of x-axis along the range of predicted values in both datasets, suggesting that there is no systematic error associated with the HSVR model [102]. The unbiased predictions can be further exhibited by its almost negligible average residuals that were −0.02 and −0.02 in the training set and test set, respectively (Table S1).

The predictivity of generated HSVR model was further evaluated by the validation requirements proposed by Golbraikh et al. [109], Ojha et al. [110], Roy et al. [111], and Chirico and Gramatica [112] (Equations (18)–(21)) in the training set and test set. Table 4 summarizes the results, from which it can be observed that HSVR maintained similar high levels of performance in the training set and test set. Additionally, HSVR fulfilled all validation requirements, indicating that this predictive model is highly accurate and predictive.

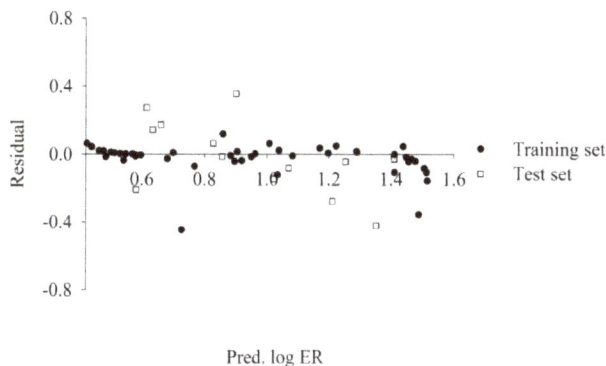

Figure 3. Residual vs. the log ER predicted by HSVR in the training set (solid circle) and test set (open square).

Table 4. Validation verification of HSVR based on prediction performance of those molecules in the training set and test set.

	Training Set	Test Set
n	50	13
r_o^2	0.95	0.77
k	1.03	1.05
$r_o'^2$	0.94	0.52
r_m^2	0.90	0.72
$r_m'^2$	0.85	0.60
$\langle r_m^2 \rangle$	0.88	0.66
Δr_m^2	0.05	0.12
$r^2, q_{CV}^2, q_{.}^2, q_{Fn}^2 \geq 0.70$	x	x
$\left| r^2 - q_{CV}^2 \right| < 0.10$	x	N/A
$(r^2 - r_o^2)/r^2 < 0.10$ and $0.85 \leq k \leq 1.15$	x	x
$\left| r_o^2 - r_o'^2 \right| < 0.30$	x	x
$r_m^2 \geq 0.65$	x	x
$\langle r_m^2 \rangle \geq 0.65$ and $\Delta r_m^2 < 0.20$	x	x
CCC ≥ 0.85	N/A [†]	x

[†] Not applicable.

2.6. Mock Test

To mimic real world challenges, the developed HSVR model was further tested on the P-gp substrates assayed by Crivori et al. [51]. Of all marketed drugs measured by Crivori et al., 12 were also enrolled in this study, yielding a good way to calibrate the testing system. However, these molecules were measured in Caco-2 cells, whereas all of the molecules adopted in this study were tested in MDCK cells, suggesting that those compounds assayed by Crivori et al. are not qualified as the second external or test set since those validation criteria (vide supra) are not applicable to these compounds. To eliminate the discrepancy between both assay systems, the linear correlation between both assay systems for those common molecules was first inspected and the obtained scatter plot is illustrated in Figure 4. It can be observed that the experimental values in both systems were modestly correlated with each other well with an *r* value of 0.78. Thus, it is plausible to examine the HSVR model with those novel P-gp substrates assayed in Caco-2 cells.

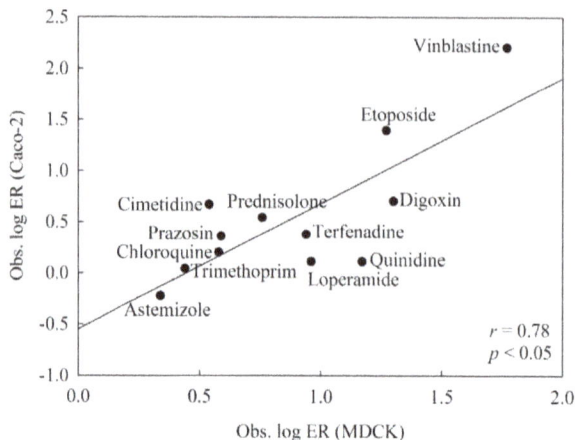

Figure 4. The observed log ER values (Caco-2) vs. the observed log ER values (MDCK).

Figure 5 displays the tested results of the nine novel drugs. It can be observed that the r value between experimental log ER obtained in the Caco-2 cells and predicted log ER in the MDCK cells was 0.77. The negligible difference between both numbers (0.78 vs. 0.77) suggests that the predictions by the HSVR model can almost reproduce the experimental observations and this mock test unequivocally assured the predictive capability of HSVR.

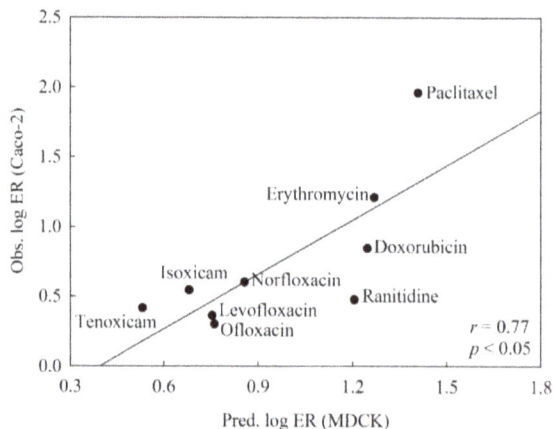

Figure 5. The observed log ER values (Caco-2) vs. the predicted log ER values (MDCK).

3. Discussion

Collectively, seven descriptors were adopted in this study. Intrinsically, the sample-to-descriptor ratio was ca. 7:1, which is significantly larger than 5, viz. the minimal requirement to lessen the probability of chance correlations in a predictive model [113]. However, the process of P-gp substrate efflux is complex since it can take place thought various routes (vide supra). As such, different descriptors were adopted by different classification models. Of various descriptors selected by qualitative predictive models, hydrophobic, HBA, and HBD are the most frequently selected chemical features, as illustrated by the model proposed by Penzotti et al. [47]. However, the analysis of Amgen in-house compounds can reveal that HBD and topological PSA (tPSA) are the predominant factors associated with ER [105].

Figure 6 displays the average log ER for each histogram bin of HBD for all molecules selected in this study. It can be observed that the average log ER value initially increased with HBD when HBD was no more than 6 and then subsequently decreased when HBD was more than 6. Such positive dependence of log ER on HBD is, in fact, consistent with the analysis made by Hitchcock et al. [105]. However, those Amgen in-house compounds had HBD of no more than 5, leading to an only positive relationship between log ER and HBD. Such discrepancy in both systems can be conceivably attributed to the fact that the initial P-gp substrate binding can be enhanced by HBD as illustrated by the pharmacophore models of Penzotti et al. [47], whereas the consequent transport of the substrates into the extracellular environment can be hampered by too many HBDs, plausibly because of the increase in water desolvation energy [114] and the decrease in membrane fluidity [115]. As such, a nonlinear relationship between HBD and log ER was yielded consequently.

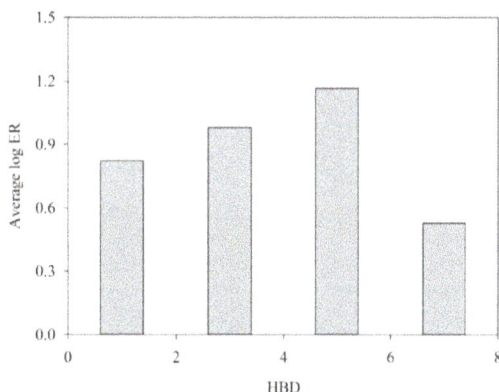

Figure 6. Average log ER vs. the distribution of HBD.

It has been observed that hydrophobicity, which normally can be represented by log P, plays an important role in P-gp–substrate interaction due to the hydrophobic nature of the substrate binding pocket, resulting in stronger P-gp substrate binding for those more hydrophobic substrates [116]. Nevertheless, the interaction between substrates and lipid bilayer as well as the release of substrates into the extracellular environment also depend on the hydrophobicity of substrates (vide supra), leading to a nonlinear relationship between log P and log ER. Figure 7 displays the average log ER for histogram bin of log P for all molecules enlisted in this study. It can be observed that the average log ER initially increased with log P when log P was smaller and decreased with log P when log P became higher. Such observation is qualitatively similar to the trend of P_{app} (A→B) found by Hitchcock et al. [105].

Nevertheless, it is unusual to observe that log P was not included in this study, whereas the number of aromatic rings (n_{Ar}) was enlisted in this study. Such inconsistency can be realized by the fact that the average log P values increased with n_{Ar} for all of molecules included in this study, as illustrated in Figure 8, which displays the average log P versus the distribution of n_{Ar}. As such, it is plausible to replace log P by n_{Ar}. Furthermore, it has been found that aromatic ring moieties are important in substrate recognition and efflux modulation [117,118]. More importantly, the empirical observation has indicated that models with the selection of n_{Ar} unanimously showed better performance than those with the selection of log P (data not shown).

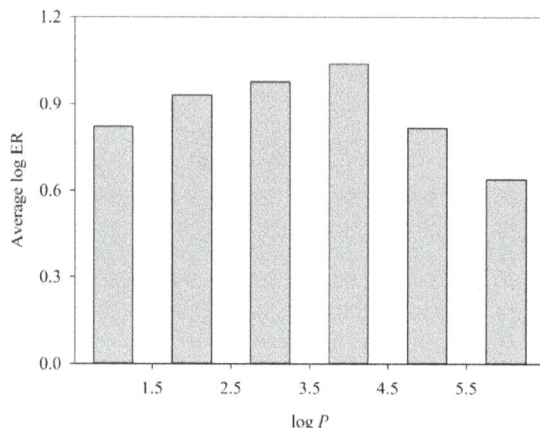

Figure 7. Average log ER vs. the distribution of log *P*.

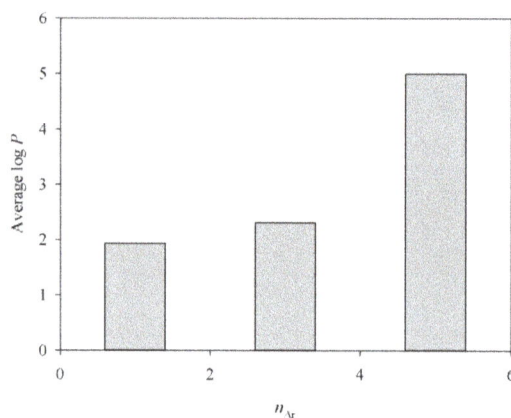

Figure 8. Average log *P* vs. the distribution of number of aromatic ring (n_{Ar}).

The significant role of HBA in the P-gp–substrate interaction has been manifested by molecular docking simulations [71] as well as numerous qualitative models. Additionally, it has been suggested that HBA can enhance P-gp-mediated efflux [56]. Nevertheless, it is unusual to observe that none of SVR models in the ensemble has adopted HBA, plausibly because the descriptor number of nitrogen and oxygen atoms (n_{N+O}) correlated well with HBA as demonstrated by Figure 9, which displays n_{N+O} versus HBA. In fact, Desai et al. [56] adopted n_{N+O} instead of HBA as the substrate classification criterion. Furthermore, empirical model development has shown that models with the selection of n_{N+O} executed better than those with the selection of HBA (data not shown). As a result, the descriptor n_{N+O} was selected in lieu of HBA.

The descriptor tPSA is a modified version to swiftly calculate the polar surface area only based on the additive polar surface areas [119]. The recursive partitioning (RP) model of Joung et al. [68] indicated the significant role of PSA in classifying molecules as P-gp substrates/non-substrates. Moreover, Hitchcock et al. also found the profound contribution of tPSA to P-gp mediated efflux (vide supra). Accordingly, the more sophisticated version of PSA was adopted in this study since it can function as polarity and hydrogen-bonding features [66].

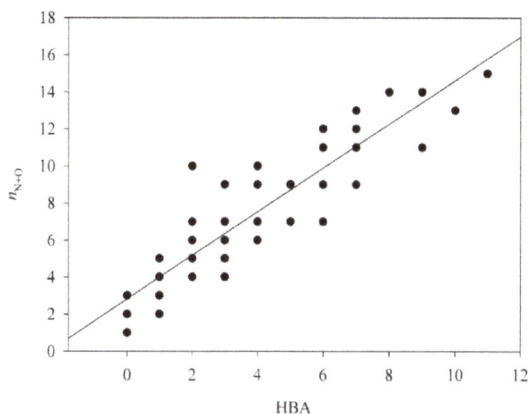

Figure 9. The number of nitrogen and oxygen (n_{N+O}) vs. HBA.

It has been observed that the substrate size, which can be characterized by molecular weight (MW), molecular volume (V_m), and total surface area (SA), can have a large impact on P-gp–substrate interaction as well as passive permeability [120]. Nevertheless, it has been suggested that both V_m and SA can be better metrics to estimate the actual molecular size [121], and MW, conversely, was closely associated with V_m with an r^2 values of 0.98 for the molecules enlisted in this study. In fact, it has been postulated that V_m rather than MW is a better metric to associate with ER [122]. Accordingly, V_m and SA were adopted to render the size effects, whereas MW was discarded to reduce the probability of spurious correlations.

It has been found that P-gp substrates generally have more rotatable bonds than non-substrates since more flexible molecules can be more easily to adopt favorable orientation to interact with P-gp [49,66,123]. In fact, non-CNS drugs are more flexible than their CNS counterparts [23] since molecules with more conformational flexibility can favor the internal H-bond formation, which, in turn, can enhance the passive membrane permeability [124]. As such, substrate conformational flexibility, which can be characterized by the number of rotatable bond (n_{Rot}), can facilitate not only the active transport but also passive permeability of P-gp substrates, and n_{Rot} was adopted in this investigation.

Gunaydin et al. [69] only took into account the contribution of the differences between free energy in water (G_{H2O}) and that in chloroform (G_{CHCl3}), viz. $\Delta G_{H2O-CHCl3}$, since it was hypothesized that P-gp undergoes a conformation change from the intercellular-facing state to extracellular-facing state upon binding with substrates. As such, the transported substrates experience from a lipophilic environment into a hydrophilic one. In addition to $\Delta G_{H2O-CHCl3}$, the contribution of $\Delta G_{DMSO-CHCl3}$ was also computed in this study to mimic the assay conditions. Nevertheless, neither of the solvation free energy differences was selected in this study due to their insignificant contribution to ERs (data not shown), plausibly because the P-gp conformation change can only account for a small part of the whole complicated efflux process and, additionally, passive permeability is not resulted from the P-gp conformation change. The predictive model of Gunaydin et al. [102], nevertheless, was derived only based on 12 marketed drugs that cannot comprehensively render the complex efflux. As such, more descriptors will be required in case of more diverse samples.

Didziapetris et al. [63] proposed the "rule-of-fours," which states that molecules with: (i) $n_{N+O} \geq 8$; (ii) MW > 400; and (iii) acid pK_a > 4 are likely to be P-gp substrates. Of all molecule with ER > 2 selected in this study, viz. substrates, approximately 32%, 52%, and 100% can meet the criteria $n_{N+O} \geq 8$, MW > 400, and acid pK_a > 4, respectively, and only 29% can completely fulfill those three criteria. Actually, Li et al. [66] also found that only ca. 34% of samples can simultaneously meet those three criteria. Furthermore, it is not unusual to observe that different rules have been proposed to classify molecules into P-gp substrates/non-substrates. Desai et al. [56], for instance, have proposed the molecules with TPSA >100 Å^2 and most basic pK_a > 8 have higher probability to be substrates. The inconsistency in various proposed rules can be plausibly explained as those rules were derived only based on linear analyses of those P-gp substrates/non-substrates. However, such bisection is not always true, as manifested by the naïve Bayesian classifiers built by Li et al. [66]. In addition, the size and hydrophobicity of substrates can affect the substrate-membrane interactions nonlinearly [125]. Further complexity can be raised once the P-gp substrate efflux is considered instead of P-gp substrate/non-substrate classification since the P-gp substrate efflux can take place through various routes (vide supra), leading to nonlinear relationships between some descriptors and log ER, such as HBD and log *P* (Figures 6 and 7). Numerous attempts have been made in this study to develop various partial least square (PLS) models to accommodate the novel 2-QSAR scheme [86] and no satisfactory models were produced (data not shown). Conversely, the accurate and predictive HSVR can comprehensively describe such nonlinear dependence of log ER on descriptors.

Moreover, it has been observed that P-gp and other ABC members, namely breast cancer resistant protein (BCRP/ABCG2) and multidrug resistance-associated protein 4 (ABCC4/MRP4), play a critical role in BBB permeability [126], which can take place via various routes [127] in addition to the already complicated P-gp mediated efflux. As such, it is plausible to expect that it is extremely difficult to develop a sound in silico model to predict BBB permeability if not entirely impossible [128]. The development of an accurate in silico model in this study to predict the P-gp substrate efflux can pave the way to establish a sound theoretical model to predict the BBB permeability in the future. Most molecules adopted in this study are marketed drugs for treating various illnesses, such as HIV infection, allergy symptoms, rheumatoid arthritis, hypertension, diarrhea, and different types of cancer in addition to assorted CNS-related disorders (Supplementary Materials Table S1). The broad spectrum of therapeutic agents unequivocally indicates that the data samples are structurally diverse, which can be further manifested by the fact that the average minimum distance between two molecules, viz. the distance between two nearest neighbors, in the chemical space was 2.06 with an standard deviation of 1.39 and the maximum distance between two collected samples was 29.57 (Supplementary Materials Figure S1), giving rise to an ratio of ca. 1:14. As such, it is plausible to expect that developed HSVR should have a larger coverage of applicability domain accordingly, which is an important characteristic for a predictive model in practical application. More importantly, the derived HSVR model and published P-gp substrate/non-substrate classification models can work in a synergistic fashion, in which the latter can be used to identify those P-gp substrates and the former can be deployed to predict their efflux ratios.

4. Materials and Methods

4.1. Data Compilation

A sound predictive model can only be built based on good quality of sample data [102]. To compile quality data for this study, a comprehensive literature search was conducted to retrieve efflux ratio values from various sources to maximize the structural diversity. If there were two or more available efflux ratio data for a given compound and in close range, the average values were then taken to warrant better consistency. Further data curation was carried out by cautiously inspecting molecular structures to remove those molecules without definite stereochemistry.

4.2. Molecular Descriptors

All of the molecules enlisted in this study were subjected to full geometry optimization using the density functional theory (DFT) B3LYP method with the basis set 6-31G(d,p) by the *Gaussian 09* package (Gaussian, Wallingford, CT) in the dimethyl sulfoxide (DMSO) solvent system using the polarizable continuum model (PCM) [129,130] to mimic the experimental conditions. These geometries were confirmed to be real minima on the potential energy surface by force calculations when no imaginary frequency was obtained. Additionally, atomic charges were also calculated by the molecular electrostatic potential-based method of Merz and Kollman [131] and the highest occupied molecular orbital energy (E_{HOMO}), lowest unoccupied molecular orbital energy (E_{LUMO}), free energy (ΔG), and dipole (μ) were also retrieved from the optimization calculations since those quantum mechanics descriptors have been adopted previously. As such, it is of necessity to employ a more sophisticated quantum mechanics method to optimize those selected molecules and to calculate their associated descriptors.

The *Discovery Studio* package (BIOVIA, San Diego, CA) and *E-Dragon* (available at the web site http://www.vcclab.org/lab/edragon/) were also utilized to calculate more than 200 one-, two-, and three-dimensional molecular descriptors of those optimized molecules. These descriptors can be classified as electronic descriptors, spatial descriptors, structural descriptors, thermodynamic descriptors, topological descriptors, and E-state indices.

Data filtering was initially performed by removing those descriptors missing for at least one sample or showing little or no discrimination against all samples. Furthermore, only one descriptor should be kept among those descriptors with intercorrelation values of $r^2 > 0.8$ to reduce the probability of spurious correlations as postulated by Topliss and Edwards [113]. It is not uncommon to observe that certain descriptors with broader ranges outweigh those with narrower ranges because of substantial variations in magnitudes. Nevertheless, such problem can be resolved when the non-descriptive descriptors, viz. real variable descriptors, are normalized with the following equation [132]

$$\chi_{ij} = \left(x_{ij} - \langle x_j \rangle\right) / \left[\sum_{i=1}^{n}\left(x_{ij} - \langle x_j \rangle\right)^2 / (n-1)\right]^{1/2} \tag{3}$$

where x_{ij} and χ_{ij} represent the original and normalized jth descriptors of the ith compound, respectively; $\langle x_j \rangle$ stands for the mean value of the original jth descriptor; and n is the number of samples.

Descriptor selection plays a pivotal role in determining the performance of predictive models [133]. More descriptors will be needed once there are more training samples with more diverse structures [102]. Conversely, it is highly possible to yield an over-trained model when there are too many selected descriptors [134]. The descriptor selection was initially executed by genetic function approximation (GFA) using the QSAR module of *Discovery Studio* due to its effectiveness and efficiency [135]. Further descriptor selection was carried out by the recursive feature elimination (RFE) method, in which the predictive model was repeatedly generated by all but one of descriptors. The descriptors were then ranked according to their contributions to the predictive performance; and the descriptor with least contribution was discarded [136].

4.3. Data Partition

The collected molecules were divided into two datasets, namely the training set and test set, to develop and to verify the predictive models using the Kennard–Stone (KS) algorithm [137] implemented in *MATLAB* (The Mathworks, Natick, MA, USA) with an approximate 4:1 ratio as suggested [106]. It has been suggested that a sound model can be derived only based on chemically and biologically similar training samples and test samples [138]. As such, the data distribution was carefully examined to ensure the high levels of biological and chemical similarity in both datasets.

4.4. Hierarchical Support Vector Regression

Support vector machine (SVM) proposed by Vapnik et al. [139] was initially designated for use in classification and consequently modified for regression problems by nonlinearly mapping the input data into a higher-dimension space, in which a linear regression is performed [140]. SVM regression takes into account both the training error and the model complexity as compared with the traditional regression algorithms, which develop predictive models by minimizing the training error. As such, SVM performs better than traditional regression methods because of its advantageous characteristics, namely dimensional independence, limited number of freedom, excellent generalization capability, global optimum, and easy to implement [141].

Similar to other linear or machine learning (ML)-based QSAR techniques, SVM has to tradeoff between the characteristics of a global model, viz. broader coverage of applicability domain (AD), and a local model, viz. higher level of predictivity [108]. This seeming dilemma, nevertheless, can be plausibly resolved using the hierarchical support vector regression (HSVR) scheme, which was initially proposed by Leong et al. and was derived from SVM [83], because HSVR can simultaneously take into consideration both seemingly mutually exclusive characteristics. Practically speaking, it has been demonstrated that HSVR outperformed a number of ML-based models, namely artificial neural network (ANN), genetic algorithm (GA), and SVM [85].

The detail of HSVR has been mentioned elsewhere [83]. Briefly, a panel of SVR models was built by the *LIBSVM* package (software available at http://www.csie.ntu.edu.tw/~cjlin/libsvm) based on various descriptor combinations, and each SVR model represented a local model. The model generation and verification were executed using the modules *svm-train* and *svm-predict*, respectively, implemented in the *LIBSVM* package. The regression modes, namely, ε-SVR and γ-SVR, were adopted, and radial basis function (RBF) was employed as the kernel due to its simplicity and better performance when compared with the others [142]. The runtime parameters, namely regression modes ε-SVR and ν-SVR, the associated ε and ν, cost C, and the kernel width γ, were scanned by the systemic grid search algorithm using an in-house Perl script [143], in which all parameters were changed independently in a parallel fashion.

Two SVR models were initially adopted to develop an SVR ensemble (SVRE), which, in turn, was further subjected to regression by another SVR to yield the final HSVR model. The two-member SVREs were continuously assembled until the HSVR model performed well. Otherwise, the three- or even four-member ensembles were built by adding one or more SVR models, respectively, if all two-member ensembles failed to perform well. The descriptor selection and ensemble assembly were predominantly governed by the principle of Occam's razor [144] by adopting the fewest descriptors and SVR models.

4.5. Predictive Evaluation

The predictivity of a generated model was evaluated by several statistic metrics. The coefficients r^2 and q^2 in the training set and external set, respectively, for the linear least square regression were computed by the following equation

$$r^2, q^2 = 1 - \sum_{i=1}^{n}(\hat{y}_i - y_i)^2 / \sum_{i=1}^{n}(y_i - \langle \hat{y} \rangle)^2 \tag{4}$$

where $\hat{y}_i$ and y_i are the predicted and observed values, respectively; and $\langle \hat{y} \rangle$ and n stand for the average predicted value and the number of samples in the dataset, respectively.

Furthermore, the residual Δ_i, which is the difference between y_i and $\hat{y}_i$, was calculated

$$\Delta_i = y_i - \hat{y}_i \tag{5}$$

The root mean square error (RMSE) and the mean absolute error (MAE) for n samples in the dataset were computed

$$\text{RMSE} = \left[\sum_{i=1}^{n}\Delta_i^2/n\right]^{1/2}, \tag{6}$$

$$\text{MAE} = \frac{1}{n}\sum_{i=1}^{n}|\Delta_i|, \tag{7}$$

The produced model was further subjected to 10-fold cross-validation instead of the widely used leave-one-out due to its better performance [145], giving rise to the correlation coefficient of 10-fold cross validation q_{CV}^2. In addition to cross-validation, the developed models were also internally validated by the *Y*-scrambling test [102], which was carried out by randomly permuting the log ER values, viz. *Y* values, to refit the previously developed models while the descriptors were remained unaltered, giving rise to the correlation coefficient r_s^2. The observed log ER values were scrambled 25 times as suggested [107] to produce the average correlation coefficient $\langle r_s^2 \rangle$. Furthermore, various modified versions of r^2 proposed by Ojha et al. [110] were also computed

$$r_m^2 = r^2 \left(1 - \sqrt{|r^2 - r_0^2|} \right), \tag{8}$$

$$r'^2_m = r^2 \left(1 - \sqrt{|r^2 - r'^2_0|} \right), \tag{9}$$

$$\langle r_m^2 \rangle = \left(r_m^2 + r'^2_m \right) / 2, \tag{10}$$

$$\Delta r_m^2 = \left| r_m^2 - r'^2_m \right|, \tag{11}$$

where the correlation coefficient r_0^2 and the slope of the regression line k were calculated from the regression line (predicted vs. observed values) through the origin, whereas r'^2_0 was calculated from the regression line (observed vs. predicted values) through the origin.

Moreover, the correlation coefficients q_{F1}^2, q_{F2}^2, and q_{F3}^2 and concordance correlation coefficient (CCC) proposed by Shi et al. [146], Schüürmann et al. [147], Consonni et al. [148], and Chirico and Gramatica [149] were also computed by *QSARINS* [150,151] to evaluate the model performance in the external dataset

$$q_{F1}^2 = 1 - \sum_{i=1}^{n_{EXT}} (y_i - \hat{y}_i)^2 / \sum_{i=1}^{n_{EXT}} (y_i - \langle y_{TR} \rangle)^2, \tag{12}$$

$$q_{F2}^2 = 1 - \sum_{i=1}^{n_{EXT}} (y_i - \hat{y}_i)^2 / \sum_{i=1}^{n_{EXT}} (y_i - \langle y_{EXT} \rangle)^2, \tag{13}$$

$$q_{F3}^2 = 1 - \left[\sum_{i=1}^{n_{EXT}} (y_i - \hat{y}_i)^2 / n_{EXT} \right] / \left[\sum_{i=1}^{n_{TR}} (y_i - \langle y_{TR} \rangle)^2 / n_{TR} \right], \tag{14}$$

$$CCC = \frac{2 \sum_{i=1}^{n_{EXT}} (y_i - \langle y_{EXT} \rangle)(\hat{y}_i - \langle \hat{y}_{EXT} \rangle)}{\sum_{i=1}^{n_{EXT}} (y_i - \langle y_{EXT} \rangle)^2 + (\hat{y}_i - \langle \hat{y}_{EXT} \rangle)^2 + n_{EXT}(\langle y_{EXT} \rangle - \langle \hat{y}_{EXT} \rangle)^2} \tag{15}$$

where n_{TR} and n_{EXT} are the numbers of samples in the training set and external set, respectively; $\langle \hat{y}_{TR} \rangle$ is the average predicted value in the training set; and $\langle y_{EXT} \rangle$ and $\langle \hat{y}_{EXT} \rangle$ are the average observed and predicted values in the external set, respectively.

Various criteria for those statistical parameters have been proposed to gauge the model predictivity [152]. For instance, Chirico and Gramatica considered that both q_{F3}^2 and CCC are the best validation parameters to measure the predictivity [149], whereas Roy et al. suggested that $\langle r_m^2 \rangle$ and Δr_m^2 are the most stringent metrics [111]. Recently, Todeschini et al. demonstrated that q_{F3}^2 is the most reliable metric [112]. The parameter q_{F2}^2 has been adopted by Organization for Economic Co-operation and Development (OECD) to assess the performance of QSAR models [147].

More importantly, a model can be considered as predictive if it can meet the most stringent criteria collectively proposed by Golbraikh et al. [109], Ojha et al. [110], Roy et al. [111], and Chirico and Gramatica [112].

$$r^2, q_{CV}^2, q^2, q_{Fn}^2 \geq 0.70, \tag{16}$$

$$\left| r^2 - q_{CV}^2 \right| < 0.10, \tag{17}$$

$$\left(r^2 - r_0^2 \right) / r^2 < 0.10 \text{ and } 0.85 \leq k \leq 1.15, \tag{18}$$

$$\left| r_o^2 - r'^2_o \right| < 0.30, \tag{19}$$

$$r_m^2 \geq 0.65, \tag{20}$$

$$\left\langle r_m^2 \right\rangle \geq 0.65 \text{ and } \Delta r_m^2 < 0.20, \tag{21}$$

$$CCC \geq 0.85, \tag{22}$$

where r in Equations (18)–(21) represents the parameters r and q in the training set and external set, respectively; and q_{Fn} stands for q_{F1}, q_{F2}, and q_{F3}.

5. Conclusions

P-gp substrate efflux can be a major obstacle in the success of CNS-targeted therapeutic delivery as well as a critical pharmacokinetic factor for causing DDIs. On the other hand, the CNS-related side-effects of non-CNS drugs can be reduced by P-gp mediated efflux. As such, P-gp substrate efflux is of critical importance to drug discovery and development regardless of CNS drugs or non-CNS drugs. An in silico model to predict the P-gp substrate efflux can be valuable to drug discovery and development. Nevertheless, P-gp substrate efflux is a complex process that can take place through various routes, namely active transport and passive permeability, leading to different descriptor combinations as well as different relationships to render these variations in different mechanisms. In this study, a QSAR predictive model derived from the novel hierarchical support vector regression (HSVR) scheme, which can simultaneously possess the advantageous characteristics of a local model and a global model, viz. broader coverage of applicability domain and higher level of predictivity, respectively, was developed to envisage the P-gp substrate efflux ratio. The developed HSVR showed great prediction accuracy for the 50 and 13 molecules in the training set and test set, respectively, with excellent predictivity and statistical significance. When mock tested by a group of molecules to mimic real challenges, the derived HSVR model also executed accordantly well. Furthermore, the HSVR model can elucidate the discrepancies among all published P-gp substrate classifiers, indicating its superiority. Hence, it can be affirmed that this HSVR model can be adopted as an accurate and reliable predictive tool, even in the high throughput fashion, to facilitate drug discovery and development by designing drug candidates with a more desirable pharmacokinetic profile.

Supplementary Materials: Table S1. Selected compounds for this study, their names, SMILES strings, CAS numbers, observed log ER values and predicted values by SVR A, SVR B, and HSVR, data partitions, and references; Table S2. Optimal runtime parameters for the SVR models; Figure S1. Molecular distribution for the samples in the training set (solid circle) and test set (open square) in the chemical space spanned by three principal components; Figure S2. Histograms of: (A) observed log ER; (B) molecular weight (MW); (C) polar surface area (PSA); (D) number of hydrogen bond acceptor (HBA); and (E) number of hydrogen bond donor (HBD) in density form for all molecules in the training set and test set.

Author Contributions: C.C., C.F.W., and M.K.L. conceived and designed the study; C.C., M.H.L., and M.K.L. performed the experiments and analyzed the data; and C.C., C.F.W., and M.K.L. wrote the paper.

Acknowledgments: This work was financially supported by the Ministry of Science and Technology, Taiwan. Some calculations were performed at the National Center for High-Performance Computing, Taiwan. The authors are grateful to Prof. Paola Gramatica for providing free license of *QSARINS* and Yi-Lung Ding for helping data analysis.

Conflicts of Interest: The authors declare that they have no conflict of interest.

References

1. Schinkel, A.H.; Jonker, J.W. Mammalian drug efflux transporters of the atp binding cassette (abc) family: An overview. *Adv. Drug Deliv. Rev.* **2003**, *55*, 3–29. [CrossRef]
2. Thiebaut, F.; Tsuruo, T.; Hamada, H.; Gottesman, M.M.; Pastan, I.; Willingham, M.C. Cellular localization of the multidrug-resistance gene product p-glycoprotein in normal human tissues. *Proc. Natl. Acad. Sci. USA* **1987**, *84*, 7735–7738. [CrossRef] [PubMed]

3. Kim, R.B.; Fromm, M.F.; Wandel, C.; Leake, B.; Wood, A.J.; Roden, D.M.; Wilkinson, G.R. The drug transporter p-glycoprotein limits oral absorption and brain entry of hiv-1 protease inhibitors. *J. Clin. Investig.* **1998**, *101*, 289–294. [CrossRef] [PubMed]

4. Cordon-Cardo, C.; O'Brien, J.P.; Casals, D.; Rittman-Grauer, L.; Biedler, J.L.; Melamed, M.R.; Bertino, J.R. Multidrug-resistance gene (p-glycoprotein) is expressed by endothelial cells at blood-brain barrier sites. *Proc. Natl. Acad. Sci. USA* **1989**, *86*, 695–698. [CrossRef] [PubMed]

5. Schinkel, A.H. P-glycoprotein, a gatekeeper in the blood-brain barrier. *Adv. Drug Deliv. Rev.* **1999**, *36*, 179–194. [CrossRef]

6. Vähäkangas, K.; Myllynen, P. Drug transporters in the human blood-placental barrier. *Br. J. Pharmacol.* **2009**, *158*, 665–678. [CrossRef] [PubMed]

7. Gosselet, F.; Saint-Pol, J.; Candela, P.; Fenart, L. Amyloid-β peptides, alzheimer's disease and the blood-brain barrier. *Curr. Alzheimer Res.* **2013**, *10*, 1015–1033. [CrossRef] [PubMed]

8. Mawuenyega, K.G.; Sigurdson, W.; Ovod, V.; Munsell, L.; Kasten, T.; Morris, J.C.; Yarasheski, K.E.; Bateman, R.J. Decreased clearance of cns β-amyloid in alzheimer's disease. *Science* **2010**, *330*, 1774. [CrossRef] [PubMed]

9. van Assema, D.M.E.; Lubberink, M.; Bauer, M.; van der Flier, W.M.; Schuit, R.C.; Windhorst, A.D.; Comans, E.F.I.; Hoetjes, N.J.; Tolboom, N.; Langer, O.; et al. Blood–brain barrier p-glycoprotein function in alzheimer's disease. *Brain* **2012**, *135*, 181–189. [CrossRef] [PubMed]

10. Jedlitschky, G.; Vogelgesang, S.; Kroemer, H.K. Mdr1-p-glycoprotein (abcb1)-mediated disposition of amyloid-β peptides: Implications for the pathogenesis and therapy of alzheimer's disease. *Clin. Pharmacol. Ther.* **2010**, *88*, 441–443. [CrossRef] [PubMed]

11. Cascorbi, I.; Flüh, C.; Remmler, C.; Haenisch, S.; Faltraco, F.; Grumbt, M.; Peters, M.; Brenn, A.; Thal, D.R.; Warzok, R.W.; et al. Association of atp-binding cassette transporter variants with the risk of alzheimer's disease. *Pharmacogenomics* **2013**, *14*, 485–494. [CrossRef] [PubMed]

12. Brenn, A.; Grube, M.; Peters, M.; Fischer, A.; Jedlitschky, G.; Kroemer, H.K.; Warzok, R.W.; Vogelgesang, S. Beta-amyloid downregulates mdr1-p-glycoprotein (abcb1) expression at the blood-brain barrier in mice. *Int. J. Alzheimers Dis.* **2011**, *2011*. [CrossRef] [PubMed]

13. Neuwelt, E.A.; Bauer, B.; Fahlke, C.; Fricker, G.; Iadecola, C.; Janigro, D.; Leybaert, L.; Molnár, Z.; O'Donnell, M.E.; Povlishock, J.T.; et al. Engaging neuroscience to advance translational research in brain barrier biology. *Nat. Rev. Neurosci.* **2011**, *12*, 169–182. [CrossRef] [PubMed]

14. Wolf, A.; Bauer, B.; Hartz, A. Abc transporters and the alzheimer's disease enigma. *Front. Psychiatry* **2012**, *3*. [CrossRef] [PubMed]

15. Selick, H.E.; Beresford, A.P.; Tarbit, M.H. The emerging importance of predictive adme simulation in drug discovery. *Drug Discov. Today* **2002**, *7*, 109–116. [CrossRef]

16. Montanari, F.; Ecker, G.F. Prediction of drug–abc-transporter interaction—recent advances and future challenges. *Adv. Drug Deliv. Rev.* **2015**, *86*, 17–26. [CrossRef] [PubMed]

17. Greiner, B.; Eichelbaum, M.; Fritz, P.; Kreichgauer, H.P.; von Richter, O.; Zundler, J.; Kroemer, H.K. The role of intestinal p-glycoprotein in the interaction of digoxin and rifampin. *J. Clin. Investig.* **1999**, *104*, 147–153. [CrossRef] [PubMed]

18. Padowski, J.M.; Pollack, G.M. Influence of time to achieve substrate distribution equilibrium between brain tissue and blood on quantitation of the blood–brain barrier p-glycoprotein effect. *Brain Res.* **2011**, *1426*, 1–17. [CrossRef] [PubMed]

19. Bagal, S.; Bungay, P. Restricting cns penetration of drugs to minimise adverse events: Role of drug transporters. *Drug Discov. Today Technol.* **2014**, *12*, e79–e85. [CrossRef] [PubMed]

20. Hochman, J.H.; Ha, S.N.; Sheridan, R.P. Establishment of p-glycoprotein structure–transport relationships to optimize cns exposure in drug discovery. In *Blood-Brain Barrier in Drug Discovery: Optimizing Brain Exposure of Cns Drugs and Minimizing Brain Side Effects for Peripheral Drugs*; Di, L., Kerns, E.H., Eds.; John Wiley & Sons, Inc.: Hoboken, NJ, USA, 2015; pp. 113–124.

21. Schinkel, A.H.; Wagenaar, E.; Mol, C.A.; van Deemter, L. P-glycoprotein in the blood-brain barrier of mice influences the brain penetration and pharmacological activity of many drugs. *J. Clin. Investig.* **1996**, *97*, 2517–2524. [CrossRef] [PubMed]

22. Aszalos, A. Drug–drug interactions affected by the transporter protein, p-glycoprotein (abcb1, mdr1): II. Clinical aspects. *Drug Discov. Today* **2007**, *12*, 838–843. [CrossRef] [PubMed]

23. Doan, K.M.M.; Humphreys, J.E.; Webster, L.O.; Wring, S.A.; Shampine, L.J.; Serabjit-Singh, C.J.; Adkison, K.K.; Polli, J.W. Passive permeability and p-glycoprotein-mediated efflux differentiate central nervous system (cns) and non-cns marketed drugs. *J. Pharmacol. Exp. Ther.* **2002**, *303*, 1029–1037. [CrossRef] [PubMed]

24. Hennessy, M.; Spiers, J.P. A primer on the mechanics of p-glycoprotein the multidrug transporter. *Pharmacol. Res.* **2007**, *55*, 1–15. [CrossRef] [PubMed]

25. Gottesman, M.M.; Pastan, I. Biochemistry of multidrug resistance mediated by the multidrug transporter. *Ann. Rev. Biochem.* **2003**, *62*, 385–427. [CrossRef] [PubMed]

26. Breier, A.; Gibalova, L.; Seres, M.; Barancik, M.; Sulova, Z. New insight into p-glycoprotein as a drug target. *Anticancer Agents Med. Chem.* **2013**, *13*, 159–170. [CrossRef] [PubMed]

27. Ambudkar, S.V.; Dey, S.; Hrycyna, C.A.; Ramachandra, M.; Pastan, I.; Gottesman, M.M. Biochemical, cellular, and pharmacological aspects of the multidrug transporter. *Annu. Rev. Pharmacol. Toxicol.* **1999**, *39*, 361–398. [CrossRef] [PubMed]

28. Siegel, R.L.; Miller, K.D.; Jemal, A. Cancer statistics, 2017. *CA Cancer J. Clin.* **2017**, *67*, 7–30. [CrossRef] [PubMed]

29. Clarke, J.; Penas, C.; Pastori, C.; Komotar, R.J.; Bregy, A.; Shah, A.H.; Wahlestedt, C.; Ayad, N.G. Epigenetic pathways and glioblastoma treatment. *Epigenetics* **2013**, *8*, 785–795. [CrossRef] [PubMed]

30. Wang, T.; Agarwal, S.; Elmquist, W.F. Brain distribution of cediranib is limited by active efflux at the blood-brain barrier. *J. Pharmacol. Exp. Ther.* **2012**, *341*, 386–395. [CrossRef] [PubMed]

31. Palmeira, A.; Sousa, E.H.; Vasconcelos, M.M.; Pinto, M. Three decades of p-gp inhibitors: Skimming through several generations and scaffolds. *Curr. Med. Chem.* **2012**, *19*, 1946–2025. [CrossRef] [PubMed]

32. van Hoppe, S.; Schinkel, A.H. What next? Preferably development of drugs that are no longer transported by the abcb1 and abcg2 efflux transporters. *Pharmacol. Res.* **2017**, 122–144. [CrossRef] [PubMed]

33. Crivori, P. Computational models for p-glycoprotein substrates and inhibitors. In *Antitargets: Prediction and Prevention of Drug Side Effects*; Vaz, R.J., Klabunde, T., Eds.; Wiley-VCH Verlag GmbH & Co. KGaA: Weinheim, Germany, 2008; Volume 38, pp. 367–397.

34. Terasaki, T.; Hosoya, K.I. The blood-brain barrier efflux transporters as a detoxifying system for the brain. *Adv. Drug Deliv. Rev.* **1999**, *36*, 195–209. [CrossRef]

35. Garg, P.; Verma, J. In silico prediction of blood brain barrier permeability: An artificial neural network model. *J. Chem. Inf. Model.* **2006**, *46*, 289–297. [CrossRef] [PubMed]

36. Kalvass, J.C.; Maurer, T.S.; Pollack, G.M. Use of plasma and brain unbound fractions to assess the extent of brain distribution of 34 drugs: Comparison of unbound concentration ratios to in vivo p-glycoprotein efflux ratios. *Drug Metab. Dispos.* **2007**, *35*, 660–666. [CrossRef] [PubMed]

37. Di, L.; Rong, H.; Feng, B. Demystifying brain penetration in central nervous system drug discovery. *J. Med. Chem.* **2013**, *56*, 2–12. [CrossRef] [PubMed]

38. Inoue, T.; Osada, K.; Tagawa, M.; Ogawa, Y.; Haga, T.; Sogame, Y.; Hashizume, T.; Watanabe, T.; Taguchi, A.; Katsumata, T.; et al. Blonanserin, a novel atypical antipsychotic agent not actively transported as substrate by p-glycoprotein. *Prog. Neuropsychopharmacol. Biol. Psychiatry* **2012**, *39*, 156–162.

39. Polli, J.W.; Wring, S.A.; Humphreys, J.E.; Huang, L.; Morgan, J.B.; Webster, L.O.; Serabjit-Singh, C.S. Rational use of in vitro p-glycoprotein assays in drug discovery. *J. Pharmacol. Exp. Ther.* **2001**, *299*, 620–628. [PubMed]

40. Hochman, J.H.; Yamazaki, M.; Ohe, T.; Lin, J.H. Evaluation of drug interactions with p-glycoprotein in drug discovery: In vitro assessment of the potential for drug-drug interactions with p-glycoprotein. *Curr. Drug MeTable* **2002**, *3*, 257–273. [CrossRef]

41. Schwab, D.; Fischer, H.; Tabatabaei, A.; Poli, S.; Huwyler, J. Comparison of in vitro p-glycoprotein screening assays: Recommendations for their use in drug discovery. *J. Med. Chem.* **2003**, *46*, 1716–1725. [CrossRef] [PubMed]

42. Zhang, Y.; Bachmeier, C.; Miller, D.W. In vitro and in vivo models for assessing drug efflux transporter activity. Adv. *Drug Deliv. Rev.* **2003**, *55*, 31–51. [CrossRef]

43. Sugano, K.; Shirasaka, Y.; Yamashita, S. Estimation of michaelis–menten constant of efflux transporter considering asymmetric permeability. *Int. J. Pharm.* **2011**, *418*, 161–167. [CrossRef] [PubMed]

44. Storch, C.H.; Nikendei, C.; Schild, S.; Haefeli, W.E.; Weiss, J.; Herzog, W. Expression and activity of p-glycoprotein (mdr1/abcb1) in peripheral blood mononuclear cells from patients with anorexia nervosa compared with healthy controls. *Int. J. Eating Disord.* **2008**, *41*, 432–438. [CrossRef] [PubMed]

45. Egan, W.J. Computational models for adme. In *Annual Reports in Medicinal Chemistry*; John, E.M., Ed.; Academic Press: San Diego, CA, USA, 2007; Volume 42, pp. 449–467.

46. Sliwoski, G.; Kothiwale, S.; Meiler, J.; Lowe, E.W. Computational methods in drug discovery. *Pharmacol. Rev.* **2014**, *66*, 334–395. [CrossRef] [PubMed]

47. Penzotti, J.E.; Lamb, M.L.; Evensen, E.; Grootenhuis, P.D.J. A computational ensemble pharmacophore model for identifying substrates of p-glycoprotein. *J. Med. Chem.* **2002**, *45*, 1737–1740. [CrossRef] [PubMed]

48. Gombar, V.K.; Polli, J.W.; Humphreys, J.E.; Wring, S.A.; Serabjit-Singh, C.S. Predicting p-glycoprotein substrates by a quantitative structure-activity relationship model. *J. Pharm. Sci.* **2004**, *93*, 957–968. [CrossRef] [PubMed]

49. Xue, Y.; Yap, C.W.; Sun, L.Z.; Cao, Z.W.; Wang, J.F.; Chen, Y.Z. Prediction of p-glycoprotein substrates by a support vector machine approach. *J. Chem. Inf. Comput. Sci.* **2004**, *44*, 1497–1505. [CrossRef] [PubMed]

50. Wang, Y.H.; Li, Y.; Yang, S.L.; Yang, L. Classification of substrates and inhibitors of p-glycoprotein using unsupervised machine learning approach. *J. Chem. Inf. Model.* **2005**, *45*, 750–757. [CrossRef] [PubMed]

51. Crivori, P.; Reinach, B.; Pezzetta, D.; Poggesi, I. Computational models for identifying potential p-glycoprotein substrates and inhibitors. *Mol. Pharma.* **2006**, *3*, 33–44. [CrossRef]

52. de Cerqueira Lima, P.; Golbraikh, A.; Oloff, S.; Xiao, Y.; Tropsha, A. Combinatorial qsar modeling of p-glycoprotein substrates. *J. Chem. Inf. Model.* **2006**, *46*, 1245–1254. [CrossRef] [PubMed]

53. Huang, J.; Ma, G.; Muhammad, I.; Cheng, Y. Identifying p-glycoprotein substrates using a support vector machine optimized by a particle swarm. *J. Chem. Inf. Model.* **2007**, *47*, 1638–1647. [CrossRef] [PubMed]

54. Li, W.-X.; Li, L.; Eksterowicz, J.; Ling, X.B.; Cardozo, M. Significance analysis and multiple pharmacophore models for differentiating p-glycoprotein substrates. *J. Chem Inf. Model.* **2007**, *47*, 2429–2438. [CrossRef] [PubMed]

55. Wang, Z.; Chen, Y.; Liang, H.; Bender, A.; Glen, R.C.; Yan, A. P-glycoprotein substrate models using support vector machines based on a comprehensive data set. *J. Chem. Inf. Model.* **2011**, *51*, 1447–1456. [CrossRef] [PubMed]

56. Desai, P.V.; Sawada, G.A.; Watson, I.A.; Raub, T.J. Integration of in silico and in vitro tools for scaffold optimization during drug discovery: Predicting p-glycoprotein efflux. Mol. *Pharmaceutics* **2013**, *10*, 1249–1261. [CrossRef] [PubMed]

57. Ecker, G.F.; Stockner, T.; Chiba, P. Computational models for prediction of interactions with abc-transporters. *Drug Discov. Today* **2008**, *13*, 311–317. [CrossRef] [PubMed]

58. Adenot, M. A practical approach to computational models of the blood–brain barrier. In *Handbook of Neurochemistry and Molecular Neurobiology: Neural Membranes and Transport*; Lajtha, A., Reith, M.E.A., Eds.; Springer: New York, NY, USA, 2007; pp. 109–150.

59. Ivanciuc, O. Artificial immune systems in drug design: Recognition of p-glycoprotein substrates with airs (artificial immune recognition system). Internet Electron. *J. Mol. Des.* **2006**, *5*, 542–554.

60. Bikadi, Z.; Hazai, I.; Malik, D.; Jemnitz, K.; Veres, Z.; Hari, P.; Ni, Z.; Loo, T.W.; Clarke, D.M.; Hazai, E.; et al. Predicting p-glycoprotein-mediated drug transport based on support vector machine and three-dimensional crystal structure of p-glycoprotein. *PLoS ONE* **2011**, *6*, e25815. [CrossRef] [PubMed]

61. Erić, S.; Kalinić, M.; Ilić, K.; Zloh, M. Computational classification models for predicting the interaction of drugs with p-glycoprotein and breast cancer resistance protein. *SAR QSAR Environ. Res.* **2014**, *25*, 939–966. [CrossRef] [PubMed]

62. Pan, X.; Mei, H.; Qu, S.; Huang, S.; Sun, J.; Yang, L.; Chen, H. Prediction and characterization of p-glycoprotein substrates potentially bound to different sites by emerging chemical pattern and hierarchical cluster analysis. *Int. J. Pharm.* **2016**, *502*, 61–69. [CrossRef] [PubMed]

63. Didziapetris, R.; Japertas, P.; Avdeef, A.; Petrauskas, A. Classification analysis of p-glycoprotein substrate specificity. *J. Drug Target.* **2003**, *11*, 391–406. [CrossRef] [PubMed]

64. Broccatelli, F. Qsar models for p-glycoprotein transport based on a highly consistent data set. *J. Chem. Inf. Model.* **2012**, *2*, 2462–2470. [CrossRef] [PubMed]

65. Poongavanam, V.; Haider, N.; Ecker, G.F. Fingerprint-based in silico models for the prediction of p-glycoprotein substrates and inhibitors. *Bioorg. Med. Chem.* **2012**, *20*, 5388–5395. [CrossRef] [PubMed]

66. Li, D.; Chen, L.; Li, Y.; Tian, S.; Sun, H.; Hou, T. Admet evaluation in drug discovery. 13. Development of in silico prediction models for p-glycoprotein substrates. *Mol. Pharm.* **2014**, *11*, 716–726. [PubMed]

67.   Estrada, E.; Molina, E.; Nodarse, D.; Uriarte, E. Structural contributions of substrates to their binding to p-glycoprotein. A topsmode approach. *Curr. Pharm. Des.* **2010**, *16*, 2676–2709. [CrossRef] [PubMed]
68.   Joung, J.Y.; Kim, H.; Kim, H.M.; Ahn, S.K.; Nam, K.-Y.; No, K.T. Prediction models of p-glycoprotein substrates using simple 2d and 3d descriptors by a recursive partitioning approach. *Bull. Korean Chem. Soc.* **2012**, *33*, 1123–1127. [CrossRef]
69.   Gunaydin, H.; Weiss, M.M.; Sun, Y. De novo prediction of p-glycoprotein-mediated efflux liability for druglike compounds. *ACS Med. Chem. Lett.* **2013**, *4*, 108–112. [CrossRef] [PubMed]
70.   Dolghih, E.; Jacobson, M.P. Predicting efflux ratios and blood-brain barrier penetration from chemical structure: Combining passive permeability with active efflux by p-glycoprotein. *ACS Chem. Neurosci.* **2012**, *4*, 361–367. [CrossRef] [PubMed]
71.   Dolghih, E.; Bryant, C.; Renslo, A.R.; Jacobson, M.P. Predicting binding to p-glycoprotein by flexible receptor docking. *PLoS Comput. Biol.* **2011**, *7*, e1002083. [CrossRef] [PubMed]
72.   Subramanian, N.; Condic-Jurkic, K.; O'Mara, M.L. Structural and dynamic perspectives on the promiscuous transport activity of p-glycoprotein. *Neurochem. Int.* **2016**, *98*, 146–152. [CrossRef] [PubMed]
73.   Leong, M.K.; Chen, H.B.; Shih, Y.H. Prediction of promiscuous p-glycoprotein inhibition using a novel machine learning scheme. *PLoS ONE* **2012**, *7*, e33829. [CrossRef] [PubMed]
74.   Garrigues, A.; Loiseau, N.; Delaforge, M.; Ferté, J.; Garrigos, M.; André, F.; Orlowski, S. Characterization of two pharmacophores on the multidrug transporter p-glycoprotein. *Mol. Pharmacol.* **2002**, *62*, 1288–1298. [CrossRef] [PubMed]
75.   Chufan, E.E.; Sim, H.M.; Ambudkar, S.V. Molecular basis of the polyspecificity of p-glycoprotein (abcb1): Recent biochemical and structural studies. In *Advances in Cancer Research: Abc Transporters and Cancer*; John, D.S., Toshihisa, I., Eds.; Academic Press: San Diego, CA, USA, 2015; Volume 125, pp. 71–96.
76.   Ferreira, R.J.; Ferreira, M.J.U.; dos Santos, D.J.V.A. Molecular docking characterizes substrate-binding sites and efflux modulation mechanisms within p-glycoprotein. *J. Chem. Inf. Model.* **2013**, *53*, 1747–1760. [CrossRef] [PubMed]
77.   Aller, S.G.; Yu, J.; Ward, A.; Weng, Y.; Chittaboina, S.; Zhuo, R.; Harrell, P.M.; Trinh, Y.T.; Zhang, Q.; Urbatsch, I.L.; et al. Structure of p-glycoprotein reveals a molecular basis for poly-specific drug binding. *Science* **2009**, *323*, 1718–1722. [CrossRef] [PubMed]
78.   Edwards, G. Ivermectin: Does p-glycoprotein play a role in neurotoxicity? *Filaria J.* **2003**, *2* (Suppl. 1), S8. [CrossRef] [PubMed]
79.   Balimane, P.V.; Han, Y.H.; Chong, S. Current industrial practices of assessing permeability and p-glycoprotein interaction. *AAPS J.* **2006**, *8*, E1–E13. [CrossRef] [PubMed]
80.   Roger, P.; Sahla, M.E.; Mäkelä, S.; Gustafsson, J.Å.; Baldet, P.; Rochefort, H. Decreased expression of estrogen receptor β protein in proliferative preinvasive mammary tumors. *Cancer Res.* **2001**, *61*, 2537–2541. [PubMed]
81.   Speck-Planche, A.; Cordeiro, M.N.D.S. Multi-target qsar approaches for modeling protein inhibitors. Simultaneous prediction of activities against biomacromolecules present in gram-negative bacteria. *Curr. Top. Med. Chem.* **2015**, *15*, 1801–1813. [CrossRef] [PubMed]
82.   Ferreira, R.J.; dos Santos, D.J.V.A.; Ferreira, M.J.U.; Guedes, R.C. Toward a better pharmacophore description of p-glycoprotein modulators, based on macrocyclic diterpenes from euphorbia species. *J. Chem. Inf. Model.* **2011**, *51*, 1315–1324. [CrossRef] [PubMed]
83.   Leong, M.K.; Chen, Y.M.; Chen, T.H. Prediction of human cytochrome p450 2b6-substrate interactions using hierarchical support vector regression approach. *J. Comput. Chem.* **2009**, *30*, 1899–1909. [CrossRef] [PubMed]
84.   Caudill, M. Using neural networks: Hybrid expert networks. *AI Expert* **1990**, *5*, 49–54.
85.   Leong, M.K.; Lin, S.W.; Chen, H.B.; Tsai, F.Y. Predicting mutagenicity of aromatic amines by various machine learning approaches. *Toxicol. Sci.* **2010**, *116*, 498–513. [CrossRef] [PubMed]
86.   Ding, Y.L.; Lyu, Y.C.; Leong, M.K. In silico prediction of the mutagenicity of nitroaromatic compounds using a novel two-qsar approach. *Toxicol. In Vitro* **2017**, *40*, 102–114. [CrossRef] [PubMed]
87.   Gnanadesikan, R.; Kettenring, J.R. Robust estimates, residuals, and outlier detection with multiresponse data. *Biometrics* **1972**, *28*, 81–124. [CrossRef]
88.   Carrara, S.; Reali, V.; Misiano, P.; Dondio, G.; Bigogno, C. Evaluation of in vitro brain penetration: Optimized pampa and mdckii-mdr1 assay comparison. *Int. J. Pharm.* **2007**, *345*, 125–133. [CrossRef] [PubMed]
89.   Chen, C.; Hanson, E.; Watson, J.W.; Lee, J.S. P-glycoprotein limits the brain penetration of nonsedating but not sedating h1-antagonists. *Drug Metab. Dispos.* **2003**, *31*, 312–318. [CrossRef] [PubMed]

90. Eriksson, U.G.; Dorani, H.; Karlsson, J.; Fritsch, H.; Hoffmann, K.-J.; Olsson, L.; Sarich, T.C.; Wall, U.; Schützer, K.-M. Influence of erythromycin on the pharmacokinetics of ximelagatran may involve inhibition of p-glycoprotein-mediated excretion. *Drug Metab. Dispos.* **2006**, *34*, 775–782. [CrossRef] [PubMed]

91. Feng, B.; Mills, J.B.; Davidson, R.E.; Mireles, R.J.; Janiszewski, J.S.; Troutman, M.D.; de Morais, S.M. In vitro p-glycoprotein assays to predict the in vivo interactions of p-glycoprotein with drugs in the central nervous system. *Drug Metab. Dispos.* **2008**, *36*, 268–275. [CrossRef] [PubMed]

92. Gertz, M.; Harrison, A.; Houston, J.B.; Galetin, A. Prediction of human intestinal first-pass metabolism of 25 cyp3a substrates from in vitro clearance and permeability data. *Drug Metab. Dispos.* **2010**, *38*, 1147–1158. [CrossRef] [PubMed]

93. Huang, L.; Wang, Y.; Grimm, S. Atp-dependent transport of rosuvastatin in membrane vesicles expressing breast cancer resistance protein. *Drug Metab. Dispos.* **2006**, *34*, 738–742. [CrossRef] [PubMed]

94. Luo, S.; Pal, D.; Shah, S.J.; Kwatra, D.; Paturi, K.D.; Mitra, A.K. Effect of hepes buffer on the uptake and transport of p-glycoprotein substrates and large neutral amino acids. *Mol. Pharm.* **2010**, *7*, 412–420. [CrossRef] [PubMed]

95. Taub, M.E.; Podila, L.; Ely, D.; Almeida, I. Functional assessment of multiple p-glycoprotein (p-gp) probe substrates: Influence of cell line and modulator concentration on p-gp activity. *Drug Metab. Dispos.* **2005**, *33*, 1679–1687. [CrossRef] [PubMed]

96. Troutman, M.D.; Thakker, D.R. Novel experimental parameters to quantify the modulation of absorptive and secretory transport of compounds by p-glycoprotein in cell culture models of intestinal epithelium. *Pharm. Res.* **2003**, *20*, 1210–1224. [CrossRef] [PubMed]

97. Wager, T.T.; Chandrasekaran, R.Y.; Hou, X.; Troutman, M.D.; Verhoest, P.R.; Villalobos, A.; Will, Y. Defining desirable central nervous system drug space through the alignment of molecular properties, in vitro adme, and safety attributes. *ACS Chem. Neurosci.* **2010**, *1*, 420–434. [CrossRef] [PubMed]

98. Callegari, E.; Malhotra, B.; Bungay, P.J.; Webster, R.; Fenner, K.S.; Kempshall, S.; LaPerle, J.L.; Michel, M.C.; Kay, G.G. A comprehensive non-clinical evaluation of the cns penetration potential of antimuscarinic agents for the treatment of overactive bladder. *Br. J. Clin. Pharmacol.* **2011**, *72*, 235–246. [CrossRef] [PubMed]

99. Obradovic, T.; Dobson, G.; Shingaki, T.; Kungu, T.; Hidalgo, I. Assessment of the first and second generation antihistamines brain penetration and role of p-glycoprotein. *Pharm. Res.* **2007**, *24*, 318–327. [CrossRef] [PubMed]

100. Liu, Q.; Wang, C.; Meng, Q.; Huo, X.; Sun, H.; Peng, J.; Ma, X.; Sun, P.; Liu, K. Mdr1 and oat1/oat3 mediate the drug-drug interaction between puerarin and methotrexate. *Pharm. Res.* **2014**, *31*, 1120–1132. [CrossRef] [PubMed]

101. Kim, W.Y.; Benet, L.Z. P-glycoprotein (p-gp/mdr1)-mediated efflux of sex-steroid hormones and modulation of p-gp expression in vitro. *Pharm. Res.* **2004**, *21*, 1284–1293. [CrossRef] [PubMed]

102. Cherkasov, A.; Muratov, E.N.; Fourches, D.; Varnek, A.; Baskin, I.I.; Cronin, M.; Dearden, J.; Gramatica, P.; Martin, Y.C.; Todeschini, R.; et al. Qsar modeling: Where have you been? Where are you going to? *J. Med. Chem.* **2014**, *57*, 4977–5010. [CrossRef] [PubMed]

103. McIver, Z.A.; Kryman, M.W.; Choi, Y.; Coe, B.N.; Schamerhorn, G.A.; Linder, M.K.; Davies, K.S.; Hill, J.E.; Sawada, G.A.; Grayson, J.M.; et al. Selective photodepletion of malignant t cells in extracorporeal photopheresis with selenorhodamine photosensitizers. *Bioorg. Med. Chem.* **2016**, *24*, 3918–3931. [CrossRef] [PubMed]

104. Lee, W.; Crawford, J.J.; Aliagas, I.; Murray, L.J.; Tay, S.; Wang, W.; Heise, C.E.; Hoeflich, K.P.; La, H.; Mathieu, S.; et al. Synthesis and evaluation of a series of 4-azaindole-containing p21-activated kinase-1 inhibitors. *Bioorg. Med. Chem. Lett.* **2016**, *26*, 3518–3524. [CrossRef] [PubMed]

105. Hitchcock, S.A. Structural modifications that alter the p-glycoprotein efflux properties of compounds. *J. Med. Chem.* **2012**, *55*, 4877–4895. [CrossRef] [PubMed]

106. Tropsha, A.; Gramatica, P.; Gombar, V.K. The importance of being earnest: Validation is the absolute essential for successful application and interpretation of qspr models. *QSAR Comb. Sci.* **2003**, *22*, 69–77. [CrossRef]

107. Rücker, C.; Rücker, G.; Meringer, M. Y-randomization and its variants in qspr/qsar. *J. Chem. Inf. Model.* **2007**, *47*, 2345–2357. [CrossRef] [PubMed]

108. Netzeva, T.I.; Worth, A.; Aldenberg, T.; Benigni, R.; Cronin, M.T.D.; Gramatica, P.; Jaworska, J.S.; Kahn, S.; Klopman, G.; Marchant, C.A.; et al. Current status of methods for defining the applicability domain of (quantitative) structure-activity relationships : The report and recommendations of ecvam workshop 52. *Altern. Lab. Anim.* **2005**, *33*, 1–19.

109. Golbraikh, A.; Shen, M.; Xiao, Z.Y.; Xiao, Y.D.; Lee, K.H.; Tropsha, A. Rational selection of training and test sets for the development of validated qsar models. *J. Comput.-Aided Mol. Des.* **2003**, *17*, 241–253. [CrossRef] [PubMed]

110. Ojha, P.K.; Mitra, I.; Das, R.N.; Roy, K. Further exploring rm2 metrics for validation of qspr models. *Chemometr. Intell. Lab. Syst.* **2011**, *107*, 194–205. [CrossRef]

111. Roy, K.; Mitra, I.; Kar, S.; Ojha, P.K.; Das, R.N.; Kabir, H. Comparative studies on some metrics for external validation of qspr models. *J. Chem. Inf. Model.* **2012**, *52*, 396–408. [CrossRef] [PubMed]

112. Chirico, N.; Gramatica, P. Real external predictivity of qsar models. Part 2. New intercomparable thresholds for different validation criteria and the need for scatter plot inspection. *J. Chem. Inf. Model.* **2012**, *52*, 2044–2058. [CrossRef] [PubMed]

113. Topliss, J.G.; Edwards, R.P. Chance factors in studies of quantitative structure-activity relationships. *J. Med. Chem.* **1979**, *22*, 1238–1244. [CrossRef] [PubMed]

114. Desai, P.V.; Raub, T.J.; Blanco, M.J. How hydrogen bonds impact p-glycoprotein transport and permeability. *Bioorg. Med. Chem. Lett.* **2012**, *22*, 6540–6548. [CrossRef] [PubMed]

115. Teixeira, V.H.; Vila-Viçosa, D.; Baptista, A.M.; Machuqueiro, M. Protonation of dmpc in a bilayer environment using a linear response approximation. *J. Chem. Theory Comput.* **2014**, *10*, 2176–2184. [CrossRef] [PubMed]

116. Clay, A.T.; Sharom, F.J. Lipid bilayer properties control membrane partitioning, binding, and transport of p-glycoprotein substrates. *Biochemistry* **2013**, *52*, 343–354. [CrossRef] [PubMed]

117. Raub, T.J. P-glycoprotein recognition of substrates and circumvention through rational drug design. *Mol. Pharmaceutics* **2006**, *3*, 3–25. [CrossRef]

118. Suzuki, T.; Fukazawa, N.; San-nohe, K.; Sato, W.; Yano, O.; Tsuruo, T. Structure-activity relationship of newly synthesized quinoline derivatives for reversal of multidrug resistance in cancer. *J. Med. Chem.* **1997**, *40*, 2047–2052. [CrossRef] [PubMed]

119. Prasanna, S.; Doerksen, R.J. Topological polar surface area: A useful descriptor in 2d-qsar. *Curr. Med. Chem.* **2009**, *16*, 21–41. [CrossRef] [PubMed]

120. Ferté, J. Analysis of the tangled relationships between p-glycoprotein-mediated multidrug resistance and the lipid phase of the cell membrane. *Eur. J. Biochem.* **2000**, *267*, 277–294. [CrossRef] [PubMed]

121. Johnson, T.W.; Dress, K.R.; Edwards, M. Using the golden triangle to optimize clearance and oral absorption. *Bioorg. Med. Chem. Lett.* **2009**, *19*, 5560–5564. [CrossRef] [PubMed]

122. Pettersson, M.; Hou, X.; Kuhn, M.; Wager, T.T.; Kauffman, G.W.; Verhoest, P.R. Quantitative assessment of the impact of fluorine substitution on p-glycoprotein (p-gp) mediated efflux, permeability, lipophilicity, and metabolic stability. *J. Med. Chem.* **2016**, *59*, 5284–5296. [CrossRef] [PubMed]

123. Jabeen, I.; Wetwitayaklung, P.; Klepsch, F.; Parveen, Z.; Chiba, P.; Ecker, G.F. Probing the stereoselectivity of p-glycoprotein-synthesis, biological activity and ligand docking studies of a set of enantiopure benzopyrano[3,4-b][1,4]oxazines. *Chem. Commun.* **2011**, *47*, 2586–2588. [CrossRef] [PubMed]

124. Rezai, T.; Bock, J.E.; Zhou, M.V.; Kalyanaraman, C.; Lokey, R.S.; Jacobson, M.P. Conformational flexibility, internal hydrogen bonding, and passive membrane permeability: Successful in silico prediction of the relative permeabilities of cyclic peptides. *J. Am. Chem. Soc.* **2006**, *128*, 14073–14080. [CrossRef] [PubMed]

125. Rauch, C.; Paine, S.W.; Littlewood, P. Can long range mechanical interaction between drugs and membrane proteins define the notion of molecular promiscuity? Application to p-glycoprotein-mediated multidrug resistance (mdr). Biochim. Biophys. *Acta-Gen. Subj.* **2013**, *1830*, 5112–5118. [CrossRef] [PubMed]

126. Declèves, X.; Jacob, A.; Yousif, S.; Shawahna, R.; Potin, S.; Scherrmann, J.-M. Interplay of drug metabolizing cyp450 enzymes and abc transporters in the blood-brain barrier. *Curr. Drug MeTable* **2011**, *12*, 732–741. [CrossRef]

127. Passeleu-Le Bourdonnec, C.; Carrupt, P.-A.; Scherrmann, J.; Martel, S. Methodologies to assess drug permeation through the blood–brain barrier for pharmaceutical research. *Pharm. Res.* **2013**, *30*, 2729–2756. [CrossRef] [PubMed]

128. Leong, M.K. In silico prediction of the blood-brain barrier permeation: Are we there yet? *Med. Chem.* **2015**, *5*, 130. [CrossRef]

129. Cammi, R.; Tomasi, J. Remarks on the use of the apparent surface charges (asc) methods in solvation problems: Iterative versus matrix-inversion procedures and the renormalization of the apparent charges. *J. Comput. Chem.* **1995**, *16*, 1449–1458. [CrossRef]

130. Miertuš, S.; Scrocco, E.; Tomasi, J. Electrostatic interaction of a solute with a continuum. A direct utilizaion of ab initio molecular potentials for the prevision of solvent effects. *Chem. Phys.* **1981**, *55*, 117–129. [CrossRef]

131. Besler, B.H.; Merz, K.M.J.; Kollman, P.A. Atomic charges derived from semiempirical methods. *J. Comput. Chem.* **1990**, *11*, 431–439. [CrossRef]

132. Kettaneh, N.; Berglund, A.; Wold, S. Pca and pls with very large data sets. *Comput. Stat. Data Anal.* **2005**, *48*, 69–85. [CrossRef]

133. Tseng, Y.J.; Hopfinger, A.J.; Esposito, E.X. The great descriptor melting pot: Mixing descriptors for the common good of qsar models. *J. Comput. Aided Mol. Des.* **2012**, *26*, 39–43. [CrossRef] [PubMed]

134. Burden, F.R.; Ford, M.G.; Whitley, D.C.; Winkler, D.A. Use of automatic relevance determination in qsar studies using bayesian neural networks. *J. Chem. Inf. Comput. Sci.* **2000**, *40*, 1423–1430. [CrossRef] [PubMed]

135. Rogers, D.; Hopfinger, A.J. Application of genetic function approximation to quantitative structure-activity relationships and quantitative structure-property relationships. *J. Chem. Inf. Comput. Sci.* **1994**, *34*, 854–866. [CrossRef]

136. Guyon, I.; Weston, J.; Barnhill, S.; Vapnik, V. Gene selection for cancer classification using support vector machines. *Mach. Learn.* **2002**, *46*, 389–422. [CrossRef]

137. Kennard, R.W.; Stone, L.A. Computer aided design of experiments. *Technometrics* **1969**, *11*, 137–148. [CrossRef]

138. Tropsha, A. Recent trends in statistical qsar modeling of environmental chemical toxicity. In *Molecular, Clinical and Environmental Toxicology. Volume 3: Environmental Toxicology*; Luch, A., Ed.; Springer Basel: New York, NY, USA, 2012; Volume 101, pp. 381–411.

139. Cortes, C.; Vapnik, V. Support-vector networks. *Mach. Learn.* **1995**, *20*, 273–297. [CrossRef]

140. Vapnik, V.; Golowich, S.; Smola, A. *Support Vector Method for Function Approximation, Regression Estimation, and Signal Processing, Advances in Neural Information Processing Systems 9*; Mozer, M., Jordan, M.I., Petsche, T., Eds.; MIT Press: Cambridge, MA, USA, 1997; pp. 281–287.

141. Schölkopf, B.; Smola, A. *Learning with Kernels: Support Vector Machines, Regularization, Optimization, and Beyond*, 1st ed.; MIT Press: Cambridge, MA, USA, 2002.

142. Kecman, V. *Learning and Soft Computing :Support Vector Machines, Neural Networks, and Fuzzy Logic Models*; MIT Press: Cambridge, MA, USA, 2001; p. 576.

143. Leong, M.K.; Syu, R.G.; Ding, Y.L.; Weng, C.F. Prediction of n-methyl-d-aspartate receptor glun1-ligand binding affinity by a novel svm-pose/svm-score combinatorial ensemble docking scheme. *Sci. Rep.* **2017**, *7*, 40053. [CrossRef] [PubMed]

144. Dearden, J.C.; Cronin, M.T.D.; Kaiser, K.L.E. How not to develop a quantitative structure–activity or structure–property relationship (qsar/qspr). *SAR QSAR Environ. Res.* **2009**, *20*, 241–266. [CrossRef] [PubMed]

145. Breiman, L.; Spector, P. Submodel selection and evaluation in regression. The x-random case. *Int. Stat. Rev.* **1992**, *60*, 291–319. [CrossRef]

146. Shi, L.M.; Fang, H.; Tong, W.; Wu, J.; Perkins, R.; Blair, R.M.; Branham, W.S.; Dial, S.L.; Moland, C.L.; Sheehan, D.M. Qsar models using a large diverse set of estrogens. *J. Chem. Inf. Comput. Sci.* **2001**, *41*, 186–195. [CrossRef] [PubMed]

147. Schüürmann, G.; Ebert, R.U.; Chen, J.; Wang, B.; Kühne, R. External validation and prediction employing the predictive squared correlation coefficient-test set activity mean vs training set activity mean. *J. Chem. Inf. Model.* **2008**, *48*, 2140–2145. [CrossRef] [PubMed]

148. Consonni, V.; Ballabio, D.; Todeschini, R. Comments on the definition of the q2 parameter for qsar validation. *J. Chem. Inf. Model.* **2009**, *49*, 1669–1678. [CrossRef] [PubMed]

149. Chirico, N.; Gramatica, P. Real external predictivity of qsar models: How to evaluate it? Comparison of different validation criteria and proposal of using the concordance correlation coefficient. *J. Chem. Inf. Model.* **2011**, *51*, 2320–2335. [CrossRef] [PubMed]

150. Gramatica, P.; Chirico, N.; Papa, E.; Cassani, S.; Kovarich, S. Qsarins: A new software for the development, analysis, and validation of qsar mlr models. *J. Comput. Chem.* **2013**, *34*, 2121–2132. [CrossRef]

151. Gramatica, P.; Cassani, S.; Chirico, N. Qsarins-chem: Insubria datasets and new qsar/qspr models for environmental pollutants in qsarins. *J. Comput. Chem.* **2014**, *35*, 1036–1044. [CrossRef] [PubMed]

152. Gramatica, P.; Sangion, A. A historical excursus on the statistical validation parameters for qsar models: A clarification concerning metrics and terminology. *J. Chem. Inf. Model.* **2016**, *56*, 1127–1131. [CrossRef] [PubMed]

Sample Availability: Not available.

molecules

Article

The Cartesian Product and Join Graphs on Edge-Version Atom-Bond Connectivity and Geometric Arithmetic Indices

Xiujun Zhang [1], Huiqin Jiang [1], Jia-Bao Liu [2] and Zehui Shao [3,*]

[1]   Key Laboratory of Pattern Recognition and Intelligent Information Processing, Institutions of Higher Education of Sichuan Province, Chengdu University, Chengdu 610106, China; woodszhang@cdu.edu.cn (X.Z.); hq.jiang@hotmail.com (H.J.)
[2]   School of Mathematics and Physics, Anhui Jianzhu University, Hefei 230601, China; liujiabaoad@163.com
[3]   Institute of Computing Science and Technology, Guangzhou University, Guangzhou 510006, China
*   Correspondence: zshao@gzhu.edu.cn; Tel.: +86-135-5186-2078

Received: 25 May 2018; Accepted: 6 July 2018; Published: 16 July 2018

Abstract: The Cartesian product and join are two classical operations in graphs. Let $d_{L(G)}(e)$ be the degree of a vertex e in line graph $L(G)$ of a graph G. The edge versions of atom-bond connectivity (ABC_e) and geometric arithmetic (GA_e) indices of G are defined as

$$\sum_{ef \in E(L(G))} \sqrt{\frac{d_{L(G)}(e) + d_{L(G)}(f) - 2}{d_{L(G)}(e) \times d_{L(G)}(f)}} \quad \text{and} \quad \sum_{ef \in E(L(G))} \frac{2\sqrt{d_{L(G)}(e) \times d_{L(G)}(f)}}{d_{L(G)}(e) + d_{L(G)}(f)}, \text{ respectively. In this}$$

paper, ABC_e and GA_e indices for certain Cartesian product graphs (such as $P_n \Box P_m$, $P_n \Box C_m$ and $P_n \Box S_m$) are obtained. In addition, ABC_e and GA_e indices of certain join graphs (such as $C_m + P_n + S_r$, $P_m + P_n + P_r$, $C_m + C_n + C_r$ and $S_m + S_n + S_r$) are deduced. Our results enrich and revise some known results.

Keywords: line graph; Cartesian product graph; join graph; atom-bond connectivity index; geometric arithmetic index

1. Introduction

The invariants based on the distance or degree of vertices in molecules are called topological indices. In theoretical chemistry, physics and graph theory, topological indices are the molecular descriptors that describe the structures of chemical compounds, and they help us to predict certain physico-chemical properties. The first topological index, Wiener index, was published in 1947 [1], and the edge version of the Wiener index was proposed by Iranmanesh et al. in 2009 [2]. Because the important effects of the topological indices are proved in chemical research, more and more topological indices are studied, including the classical atom-bond connectivity index and the geometric arithmetic index.

Let G be a simple connected graph. Denote by $V(G)$ and $E(G)$ the vertex set and edge set of G, respectively. Let P_n, C_n, K_n and S_n be a path, a cycle, a complete graph and a star, respectively, on n vertices. $e = uv$ represents edge-connecting vertices u and v. $N(v)$ is an open neighborhood of vertex v, i.e., $N(v) = \{u | uv \in E(G)\}$. Denote by $d_G(v)$ (simply $d(v)$) the degree of vertex v of graph G, i.e., $d(v) = |N(v)|$. Let $L(G)$ or G^L be the line graph of G such that each vertex of $L(G)$ represents an edge of G and two vertices of $L(G)$ are adjacent if and only if their corresponding edges share a common endpoint in G [3]. It is known that the line graph $L(G)$ of any graph G is claw-free. Denote by $d_{L(G)}(e)$ the degree of edge e in G, which is the number of edges sharing a common endpoint with edge e in G, or the degree of vertex e in $L(G)$. We denote by $E_{n,m}$ (or $E_{n,m}^L$) the set of edges uv with degrees n and m of end vertices u and v in G (or in G^L), i.e., $E_{n,m} = \{uv | \{n,m\} = \{d(u), d(v)\}, \ u \in G, v \in G\}$

or $E_{n,m}^L = \{uv | \{n, m\} = \{d(u), d(v)\}, u \in L(G), v \in L(G)\}$. The distance $d_G(u, v)$ (or $d(u, v)$ for short) between u and v in G is the length of a shortest $u - v$ path.

The atom-bond connectivity (ABC) index was proposed by Estrada et al. in 1998 [4]. The ABC index is defined as:

$$ABC(G) = \sum_{uv \in E(G)} \sqrt{\frac{d_G(u) + d_G(v) - 2}{d_G(u) \times d_G(v)}} \tag{1}$$

where $d_G(u)$ and $d_G(v)$ are the degrees of the vertices u and v in G. Meanwhile, the edge version of the ABC index is:

$$ABC_e(G) = \sum_{ef \in E(L(G))} \sqrt{\frac{d_{L(G)}(e) + d_{L(G)}(f) - 2}{d_{L(G)}(e) \times d_{L(G)}(f)}} \tag{2}$$

where $d_{L(G)}(e)$ and $d_{L(G)}(f)$ are the degrees of the edges e and f, respectively, in G. The recent research on edge version ABC index can be referred to Gao et al. [5].

The geometric arithmetic (GA) index was proposed by Vukicevic and Furthla in 2009 [6]. The GA index is defined as

$$GA(G) = \sum_{uv \in E(G)} \frac{2\sqrt{d_G(u)d_G(v)}}{d_G(u) + d_G(v)} \tag{3}$$

The edge version of the GA index was proposed by Mahmiani et al. [7] and is

$$GA_e(G) = \sum_{ef \in E(L(G))} \frac{2\sqrt{d_{L(G)}(e)d_{L(G)}(f)}}{d_{L(G)}(e) + d_{L(G)}(f)} \tag{4}$$

Recent research on the edge-version GA index can be referred to the articles [5,8–16]. In addition, Das [17] obtained the upper and lower bounds of the ABC index of trees. Furtula et al. [18] found the chemical trees with extremal ABC values. Fath-Tabar et al. [19] obtained some inequalities for the ABC index of a series of graph operations. Chen et al. [20] obtained some upper bounds for the ABC index of graphs with given vertex connectivity. Das and Trinajstić [21] compared the GA and ABC indices for chemical trees and molecular graphs. Xing et al. [22] gave the upper bound for the ABC index of trees with perfect matching and characterized the unique extremal tree.

Based on the results, ABC_e and GA_e indices for certain Cartesian product graphs (such as $P_n \Box P_m$, $P_n \Box C_m$ and $P_n \Box S_m$) are obtained. In addition, ABC_e and GA_e indices of certain join graphs (such as $C_m + P_n + S_r$, $P_m + P_n + P_r$, $C_m + C_n + C_r$ and $S_m + S_n + S_r$) are deduced. Our results extend and enrich some known results [5,23,24].

2. Main Results

It is known that the Cartesian product and join operation are very complicated. In this section, we present these two classical type of graphs.

2.1. Cartesian Product Graphs

In graph theory, the Cartesian product $G \Box H$ of graphs G and H is a graph such that the vertex set of $G \Box H$ is the Cartesian product $V(G) \times V(H)$; and any two vertices (u, u') and (v, v') are adjacent in $G \Box H$ if and only if either $u = v$ and u' are adjacent with v' in H or $u' = v'$ and u are adjacent with v in G. The graph $P_n \Box P_m$ and the line graph of $P_n \Box P_m$ are illustrated in Figure 1. In the following, we discuss the edge-version ABC and GA indices of some Cartesian product graphs.

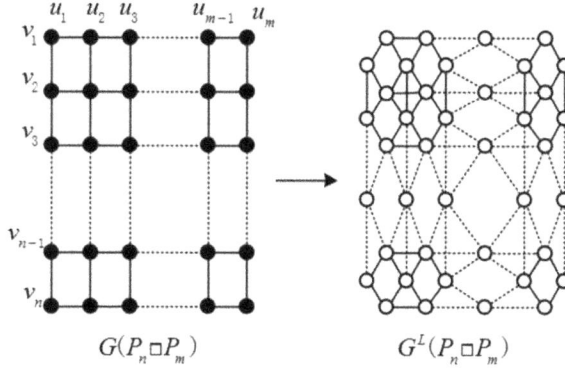

Figure 1. $P_n \Box P_m$ and the line graph of $P_n \Box P_m$.

Theorem 1. *If* $n, m \geq 4$, *then*

$$
\begin{aligned}
ABC_e(P_n \Box P_m) &= \frac{1}{2}\sqrt{\frac{3}{2}}(2n + 2m - 16) + \frac{1}{2}\sqrt{\frac{7}{5}}(4n + 4m - 24) + \sqrt{\frac{3}{10}}(6n + 6m - 32) \\
&\quad + \frac{\sqrt{10}}{6}(6nm - 18n - 18m + 52) + \frac{8\sqrt{2}}{5} + 4\sqrt{\frac{5}{3}} + 8\sqrt{\frac{2}{5}} + \frac{8}{3};
\end{aligned}
\tag{5}
$$

$$
\begin{aligned}
GA_e(P_n \Box P_m) &= 6nm - 16n - 16m + \frac{4\sqrt{5}}{9}(4n + 4m - 24) + \frac{2\sqrt{30}}{11}(6n + 6m - 32) \\
&\quad + 44 + \frac{16\sqrt{12}}{7} + 2\sqrt{15}.
\end{aligned}
\tag{6}
$$

Proof. Let $G = P_n \Box P_m$, we have $L(G)$ has $6nm - 6n - 6m + 4$ edges. Moreover, $\left|E_{3,3}^L\right| = 4$, $\left|E_{3,4}^L\right| = 8$, $\left|E_{3,5}^L\right| = 8$, $\left|E_{4,4}^L\right| = 2n + 2m - 16$, $\left|E_{4,5}^L\right| = 4n + 4m - 24$, $\left|E_{5,5}^L\right| = 4$, $\left|E_{5,6}^L\right| = 6n + 6m - 32$ and $\left|E_{6,6}^L\right| = 6nm - 18n - 18m + 52$.

$$
\begin{aligned}
ABC_e(P_n \Box P_m) &= (4)\left(\sqrt{\frac{3 + 3 - 2}{3 \times 3}}\right) + (8)\left(\sqrt{\frac{3 + 4 - 2}{3 \times 4}}\right) + (8)\left(\sqrt{\frac{3 + 5 - 2}{3 \times 5}}\right) \\
&\quad + (2n + 2m - 16)\left(\sqrt{\frac{4 + 4 - 2}{4 \times 4}}\right) \\
&\quad + (4n + 4m - 24)\left(\sqrt{\frac{4 + 5 - 2}{4 \times 5}}\right) + (4)\left(\sqrt{\frac{5 + 5 - 2}{5 \times 5}}\right) \\
&\quad + (6n + 6m - 32)\left(\sqrt{\frac{5 + 6 - 2}{5 \times 6}}\right) \\
&\quad + (6nm - 18n - 18m + 52)\left(\sqrt{\frac{6 + 6 - 2}{6 \times 6}}\right) \\
&= \frac{1}{2}\sqrt{\frac{3}{2}}(2n + 2m - 16) + \frac{1}{2}\sqrt{\frac{7}{5}}(4n + 4m - 24) \\
&\quad + \sqrt{\frac{3}{10}}(6n + 6m - 32) + \frac{\sqrt{10}}{6}(6nm - 18n - 18m + 52) \\
&\quad + \frac{8\sqrt{2}}{5} + 4\sqrt{\frac{5}{3}} + 8\sqrt{\frac{2}{5}} + \frac{8}{3};
\end{aligned}
\tag{7}
$$

$$GA_e(P_n \square P_m) = (4)(\frac{2\sqrt{3 \times 3}}{3+3}) + (8)(\frac{2\sqrt{3 \times 4}}{3+4}) + (8)(\frac{2\sqrt{3 \times 5}}{3+5})$$

$$+ (2n + 2m - 16)(\frac{2\sqrt{4 \times 4}}{4+4})$$

$$+ (4n + 4m - 24)(\frac{2\sqrt{4 \times 5}}{4+5}) + (4)(\frac{2\sqrt{5 \times 5}}{5+5})$$

$$+ (6n + 6m - 32)(\frac{2\sqrt{5 \times 6}}{5+6}) \tag{8}$$

$$+ (6nm - 18n - 18m + 52)(\frac{2\sqrt{6 \times 6}}{6+6})$$

$$= 6nm - 16n - 16m + \frac{4\sqrt{5}}{9}(4n + 4m - 24)$$

$$+ \frac{2\sqrt{30}}{11}(6n + 6m - 32) + 44 + \frac{16\sqrt{12}}{7} + 2\sqrt{15}.$$

By now, the proof is complete.

Theorem 2. *If* $n \geq 4, m \geq 3$, *then*

$$ABC_e(P_n \square C_m) = \sqrt{10}nm + (\frac{\sqrt{6}}{2} + 2\sqrt{\frac{7}{5}} + \frac{3\sqrt{30}}{5} - \frac{9\sqrt{10}}{3})m \tag{9}$$

$$GA_e(P_n \square C_m) = 6nm + (2 + \frac{16\sqrt{5}}{9} + \frac{12\sqrt{30}}{11} - 18)m \tag{10}$$

Proof. Let $G = P_n \square C_m$, we have $L(G)$ has $6nm - 6m$ edges. Moreover, $\left| E_{4,4}^L \right| = 2m$, $\left| E_{4,5}^L \right| = 4m$, $\left| E_{5,6}^L \right| = 6m$ and $\left| E_{6,6}^L \right| = 6nm - 18m$. In Figure 2, the degrees of vertices in line graph $G^L(P_n \square C_m)$ are displayed near the corresponding vertices.

$$ABC_e(P_n \square C_m) = (2m)(\sqrt{\frac{4+4-2}{4 \times 4}}) + (4m)(\sqrt{\frac{4+5-2}{4 \times 5}})$$

$$+ (6m)(\sqrt{\frac{5+6-2}{5 \times 6}}) + (6nm - 18m)(\sqrt{\frac{6+6-2}{6 \times 6}}) \tag{11}$$

$$= \sqrt{10}nm + (\frac{\sqrt{6}}{2} + 2\sqrt{\frac{7}{5}} + \frac{3\sqrt{30}}{5} - \frac{9\sqrt{10}}{3})m;$$

$$GA_e(P_n \square C_m) = (2m)(\frac{2\sqrt{4 \times 4}}{4+4}) + (4m)(\frac{2\sqrt{4 \times 5}}{4+5}) + (6m)(\frac{2\sqrt{5 \times 6}}{5+6})$$

$$+ (6nm - 18m)(\frac{2\sqrt{6 \times 6}}{6+6}) \tag{12}$$

$$= 6nm + (2 + \frac{16\sqrt{5}}{9} + \frac{12\sqrt{30}}{11} - 18)m.$$

In the end, the proof is complete.

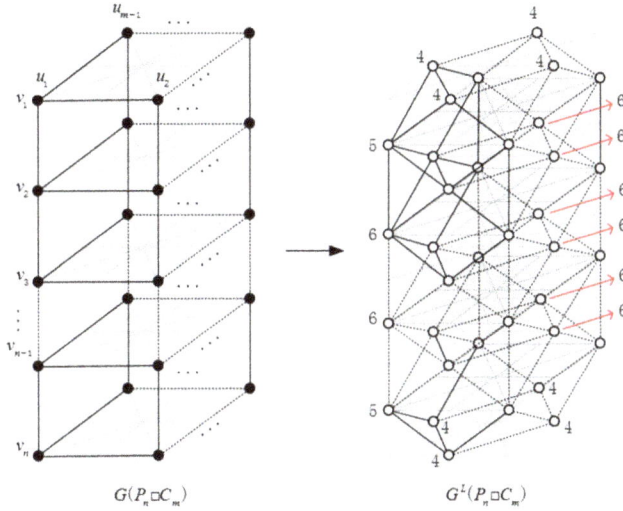

Figure 2. $G(P_n \square C_m)$ and $G^L(P_n \square C_m)$.

Theorem 3. *If $n \geq 5, m \geq 1$, then*

$$
\begin{aligned}
ABC_e(P_n \square S_m) &= \frac{(n-2)(m-1)(m-2)}{2(m+2)}\sqrt{2m+2} \\
&\quad +(n-3)(m-1)\left(\sqrt{\frac{m+4}{m+2}}+2\sqrt{\frac{3}{2(m+2)}}\right) \\
&\quad +2(m-1)\left(\sqrt{\frac{m+3}{3(m+2)}}+\sqrt{\frac{3m-1}{(m+2)(2m-1)}}+\sqrt{\frac{m+1}{3m}}\right. \\
&\quad \left.+\sqrt{\frac{3m-3}{m(2m-1)}}\right)+\frac{(m-1)(m-2)}{m}\sqrt{2m-2} \\
&\quad +\frac{1}{4}(m-1)(n-4)\sqrt{6}+(m-1)\sqrt{\frac{5}{3}} \\
&\quad +\frac{n-4}{2m}\sqrt{4m-2}+2\sqrt{\frac{4m-3}{2m(2m-1)}};
\end{aligned}
\tag{13}
$$

$$
\begin{aligned}
GA_e(P_n \square S_m) &= \frac{(n-2)(m-1)(m-2)}{2}+8(n-3)(m-1)\frac{\sqrt{(m+2)}}{m+6} \\
&\quad +4(m-1)\frac{\sqrt{3(m+2)}}{m+5}+4(n-3)(m-1)\frac{\sqrt{2m(m+2)}}{3m+2} \\
&\quad +4(m-1)\frac{\sqrt{(m+2)(2m-1)}}{3m+1}+(m-1)(m-2) \\
&\quad +4(m-1)\frac{\sqrt{3m}}{m+3}+4(m-1)\frac{\sqrt{m(2m-1)}}{3m-1} \\
&\quad +(m-1)(n-4)+8(m-1)\frac{\sqrt{3}}{7} \\
&\quad +(n-4)+4\frac{\sqrt{2m(2m-1)}}{4m-1}.
\end{aligned}
\tag{14}
$$

Proof. Let $G = P_n \square S_m$, we have $L(G)$ has $\frac{1}{2}(m^2 n + m(7n - 10) - 8n + 8)$ edges.

Moreover, $\left|E^L_{m+2,m+2}\right| = \dfrac{(n-2)(m-1)(m-2)}{2}$, $\left|E^L_{m+2,4}\right| = 2(n-3)(m-1)$, $\left|E^L_{m+2,3}\right| = 2(m-1)$,

$\left|E^L_{m+2,2m}\right| = 2(n-3)(m-1)$, $\left|E^L_{m+2,2m-1}\right| = 2(m-1)$, $\left|E^L_{m,m}\right| = (m-1)(m-2)$, $\left|E^L_{m,3}\right| = 2(m-1)$,

$\left|E^L_{m,2m-1}\right| = 2(m-1)$, $\left|E^L_{4,4}\right| = (m-1)(n-4)$, $\left|E^L_{3,4}\right| = 2(m-1)$, $\left|E^L_{2m,2m}\right| = (n-4)$ and

$\left|E^L_{2m-1,2m}\right| = 2$. In Figure 3, the degrees of vertices in line graph $G^L(P_n \square S_m)$ are displayed near by the corresponding vertices.

$$
\begin{aligned}
ABC_e(P_n \square S_m) &= \frac{(n-2)(m-1)(m-2)}{2}\left(\sqrt{\frac{m+2+m+2-2}{(m+2)\times(m+2)}}\right) \\
&+ 2(n-3)(m-1)\left(\sqrt{\frac{m+2+4-2}{(m+2)\times 4}}\right) \\
&+ 2(m-1)\left(\sqrt{\frac{m+2+3-2}{(m+2)\times 3}}\right) \\
&+ 2(n-3)(m-1)\left(\sqrt{\frac{m+2+2m-2}{(m+2)\times 2m}}\right) \\
&+ 2(m-1)\left(\sqrt{\frac{m+2+2m-1-2}{(m+2)\times(2m-1)}}\right) \\
&+ (m-1)(m-2)\left(\sqrt{\frac{m+m-2}{m\times m}}\right) \\
&+ 2(m-1)\left(\sqrt{\frac{m+3-2}{m\times 3}}\right) + 2(m-1)\left(\sqrt{\frac{m+2m-1-2}{m\times(2m-1)}}\right) \\
&+ (m-1)(n-4)\left(\sqrt{\frac{4+4-2}{4\times 4}}\right) + 2(m-1)\left(\sqrt{\frac{3+4-2}{3\times 4}}\right) \\
&+ (n-4)\left(\sqrt{\frac{2m+2m-2}{2m\times 2m}}\right) + 2\left(\sqrt{\frac{2m-1+2m-2}{(2m-1)\times 2m}}\right) \\
&= \frac{(n-2)(m-1)(m-2)}{2(m+2)}\sqrt{2m+2} \\
&+ (n-3)(m-1)\left(\sqrt{\frac{m+4}{m+2}} + 2\sqrt{\frac{3}{2(m+2)}}\right) \\
&+ 2(m-1)\left(\sqrt{\frac{m+3}{3(m+2)}} + \sqrt{\frac{3m-1}{(m+2)(2m-1)}}\right. \\
&\left. + \sqrt{\frac{m+1}{3m}} + \sqrt{\frac{3m-3}{m(2m-1)}}\right) \\
&+ \frac{(m-1)(m-2)}{m}\sqrt{2m-2} \\
&+ \frac{1}{4}(m-1)(n-4)\sqrt{6} + (m-1)\sqrt{\frac{5}{3}} \\
&+ \frac{n-4}{2m}\sqrt{4m-2} + 2\sqrt{\frac{4m-3}{2m(2m-1)}};
\end{aligned}
$$

(15)

$$GA_e(P_n \Box S_m) = \frac{(n-2)(m-1)(m-2)}{2}(\frac{2\sqrt{(m+2) \times (m+2)}}{m+2+m+2})$$

$$+2(n-3)(m-1)(\frac{2\sqrt{(m+2) \times 4}}{m+2+4})$$

$$+2(m-1)(\frac{2\sqrt{(m+2) \times 3}}{m+2+3})$$

$$+2(n-3)(m-1)(\frac{2\sqrt{(m+2) \times 2m}}{m+2+2m})$$

$$+2(m-1)(\frac{2\sqrt{(m+2) \times (2m-1)}}{m+2+2m-1})$$

$$+(m-1)(m-2)(\frac{2\sqrt{m \times m}}{m+m})$$

$$+2(m-1)(\frac{2\sqrt{m \times 3}}{m+3})+2(m-1)(\frac{2\sqrt{m \times (2m-1)}}{m+2m-1})$$

$$+(m-1)(n-4)(\frac{2\sqrt{4 \times 4}}{4+4})+2(m-1)(\frac{2\sqrt{3 \times 4}}{3+4}) \qquad (16)$$

$$+(n-4)(\frac{2\sqrt{2m \times 2m}}{2m+2m})+2(\frac{2\sqrt{(2m-1) \times (2m)}}{2m-1+2m})$$

$$= \frac{(n-2)(m-1)(m-2)}{2}+8(n-3)(m-1)\frac{\sqrt{(m+2)}}{m+6}$$

$$+4(m-1)\frac{\sqrt{3(m+2)}}{m+5}+4(n-3)(m-1)\frac{\sqrt{2m(m+2)}}{3m+2}$$

$$+4(m-1)\frac{\sqrt{(m+2)(2m-1)}}{3m+1}+(m-1)(m-2)$$

$$+4(m-1)\frac{\sqrt{3m}}{m+3}+4(m-1)\frac{\sqrt{m(2m-1)}}{3m-1}$$

$$+(m-1)(n-4)+8(m-1)\frac{\sqrt{3}}{7}$$

$$+(n-4)+4\frac{\sqrt{2m(2m-1)}}{4m-1}.$$

Until now, the proof is complete.

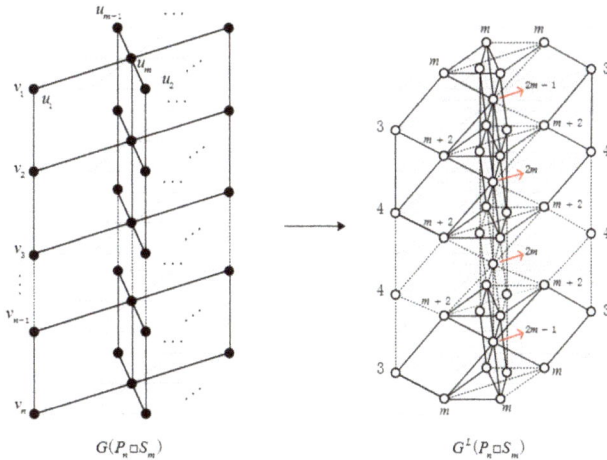

Figure 3. $G(P_n \Box S_m)$ and $G^L(P_n \Box S_m)$.

2.2. Join Graph

The results of ABC_e and GA_e indices of P_n, S_n, K_n and C_n, which were first established by [7], as well as the ABC_e and GA_e indices of some join graphs, such as $P_n + C_m$, $P_n + S_m$, $C_m + P_n + C_m$, $S_m + P_n + S_m$ and $C_m + P_n + S_r$, created by P_n, C_n and S_n were obtained by [5]. However, there are some problems in the calculation of the ABC_e and GA_e indices of join graph $C_m + P_n + S_r$ in [5].

The join graph operation's definition is given as follows: If we are given two graphs G and H and two vertices $v_i \in V(G)$, $u_j \in V(H)$, the join graph is obtained by merging v_i and u_j into one vertex. The certain join graphs $P_n + C_m$ and $P_n + S_m$ are illustrated in Figures 4 and 5, respectively.

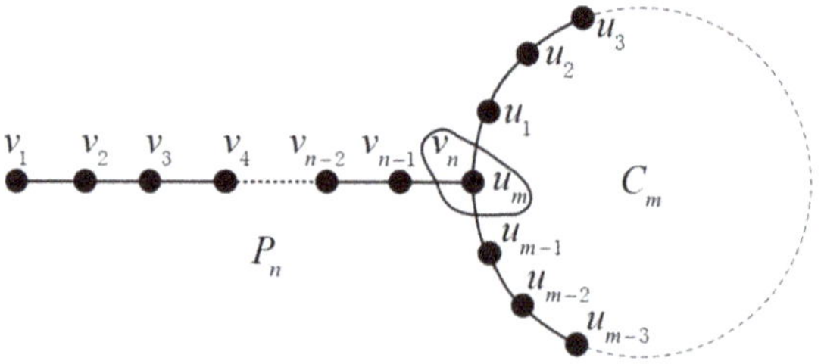

Figure 4. The join graph of $P_n + C_m$.

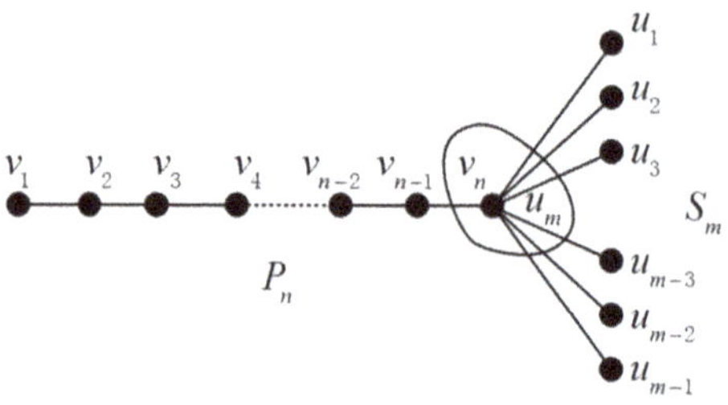

Figure 5. The join graph of $P_n + S_m$.

Theorem A is stated in [5]. However, the result is not correct. In this paper, we correct the result of Theorem A and restate it in Theorem 4 as follows:

Theorem A. *If $n, r \geq 4, m \geq 3$, then*

$$ABC_e(C_m + P_n + S_r) = \frac{r-2}{2}\sqrt{2r-4} + (r-1)\sqrt{\frac{2r-3}{r(r-1)}} + \frac{\sqrt{2}}{2}(n+m-3) + 2 \tag{17}$$

$$GA_e(C_m + P_n + S_r) = \frac{2\sqrt{2(r-1)}}{r+1} + (r-1)(\frac{r-2}{2} + \frac{2\sqrt{r(r-1)}}{2r-1}) + n + m + \frac{6\sqrt{6}}{5} - 4 \tag{18}$$

The join graph of $C_m + P_n + S_r$ is illustrated in Figure 6. It can be seen that $d_{L(G)}(v_{n-2}v_{n-1})$ is 2 and $d_{L(G)}(v_{n-1}v_n)$ is r in $C_m + P_n + S_r$, so we have one edge of types $d_{L(G)}(e) = 2$ and $d_{L(G)}(f) = r$ in $G^L(C_m + P_n + S_r)$.

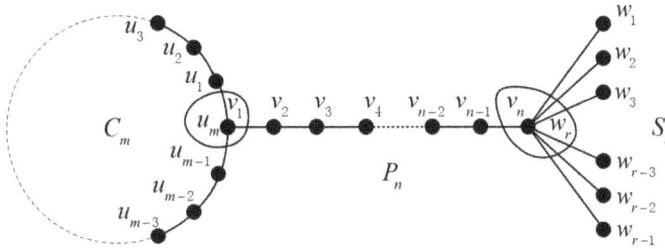

Figure 6. The join graph of $C_m + P_n + S_r$.

Theorem 4. *If $n \geq 4, r \geq 4, m \geq 3$, then we have*

$$ABC_e(C_m + P_n + S_r) = \frac{r-2}{2}\sqrt{2r-4} + (r-1)\sqrt{\frac{2r-3}{r(r-1)}} + \frac{\sqrt{2}}{2}(n+m-3) + 2 \qquad (19)$$

$$GA_e(C_m + P_n + S_r) = \frac{2\sqrt{2r}}{r+2} + (r-1)(\frac{r-2}{2} + \frac{2\sqrt{r(r-1)}}{2r-1}) + n + m + \frac{6\sqrt{6}}{5} - 4 \qquad (20)$$

Proof.   Let $G = C_m + P_n + S_r$, we have $\left|E_{2,2}^L\right| = n+m-7$, $\left|E_{2,3}^L\right| = 3$, $\left|E_{2,r}^L\right| = 1$, $\left|E_{3,3}^L\right| = 3$, $\left|E_{r-1,r-1}^L\right| = \dfrac{(r-1)(r-2)}{2}$ and $\left|E_{r-1,r}^L\right| = r-1$.

$$\begin{aligned} ABC_e(C_m + P_n + S_r) &= (n+m-7)ABC_e(E_{2,2}^L) + (3)ABC_e^L(E_{2,3}) \\ &\quad + (1)ABC_e(E_{2,r}^L) + (3)ABC_e(E_{3,3}^L) \\ &\quad + \frac{(r-1)(r-2)}{2}ABC_e(E_{r-1,r-1}^L) \\ &\quad + (r-1)ABC_e(E_{r-1,r}^L) \\ &= (n+m-7)(\sqrt{\frac{2+2-2}{2\times 2}}) + (3)(\sqrt{\frac{2+3-2}{2\times 3}}) \\ &\quad + (1)(\sqrt{\frac{2+r-2}{2\times r}}) + (3)(\sqrt{\frac{3+3-2}{3\times 3}}) \\ &\quad + \frac{(r-1)(r-2)}{2}(\sqrt{\frac{(r-1)+(r-1)-2}{(r-1)\times(r-1)}}) \\ &\quad + (r-1)(\sqrt{\frac{(r-1)+r-2}{(r-1)\times r}}) \\ &= \frac{r-2}{2}\sqrt{2r-4} + (r-1)\sqrt{\frac{2r-3}{r(r-1)}} \\ &\quad + \frac{\sqrt{2}}{2}(n+m-3) + 2. \end{aligned} \qquad (21)$$

Remark: The result of $ABC_e(C_m + P_n + S_r)$ is the same as that of [5], only because the $ABC_e(E^L_{2,r-1}) = ABC_e(E^L_{2,r})$. We must note $GA_e(E^L_{2,r-1}) \neq GA_e(E^L_{2,r})$.

$$
\begin{aligned}
GA_e(C_m + P_n + S_r) &= (n + m - 7)GA_e(E^L_{2,2}) + (3)GA_e(E^L_{2,3}) + (1)GA_e(E^L_{2,r}) \\
&\quad + (3)GA_e(E^L_{3,3}) + \frac{(r-1)(r-2)}{2}GA_e(E^L_{r-1,r}) \\
&= (n + m - 7)(\frac{2\sqrt{2 \times 2}}{2+2}) + (3)(\frac{2\sqrt{2 \times 3}}{2+3}) \\
&\quad + (1)(\frac{2\sqrt{2 \times r}}{2+r}) + (3)(\frac{2\sqrt{3 \times 3}}{3+3}) \\
&\quad + \frac{(r-1)(r-2)}{2}(\frac{2\sqrt{(r-1) \times (r-1)}}{(r-1)+(r-1)}) \\
&\quad + (r-1)(\frac{2\sqrt{(r-1) \times r}}{(r-1)+r}) \\
&= \frac{2\sqrt{2r}}{r+2} + (r-1)(\frac{r-2}{2} + \frac{2\sqrt{r(r-1)}}{2r-1}) \\
&\quad + n + m + \frac{6\sqrt{6}}{5} - 4.
\end{aligned}
\tag{22}
$$

Now the proof is complete.

Theorem 5. *If $m \geq 2, n \geq 2, r \geq 2$ and $P_m + P_n + P_r$ be the join graphs depicted in Figure 7, then*

$$
ABC_e(P_m + P_n + P_r) = \frac{\sqrt{2}}{2}(m + n + r - 4)
\tag{23}
$$

$$
GA_e(P_m + P_n + P_r) = m + n + r - 6 + \frac{4}{3}\sqrt{2}
\tag{24}
$$

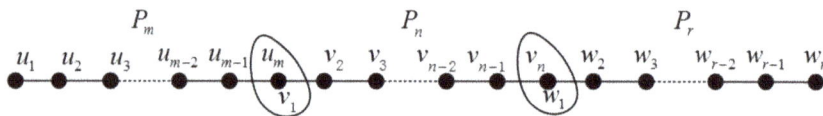

Figure 7. The join graph of $P_m + P_n + P_r$.

Proof. Let $G = P_m + P_n + P_r$, we have $\left|E^L_{2,2}\right| = m + n + r - 6$ and $\left|E^L_{1,2}\right| = 2$.

$$
\begin{aligned}
ABC_e(P_m + P_n + P_r) &= (m + n + r - 6)ABC_e(E^L_{2,2}) + 2ABC_e(E^L_{1,2}) \\
&= (m + n + r - 6)(\sqrt{\frac{2+2-2}{2 \times 2}}) + 2(\sqrt{\frac{1+2-2}{1 \times 2}}) \\
&= \frac{\sqrt{2}}{2}(m + n + r - 4).
\end{aligned}
\tag{25}
$$

$$
\begin{aligned}
GA_e(P_m + P_n + P_r) &= (m + n + r - 6)GA_e(E^L_{2,2}) + 2GA_e(E^L_{1,2}) \\
&= (m + n + r - 6)(\frac{2\sqrt{2 \times 2}}{2+2}) + 2(\frac{2\sqrt{1 \times 2}}{1+2}) \\
&= m + n + r - 6 + \frac{4}{3}\sqrt{2}.
\end{aligned}
\tag{26}
$$

Now the proof is complete.

Theorem 6. *Let* $m \geq 3, r \geq 3, n \geq 6$ *and* $C_m + C_n + C_r$ *be the join graphs depicted in Figure 8. If* $d(u_m, v_n) \geq 3$, *then*

$$ABC_e(C_m + C_n + C_r) = \frac{\sqrt{2}}{2}(m + n + r) - 2\sqrt{2} + 3\sqrt{6} \tag{27}$$

$$GA_e(C_m + C_n + C_r) = m + n + r + \frac{16\sqrt{2}}{3} \tag{28}$$

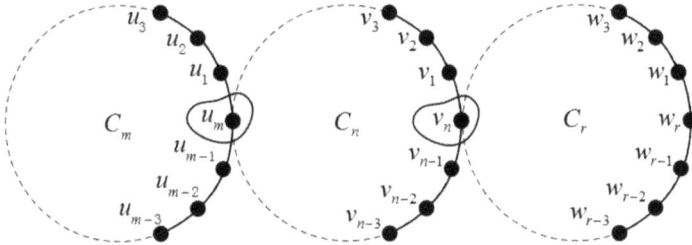

Figure 8. The join graph of $C_m + C_n + C_r$.

Proof. Let $G = C_m + C_n + C_r$, we have $\left|E_{2,2}^L\right| = m + n + r - 12$, $\left|E_{2,4}^L\right| = 8$ and $\left|E_{4,4}^L\right| = 12$.

$$
\begin{aligned}
ABC_e(C_m + C_n + C_r) &= (m + n + r - 12)ABC_e(E_{2,2}^L) \\
&\quad + 8ABC_e(E_{2,4}^L) + 12ABC_e(E_{4,4}^L) \\
&= (m + n + r - 12)\left(\sqrt{\frac{2 + 2 - 2}{2 \times 2}}\right) \\
&\quad + 8\left(\sqrt{\frac{2 + 4 - 2}{2 \times 4}}\right) + 12\left(\sqrt{\frac{4 + 4 - 2}{4 \times 4}}\right) \\
&= \frac{\sqrt{2}}{2}(m + n + r) - 2\sqrt{2} + 3\sqrt{6}.
\end{aligned} \tag{29}
$$

$$
\begin{aligned}
GA_e(P_m + P_n + P_r) &= (m + n + r - 12)GA_e(E_{2,2}^L) \\
&\quad + 8GA_e(E_{2,4}^L) + 12GA_e(E_{4,4}^L) \\
&= (m + n + r - 12)\left(\frac{2\sqrt{2 \times 2}}{2 + 2}\right) \\
&\quad + 8\left(\frac{2\sqrt{2 \times 4}}{2 + 4}\right) + 12\left(\frac{2\sqrt{4 \times 4}}{4 + 4}\right) \\
&= m + n + r + \frac{16\sqrt{2}}{3}.
\end{aligned} \tag{30}
$$

Now the proof is complete.

Theorem 7. *Let* $m \geq 2, n \geq 3, r \geq 3$ *and* $S_m + S_n + S_r$ *be the join graphs depicted in Figure 9; then, we have*

$$
\begin{aligned}
ABC_e(S_m + S_n + S_r) &= (m - 1)\sqrt{\frac{2m + n - 5}{(m - 1)(m + n - 2)}} + (n - 2)\sqrt{\frac{m + 2n - 5}{(n - 1)(m + n - 2)}} \\
&\quad + (n - 2)\sqrt{\frac{2n + r - 6}{(n - 1)(n + r - 3)}} + (r - 2)\sqrt{\frac{n + 2r - 7}{(r - 2)(n + r - 3)}} \\
&\quad + \frac{(m - 2)}{2}\sqrt{2m - 4} + \frac{(n - 2)(n - 3)}{2(n - 1)}\sqrt{2n - 4} + \frac{(r - 3)}{2}\sqrt{2r - 6} \\
&\quad + \sqrt{\frac{m + 2n + r - 7}{(m + n - 2)(n + r - 3)}};
\end{aligned} \tag{31}
$$

$$
\begin{aligned}
GA_e(S_m + S_n + S_r) \quad &= 2(m-1)\frac{\sqrt{(m-1)(m+n-2)}}{2m+n-3} + 2(n-2)\frac{\sqrt{(n-1)(m+n-2)}}{m+2n-3} \\
&+ 2(n-2)\frac{\sqrt{(n-1)(n+r-3)}}{2n+r-4} + 2(r-2)\frac{\sqrt{(r-2)(n+r-3)}}{n+2r-5} \\
&+ \frac{(m-1)(m-2)}{2} + \frac{(n-2)(n-3)}{2} + \frac{(r-2)(r-3)}{2} \\
&+ \frac{2\sqrt{(m+n-2)(n+r-3)}}{m+2n+r-5}.
\end{aligned}
\tag{32}
$$

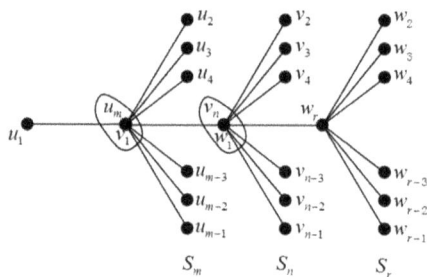

Figure 9. The join graph of $S_m + S_n + S_r$.

Proof. Let $G = S_m + S_n + S_r$, we have $\left|E^L_{m-1,m+n-2}\right| = m-1$, $\left|E^L_{n-1,m+n-2}\right| = n-2$, $\left|E^L_{n-1,n+r-3}\right| = n-2$, $\left|E^L_{r-2,n+r-3}\right| = r-2$, $\left|E^L_{m-1,m-1}\right| = \frac{(m-1)(m-2)}{2}$, $\left|E^L_{n-1,n-1}\right| = \frac{(n-2)(n-3)}{2}$, $\left|E^L_{r-2,r-2}\right| = \frac{(r-2)(r-3)}{2}$ and $\left|E^L_{m+n-2,n+r-3}\right| = 1$.

$$
\begin{aligned}
ABC_e(S_m + S_n + S_r) &= (m-1)ABC_e(E^L_{m-1,m+n-2}) \\
&+ (n-2)ABC_e(E^L_{n-1,m+n-2}) \\
&+ (n-2)ABC_e(E^L_{n-1,n+r-3}) \\
&+ (r-2)ABC_e(E^L_{r-2,n+r-3}) \\
&+ \frac{(m-1)(m-2)}{2}ABC_e(E^L_{m-1,m-1}) \\
&+ \frac{(n-2)(n-3)}{2}ABC_e(E^L_{n-1,n-1}) \\
&+ \frac{(r-2)(r-3)}{2}ABC_e(E^L_{r-2,r-2}) \\
&+ (1)ABC_e(E^L_{m+n-2,n+r-3}) \\[4pt]
&= (m-1)\sqrt{\frac{(m-1)+(m+n-2)-2}{(m-1)(m+n-2)}} \\
&+ (n-2)\sqrt{\frac{(n-1)+(m+n-2)-2}{(n-1)(m+n-2)}} \\
&+ (n-2)\sqrt{\frac{(n-1)+(n+r-3)-2}{(n-1)(n+r-3)}} \\
&+ (r-2)\sqrt{\frac{(r-2)+(n+r-3)-2}{(r-2)(n+r-3)}}
\end{aligned}
\tag{33}
$$

$$+\frac{(m-1)(m-2)}{2}\sqrt{\frac{(m-1)+(m-1)-2}{(m-1)(m-1)}}$$

$$+\frac{(n-2)(n-3)}{2}\sqrt{\frac{(n-1)+(n-1)-2}{(n-1)(n-1)}}$$

$$+\frac{(r-2)(r-3)}{2}\sqrt{\frac{(r-2)+(r-2)-2}{(r-2)(r-2)}}$$

$$+(1)\sqrt{\frac{(m+n-2)+(n+r-3)-2}{(m+n-2)(n+r-3)}}$$

$$=(m-1)\sqrt{\frac{2m+n-5}{(m-1)(m+n-2)}}$$

$$+(n-2)\sqrt{\frac{m+2n-5}{(n-1)(m+n-2)}}$$

$$+(n-2)\sqrt{\frac{2n+r-6}{(n-1)(n+r-3)}}$$

$$+(r-2)\sqrt{\frac{n+2r-7}{(r-2)(n+r-3)}}$$

$$+\frac{(m-2)}{2}\sqrt{2m-4}+\frac{(n-2)(n-3)}{2(n-1)}\sqrt{2n-4}$$

$$+\frac{(r-3)}{2}\sqrt{2r-6}+\sqrt{\frac{m+2n+r-7}{(m+n-2)(n+r-3)}}.$$

$$
\begin{aligned}
GA_e(S_m+S_n+S_r) \;=\;& (m-1)GA_e(E^L_{m-1,m+n-2})\\
&+(n-2)GA_e(E^L_{n-1,m+n-2})\\
&+(n-2)GA_e(E^L_{n-1,n+r-3})\\
&+(r-2)GA_e(E^L_{r-2,n+r-3})\\
&+\frac{(m-1)(m-2)}{2}GA_e(E^L_{m-1,m-1})\\
&+\frac{(n-2)(n-3)}{2}GA_e(E^L_{n-1,n-1})\\
&+\frac{(r-2)(r-3)}{2}GA_e(E^L_{r-2,r-2})\\
&+(1)ABC_e(E^L_{m+n-2,n+r-3})\\
=\;& (m-1)\frac{2\sqrt{(m-1)(m+n-2)}}{(m-1)+(m+n-2)}\\
&+(n-2)\frac{2\sqrt{(n-1)(m+n-2)}}{(n-1)+(m+n-2)}\\
&+(n-2)\frac{2\sqrt{(n-1)(n+r-3)}}{(n-1)+(n+r-3)}\\
&+(r-2)\frac{2\sqrt{(r-2)(n+r-3)}}{(r-2)+(n+r-3)}\\
&+\frac{(m-1)(m-2)}{2}\frac{2\sqrt{(m-1)(m-1)}}{(m-1)+(m-1)}\\
&+\frac{(n-2)(n-3)}{2}\frac{2\sqrt{(n-1)(n-1)}}{(n-1)+(n-1)}\\
&+\frac{(r-2)(r-3)}{2}\frac{2\sqrt{(r-2)(r-2)}}{(r-2)+(r-2)}
\end{aligned}
$$

(34)

$$+(1)\frac{2\sqrt{(m+n-2)(n+r-3)}}{(m+n-2)+(n+r-3)}$$

$$= 2(m-1)\frac{\sqrt{(m-1)(m+n-2)}}{2m+n-3}$$

$$+2(n-2)\frac{\sqrt{(n-1)(m+n-2)}}{m+2n-3}$$

$$+2(n-2)\frac{\sqrt{(n-1)(n+r-3)}}{2n+r-4}$$

$$+2(r-2)\frac{\sqrt{(r-2)(n+r-3)}}{n+2r-5}$$

$$+\frac{(m-1)(m-2)}{2}+\frac{(n-2)(n-3)}{2}$$

$$+\frac{(r-2)(r-3)}{2}$$

$$+\frac{2\sqrt{(m+n-2)(n+r-3)}}{m+2n+r-5}.$$

Now the proof is complete.

3. Conclusions

The physical and chemical properties of proteins, DNAs and RNAs are very important for human disease and various approaches have been proposed to predict, validate and identify their structures and features [25,26]. Among these, topological indices were proved to be very helpful in testing the chemical properties of new chemical or physical materials such as new drugs or nanomaterials. Topological indices play an important role in studying the topological properties of chemical compounds, especially organic materials i.e., carbon containing molecular structures.

Various topological indices provide a better correlation for certain physico-chemical properties. Hence, the edge version ABC and GA indices for some special Cartesian product graphs and certain join graphs are described by graph structure analysis and a mathematical derivation method in this paper. The results of the current study also have promising prospects for applications in chemical and material engineering. The conclusions we draw here will not work for other classes of indices such as distance-based and distance adjacency-based topological indices. Thus a similar kind of study is needed for other classes of indices which might be a future direction in this area of mathematical chemistry.

Author Contributions: X.Z. contributes for conceptualization, designing the experiments and analyzed the data curation, he wrote the initial draft of the paper which were investigated and approved by Z.S. and J.-B.L. H.J. contribute for computing and performed experiments. J.-B.L. contributes for validation and formal analyzing. Z.S. contributes for supervision, funding and methodology and wrote the final draft. All authors read and approved the final version of the paper.

Funding: This work was supported by Applied Basic Research (Key Project) of the Sichuan Province under grant 2017JY0095, the key project of the Sichuan Provincial Department of Education under grant 17ZA0079 and 18ZA0118, the Soft Scientific Research Foundation of Sichuan Provincial Science and Technology Department (18RKX1048).

Conflicts of Interest: The authors declare no conflict of interest.

References

1. Wiener, H. Structural determination of paraffin boiling points. *J. Am. Chem. Soc.* **1947**, *69*, 7–20. [CrossRef]
2. Iranmanesh, A.; Gutman, I.; Khormali, O.; Mahmiani, A. The edge versions of the wiener index. *MATCH Commun. Math. Comput. Chem.* **2009**, *61*, 663–672.
3. Harary, F.; Norman, R.Z. Some properties of line digraphs. *Rendiconti del Circolo Matematico di Palermo* **1960**, *9*, 161–169. [CrossRef]

4. Estrada, E.; Torres, L.; Rodriguez, L.; Gutman, I. An atom-bond connectivity index: Modelling the enthalpy of formation of Alkanes. *Indian J. Chem.* **1998**, *37A*, 849–855.
5. Gao, W.; Farahani, M.; Wang, S.; Husin, M.N. On the edge-version atom-bond connectivity and geometric arithmetic indices of certain graph operations. *Appl. Math. Comput.* **2017**, *308*, 11–17. [CrossRef]
6. Vukicevic, D.; Furtula, B. Topological index based on the ratios of geometric and arithmetical means of end-vertex degrees of edges. *J. Math. Chem.* **2009**, *46*, 1369–1376. [CrossRef]
7. Mahmiani, A.; Khormali, O.; Iranmanesh, A. On the edge version of geometric-arithmetic index. *Digest J. Nanomater. Biostruct.* **2012**, *7*, 411–414.
8. Liu, J.; Pan, X.; Yu, L.; Li, D. Complete characterization of bicyclic graphs with minimal Kirchhoff index. *Discrete Appl. Math.* **2016**, *200*, 95–107. [CrossRef]
9. Liu, J.; Wang, W.; Zhang, Y.; Pan, X. On degree resistance distance of cacti. *Discrete Appl. Math.* **2016**, *203*, 217–225. [CrossRef]
10. Li, X.L.; Shi, Y.T. A survey on the Randic' index. *MATCH Commun. Math. Comput. Chem.* **2008**, *59*, 127–156.
11. Randić, M. The connectivity index 25 years after. *J. Mol. Graph. Modell.* **2001**, *20*, 19–35. [CrossRef]
12. Ma, J.; Shi, Y.; Wang, Z.; Yue, J. On Wiener polarity index of bicyclic networks. *Sci. Rep.* **2016**, *6*, 19066. [CrossRef] [PubMed]
13. Shi, Y. Note on two generalizations of the Randić index. *Appl. Math. Comput.* **2015**, *265*, 1019–1025. [CrossRef]
14. Estes, J.; Wei, B. Sharp bounds of the Zagreb indices of k-trees. *J. Comb. Optim.* **2014**, *27*, 271–291. [CrossRef]
15. Ji, S.; Wang, S. On the sharp lower bounds of Zagreb indices of graphs with given number of cut vertices. *J. Math. Anal. Appl.* **2018**, *458*, 21–29. [CrossRef]
16. Zafar, S.; Nadeem, M.F.; Zahid, Z. On the edge version of geometric-arithmetic index of nanocones. *Stud. Univ. Babes-Bolyai Chem.* **2016**, *61*, 273–282.
17. Das, K.C.; Trinajstić, N. Comparison between first geometric-arithmetic index and atom-bond connectivity index. *Chem. Phys. Lett.* **2010**, *497*, 149–151. [CrossRef]
18. Furtula, B.; Graovac, A. Atom-bond connectivity index of trees. *Discrete Appl. Math.* **2009**, *157*, 2828–2835. [CrossRef]
19. Fath-Tabar, G.H.; Vaez-Zadeh, B.; Ashrafi, A.R.; Graova, A. Some inequalities for the atom-bond connectivity index of graph operations. *Discrete Appl. Math.* **2011**, *159*, 1323–1330. [CrossRef]
20. Chen, J.; Liu, J.; Guo, X. Some upper bounds for the atom-bond connectivity index of graphs. *Appl. Math. Lett.* **2012**, *25*, 1077–1081. [CrossRef]
21. Das, K.C. Atom-bond connectivity index of graphs. *Discrete Appl. Math.* **2010**, *158*, 1181–1188. [CrossRef]
22. Xing, R.; Zhou, B.; Du, Z. Further results on atom-bond connectivity index of trees. *Discrete Appl. Math.* **2010**, *158*, 1536–1545. [CrossRef]
23. Shao, Z.; Wu, P.; Gao, Y.; Gutman, I.; Zhang, X. On the maximum ABC index of graphs without pendent vertices. *Appl. Math. Comp.* **2017**, *315*, 298–312. [CrossRef]
24. Shao, Z.; Wu, P.; Zhang, X.; Dimitrov, D.; Liu, J. On the maximum ABC index of graphs with prescribed size and without pendent vertices. *IEEE Access* **2018**, *6*, 27604–27616. [CrossRef]
25. Zeng, X.; Lin, W.; Guo, M.; Zou, Q. A comprehensive overview and evaluation of circular RNA detection tools. *PLoS Comput. Biol.* **2017**, *13*, e1005420. [CrossRef] [PubMed]
26. Zeng, X.; Liao, Y.; Liu, Y.; Zou, Q. Prediction and validation of disease genes using HeteSim Scores. *IEEE/ACM Trans. Comput. Biol. Bioinform.* **2017**, *14*, 687–695. [CrossRef] [PubMed]

Sample Availability: Not available.

molecules

MDPI

Article

Scoring Amino Acid Mutations to Predict Avian-to-Human Transmission of Avian Influenza Viruses

Xiaoli Qiang [1], Zheng Kou [1,*], Gang Fang [1] and Yanfeng Wang [2]

[1]   Institute of Computing Science and Technology, Guangzhou University, Guangzhou 510006, China;
    qiangxl@mail.scuec.edu.cn (X.Q.); yuxiangqd@163.com (G.F.)
[2]   Henan Key Lab of Information-Based Electrical Appliances, College of Electrical and Electronic Engineering,
    Zhengzhou University of Light Industry, Zhengzhou 450002, China; wangyanfeng@zzuli.edu.cn
*   Correspondence: kouzhengcn@foxmail.com; Tel.: +86-20-3936-6191

Received: 17 May 2018; Accepted: 19 June 2018; Published: 29 June 2018

Abstract: Avian influenza virus (AIV) can directly cross species barriers and infect humans with high fatality. Using machine learning methods, the present paper scores the amino acid mutations and predicts interspecies transmission. Initially, 183 signature positions in 11 viral proteins were screened by the scores of five amino acid factors and their random forest rankings. The most important amino acid factor (Factor 3) and the minimal range of signature positions (50 amino acid residues) were explored by a supporting vector machine (the highest-performing classifier among four tested classifiers). Based on these results, the avian-to-human transmission of AIVs was analyzed and a prediction model was constructed for virology applications. The distributions of human-origin AIVs suggested that three molecular patterns of interspecies transmission emerge in nature. The novel findings of this paper provide important clues for future epidemic surveillance.

Keywords: avian influenza virus; interspecies transmission; amino acid mutation; machine learning

1. Introduction

Wild birds are regarded as the natural reservoir of avian influenza virus (AIV) [1]. Interspecies transmission might have been enabled long ago, when wild birds were domesticated by humans. A highly pathogenic subtype of AIV, avian influenza H5N1, originated in Asia in 1996 [2]. Human-origin H5N1 virus was first isolated from clinical samples in 1997, confirming that the H5N1 virus can directly cross species barriers and fatally infect the respiratory system [3,4]. Human infection by H5N1 has been continuously reported since 2003, attracting the attention of both researchers and wider society [5–8]. Moreover, viral subtypes other than H5N1 can infect humans by direct interspecies transmission. Two infectious cases of H9N2 virus have been reported; one in 1999, the other in 2003 [9,10]. H7N7 virus infected farmers in the Netherlands in 2003 [11], and H7N9 has continuously infected China's population since 2013 [12,13].

Interspecies transmission of AIV from its natural reservoir occurs in three steps: (1) the residence of AIVs in their wild animal hosts; (2) AIV contact with humans and direct infection with low probability; and (3) adaptation of AIVs to their new host and efficient human-to-human transmission thereafter. Thus far, AIV has not progressed beyond step 2, which represents initial adaption to the new host and low efficiency of transmission among the new host. The subtype viruses that can cross the species barrier and cause epidemics should be identified. Approximately twenty years has passed since human-originated AIV was first isolated from human samples in 1997. During this period, vast amounts of genomic data have accumulated in public databases. Therefore, after screening the important amino acid sites in the 11 viral proteins, the AIV risk can be predicted by machine learning methods and other mathematic models in the field of bioinformatics [14–18].

AIV transmission relies on amino acid mutations [19–21]. In a previous study, five amino acid factors (AA factors) summarized from 491 highly redundant amino acid attributes were associated with specific physiochemical amino acid properties, namely, polarity, secondary structure, molecular volume, codon diversity, and electrostatic charge [22]. In this paper, we use five AA factors to transform viral proteins and use the random forest (RF) method to select features from high-dimensional protein data and score them by their contributions to the data category. After ranking and screening the positions containing important mutation information, the classifier can predict the interspecies transmission phenotypes.

Two prediction models of AIVs have been published in the literature [23–25]. However, both of these models lack the protein data of hemagglutinin (HA) and neuraminidase (NA), and the biological meanings of the features were not clarified. To construct a more robust and meaningful model, we revise these models and screen the signature amino acid positions in HA, NA, and nine other viral proteins. To this end, we first identify 183 signature mutation positions by RF scoring, then predict AIV occurrence by four popular machine learning methods. Using the most effective classifier, we seek the important amino acid factors and the minimal range of signature positions. The study results will benefit epidemic surveillance and future studies on interspecies AIV transmission.

2. Results

2.1. Dataset

The cleaned dataset contained 869 high-quality AIV strains: 440 avian-origin AIVs (negative samples; H1–H14, H16 subtypes) and 429 human-origin AIVs (positive samples; H5N1, H5N6, H7N3, H7N7, H7N9, and H9N2 subtypes). The information related to these strains is summarized in Table S1.

2.2. Signature Amino Acid Residues

The importance score at each position in the 11 viral proteins was computed by RF. As shown in Figure 1a, the slope of the curve suddenly changes at an importance score of 9. Therefore, 9 was selected as the cutoff score, providing 183 signature positions for further machine learning.

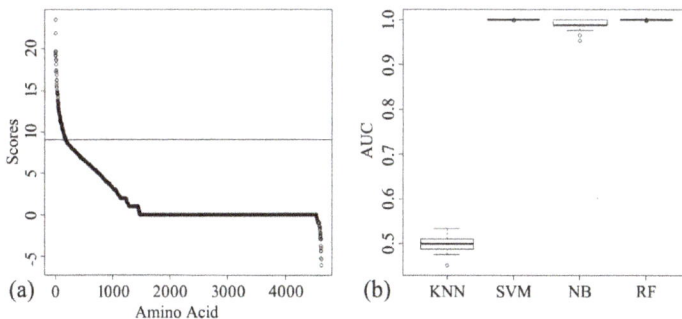

Figure 1. Importance score curve and the performances of *k*-nearest neighbor (KNN), support vector machine (SVM), naïve Bayes (NB), and random forest (RF) classifiers. (**a**) The ranked scores were calculated from five AA factors using the random forest method. The *x* and *y* coordinates denote the total length of the 11 protein alignments and the importance scores, respectively. The cutoff value (9) is indicated by the thin horizontal line. (**b**) Performances of the four classifiers were evaluated from 100 repeats of 10-fold cross-validation. The area under the curve (AUC) ranges from 0 to 1.

As shown in Table 1, the HA protein contained the largest number of signature positions (65 amino acid residues), suggesting that HA is very important for interspecies transmission of AIVs. HA is mainly involved in receptor-binding and fusion activities. Positions HA102–HA290 (Table 1) locate in or close to the region of host receptor binding [26,27], and H163 is reportedly related to the specificity of receptor binding [28]. HA91, HA96, HA328, HA377, and HA397 locate at or near the fusion peptide [29], which triggers fusion activity in acidic environments and favors transmission to humans. The four HA327 positions located in the cleavage site are important virulence sites [30].

NA protein contains 44 signature positions (Table 1). The three NA52s located in the stalk deletion region are related to the virulence and pathogenesis of H5N1 influenza A virus [31]. NA19–NA37 located in the N-terminal are associated with structural stability and enzyme activity [32]. The PB2 627 position has been implicated in increased replication or virulence of AIVs in mammals [33]. PB1 14, located in the binding region of polymerase, is related to viral genome replication [34]. M2 97, which is affiliated with viral particle ensembles [35], was also screened. NEP 14, NP 373, and NP 377 are reportedly involved in intracellular transport of viral proteins [36,37].

Table 1. Scores for the 183 signature amino acids of avian influenza viruses (AIVs).

Num	Pro [1]	Pos [2]	Score	Num	Pro	Pos	Score	Num	Pro	Pos	Score
1	PB2	389	11.95	62	HA	176	13.61	123	NA	65	10.98
2	PB2	478	9.81	63	HA	179	10.08	124	NA	66	9.93
3	PB2	598	17.36	64	HA	185	14.73	125	NA	72	10.96
4	PB2	627	9.83	65	HA	189	14.55	126	NA	79	11.38
5	PB2	648	15.55	66	HA	207	9.49	127	NA	85	9.57
6	PB2	676	9.94	67	HA	211	11.15	128	NA	88	10.13
7	PB1	14	19.16	68	HA	213	11.40	129	NA	100	11.34
8	PB1	48	18.13	69	HA	216	12.17	130	NA	187	10.48
9	PB1	113	18.58	70	HA	221	10.57	131	NA	205	9.62
10	PB1	149	11.09	71	HA	222	9.02	132	NA	233	10.13
11	PB1	257	13.74	72	HA	240	17.36	133	NA	249	9.05
12	PB1	383	12.14	73	HA	251	16.26	134	NA	257	17.24
13	PB1	384	9.34	74	HA	266	10.96	135	NA	265	9.29
14	PB1	387	11.50	75	HA	273	12.53	136	NA	285	10.46
15	PB1	525	9.95	76	HA	274	9.23	137	NA	287	10.65
16	PB1	573	13.38	77	HA	275	9.38	138	NA	288	10.28
17	PB1	628	9.59	78	HA	289	10.36	139	NA	333	10.07
18	PB1-F2	4	9.38	79	HA	290	11.74	140	NA	338	9.02
19	PB1-F2	26	9.24	80	HA	297	10.48	141	NA	347	9.82
20	PB1-F2	48	13.50	81	HA	315	11.98	142	NA	359	10.08
21	PB1-F2	50	11.81	82	HA	323	13.04	143	NA	368	11.05
22	PB1-F2	57	16.85	83	HA	327	12.84	144	NA	369	10.82
23	PB1-F2	77	11.29	84	HA	327	16.23	145	NA	399	11.71
24	PA	37	18.74	85	HA	327	19.25	146	NA	415	9.43
25	PA	61	12.34	86	HA	327	10.41	147	NA	416	13.74
26	PA	63	9.70	87	HA	328	16.24	148	NA	418	9.09
27	PA	129	9.34	88	HA	377	13.91	149	NA	445	12.13
28	PA	337	11.25	89	HA	397	16.18	150	NA	468	9.66
29	PA	356	12.77	90	HA	407	9.49	151	M1	15	9.79
30	PA	367	14.56	91	HA	431	13.52	152	M1	27	12.16
31	PA	405	10.01	92	HA	492	9.49	153	M1	37	14.66
32	PA	554	14.67	93	HA	495	11.15	154	M1	46	14.96
33	PA	607	11.97	94	HA	496	10.62	155	M1	101	13.28
34	PA	684	12.20	95	HA	500	11.88	156	M1	140	12.40
35	PA	712	9.25	96	HA	503	12.76	157	M1	142	11.31

Table 1. *Cont.*

Num	Pro [1]	Pos [2]	Score	Num	Pro	Pos	Score	Num	Pro	Pos	Score
36	HA	40	9.42	97	HA	526	11.91	158	M1	166	17.35
37	HA	42	9.21	98	HA	530	11.26	159	M1	205	11.09
38	HA	45	11.92	99	HA	531	11.67	160	M1	219	13.18
39	HA	46	16.27	100	HA	534	12.77	161	M1	224	23.52
40	HA	53	9.87	101	NP	34	17.45	162	M1	232	14.80
41	HA	57	9.42	102	NP	77	12.39	163	M1	242	19.59
42	HA	65	10.99	103	NP	105	10.61	164	M1	248	11.25
43	HA	66	11.13	104	NP	373	14.73	165	M2	13	13.66
44	HA	79	12.71	105	NP	377	21.88	166	M2	21	10.53
45	HA	81	12.03	106	NP	482	19.71	167	M2	97	15.79
46	HA	84	10.27	107	NA	19	9.20	168	NS1	77	10.59
47	HA	91	17.33	108	NA	23	11.02	169	NS1	80	12.48
48	HA	96	14.98	109	NA	37	9.57	170	NS1	81	12.55
49	HA	102	9.04	110	NA	41	11.30	171	NS1	82	12.01
50	HA	112	12.67	111	NA	42	9.33	172	NS1	83	14.52
51	HA	114	19.46	112	NA	47	10.12	173	NS1	84	10.21
52	HA	115	9.66	113	NA	48	11.23	174	NS1	172	14.21
53	HA	121	10.42	114	NA	49	10.85	175	NS1	179	11.18
54	HA	124	10.28	115	NA	50	9.14	176	NS1	197	9.32
55	HA	131	12.31	116	NA	52	12.38	177	NS1	212	14.19
56	HA	142	12.01	117	NA	52	10.34	178	NEP	14	13.01
57	HA	163	10.07	118	NA	52	9.75	179	NEP	22	15.38
58	HA	164	9.03	119	NA	53	9.03	180	NEP	40	10.28
59	HA	167	14.22	120	NA	58	11.05	181	NEP	60	9.17
60	HA	173	12.81	121	NA	60	9.34	182	NEP	100	10.58
61	HA	174	10.16	122	NA	63	9.44	183	NEP	115	11.10

[1] Viral protein; [2] Position of amino acid residue as H3 subtype numbering.

The AA factors and RF method screened 183 signature positions, some of which are reported to be associated with the mechanism of interspecies transmission. All of the residues were useful, not only for constructing the prediction model but also for further investigating the molecular mechanisms underlying the interspecies transmission of AIVs.

2.3. Performance of the Prediction Model

The performances of the four classifiers are presented as boxplots in Figure 1b. The results were obtained from 100 repeats of 10-fold cross-validation. The area under the curve (AUC) medians in the support vector machine (SVM) and RF classifiers were almost 1. The AUC was clearly lower in the *k*-nearest neighbor (KNN) classifier, possibly because of the nonlinear prediction rules. Although the naïve Bayes (NB) classifier achieved a similar AUC score to the SVM classifier, its performance was poorer and less stable than those of the SVM and RF classifiers. Considering the complexity of the computation, the SVM classifier was selected as the optimal machine learning model for predicting avian-to-human transmission of AIVs.

2.4. Contributions of the AA Factors

The AIV strains were characterized by five AA factors. To understand the mechanism of interspecies transmission, the performance of the SVM classifier was calculated for all combination patterns of these AA factors. The result reveals the importance of the five AA factors. Most of the stable performances of the SVM classifier were contributed by AA Factor 3 or AA Factor 4 (Figure 2a). Notably, the median AUC values were almost 1 and remained stable under AA Factor 3 or AA Factor 4 alone. The SVM classifiers were unstable under AA Factor 1, AA Factor 2, and AA Factor 5 alone. Moreover, AA Factor 3 yielded a slightly better result than AA Factor 4. These results indicate an important role for AA Factor 3 in the avian-to-human transmission of AIVs. Therefore, AA Factor 3 was employed in further analysis.

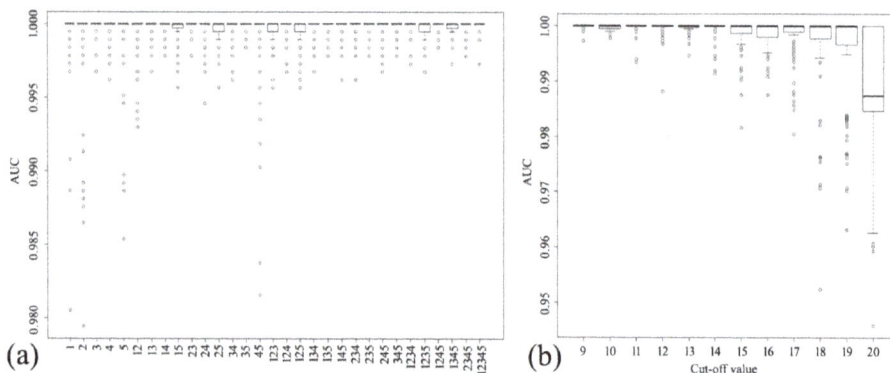

Figure 2. Contributions of AA factors and different mutation sets. (**a**) Performance of SVM classifier for different combinations of the five AA factors. The *x* and *y* coordinates denote the 31 combination patterns and the AUC values (from 0 to 1), respectively. Along the x axis, '13' denotes that the set of 183 amino acid residues was transformed using AA Factor 1 and AA Factor 3 together, for example. (**b**) Contributions of mutation positions for different cutoff values (range 9–20). The *y* coordinate shows the AUC values.

2.5. Contributions of Mutation Positions at Different Cutoff Values

As mentioned above, 183 mutation sites survived a cutoff value of 9. To further explore the mechanism of interspecies transmission, we should reduce the range of crucial positions. To this end, the cutoff value was incremented in steps of 1 (thereby decreasing the number of mutation sites), and the performance of the SVM classifier was calculated with the five AA factors. As shown in Figure 2b, the SVM classifier achieved stable and high performance at cutoffs up to 14. The SVM classifier destabilized at higher cutoffs.

Considering the results under AA factor combinations and cutoff values, the performance of the SVM classifier with AA Factor 3 alone was assessed for different cutoffs. In this situation, the SVM classifier performed stably and well up to a cutoff of 13 (Figure 3a). The analysis results confirm that 13 is the extreme cutoff, giving 50 signature positions (Figure 3b). This set was regarded as the minimal mutation position set for predicting AIVs. We transformed these 50 signature residues using AA Factor 3 alone, and obtained the patterns of the human-origin AIVs (positive samples) by the multidimensional scaling method (Table S2). The resulting clusters are shown in Figure 3c. Cluster 1 comprises three H9N2 viruses (A/Hong Kong/1073/1999; A/Korea/KBNP-0028/2000; A/Bangladesh/0994/2011), two H7N3 viruses (A/Canada/rv504/2004; A/Mexico/InDRE7218/2012), two H7N7 viruses (A/Netherlands/219/2003; A/Italy/3/2013), and one H5N1 virus (A/Hong Kong/482/1997). Cluster 2 includes only H5N1 viruses isolated from 2003 to 2015. Cluster 3 is composed of H7N9 viruses, two H5N6 viruses (A/Yunnan/14563/2015; A/Yunnan/0127/2015), and two H9N2 viruses (A/Hong Kong/308/2014; A/Hunan/44558/2015). The distribution of the human-origin AIVs suggests that three molecular patterns of avian-to-human interspecies transmission emerge in nature.

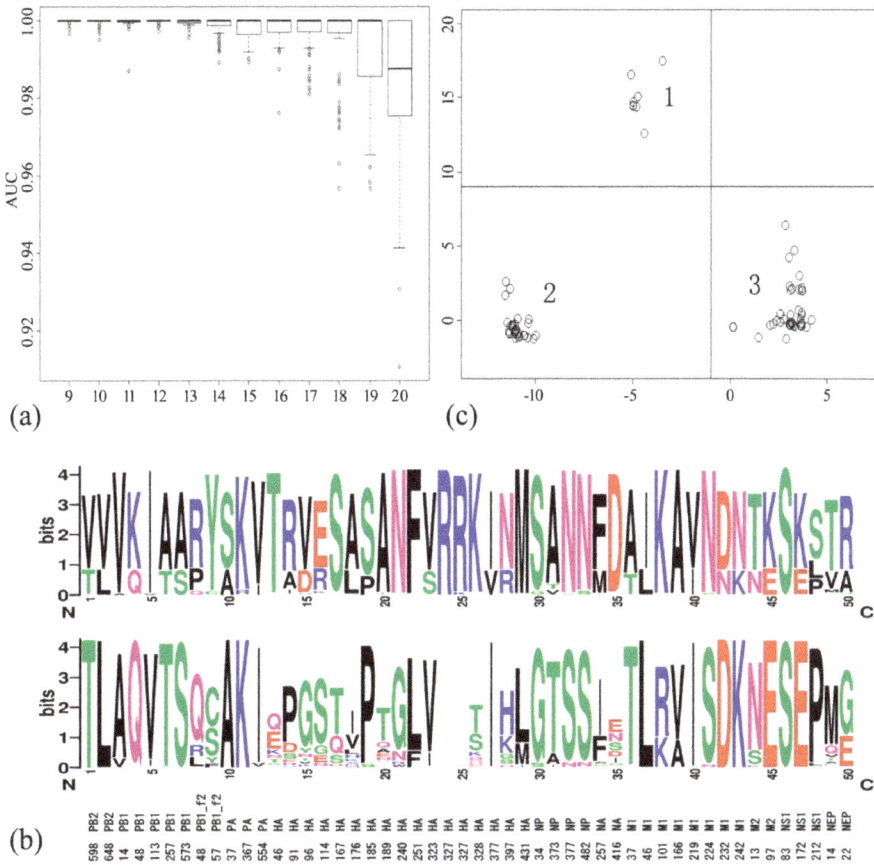

Figure 3. Minimal amino acid set for predicting AIVs. (**a**) Contributions of reduced mutation position sets. The *x* and *y* coordinates denote the cutoff (range 9–20) and the AUC values (range 0–1), respectively. (**b**) Profiles of 50 signature positions from human-origin (top) and avian-origin (bottom) AIVs. (**c**) Three patterns of human-origin AIVs clustered by the multidimensional scaling (MDS) method.

3. Discussion

Avian influenza viruses can cross the species barrier, potentially causing a human pandemic. In this paper, human AIV transmission was predicted by a machine learning model with excellent performance (namely, SVM). We firstly screened 183 mutation positions in 11 viral proteins after ranking them by random forest (RF). Most of the screened amino acid positions locate in the important functional regions of receptor binding, fusion peptides, intracellular transport, protein active sites, or virus assembly [26–37]. Some of the residues at these positions have been related to interspecies transmission in earlier reports, such as HA102–HA290 [26,27], H163 [28], HA91, HA96, HA328, HA377 and HA397 [29], HA327 [30], NA52 [31], and PB2 627 [33]. The signature positions guarantee the accuracy of the classifier and are biologically meaningful, which will benefit epidemic surveillance and further studies on interspecies AIV transmission. The proposed method provides important clues for future surveillance and is a useful pre-screening tool for phenotype screening in high-level biological safety laboratories.

The signature positions related with the phenotype of interspecies transmission were screened by the method of random forest. Some yielded a modest score (PB2 627, for example). PB2 E627K was firstly identified in a mouse model [33] and found in the protein of other human-origin avian influenza viruses [12]. In part of the PB2 protein of the human seasonal influenza virus from the public database, PB2 E627 still existed. It is possible that the mutation PB2 E627K is not a strong marker for interspecies transmission, which is consistent with our results. In the future, we need to update the model with new molecular evidence in the field of virology and with more powerful technology in the field of machine learning.

Amino acid mutations in the HA protein are essential for AIV transmission in mammals [21], but mutations in other viral proteins are also necessary [19,20]. Mutations of different proteins introduce synergy and nonlinearity in interspecies transmission. This concept was supported by the present study. Specifically, the linear classifier (the KNN model) showed poor predictive performance on the initial set of 183 signature positions. Moreover, the minimal signature position set was 50 amino acid long and distributed among different viral proteins. This synergistic effect should be notable in further study.

The molecular characteristics of AA Factor 3 are related to molecular size or volume with high factor coefficients of bulkiness, residue volume, average volume of buried residues, side chain volume, and molecular weight [22]. Molecular size or volume is strong related with the binding of biology molecules, such as viral surface protein, host receptor, enzyme, and substrate. In this paper, the AA Factor 3 makes an important contribution to the prediction in terms of high accuracy, which agrees with previous results concerning the receptor binding of viral surface protein [26–28], enzyme activity of viral neuraminidase [32], and RNA binding of viral polymerase [34]. The slightly poor performance of other factors may suggest that host receptor binding, virus partial release triggered by viral neuraminidase, and viral polymerase activity play key roles for the interspecies transmission of avian influenza virus.

The patterns of human-origin AIVs were clarified by the MDS method. Cluster 1 was composed of one H5N1 virus from 1997; three H9N2 viruses from 1999, 2000, and 2011; two H7N3 viruses from 2004 and 2012; and two H7N7 viruses from 2003 and 2013. Cluster 2 contained only H5N1 viruses isolated from 2003 to 2015. Cluster 3 contained H7N9 viruses, two H5N6 viruses from 2015, and two H9N2 viruses from 2014 and 2015. The distribution of human-origin AIVs implies that three molecular patterns of avian-to-human interspecies transmission have emerged. Further investigations on the appearance of novel patterns should be undertaken in future.

The proposed method is applicable to other infectious pathogens that can cross species barriers. As deep learning technology develops, powerful methods that omit feature selection and complex computations might emerge. To better understand the interspecies transmission mechanism of AIVs, the prediction model could be supplemented with information on the host's genetic background [38].

4. Materials and Methods

4.1. Dataset

The avian- and human-origin AIVs were collected from the EpiFlu public database (http://platform.gisaid.org/epi3/frontend) and processed using multiple public bioinformatics tools and algorithms (Figure 4). The details of each procedure are described below.

Step 1: In total, 6305 avian-origin and 644 human-origin AIV strains were obtained from the public influenza virus database. The strains were isolated between January 1996 and February 2016. GISAID deposits high-quality genomic sequences along with their clinical information.

Step 2: Our prediction classifiers were based on eleven viral proteins (PB2, PB1, PBI-F2, PA, HA, NP, NA, M1, M2, NS1, and NEP) with reported roles in interspecies transmission. AIV strains lacking any of these 11 protein sequences in the GISAID database, and strains without subtype information, were excluded in this step.

Molecules **2018**, 23, 1584

Step 3: The amino acid residues in the 11 proteins were numbered by the multiple sequence alignment tool MUSCLE [39], using the seasonal human H3 subtype virus as the reference. This step eliminated strains lacking more than 3 amino acids at any protein terminal. The missing residues were replaced by the corresponding residues in the protein sequence with highest identity.

Step 4: To reduce redundancy in the dataset, the AIV strains should differ by at least one amino acid. The amino acid sequences were compared using the CD-Hit tool [40].

Step 5: If the genome sequences of the avian-origin and human-origin AIV strains share high identity, the interspecies transmission capabilities of the avian-origin strains are ambiguous. Therefore, this step eliminated avian-origin strains in which any nucleotide sequence of the eight genome segments shared > 97% identity with that of the human-origin strains. The elimination was performed by the BLAST + tool [41].

Step 6: Ambiguous amino acid residues such as 'X' and 'B' were replaced by the corresponding residues in the protein sequence with highest identity.

The final dataset for predicting AIV interspecies transmission contained 429 positive samples (human-origin AIVs) and 440 negative samples (avian-origin AIVs). All of these strains are listed in Table S1.

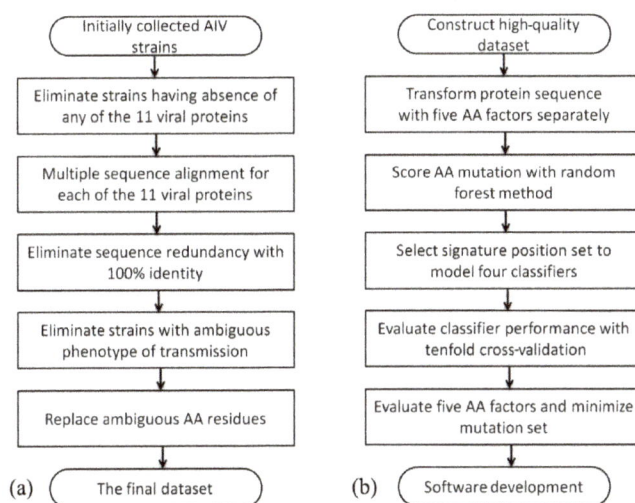

Figure 4. Flowchart of methods used in this paper. (**a**) High-quality dataset construction; (**b**) Machine learning algorism.

4.2. Recognition of Signature Positions

The random forest method is very popularly used for feature selection of prediction problems and can rank the importance of the features in a large scale to discriminate the different categories. The signature positions in the 11 viral proteins were recognized by the RF method (RF, https://cran.r-project.org/web/packages/randomForest/index.html). In each strain, the 11 proteins were concentrated in the following order: PB2 > PB1 > PB1-F2 > PA > HA > NP > NA > M1 > M2 > NS1 > NEP. The proteins with the length of 4620 amino acids were then transformed into numerical sequences of the amino acid factor. Any deletions or insertions in the protein were replaced by zeros. The strains were processed sequentially and accumulated into the total dataset, which was input to the RF. The positive samples (human-origin AIVs) and negative samples (avian-origin AIVs) were classified by their importance scores at each amino acid position. As the classification was based on five factors, the final importance score at each position was the sum of five calculations. Therefore, highly scoring positions were important for distinguishing positive and negative samples. These high scorers were regarded as important amino acid mutations in the interspecies transmission of AIVs. Breiman's random forest algorithm was used as default.

4.3. Constructing the Classifier Model

The machine learning method can solve the classification problem and the numeric features of the positive and negative samples are essential for classification. After screening the signature positions as mentioned above, each strain was represented by an amino acid residue set. These amino acid sets were again transformed into numerical sequences of the five AA factors. Each strain was represented as a numeric vector of length 5N, where N is the number of amino acids in an amino residue set. The interspecies AIV transmission was then predicted by four popular machine learning models that are widely used in bioinformatics and computational biology: (1) support vector machine (SVM, https://cran.r-project.org/web/packages/e1071/index.html), (2) random forest (RF, https://cran.r-project.org/web/packages/randomForest/index.html), (3) naïve Bayes (NB, https://cran.r-project.org/web/packages/e1071/index.html), and (4) *k*-nearest neighbor (KNN, https://cran.r-project.org/web/packages/class/index.html). The present prediction task is a two-class classification problem (in which human-origin and avian-origin AIVs are classified as positive and negative, respectively). The four classifiers were implemented in the R environment and related packages.

The SVM classifier performs the classification in a high-dimensional feature space, which was transferred from the input feature vector with the kernel function. If the samples from two categories were partly overlapped in the original feature space, the SVM will have good performance. In this paper, the optimal hyperplane is determined with the regularization parameter C (C = 1) and the radial basis function (RBF) as default. The RF classifier is an ensemble of many decision trees. Each tree is fully grown using part of the samples in the training dataset selected with the bootstrap technique. The NB is constructed based on the Bayes theorem. Both RF and NB were implemented with the default parameter in the package. The KNN classifier is a nonparametric method to determine a sample category by a majority vote of its neighbors; the number of neighbors in this paper was set to be 3 (k = 3).

4.4. Evaluating the Performance of Different Classifiers

The four classifiers were trained on 387 positive samples and 396 negative samples randomly selected from the AIV dataset. The remaining 10% of samples (42 positive and 44 negative samples) were reserved as an independent test dataset for assessing the performances of the classifiers. The classifier performances were evaluated by 10-fold cross-validation and the receiver operating characteristic (ROC) curve. The area under the ROC curve (AUC) reveals the optimal parameters in the four classifiers. To compare the classifier performances, we repeated the evaluation process 100 times and plotted the distributions of the resulting AUC values. The ROC curve relates the true and false positive rates, where both rates range from 0 to 1. The AUC was calculated by the 'ROCR' package in R (https://cran.r-project.org/web/packages/ROCR/index.html). As both rates range from 0 to 1, AUC also ranges from 0 to 1. A higher AUC value denotes a higher performance of the classifier. The human-origin AIVs were shown by the multidimensional scaling method in R (MDS, https://cran.r-project.org/web/packages/MASS/index.html) and the amino acid profile was drawn by the WebLogo server (http://weblogo.berkeley.edu/logo.cgi).

4.5. Prediction Software

By integrating the features at the signature positions with the best-performing classifier, we constructed a software program for predicting avian-to-human transmission of AIVs (delivery by request).

Supplementary Materials: Table S1: AIV Strains in the final dataset, Table S2: Human-origin AIVs clustered by the MDS method.

Author Contributions: Conceptualization, X.Q. and Z.K.; Methodology, X.Q. and G.F.; Software, Y.W.; Writing-review and editing, Z.K.

Funding: This research was funded by the Chinese National Natural Science Foundation (61379059, 61472372, 61632002).

Acknowledgments: We would like to acknowledge the originating and submitting laboratories of the viral sequences from GISAID's EpiFlu Database. We thank Leonie Pipe for manuscript editing.

Conflicts of Interest: The authors declare no conflict of interest.

References

1. Webster, R.G.; Bean, W.J.; Gorman, O.T.; Chambers, T.M.; Kawaoka, Y. Evolution and ecology of influenza A viruses. *Microbiol. Rev.* **1992**, *56*, 152–179. [CrossRef] [PubMed]
2. Xu, X.; Subbarao, K.; Cox, N.J.; Guo, Y. Genetic characterization of the pathogenic influenza A/Goose/Guangdong/1/96 (H5N1) virus: Similarity of its hemagglutinin gene to those of H5N1 viruses from the 1997 outbreaks in Hong Kong. *Virology* **1999**, *261*, 15–19. [CrossRef] [PubMed]
3. Claas, E.C.J.; Osterhaus, A.D.; van Beek, R.; De Jong, J.C.; Rimmelzwaan, G.F.; Senne, D.A.; Krauss, S.; Shortridge, K.F.; Webster, R.G. Human influenza A H5N1 virus related to a highly pathogenic avian influenza virus. *Lancet* **1998**, *351*, 472–477. [CrossRef]
4. Subbarao, K.; Klimov, A.; Katz, J.; Regnery, H.; Lim, W.; Hall, H.; Perdue, M.; Swayne, D.; Bender, C.; Huang, J.; et al. Characterization of an avian influenza A (H5N1) virus isolated from a child with a fatal respiratory illness. *Science* **1998**, *279*, 393–396. [CrossRef] [PubMed]
5. Chen, H.; Smith, G.J.; Li, K.S.; Wang, J.; Fan, X.H.; Rayner, J.M.; Vijaykrishna, D.; Zhang, J.X.; Zhang, L.J.; Guo, G.T.; et al. Establishment of multiple sublineages of H5N1 influenza virus in Asia: Implications for pandemic control. *Proc. Natl. Acad. Sci. USA* **2006**, *103*, 2845–2850. [CrossRef] [PubMed]
6. Li, K.S.; Guan, Y.; Wang, J.; Smith, G.J.; Xu, K.M.; Duan, L.; Rahardjo, A.P.; Puthavathana, P.; Buranathai, C.; Nguyen, T.D.; et al. Genesis of a highly pathogenic and potentially pandemic H5N1 influenza virus in eastern Asia. *Nature* **2004**, *430*, 209–213. [CrossRef] [PubMed]
7. Zhu, Q.Y.; Qin, E.D.; Wang, W.; Yu, J.; Liu, B.H.; Hu, Y.; Hu, J.F.; Cao, W.C. Fatal infection with influenza A (H5N1) virus in China. *N. Engl. J. Med.* **2006**, *354*, 2731–2732. [CrossRef] [PubMed]
8. Shu, Y.L.; Yu, H.J.; Li, D.X. Lethal avian influenza A (H5N1) infection in a pregnant woman in Anhui province, China. *N. Engl. J. Med.* **2006**, *354*, 1421–1422. [CrossRef] [PubMed]
9. Peiris, M.; Yuen, K.Y.; Leung, C.W.; Chan, K.H.; Ip, P.L.; Lai, R.W.; Orr, W.K.; Shortridge, K.F. Human infection with influenza H9N2. *Lancet* **1999**, *354*, 916–917. [CrossRef]
10. Butt, K.M.; Smith, G.J.; Chen, H.; Zhang, L.J.; Leung, Y.H.; Xu, K.M.; Lim, W.; Webster, R.G.; Yuen, K.Y.; Malik Peiris, J.S.; et al. Human infection with an avian H9N2 influenza A virus in Hong Kong in 2003. *J. Clin. Microbiol.* **2005**, *43*, 5760–5767. [CrossRef] [PubMed]
11. Fouchier, R.A.; Schneeberger, P.M.; Rozendaal, F.W.; Broekman, J.M.; Kemink, S.A.; Munster, V.; Kuiken, T.; Rimmelzwaan, G.F.; Schutten, M.; van Doornum, G.J.J.; et al. Avian influenza A virus (H7N7) associated with human conjunctivitis and a fatal case of acute respiratory distress syndrome. *Proc. Natl. Acad. Sci. USA* **2004**, *101*, 1356–1361. [CrossRef] [PubMed]
12. Gao, R.; Cao, B.; Hu, Y.; Feng, Z.; Wang, D.; Hu, W.; Chen, J.; Jie, Z.; Qiu, H.; Xu, K.; et al. Human infection with a novel avian-origin influenza A (H7N9) virus. *N. Engl. J. Med.* **2013**, *368*, 1888–1897. [CrossRef] [PubMed]
13. Cao, H.F.; Liang, Z.H.; Feng, Y.; Zhang, Z.N.; Xu, J.; He, H. A confirmed severe case of human infection with avian-origin influenza H7N9: A case report. *Exp. Ther. Med.* **2015**, *9*, 693–696. [CrossRef] [PubMed]
14. Zeng, X.; Liu, L.; Lv, L.; Zou, Q. Prediction of potential disease-associated microRNAs using structural perturbation method. *Bioinformatics* **2018**. [CrossRef] [PubMed]
15. Zeng, X.; Zhang, X.; Zou, Q. Integrative approaches for predicting microRNA function and prioritizing disease-related microRNA using biological interaction networks. *Brief. Bioinform.* **2016**, *17*, 193–203. [CrossRef] [PubMed]

16. Gang, F.; Zhang, S.; Dong, Y. Optimizing DNA assembly based on statistical language modelling. *Nucleic Acids Res.* **2017**, *45*, e182. [CrossRef]

17. Zeng, X.; Liao, Y.; Liu, Y.; Zou, Q. Prediction and validation of disease genes using HeteSim Scores. *IEEE/ACM Trans. Comput. Biol. Bioinform.* **2017**, *14*, 687–695. [CrossRef] [PubMed]

18. Zeng, X.; Lin, W.; Guo, M.; Zou, Q. A comprehensive overview and evaluation of circular RNA detection tools. *PLoS Comput. Biol.* **2017**, *13*, e1005420. [CrossRef] [PubMed]

19. Herfst, S.; Schrauwen, E.J.; Linster, M.; Chutinimitkul, S.; de Wit, E.; Munster, V.J.; Sorrell, E.M.; Bestebroer, T.M.; Burke, D.F.; Smith, D.J.; et al. Airborne transmission of influenza A/H5N1 virus between ferrets. *Science* **2012**, *336*, 1534–1541. [CrossRef] [PubMed]

20. Imai, M.; Watanabe, T.; Hatta, M.; Das, S.C.; Ozawa, M.; Shinya, K.; Zhong, G.; Hanson, A.; Katsura, J.; Watanabe, S.; et al. Experimental adaptation of an influenza H5 HA confers respiratory droplet transmission to a reassortant H5 HA/H1N1 virus in ferrets. *Nature* **2012**, *486*, 420–428. [CrossRef] [PubMed]

21. Glaser, L.; Stevens, J.; Zamarin, D.; Wilson, I.A.; García-Sastre, A.; Tumpey, T.M.; Basler, C.F.; Taubenberger, J.K.; Palese, P. A single amino acid substitution in 1918 influenza virus hemagglutinin changes receptor binding specificity. *J. Virol.* **2005**, *79*, 11533–11536. [CrossRef] [PubMed]

22. Atchley, W.R.; Zhao, J.; Fernandes, A.D.; Drüke, T. Solving the protein sequence metric problem. *Proc. Natl. Acad. Sci. USA* **2005**, *102*, 6395–6400. [CrossRef] [PubMed]

23. Kou, Z.; Lei, F.; Wang, S.; Zhou, Y.; Li, T. Molecular patterns of avian influenza A viruses. *Chin. Sci. Bull.* **2008**, *53*, 2002–2007. [CrossRef]

24. Qiang, X.; Kou, Z. Prediction of interspecies transmission for avian influenza A virus based on a back-propagation neural network. *Math. Comput. Model.* **2010**, *52*, 2060–2065. [CrossRef]

25. Wang, J.; Kou, Z.; Duan, M.; Ma, C.; Zhou, Y. Using amino acid factor scores to predict avian-to-human transmission of avian influenza viruses: A machine learning study. *Protein Pept. Lett.* **2013**, *20*, 1115–1121. [CrossRef] [PubMed]

26. Stevens, J.; Corper, A.L.; Basler, C.F.; Taubenberger, J.K.; Palese, P.; Wilson, I.A. Structure of the uncleaved human H1 hemagglutinin from the extinct 1918 influenza virus. *Science* **2004**, *303*, 1866–1870. [CrossRef] [PubMed]

27. Hulse, D.J.; Webster, R.G.; Russell, R.J.; Perez, D.R. Molecular determinants within the surface proteins involved in the pathogenicity of H5N1 influenza viruses in chickens. *J. Virol.* **2004**, *78*, 9954–9964. [CrossRef] [PubMed]

28. Mishin, V.P.; Novikov, D.; Hayden, F.G.; Gubareva, L.V. Effect of hemagglutinin glycosylation on influenza virus susceptibility to neuraminidase inhibitors. *J. Virol.* **2005**, *79*, 12416–12424. [CrossRef] [PubMed]

29. Chen, J.; Skehel, J.J.; Wiley, D.C. N- and C-terminal residues combine in the fusion-pH influenza hemagglutinin HA (2) subunit to form an N cap that terminates the triple-stranded coiled coil. *Proc. Natl. Acad. Sci. USA* **1999**, *96*, 8967–8972. [CrossRef] [PubMed]

30. Schrauwen, E.J.A.; de Graaf, M.; Herfst, S.; Rimmelzwaan, G.F.; Osterhaus, A.D.M.E.; Fouchier, R.A.M. Determinants of virulence of influenza A virus. *Eur. J. Clin. Microbiol. Infect. Dis.* **2014**, *33*, 479–490. [CrossRef] [PubMed]

31. Zhou, H.; Yu, Z.; Hu, Y.; Tu, J.; Zou, W.; Peng, Y.; Zhu, J.; Li, Y.; Zhang, A.; Yu, Z.; et al. The special neuraminidase stalk-motif responsible for increased virulence and pathogenesis of H5N1 influenza A virus. *PLoS ONE* **2009**, *4*, e6277. [CrossRef] [PubMed]

32. Barman, S.; Adhikary, L.; Chakrabarti, A.K.; Bernas, C.; Kawaoka, Y.; Nayak, D.P. Role of transmembrane domain and cytoplasmic tail amino acid sequences of influenza a virus neuraminidase in raft association and virus budding. *J. Virol.* **2004**, *78*, 5258–5269. [CrossRef] [PubMed]

33. Hatta, M.; Gao, P.; Halfmann, P.; Kawaoka, Y. Molecular basis for high virulence of Hong Kong H5N1 influenza A viruses. *Science* **2001**, *293*, 1840–1842. [CrossRef] [PubMed]

34. Perez, D.R.; Donis, R.O. Functional analysis of PA binding by influenza a virus PB1: Effects on polymerase activity and viral infectivity. *J. Virol.* **2001**, *75*, 8127–8136. [CrossRef] [PubMed]

35. Iwatsuki-Horimoto, K.; Horimoto, T.; Noda, T.; Kiso, M.; Maeda, J.; Watanabe, S.; Muramoto, Y.; Fujii, K.; Kawaoka, Y. The cytoplasmic tail of the influenza A virus M2 protein plays a role in viral assembly. *J. Virol.* **2006**, *80*, 5233–5240. [CrossRef] [PubMed]

36. Bullido, R.; Gomez-Puertas, P.; Albo, C.; Portela, A. Several protein regions contribute to determine the nuclear and cytoplasmic localization of the influenza A virus nucleoprotein. *J. Gen. Virol.* **2000**, *81*, 135–142. [CrossRef] [PubMed]

37. Iwatsuki-Horimoto, K.; Horimoto, T.; Fujii, Y.; Kawaoka, Y. Generation of influenza A virus NS2 (NEP) mutants with an altered nuclear export signal sequence. *J. Virol.* **2004**, *78*, 10149–10155. [CrossRef] [PubMed]

38. Srivastava, B.; Błazejewska, P.; Heßmann, M.; Bruder, D.; Geffers, R.; Susanne, M.; Gruber, A.D.; Schughart, K. Host genetic background strongly influences the response to influenza A virus infections. *PLoS ONE* **2009**, *4*, e4857. [CrossRef] [PubMed]

39. Edgar, R.C. MUSCLE: multiple sequence alignment with high accuracy and high throughput. *Nucleic Acids Res.* **2004**, *32*, 1792–1797. [CrossRef] [PubMed]

40. Li, W.; Godzik, A. Cd-hit: A fast program for clustering and comparing large sets of protein or nucleotide sequences. *Bioinformatics* **2006**, *22*, 1658–1659. [CrossRef] [PubMed]

41. Altschul, S.; Gish, W.; Miller, W.; Myers, E.; Lipman, D. Basic local alignment search tool. *J. Mol. Biol.* **1990**, *215*, 403–410. [CrossRef]

Sample Availability: Samples of the compounds are not available from the authors.

molecules

Article

Detection of Protein Complexes Based on Penalized Matrix Decomposition in a Sparse Protein–Protein Interaction Network

Buwen Cao [1,2,*], Shuguang Deng [1,*], Hua Qin [1], Pingjian Ding [2], Shaopeng Chen [3] and Guanghui Li [2,4]

[1] College of Information and Electronic Engineering, Hunan City University, Yiyang 413000, China; qinhua_hcu@163.com

[2] College of Computer Science and Electronic Engineering, Hunan University, Changsha 410082, China; dpj@hnu.edu.cn (P.D.); ghli16@163.com (G.L.)

[3] College of Mathematics and Computer Science, Hunan Normal University, Changsha 410081, China; chenshaopeng2010@gmail.com

[4] School of Information Engineering, East China Jiaotong University, Nanchang 330013, China

* Correspondence: cbwchj@126.com (B.C.); shuguangdeng@163.com (S.D.);
Tel.: +86-0737-6353-128 (B.C. & S.D.)

Academic Editors: Xiangxiang Zeng, Alfonso Rodríguez-Patón and Quan Zou
Received: 21 May 2018; Accepted: 12 June 2018; Published: 15 June 2018

Abstract: High-throughput technology has generated large-scale protein interaction data, which is crucial in our understanding of biological organisms. Many complex identification algorithms have been developed to determine protein complexes. However, these methods are only suitable for dense protein interaction networks, because their capabilities decrease rapidly when applied to sparse protein–protein interaction (PPI) networks. In this study, based on penalized matrix decomposition (PMD), a novel method of penalized matrix decomposition for the identification of protein complexes (i.e., PMD_{pc}) was developed to detect protein complexes in the human protein interaction network. This method mainly consists of three steps. First, the adjacent matrix of the protein interaction network is normalized. Second, the normalized matrix is decomposed into three factor matrices. The PMD_{pc} method can detect protein complexes in sparse PPI networks by imposing appropriate constraints on factor matrices. Finally, the results of our method are compared with those of other methods in human PPI network. Experimental results show that our method can not only outperform classical algorithms, such as CFinder, ClusterONE, RRW, HC-PIN, and PCE-FR, but can also achieve an ideal overall performance in terms of a composite score consisting of F-measure, accuracy (ACC), and the maximum matching ratio (MMR).

Keywords: protein–protein interaction (PPI); clustering; protein complex; penalized matrix decomposition

1. Introduction

The identification of protein complexes is highly beneficial for the investigation of all kinds of organisms to understand biological processes and determine inherent organizational structures within cells [1]. The dramatic development of computational methods stimulates many protein complex identification algorithms for protein–protein interaction (PPI) networks, which are generally organized into three catalogs. The first catalog includes clustering methods that are also divided into three sub-catalogs. First, the local search approaches based on density are used to identify densely connected subgraphs in PPI networks, in which subgraphs with density above a pre-defined threshold, such as MCODE (Molecular Complex Detection) [2], CFinder (a software tool for network cluster detection) [3], DPCLus (a Density-Periphery based graph CLustering software) [4], and ICPM (Iterative

Clique Percolation Method) [5], are considered protein complexes. However, these approaches tend to neglect surrounding proteins that are connected to the kernel clusters with sparse links, which can show experimentally validated true interactions [6]. Another kind of method for detecting protein complexes uses classical hierarchy clustering techniques, which mainly depend on the distance between proteins to detect meaningful groups [6] and contain HC-PIN ((fast Hierarchical Clustering algorithm for Protein Interaction Network, agglomerative method) [7] and G-N algorithms (divisive method) [8]. Many hierarchical clustering methods employ similarities among the proteins that are calculated on the basis of network topology characteristics or biological meaning due to the further development of clustering technology. Such approaches mainly include NEMO (NEtwork MOdule identification) [9], ClusterONE (Clustering algorithm with Overlapping Neighborhood Expansion) [10], RFC (Rough Fuzzy Clustering) [11], MINE (Module Identification in Networks) [12], PageRankNibble [13], SPICi (Speed and Performance In Clustering,) [14], PCE-FR (Pseudo-Clique Extension based on Fuzzy Relation) [15], MTGO (Module detection via Topological information and GO knowledge) [16], WCOACH (Weighted COACH) [17], DCAFP (Density-based Clustering Approach for identifying overlapping protein complexes with Functional Preferences) [18], and cwMINE (Combined Weight of Module Identification in Networks) [19]. Experimental results show that these novel methods greatly outperform classical hierarchical clustering approaches. Except for the aforementioned clustering approaches, many other protein complex detection algorithms, such as RNSC (Restricted Neighborhood Search Clustering) [20], MCL [1], RRW (Repeated Random Walks algorithm) [21], CMC (Clustering-based on Maximal Cliques) [22], Coach [23], and AP (Affinity Propagation) with its variant [24] have achieved satisfactory results.

Another type of method used to detect protein complexes employs an intelligent optimization algorithm, which seeks the optimal solution of PPI based on a heuristic concept [25]. For large databases, the complexity of intelligent optimization algorithms is too high to run a correct consequence. The major weakness of the aforementioned methods is that their performance deteriorates when they are employed to sparse PPI networks [19,26]. To address this problem, matrix decomposition is proposed to improve the disadvantages of these methods. A co-clustering algorithm based on the adjacent matrix of PPI networks was proposed [6] and obtained overlapping and non-overlapping protein complexes successfully. The results show that the method reached a remarkable balance between network coverage and accuracy (ACC) and outperformed classical methods. Matrix factorization can be mainly organized into two main levels. The first level is the non-negative matrix factorization (NMF) (which integrates gene ontology (GO), gene expression data, and the PPI network to form the corresponding adjacency matrix and then decomposes it with common factors to achieve the overlapping functional modules with high ACC [27]). Zhang et al. [28] proposed sparse network-regularized multiple NMFs (SNMNMFs) to identify the microRNA regulatory modules and demonstrated the ideal performance of the proposed method in ovarian cancer dataset. The second level is the penalized matrix decomposition (PMD), which is widely applied in various datasets, such as microarray data [29], including gene expression data, and proteomic datasets [30].

Inspired by Ref. [24], PMD_{pc}, an approach used to identify the protein interaction network of protein complexes was originally proposed. First, the adjacent matrix of the protein interaction network was normalized. Second, the normalized matrix was decomposed into three factor matrices. Finally, the PMD_{pc} algorithm and several classical algorithms were executed from the well-investigated human PPI network. The experimental results show that our approach achieved satisfactory performance in terms of F-measure, ACC, and maximum matching ratio (MMR).

2. Results and Discussion

When PMD_{pc} is applied to identify the protein complexes in PPI network, the parameters of c_1, c_2, and k are crucial for the decomposition of the network. Considering that u should be sparse, we take $c_1 = 0.25 \times \sqrt{n}$ and $c_2 = 0.25 \times \sqrt{p}$ [31].

To study the parameter of k on the effect on the experimental results, we repeated the execution of algorithm and studied how the algorithm behaves in terms of F-measure and let $k \in (0, 2500]$ with a 100 increment. The detailed experimental results with different k values are presented in Figure 1. From Figure 1, we can clearly see that k is less than 1000; the experimental results fall short of satisfaction.

The value of the F-measure increases gradually until $k = 1600$ with the increase in k, such that the maximum value of 0.398, the F-measure, displays a steady state when it changes from 1600 to 2000. When k is greater than 2000, the value of F-measure shows a downward trend. Therefore, k is set to 2000.

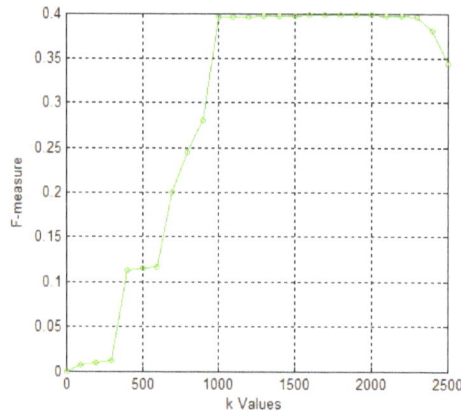

Figure 1. Values of F-measure for different values of $k \in (0, 2500]$ with a 100 increment in HPRD dataset.

Five classical protein complex algorithms, namely, CFinder [3], ClusterONE [10], RRW [21], HC-PIN [7], and PCE-FR [15], are applied on human PPI network of HPRD (Human Protein Reference Database, HPRD) to demonstrate the performance of PMD_{pc}. The complexes of the aforementioned algorithms with sizes less than 2 are filtered in our work. Moreover, the parameters of each method that is compared with our method are set using the default values recommended by the authors. The experimental result is shown in Table 1.

Table 1. Results of six protein complexes Algorithms in HPRD Dataset.

Algorithms	Number	Precision	Recall	F-Measure	ACC	Sep	MMR	MCC
CFinder	49	0.959	0.143	0.249	0.184	0.165	0.017	0.327
ClusterONE	755	0.295	0.186	0.229	0.333	0.209	0.084	0.391
RRW	167	0.671	0.190	0.296	0.236	0.231	0.034	0.209
HC-PIN	99	0.646	0.140	0.230	0.256	0.233	0.024	0.196
PCE-FR	274	0.534	0.178	0.267	0.279	0.169	0.029	0.035
PMD_{pc}	118	0.451	0.356	0.398	0.362	0.777	0.010	0.343

Table 1 shows that PMD_{pc} achieves a satisfactory performance on human PPI networks. Particularly, PMD_{pc} obtains the highest value of recall, F-measure, ACC, and Sep, which are 0.356, 0.398, 0.362, and 0.777, respectively. These results are significantly superior to the five other algorithms. Furthermore, CFinder achieves the highest precision of 0.959 and the lowest MMR of 0.017. ClusterONE identifies 755 protein complexes and achieves the highest MMR of 0.084. These values elaborate that our approach achieved an ideal result in identifying protein complexes from sparse PPI networks.

From Table 1, we can also clearly see that our method obtains the second highest value of MCC, which is 12.28% lower than that of ClusterONE. It demonstrates that our method achieved satisfactory performance in dealing imbalanced data.

To void the advantage of some evaluation metric, the composite score [24] is employed to wrap up the global performance. Interestingly, the composite comparison of our method shows absolute advantage in terms of F-measure, accuracy, and maximum matching ratio. Figure 2 presents the comparison results of the six algorithms on the HPRD dataset. The composite score of F-measure, accuracy, and maximum matching ratio is 0.770, which is 19.20% higher than the highest value of the five other methods. It further demonstrates the effectiveness of our method.

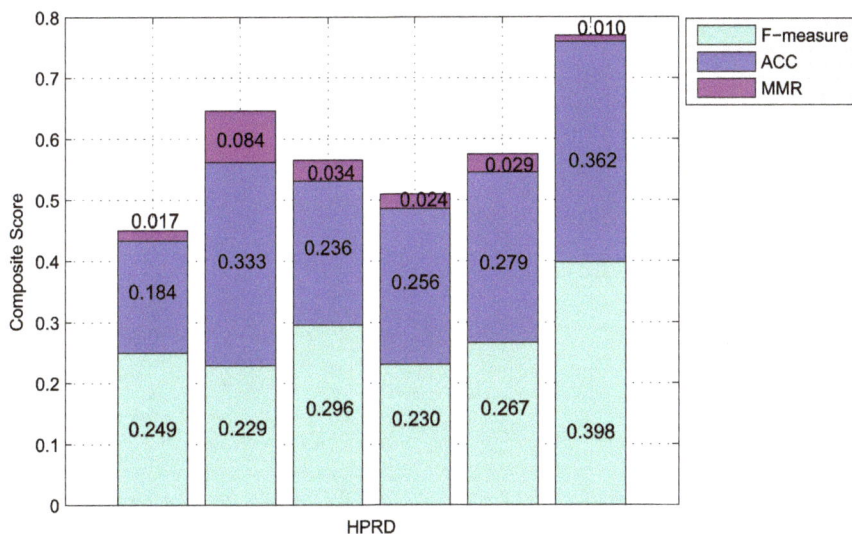

Figure 2. Results comparison of the six algorithms in HPRD dataset using CHPC2012 gold standard dataset. Columns correspond to the following algorithms, CFinder, ClusterONE, HC-PIN, PCE-FR, and PMD_{pc} from left to right. Various color of the same columns denotes the individual components of the composite score of the algorithm (cyan = F-measure, blue = ACC, and purple = MMR). The total height of each column is the value of the composite score for a special algorithm in a special dataset. Large score shows the clustering result is better.

3. Materials and Methods

3.1. Materials and Datasets

Our method is applied to detect the protein complexes in the human PPI dataset downloaded from Ref. [24], in which 9459 proteins and 36,935 interactions with the density of 0.0008 are included. The gold standard dataset is employed to evaluate the performance of the protein complexes identified in sparse PPI networks, which is CHPC2012 [32], integrating three databases, namely, CORUM [33], HPRD [34], and PINdb [35], and includes 1389 complexes and 3065 proteins.

3.2. Methods

Consider a sample dataset that consists of p eigenvectors in n samples, which is described by a matrix X with size $n \times p$ [30]. Without loss of generality, we assume that the means of column and row X are zero. The singular value decomposition of matrix X can be written as follows:

$$X = U\Delta V^T, U^T U = I_n, V^T V = I_p \tag{1}$$

The decomposition of sparse matrix is executed by imposing additional constraints on U and V. The single-factor PMD can be optimized using the following objective function, which is formulated as [30]

$$\operatorname*{argmin}_{\delta,u,v} \tfrac{1}{2}||\eta - \delta uv^T||_F^2,$$
$$s.t. ||u||_2^2 = 1, ||v||_2^2 = 1, \tag{2}$$
$$P_1(u) \le c_1, P_2(v) \le c_2, \delta \ge 0.$$

in which u is a column of U, v is a column of V, δ is a diagonal element of the matrix of η, $||\bullet||_F$ is the Frobenius norm, and P_1 and P_2 are penalty functions that have variety of forms [30].

Let U and V be $n \times R$ and $p \times R$ orthogonal matrices, respectively, and Δ a diagonal matrix with diagonal elements δ_r [30]

$$\frac{1}{2}||\eta - U\Delta V^T||_F^2 = \frac{1}{2}||\eta||_F^2 - \sum_{r=1}^{R} u_r^T \eta v_r \delta_r + \frac{1}{2}\sum_{r=1}^{R} \delta_r^2 \tag{3}$$

Therefore, when $R = 1$, we can infer that u and v satisfy Equation (7) and the following condition:

$$\operatorname*{argmax}_{u,v} u^T \eta v$$
$$s.t. ||u||_2^2 = 1, ||v||_2^2 = 1, P_1(u) \le c_1, P_2(v) \le c_1 \tag{4}$$

Moreover, δ satisfies Equation (2) when $\delta = u^T \eta v$.

The optimization problem in Equation (4) can be applied to the following biconvex optimization [30]:

$$\operatorname*{argmax}_{u,v} u^T \delta v$$
$$s.t. ||u||_2^2 \le 1, ||v||_2^2 \le 1, P_1(u) \le c_1, P_2(v) \le c_2 \tag{5}$$

Equation (5) satisfies Equation (4) based on the appropriate value of c [30]. Equation (5) is called the single factor PMD, and the iterative algorithm used to optimize it is described in Algorithm 1:

Algorithm 1. Calculating the single factor of PMD.

Step1. Initialize v and let unit $L_2 - norm$.
Step2. Interate until convergence:

(i)　　$u \leftarrow \arg\max_u u^T \delta v, s.t. ||u||_2^2 \le 1, P_1(u) \le c_1$
(ii)　　$v \leftarrow \arg\max_v u^T \delta v, s.t. ||v||_2^2 \le 1, P_2(v) \le c_2$

Step3. $d \leftarrow u^T \delta v$

Equation (2) is computed repeatedly to obtain other PMD factors. The corresponding algorithm is described in Algorithm 2.

Algorithm 2. Calculating the k factor of PMD.

Step1. $\eta^1 \leftarrow \eta$;
Step2. For $r \in 1, 2, \ldots, R$

(i)　　The single factor PMD (Algorithm 1) is executed on the matrix of η^r, computing u_r, v_r, δ_r, respectively;
(ii)　　$\eta^{r+1} \leftarrow \eta^r - \delta_r u_r v_r^T$

The constraint is imposed on u and v with $L_1 - norm$, i.e., $||u||_1 \le c_1, ||v||_1 \le c_2$. By selecting parameters c_1 and c_2 appropriately, PMD can make factors u and v sparse. Generally, c_1 and c_2

should be restricted to ranges $1 \leq c_1 \leq \sqrt{n}$ and $1 \leq c_2 \leq \sqrt{p}$. Thus, the PMD method is shaped as $PMD(L_1, L_2)$, which is described as follows:

$$\underset{u,v}{\text{argmax}} u^T \eta v$$
$$s.t. \|u\|_2^2 \leq 1, \|v\|_2^2 \leq 1, \|u\|_1 \leq c_1, \|v\|_1 \leq c_2 \tag{6}$$

Let S denote the operator of the soft threshold, i.e., $S(a,c) = \text{sgn}(a)(|a| - c)_+$, in which $c > 0$, $x_+ = \begin{cases} x & x > 0 \\ 0 & x \leq 0 \end{cases}$. The corresponding theorem is as follows:

Theorem 1. *Considering the optimization problem*

$$\underset{u}{\text{argmax}} u^T a$$
$$s.t. \|u\|_2^2 \leq 1, \|u\|_1 \leq c. \tag{7}$$

The solution is $u = \frac{S(a,\Delta)}{\|S(a,\Delta)\|_2}$. If $\|u\|_1 \leq c$, then $\Delta = 0$; otherwise, $\|u\|_1 = c$ s.t. $\Delta > 0$. The detailed proof regarding the theorem can be found in Ref. [30]. The analysis shows the solution of Equation (6) with Algorithm 1. According to Theorem 1, the single factor PMD can be optimized, as shown in Algorithm 3:

Algorithm 3. The optimization process of the single factor PMD.

Step1. Initialize v and let unit $L_2 - \text{norm}$.
Step2. Iterate until convergence:

(i) $u \leftarrow \frac{S(Xv,\Delta_1)}{\|S(Xv,\Delta_1)\|_2}$, if $\|u\|_1 \leq c_1$, then $\Delta_1 = 0$, else $\|u\|_1 = c_1, s.t., \Delta_1 > 0$

(ii) $u \leftarrow \frac{S(X^T u,\Delta_2)}{\|S(X^T u,\Delta_2)\|_2}$, if $\|v\|_1 \leq c_2$, then $\Delta_2 = 0$, else $\|v\|_1 = c_2, s.t., \Delta_2 > 0$

Step3. $d \leftarrow u^T \delta v$

To obtain the sparse factors of u and v, we let $c_1 = c\sqrt{n}, c_2 = c\sqrt{p}$, and the values of Δ_1 and Δ_2 are selected by the binary search.

For comprehensive discussion, discovered protein complexes and gold standard dataset are matched. The following evaluation measures are employed in this study.

F-measure. Two protein complexes, namely, p and g, are generated from the predicted protein complex and gold standard sets, respectively. The overlapping score $os(p, g)$ quantizes the closeness between the sets and is defined as follows [24]:

$$os(p, g) = \frac{|C_p \cap C_g|}{|C_p| \bullet |C_g|} \tag{8}$$

in which C_p, C_g denote protein complex sets p and g, respectively. If $os(p, g) \geq \theta$, then the two complexes are matched, in which θ is the threshold. θ is set as 0.2, which is consistent with many experiments for protein complex identification [24]. Let P and G represent the detected protein complex and gold standard sets, respectively; N_{cp} describes the number of identified protein complexes that match at least one gold standard set, i.e., $N_{cp} = |\{p|p \in P, \exists g \in G, os(p, g) \geq \theta\}|$; and N_{cp} presents the number of gold standard protein complexes that match at least one identified complex, that is $N_{cg} = |\{g|g \in G, \exists p \in P, os(p, g) \geq \theta\}|$. F-measure is mathematically defined as [24]

$$F - measure = \frac{2 \times Precision \times Recall}{Precision + Recall} \tag{9}$$

in which $Precision = N_{cp}/|P|$, $Recall = N_{cg}/|G|$. F-measure is defined as the harmonic mean of precision and recall, which can evaluate the overall performance of the detection methods.

ACC (Accuracy, ACC). ACC is used to quantify the quality of detected protein complexes, which is the geometric means of sensitivity and positive predictive value, PPV. The corresponding formulas are described as follows [24]:

$$ACC = \sqrt{S_n \times PPV} \tag{10}$$

in which $S_n = \frac{\sum_{i=1}^{n} \max_{j=1}^{m} t_{ij}}{\sum_{i=1}^{n} n_i}$, $PPV = \frac{\sum_{j=1}^{m} \max_{i=1}^{n} t_{ij}}{\sum_{j=1}^{m} \sum_{i=1}^{n} t_{ij}}$.

Sep (Separation, Sep). To void the case wherein proteins of a gold standard complex are matched with several identified protein complexes, Sep is used to measure the one-to-one correspondence between generated protein complexes and gold standard protein complexes. The formula is described as follows [24]:

$$Sep_g = \frac{\sum_{i=1}^{n} \sum_{j=1}^{m} Sep_{ij}}{n}, \; Sep_p = \frac{\sum_{j=1}^{m} \sum_{i=1}^{n} Sep_{ij}}{m}, \; Sep = \sqrt{Sep_g \times Sep_p}, \tag{11}$$

in which $Sep_{ij} = \frac{(t_{ij})^2}{\sum_{i=1}^{n} t_{ij} * \sum_{j=1}^{m} t_{ij}}$. In Formulas (10) and (11), n is the number of protein complexes in the gold standard dataset, m is the number of identified protein complexes, t_{ij} denotes the size of intersection between the ith gold standard complex and the jth detected complex, and n_i denotes the number of proteins included in the ith gold standard complex.

MMR (Maximum Matching Ratio). MMR is used to describe the maximum one-to-one matching between the identified and gold standard protein complexes, which are defined as follows [24]:

$$MMR(g, p) = \frac{\sum_{i=1}^{n} \max_{j=1}^{m} os(g_i, p_j)}{N_i} \tag{12}$$

in which os represents the overlapping score between two protein complexes, g_i is the ith gold standard complex, and p_j represents the jth identified protein complex.

MCC (Matthews Correlation Coefficient). MCC is widely used in bioinformatics as a performance metric that can handle imbalanced data. The formula is described as follows [24]:

$$MCC = \frac{TP \times TN - FP \times FN}{\sqrt{(TP + FN)(TP + FP)(TN + FP)(TN + FN)}} \tag{13}$$

in which TP, TN, FP, and FN mean the true positive, true negative, false positive, and false negative, respectively.

3.3. Detection of Protein Complexes Using PMD_{pc}

A PPI network is usually modeled as an undirected weight graph $G = (V, E, \omega)$, in which V represents a set of nodes (proteins), E is a set of edges (protein pairs), and ω is a set of similarity value between each protein pairs. The similarity of GO (Gene Ontology, GO) terms is mathematically expressed as follows [36]:

$$Sim(i, j) = \frac{|N(i) \cap N(j)|}{\min(N(i), N(j))} \tag{14}$$

in which $Sim(i, j)$ indicates the GO similarity of the protein pair (i, j). $N(i)$ denotes the number of GO terms that annotate the protein i. The PPI network is stocked as the matrix X with a size of $n \times n$, which is transformed into the vertex–PCA matrix X of size $n \times p$ by the principal component analysis, in which each row of X represents a protein in all n samples (protein complexes), and each column of X represents the expression level of a sample in all p proteins.

According to Section 3.2, the matrix X is decomposed into three matrices, namely, U, V, and Δ by PMD. The graphical description of PMD_{pc} is shown in Figure 3, in which u_k is the kth principal component, v_k is the kth expression model of the principal component, and u_{ik} indicates that the kth protein is projected on the kth protein complex. Therefore, matrix U is decomposed into several clusters (protein complexes) due to matrix decomposition.

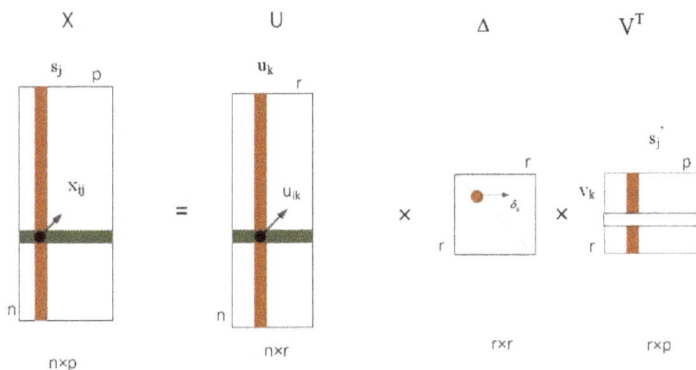

Figure 3. Graphical description of PMD_{pc}. Matrix X is decomposed into two base matrices, namely, U, V, and a diagonal matrix Δ.

PMD_{pc} is implemented in Java, and all experiments are performed on an Intel(R) Core(TM) i7-5557U CPU with 2.2 GHz and 8 GB RAM running Windows 7.0. The elapsed time is 9533 s.

4. Conclusions

The identification of protein complex helps us to discover and understand the cellular organizations and biological functions in PPI networks. Previous computational approaches mainly identified protein complexes in dense PPI networks, which had inferior performances in sparse PPI networks. In this work, PMD_{pc} is proposed on the basis of the penalized matrix decomposition to detect protein complexes in the human protein interaction network with 0.0008 density.

The performance of our method, PMD_{pc}, is compared with the performances of CFinder, ClusterONE, RRW, HC-PIN, and PCE-FR on the human PPI dataset derived from HPRD to validate the utilization of our method. The experimental results show that our proposed algorithm is better than the five classical approaches based on F-measure, ACC, and MMR. However, only the human PPI network was taken as the experimental dataset. The new method should be suitable for substructure detection with other sparse networks. Therefore, our algorithm will be used in the future to investigate other species of complex networks, such as gene regulatory and disease networks.

Supplementary Materials: The following are available online.

Author Contributions: Conceptualization: B.C. and S.D.; data curation: S.C.; formal analysis: S.D.; investigation: S.C.; methodology: B.C.; project administration: S.D.; resources: P.D.; software: S.C.; supervision: S.D.; validation: B.C., H.Q., and G.L.; writing-original draft: B.C.; writing-review & editing: P.D.

Funding: This research was funded by the National Natural Science Foundation of China grant numbers [61572180, 61472467, 61471164, 61672011, and 61602164], the Hunan Provincial Natural Science Foundation of China grant numbers [2016JJ2012 and 2018JJ2024], the Key Project of the Education Department of Hunan Province grant number [17A037], and the Scientific and Technological Research Project of Education Department in Jiangxi Province grant number [GJJ170383].

Conflicts of Interest: The authors declare no conflict of interest.

References

1. Enright, A.J.; Van Dongen, S.; Ouzounis, C.A. An efficient algorithm for large-scale detection of protein families. *Nucleic Acids Res.* **2002**, *30*, 1575–1584. [CrossRef] [PubMed]
2. Bader, G.D.; Hogue, C.W. An automated method for finding molecular complexes in large protein interaction networks. *BMC Bioinform.* **2003**, *4*, 2. [CrossRef]
3. Adamcsek, B.; Palla, G.; Farkas, I.J.; Derenyi, I.; Vicsek, T. Cfinder: Locating cliques and overlapping modules in biological networks. *Bioinformatics* **2006**, *22*, 1021–1023. [CrossRef] [PubMed]
4. Altaf-Ul-Amin, M.; Shinbo, Y.; Mihara, K.; Kurokawa, K.; Kanaya, S. Development and implementation of an algorithm for detection of protein complexes in large interaction networks. *BMC Bioinform.* **2006**, *7*, 1–13. [CrossRef] [PubMed]
5. Gao, L.; Sun, P.G.; Song, J. Clustering algorithm for detecting functional modules in protein interaction networks. *J. Bioinform. Comput. Biol.* **2011**, *7*, 217–242. [CrossRef]
6. Pizzuti, C.; Rombo, S.E. A coclustering approach for mining large protein-protein interaction networks. *IEEE ACM Trans. Comput. Biol.* **2012**, *9*, 717–730. [CrossRef] [PubMed]
7. Wang, J.X.; Li, M.; Chen, J.E.; Pan, Y. A fast hierarchical clustering algorithm for functional modules discovery in protein interaction networks. *IEEE ACM Trans. Comput. Biol.* **2011**, *8*, 607–620. [CrossRef] [PubMed]
8. Girvan, M.; Newman, M.E.J. Community structure in social and biological networks. *Proc. Natl. Acad. Sci. USA* **2002**, *99*, 7821–7826. [CrossRef] [PubMed]
9. Rivera, C.G.; Vakil, R.; Bader, J.S. Nemo: Network module identification in cytoscape. *BMC Bioinform.* **2010**, *11* (Suppl. 1), S61. [CrossRef] [PubMed]
10. Nepusz, T.; Yu, H.Y.; Paccanaro, A. Detecting overlapping protein complexes in protein-protein interaction networks. *Nat. Methods* **2012**, *9*, 471–472. [CrossRef] [PubMed]
11. Wu, H.; Gao, L.; Dong, J.H.; Yang, X.F. Detecting overlapping protein complexes by rough-fuzzy clustering in protein-protein interaction networks. *PLoS ONE* **2014**, *9*, 1856. [CrossRef] [PubMed]
12. Rhrissorrakrai, K.; Gunsalus, K.C. Mine: Module identification in networks. *BMC Bioinform.* **2011**, *12*, 192. [CrossRef] [PubMed]
13. Voevodski, K.; Teng, S.H.; Xia, Y. Finding local communities in protein networks. *BMC Bioinform.* **2009**, *10*, 297. [CrossRef] [PubMed]
14. Jiang, P.; Singh, M. Spici: A fast clustering algorithm for large biological networks. *Bioinformatics* **2010**, *26*, 1105–1111. [CrossRef] [PubMed]
15. Cao, B.W.; Luo, J.W.; Liang, C.; Wang, S.L.; Ding, P.J. Pce-fr: A novel method for identifying overlapping protein complexes in weighted protein-protein interaction networks using pseudo-clique extension based on fuzzy relation. *IEEE Trans. Nanobiosci.* **2016**, *15*, 728–738. [CrossRef] [PubMed]
16. Vella, D.; Marini, S.; Vitali, F.; di Silvestre, D.; Mauri, G.; Bellazzi, R. Mtgo: Ppi network analysis via topological and functional module identification. *Sci. Rep.* **2018**, *8*, 5499. [CrossRef] [PubMed]
17. Kouhsar, M.; Zare-Mirakabad, F.; Jamali, Y. Wcoach: Protein complex prediction in weighted ppi networks. *Genes Genet. Syst.* **2015**, *90*, 317–324. [CrossRef] [PubMed]
18. Hu, L.; Chan, K.C.C. A density-based clustering approach for identifying overlapping protein complexes with functional preferences. *BMC Bioinform.* **2015**, *16*, 174. [CrossRef] [PubMed]
19. Cao, B.; Luo, J.; Liang, C.; Wang, S. Identifying protein complexes by combining network topology and biological characteristics. *J. Comput. Theor. Nanosci.* **2016**, *13*, 1546–1955. [CrossRef]
20. King, A.D.; Przulj, N.; Jurisica, I. Protein complex prediction via cost-based clustering. *Bioinformatics* **2004**, *20*, 3013–3020. [CrossRef] [PubMed]
21. Macropol, K.; Can, T.; Singh, A.K. Rrw: Repeated random walks on genome-scale protein networks for local cluster discovery. *BMC Bioinform.* **2009**, *10*, 283. [CrossRef] [PubMed]
22. Liu, G.M.; Wong, L.; Chua, H.N. Complex discovery from weighted ppi networks. *Bioinformatics* **2009**, *25*, 1891–1897. [CrossRef] [PubMed]
23. Wu, M.; Li, X.L.; Kwoh, C.K.; Ng, S.K. A core-attachment based method to detect protein complexes in ppi networks. *BMC Bioinform.* **2009**, *10*, 169. [CrossRef] [PubMed]
24. Maulik, U.; Mukhopadhyay, A.; Bhattacharyya, M.; Kaderali, L.; Brors, B.; Bandyopadhyay, S.; Eils, R. Mining quasi-bicliques from hiv-1-human protein interaction network: A multiobjective biclustering approach. *IEEE ACM Trans. Comput. Biol.* **2013**, *10*, 423–435. [CrossRef] [PubMed]

25. Cao, B.; Luo, J.; Liang, C.; Wang, S.; Song, D. Moepga: A novel method to detect protein complexes in yeast protein-protein interaction networks based on multiobjective evolutionary programming genetic algorithm. *Comput. Biol. Chem.* **2015**, *58*, 173–181. [CrossRef] [PubMed]

26. Zhu, L.; Deng, S.-P.; You, Z.-H.; Huang, D.-S. Identifying spurious interactions in the protein-protein interaction networks using local similarity preserving embedding. *IEEE/ACM Trans. Comput. Biol. Bioinform.* **2017**, *14*, 345–352. [CrossRef] [PubMed]

27. Zhang, Y.; Du, N.; Ge, L. A collective nmf method for detecting protein functional module from multiple data sources. In Proceedings of the ACM Conference on Bioinformatics, Computational Biology and Biomedicine, Orlando, FL, USA, 7–10 October 2012; pp. 655–660. [CrossRef]

28. Zhang, S.H.; Li, Q.J.; Liu, J.; Zhou, X.J. A novel computational framework for simultaneous integration of multiple types of genomic data to identify microrna-gene regulatory modules. *Bioinformatics* **2011**, *27*, I401–I409. [CrossRef] [PubMed]

29. Zheng, C.H.; Zhang, L.; Ng, V.T.Y.; Shiu, S.C.K.; Huang, D.S. Molecular pattern discovery based on penalized matrix decomposition. *IEEE ACM Trans. Comput. Biol.* **2011**, *8*, 1592–1603. [CrossRef] [PubMed]

30. Witten, D.M.; Tibshirani, R.; Hastie, T. A penalized matrix decomposition, with applications to sparse principal components and canonical correlation analysis. *Biostatistics* **2009**, *10*, 515–534. [CrossRef] [PubMed]

31. Liu, J.-X.; Liu, J.; Gao, Y.-L.; Mi, J.-X.; Ma, C.-X.; Wang, D. A class-information-based penalized matrix decomposition for identifying plants core genes responding to abiotic stresses. *PLoS ONE* **2014**, *9*, e106097. [CrossRef] [PubMed]

32. Wu, M.; Yu, Q.; Li, X.L.; Zheng, J.; Huang, J.F.; Kwoh, C.K. Benchmarking human protein complexes to investigate drug-related systems and evaluate predicted protein complexes. *PLoS ONE* **2013**, *8*. [CrossRef] [PubMed]

33. Yang, P.; Li, X.; Wu, M.; Kwoh, C.K.; Ng, S.K. Inferring gene-phenotype associations via global protein complex network propagation. *PLoS ONE* **2011**, *6*, e21502. [CrossRef] [PubMed]

34. Peri, S.; Navarro, J.D.; Kristiansen, T.Z.; Amanchy, R.; Surendranath, V.; Muthusamy, B.; Gandhi, T.K.; Chandrika, K.N.; Deshpande, N.; Suresh, S.; et al. Human protein reference database as a discovery resource for proteomics. *Nucleic Acids Res.* **2004**, *32*, D497. [CrossRef] [PubMed]

35. Luc, P.V.; Tempst, P. Pindb: A database of nuclear protein complexes from human and yeast. *Bioinformatics* **2004**, *20*, 1413–1415. [CrossRef] [PubMed]

36. Shalgi, R.; Lieber, D.; Oren, M.; Pilpel, Y. Global and local architecture of the mammalian microrna-transcription factor regulatory network. *PLoS Comput. Biol.* **2007**, *3*, e131. [CrossRef] [PubMed]

Sample Availability: Samples of the compounds are not available from the authors.

Article

MDPI

Discovering Structural Motifs in miRNA Precursors from the *Viridiplantae* Kingdom

Joanna Miskiewicz [1] and Marta Szachniuk [1,2,*]

[1]   Institute of Computing Science, Poznan University of Technology, 60-965 Poznan, Poland;
      joanna.miskiewicz@cs.put.poznan.pl
[2]   Institute of Bioorganic Chemistry, Polish Academy of Sciences, 61-704 Poznan, Poland
*   Correspondence: marta.szachniuk@cs.put.poznan.pl; Tel.: +48-616-653-030

Received: 29 April 2018; Accepted: 4 June 2018; Published: 6 June 2018

Abstract: A small non-coding molecule of microRNA (19–24 nt) controls almost every biological process, including cellular and physiological, of various organisms' lives. The amount of microRNA (miRNA) produced within an organism is highly correlated to the organism's key processes, and determines whether the system works properly or not. A crucial factor in plant biogenesis of miRNA is the Dicer Like 1 (DCL1) enzyme. Its responsibility is to perform the cleavages in the miRNA maturation process. Despite everything we already know about the last phase of plant miRNA creation, recognition of miRNA by DCL1 in pre-miRNA structures of plants remains an enigma. Herein, we present a bioinformatic procedure we have followed to discover structure patterns that could guide DCL1 to perform a cleavage in front of or behind an miRNA:miRNA* duplex. The patterns in the closest vicinity of microRNA are searched, within pre-miRNA sequences, as well as secondary and tertiary structures. The dataset consists of structures of plant pre-miRNA from the *Viridiplantae* kingdom. The results confirm our previous observations based on *Arabidopsis thaliana* precursor analysis. Hereby, our hypothesis was tested on pre-miRNAs, collected from the miRBase database to show secondary structure patterns of small symmetric internal loops 1-1 and 2-2 at a 1–10 nt distance from the miRNA:miRNA* duplex.

Keywords: miRNA biogenesis; structural patterns; DCL1

1. Introduction

MicroRNAs (miRNAs) represent a group of small noncoding RNAs (sRNA) that consist of about 21–24 nucleotides [1–8]. They are present in animals, plants, and single-cell eukaryotes. The key role of miRNA is to regulate gene expression via degrading or blocking the targeted mRNA transcript [9,10]. With the ability to silence various genes, microRNA can modulate the homeostasis of the organism by interfering with specific mRNAs, as well as by preventing further expression of genes engaged in development, metabolism, or differentiation [3,11–14]. Mis-regulation of miRNAs, which are involved in different biological processes, is believed to be a major contributor to various diseases [15]. The recognition of targeted transcripts comes through nearly complete (in plants) or partially complete (in animals) base pair complementarity [6,16]. The multistep miRNA biogenesis differs between plants and animals, mainly in the cell location where each stage of the process is held and in the contributing proteins. The transcribed miRNA gene (pri-miRNA) in animals is cleaved into a precursor (pre-miRNA) structure by a microprocessor. The microprocessor primarily consists of two enzymes: RNAse III Drosha and DiGeorge Syndrome Critical Region 8 (DGCR8) (in several organisms DGCR8 is replaced by Pasha) [17–19]. At this phase, pre-microRNA is transported from the nucleus to the cytoplasm by Exportin 5 protein (XPO5). Next, Dicer (the other RNase III type enzyme), performs cleavages in pre-miRNA to release the duplex of microRNA (miRNA:miRNA*) [19].

In plants, all endonucleolytic cleavages of pri-miRNA and pre-miRNA are performed in the nucleus by Dicer-Like 1 (DCL1), being a homologue of Dicer. The process of plant miRNA maturation also requires engagement of HYPONASTIC LEAVES 1 (HYL1), a protein that contains a dsRNA-binding domain, and SERRATE (SE), a protein containing a zinc-finger domain. After creation, pre-miRNA is exported to the cytoplasm by the HASTY enzyme, a homologue of XPO5 [5,12,20,21]. In both, animal and plant cells, the miRNA:miRNA* duplex consists of a guide and a passenger strand. During incorporation of the duplex into the RNA-induced silencing complex (RISC), the passenger strand is discarded, while the guide strand leads the complex toward the target mRNA [22–24]. The passenger strand (miRNA*) is either degraded or used as a guide for other transcripts. Besides miRNA, which determines the targeted mRNA via base pair complementarity, RISC includes an ARGONAUTE (AGO) protein, the effector molecule with slicing activity [7,25]. The RISC enables degradation of the target mRNA or inhibition of the translation process by several mechanisms, including ARGONAUTE endonuclease activity, which enables slicing of targeted mRNA [3,5,10,25]. Biogenesis of animal miRNAs can be classified as a well-known process. The cleavages performed on animal pre-miRNA by the molecular ruler Dicer are measured from the pre-miRNA terminus, either the 3′ or the 5′ end, to the RNase III domain-dependent cleavage site [26,27]. In plants, it is still a mystery how the DCL1 enzyme recognizes miRNAs within pre-miRNA structures to perform cuts and release the miRNA:miRNA* duplex. Therefore, we have decided to analyze a set of available pre-miRNA structures and look for structural patterns occurring in miRNA vicinity. It is assumed that some motifs should exist and guide DCL1. Herein, we present a broad approach to pattern searching within pre-miRNAs. We have applied it to structures from four phyla of the *Viridiplantae* kingdom. We drew from our previous research concerning structural motifs in precursor microRNAs of *Arabidopsis thaliana*.

2. Results

2.1. A Scheme of Data Processing

Our research project has followed several steps (Figure 1). At first, the data for an analysis was collected and pre-processed. After dataset preparation, a semi-automated processing of pre-miRNAs followed. It was conducted at three structure levels. We started by investigating the sequences, and going through secondary structure studies, we ended up with a three-dimensional (3D) structure analysis. A detailed description of these steps is provided in the following paragraphs.

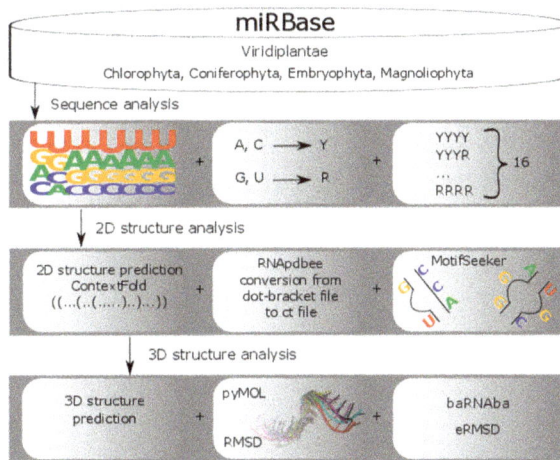

Figure 1. Precursor microRNA (pre-miRNA) analysis workflow.

2.2. Dataset Preparation

In order to find structural motifs in plant pre-miRNA, which could help understand DCL1 performance, we prepared a dataset based on sequences stored in the miRBase database [4]. We considered records under the *Viridiplantae* kingdom assigned to the following phyla: *Magnoliophyta* (6547 sequences), *Coniferophyta* (108 sequences), *Chlorophyta* (50 sequences), and *Embryophyta* (287 sequences). Altogether, our initial collection contained 6992 sequences. The Table S1 from Supplementary Materials contains number of sequences extracted from miRBase website [4] distributed by phylum, clade, family and species. Next, we extracted the relevant information of collected *Viridiplantae* from the miRBase [4] website, and shaped it to adjust to further processing. This was done using self-prepared scripts written in Python language. The prepared data files contained an accession number for each pre-miRNA (in accordance with the miRBase nomenclature) assigned to the sequence, and an miRNA position within its appropriate precursor. From the miRBase [4] database, we also collected evidence about every miRNA found within the set of 6992 sequences, which could be experimental (by similarity) or not experimental. In our research, we planned to focus the analysis on the miRNA vicinity. Thus, we needed to have the sequences and structures of miRNA precursors containing miRNAs with sufficiently large neighbouring regions. It had been decided that eight nucleotides per strand constituted a sufficient size for the vicinity sequence to be analyzed. In the initial collection of 6992 sequences, we identified 5345 pre-miRNA sequences in which miRNAs were surrounded by at least 8 nt on their 5′ and 3′ ends: 4956 from *Magnoliophyta*, 80 from *Coniferophyta*, 38 from *Chlorophyta*, and 271 from *Embryophyta*. These sequences were selected to form the basic *S8* set used in the majority of forthcoming experiments. Within this set, at least one miRNA per each sequence was confirmed experimentally (in the subset of 4388 sequences) or by similarity (within the subset of 343 sequences). In the remaining 614 sequences of the *S8* set (<11.5%), miRNAs were confirmed non-experimentally (i.e., the miRNA sequence was revealed by sequencing, and not used in any experiment yet).

Further, we found it also necessary to limit the miRNA vicinity size to 4 nt. To meet this requirement, from the initial 6992 sequences, we picked 5975 pre-miRNAs with at least 4 nucleotides on both sides of miRNA: 5555 from *Magnoliophyta*, 99 from *Coniferophyta*, 41 from *Chlorophyta*, and 280 from *Embryophyta*. These were collected in the *S4* set, which included 5345 sequences from the *S8* set (vicinity size ≥8 nt) and 630 sequences with vicinity size between 4 and 7 nt. These sequence collections allowed us to properly define the search space for our computational experiments. Within the *S4* set, at least one miRNA per sequence was confirmed experimentally (in the subset of 4890 sequences) or by similarity (within the subset of 389 sequences). In the remaining 696 sequences of the *S4* set (<12%), miRNAs were not confirmed experimentally (i.e., miRNA sequence was revealed by sequencing, and not used in any experiment yet).

2.3. Primary Structure-Based Analysis

In the first computational experiment, we have used the *S8* set of the pre-miRNA sequences. In every sequence from *S8*, either one or two miRNAs were found. We identified an 8 nt-long vicinity sequence on the 5′ and 3′ end of each of these miRNAs. These sequence fragments were extracted to form *VS8-5′* and *VS8-3′* subsets of a large *VS8* collection, including 12802 vicinity sequences with the length equal to 8 nt exactly. Subset *VS8-5′* contains 6401 vicinity sequences occurring in the miRNA vicinity on the 5′ end, and subset *VS8-3′* has 6401 sequences from the 3′ end vicinity. Both subsets, *VS8-5′* and *VS8-3′*, were processed using WebLogo tool versions 2.8.2 (https://weblogo.berkeley.edu/logo.cgi) [28] and 3.0 (http://weblogo.threeplusone.com/create.cgi) [28]. WebLogo allowed us to obtain a diagram showing the most- and the least-frequent nucleotides occurring on each of the eight positions of miRNA vicinity sequence. The first position in each sequence is the first nucleotide behind the microRNA, counting towards the 3′ end (in the *VS8-5′* subset) or towards the 5′ end (in the *VS8-3′* subset). The most frequent nucleotides are shown at the top of the stack, while the least frequent ones are at the bottom (Figure 2). Detailed information about nucleotides occupying the following positions within vicinity sequences is provided in Table 1 (for the *VS8-5'* subset) and Table 2 (for the *VS8-3'* subset).

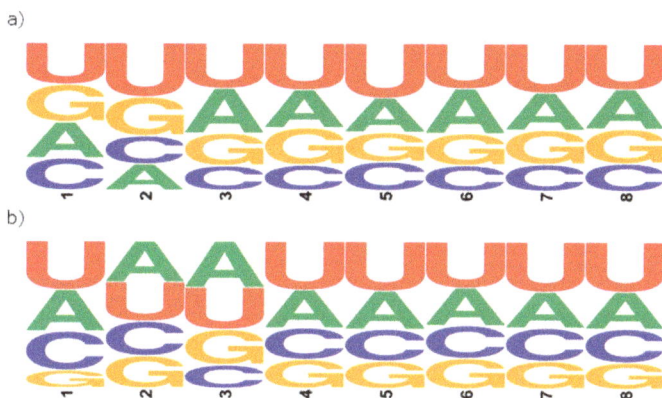

Figure 2. WebLogo 2.8.2 [28] diagram for sequences from the (**a**) *VS8-5'* and (**b**) *VS8-3'* subsets.

Table 1. WebLogo 3.0 [28] results for vicinity sequences in the *VS8-5'* subset.

Position	A [%]	C [%]	G [%]	U [%]	R [%]	Y [%]
1	24.48	21.87	25.62	28.03	50.10	49.90
2	17.78	19.28	26.56	36.38	44.34	55.66
3	30.32	16.67	22.37	30.64	52.69	47.31
4	25.68	17.31	25.46	31.54	51.15	48.85
5	23.18	19.56	20.26	36.99	43.45	56.55
6	30.21	17.23	22.11	30.45	52.32	47.68
7	25.17	18.12	23.54	33.17	48.71	51.29
8	26.71	18.75	23.76	30.78	50.48	49.52

Table 2. WebLogo 3.0 [28] results for vicinity sequences in the *VS8-3'* subset.

Position	A [%]	C [%]	G [%]	U [%]	R [%]	Y [%]
1	27.98	26.51	12.19	33.32	40.17	59.83
2	27.90	23.37	22.09	26.64	49.99	50.01
3	31.48	15.92	24.14	28.46	55.62	44.38
4	25.56	21.45	19.72	33.28	45.27	54.73
5	25.84	21.67	18.73	33.76	44.57	55.43
6	25.73	22.26	20.81	31.20	46.54	53.46
7	24.57	22.54	19.00	33.89	43.57	56.43
8	25.31	22.81	18.15	33.73	43.46	56.54

It can be observed that Uracil is the most frequent nucleotide on almost every position of each vicinity sequence. In sequences from the *VS8-5'* subset, the second position is heavily occupied by Uracil (36.38% of sequences in *VS8-5'* have Uracil on the second position), and rather poorly by Adenine (17.78%). This can indicate an unpairing in the structure, which occurs exactly on this position. In the *VS8-3'* subset, bigger differences are observed between Cytosine and Guanine occupation. The biggest difference reaches 14.22%, and concerns the first position of the vicinity sequence. In the *VS8-5'* subset, nucleotides on the first position are almost evenly distributed, while the second position seems to create an unpaired region. The *VS8-3'* subset seems to be contrary to this. It shows almost equally distributed values on the second position and highly varied distribution in the first position. Thus, it is possible that in the region of the first two positions beyond the miRNA sequence, one could find a small mismatch, revealed as a bulge or a loop in the structure. In the second experiment, aimed to search for sequential motifs in miRNA vicinity, we decided to represent each nucleotide in nucleotide ambiguity code (IUPAC) [29], based on the number of carbon-nitrogen rings, as a purine (R) or pyrimidine (Y). At first, this experiment was run on the previously created *VS8-3'* and *VS8-5'* subsets. In every vicinity sequence from these subsets, we changed the representation of adenines (A) and guanines (G) into purines (R) and uracils (U) and cytosines (C) into pyrimidines (Y). Next, we searched for exactly 8 nt-long patterns that were also encoded using the two-letter alphabet {R, Y}. All permutations for eight positions with two possible variants, purine or pyrimidine, gave us 256 possible patterns. We did not observe any significant results in this experiment. Therefore, we decided to restrict the search space and run the experiment for shorter vicinity sequences. We have taken the *S4* set of 5975 pre-miRNAs, containing miRNAs with neighbouring regions having at least 4 nucleotides on both the 5' and 3' end next to the miRNA region. From this collection, we extracted 14300 vicinity sequences 4 nt long, and divided them into two subsets, *VS4-5'* and *VS4-3'*, in the same manner as *VS8*. Each of these subsets contained 7150 short sequences. Every vicinity sequence from *VS4-5'* and *VS4-3'* was next represented with the two-letter alphabet {R, Y}, and the search for 4 nt-long patterns was performed, providing the results as presented in Table 3.

Table 3. Pattern occurrence in the *VS4-5'* and *VS4-3'* subset.

Pattern	VS4-5' [%]	VS4-3' [%]	Total [%]
RRYR	4.36	3.82	4.09
YRYR	4.41	4.57	4.49
RYYR	6.22	3.90	5.06
RRRY	5.43	5.92	5.67
RYRY	6.08	5.33	5.71
RRYY	6.13	5.30	5.71
YRYY	4.98	6.78	5.88
RYYY	6.90	4.98	5.94
YYYR	6.77	5.45	6.11
RYRR	7.50	5.29	6.39
RRRR	7.43	5.64	6.53
YYRY	6.77	6.67	6.72
YRRR	6.38	7.40	6.89
YRRY	4.83	10.10	7.46
YYYY	7.29	9.17	8.23
YYRR	8.55	9.68	9.11

The first symbol of a pattern corresponds to the nucleotide on the first position beyond miRNA sequence. From these statistics, we can observe that five of the most frequent motifs start with pyrimidine: YYRY, YRRR, YRRY, YYYY, and YYRR. This suggests that many sequences which encounter miRNA involve uracil or cytosine right before the first nucleotide of miRNA sequence.

2.4. Secondary Structure-Based Analysis

The second part of our analysis concerned the secondary structures. Since our input data collection contained sequences only, we decided to predict their secondary structures using ContextFold version 1.0 [30] installed on a local computer. The software was chosen based on the CompaRNA benchmark [31]. All 5975 sequences from the *S4* set were processed by ContextFold [30] to predict their secondary structures. Predicted structures were encoded in dot-bracket notation. For the facilitation of further analysis, we used RNApdbee program (http://rnapdbee.cs.put.poznan.pl/) [32–34] to transform two-dimensional (2D) structures from dot-bracket to CT (Connect) format. Next, we applied a script called MotifSeeker implemented in Python language. The MotifSeeker processes CT files, and searches for bulges and internal loops in the vicinity of the miRNA:miRNA* duplex (up to four nucleotides beyond the miRNA on both sides). The generated output file contains brief information about what motif has been found, on which strand, and how far it was from the microRNA. Guided by our previous study of the pre-miRNA sequences of *Arabidopsis thaliana* [5] and current WebLogo [28] results, we expected an accumulation of mismatches between the first and fourth position beyond miRNA. Although it is known that similar sequences do not always maintain the similarities at higher structural levels [35], we supposed that in our case, the analyzed structures would share some of their pattern in the short fragment beyond the miRNA:miRNA* duplex at the secondary or tertiary structural level. MotifSeeker allowed us to identify the most frequently occurring secondary structure pattern, along with its distance from the miRNA:miRNA* duplex, and a number of structures in which the motif was found. According to our assumptions, the first eight most frequent patterns had small mismatches: symmetric internal loops 1-1 (single unpaired nucleotide on every strand of the vicinity region) and 2-2 (two unpaired nucleotides on every strand of the vicinity region). We have found that in 21.56% of the 5975 secondary structures, the first nucleotides beyond the miRNA:miRNA* duplex were unpaired and formed symmetric 1-1 internal loops. The same 1-1 pattern was shared by 13.82% of the secondary structures, starting from the second position, and 16.55% of the structures starting from the third position beyond the miRNA:miRNA* duplex. This means that over 50% (exactly 51.93%) of the analyzed secondary structures contain the 1-1 motif at the maximum distance of three positions beyond miRNA. In Table 4, we present the exact number of motifs found within the structures in which we discovered the pattern. All motifs identified by MotifSeeker are represented in Figure 3, where each position is defined by the pattern type (1-1 or 2-2) and the distance between the motif and the miRNA, from 1 nt (D:1) up to 4 nt beyond miRNA (D:4). The MotifSeeker code and input files can be found here: http://bio.cs.put.poznan.pl/fileserver/.

Table 4. Motif occurrence in the *S4* set. The number of motifs was calculated based on the number of specific patterns in defined locations, referring to structures which contain at least one motif.

Motif/Distance	Number of Motifs	Number of Structures with at Least One Motif
1-1/D:1	1397	1288
1-1/D:3	1043	989
1-1/D:2	861	826
1-1/D:4	807	769
2-2/D:3	221	219
2-2/D:1	190	187
2-2/D:2	149	147
2-2/D:4	118	117

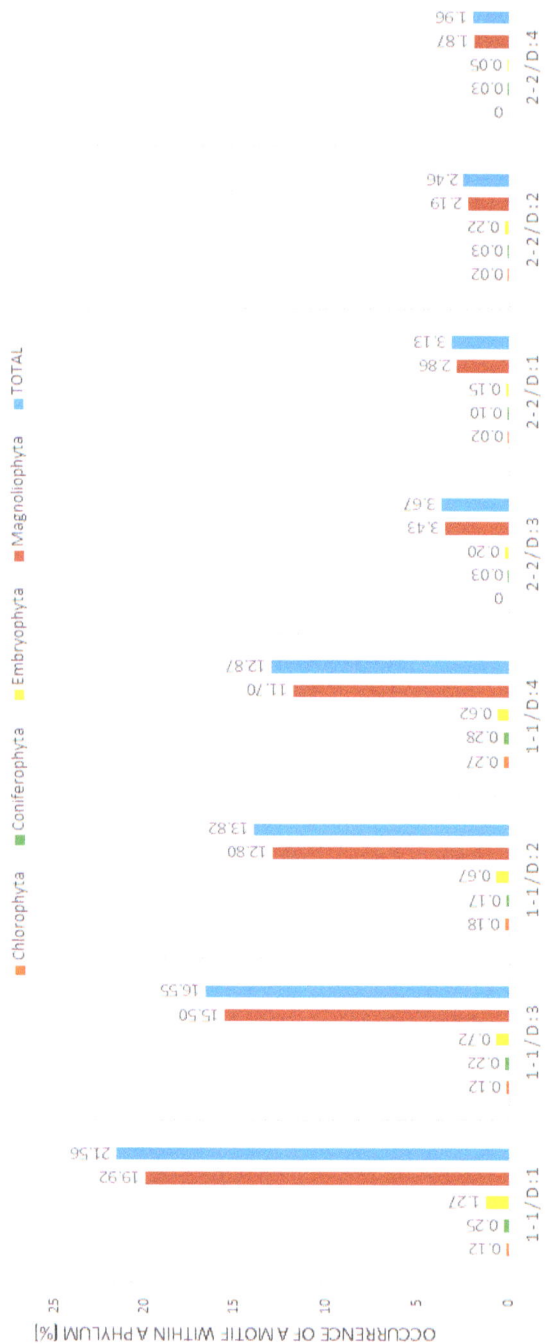

Figure 3. Distribution of eight most-occurring two-dimensional (2D) motifs in 5975 structures by phyla. The results are arranged from the least frequent motif to the most common one.

2.5. Tertiary Structure-Based Analysis

In the third stage of analysis, the tertiary structures of miRNA vicinity were analyzed using bioinformatics tools. Over many years, lots of methods for RNA 3D structure analysis have been developed [36,37]. In our experiments, we decided to focus on three of them: RNAComposer [38,39], PyMOL [40], and baRNAba [41]. First, we predicted 40 tertiary structures by using RNAComposer [38,39]. The input set for the prediction process included 10 sequences for each phylum picked randomly from S4 dataset. The obtained models were next processed by using the PyMOL program [40]. From each predicted tertiary structure, the closest vicinity regions of miRNA were cut out for alignment. Due to the shift between the 5′ and 3′ miRNA, we decided to use regions that were overlapping the miRNA:miRNA* duplex for 4 nt beyond the duplex and 4 nt within the duplex. This resulted in obtaining 8 nt-long structures from both sides of the miRNA:miRNA*. For each phylum, we have generated 20 short 3D fragments. Among them, one random structure was chosen as a reference—the remaining ones were aligned to it. Thus, we created four different alignments (Figure 4), with root mean square deviation (RMSD) values measured by PyMOL [40] and eRMSD values computed by the baRNAba software [41]. RMSD allowed us to measure the similarity between the superimposed atomic coordinates [42] whereas eRMSD facilitated to measure the distance between structures based only on the relative positions and orientations of nucleobases [41].

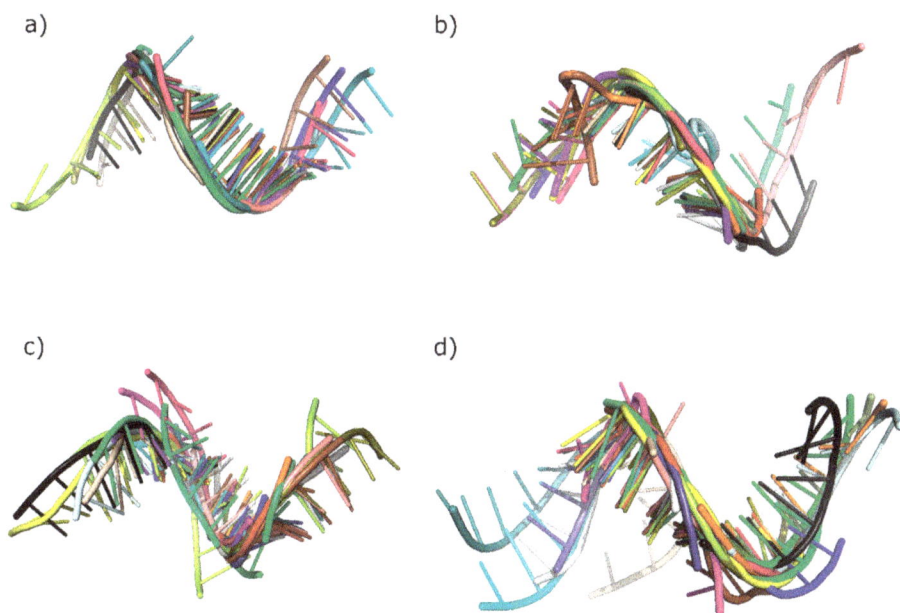

Figure 4. Aligned three-dimensional (3D) substructures within each phylum: (**a**) *Chlorophyta*, (**b**) *Coniferophyta*, (**c**) *Embryophyta*, and (**d**) *Magnoliophyta*.

The RMSD values presented in Table 5 do not exceed 2.5 Å, while the average values are not higher than 1.5 Å. Relatively low values are also found in Table 6, representing eRMSD. The highest value in Table 6 is 1.101 Å, and all four calculated averages are below 0.90 Å. In both situations, the results indicate high 3D structure similarity between the four phyla. Thus, the closest region to the miRNA:miRNA* duplex seems to be highly conserved between the phyla in *Viridiplantae* kingdom.

Table 5. RMSD values of 3D fragments from each phylum.

Fragment Id	RMSD [Å]			
	Chlorophyta	*Coniferophyta*	*Embryophyta*	*Magnoliophyta*
1	2.112	0.463	1.882	2.245
2	0.278	0.430	0.290	2.270
3	0.256	1.194	2.058	1.135
4	0.117	0.381	1.626	0.679
5	0.467	0.258	2.351	0.352
6	2.209	1.228	1.810	0.567
7	0.257	0.469	1.966	0.123
8	0.560	1.226	1.587	0.449
9	0.142	1.018	1.773	2.171
10	0.864	0.412	1.247	1.672
11	0.502	0.461	0.910	0.845
12	0.547	0.444	1.573	0.607
13	0.034	1.377	0.974	1.171
14	0.389	0.846	1.546	0.963
15	1.155	1.036	0.944	0.836
16	0.139	0.481	0.837	1.094
17	0.686	1.210	1.839	0.597
18	0.637	0.390	1.730	1.344
19	2.159	0.266	0.330	2.304
Average	0.711	0.715	1.435	1.128

Table 6. eRMSD values of 3D fragments from each phylum.

Fragment Id	eRMSD [Å]			
	Chlorophyta	*Coniferophyta*	*Embryophyta*	*Magnoliophyta*
1	0.459	0.765	0.802	0.554
2	0.788	0.771	0.434	0.503
3	0.587	0.436	0.725	0.730
4	0.291	1.047	0.776	1.101
5	0.477	1.047	0.868	0.325
6	0.432	0.746	0.858	0.444
7	0.561	1.025	0.868	0.832
8	0.442	0.799	0.817	0.365
9	0.459	0.675	0.767	0.455
10	0.438	0.800	0.842	0.643
11	0.386	0.749	1.080	0.390
12	0.251	0.753	0.841	0.398
13	0.605	0.745	0.906	0.457
14	0.410	0.680	0.791	0.447
15	0.410	0.891	0.883	0.394
16	0.463	0.729	0.901	0.467
17	0.564	1.023	0.788	0.331
18	0.528	1.058	0.764	0.604
19	0.453	0.712	0.810	0.604
Average	0.474	0.813	0.817	0.529

3. Discussion

MicroRNA research has become increasingly popular since these molecules were discovered [43,44]. Nowadays, it is not only in-vivo or in-vitro methods that are used to examine the nature of miRNAs. In-silico approaches allow us to predetermine the direction of experiments, and help to narrow the search space to answer the questions raised. Here, we focused on plant microRNAs and performed a series of computational experiments using bioinformatic methods and programs. At each level of the RNA structure, we searched for specific motifs that could guide the DCL1 enzyme to the cutting position of the miRNA:miRNA* duplex. Every analytical step we carried out led to us finding small mismatches placed in the closest vicinity of the 5′ and 3′ ends of the miRNA. Although the results of sequence analysis did not unequivocally indicate the specific unpairing in this area, the secondary structure study proved this hypothesis. In the phase of 2D structure analysis, we discovered a high number of symmetric 1-1 and 2-2 internal loops occurring no further than four nucleotides behind the miRNA:miRNA* duplex. This supports the results of our previous research on *Arabidopsis thaliana*, where we also found a significant number of such motifs in the direct vicinity of miRNA [10]. Additionally, we examined tertiary structures by aligning predicted 3D models of the miRNA neighbourhood and calculating two distance measures (RMSD and eRMSD) between them, divided by phyla. The results confirmed the appearance of a conserved region close to the duplex. In conclusion, the taken bioinformatic pathway helped us to discover potential motifs recognized by the DCL1 enzyme. By examining each structural level, we managed to extract the necessary information and draw proper conclusions. Obtained via in-silico methods, the results clearly point out the significance of closest vicinity of miRNA and mismatches occurring in this region.

4. Materials and Methods

The research focused on three structural levels of RNA architecture: sequence, secondary, and tertiary structure. Sequences were obtained from miRBase (http://www.mirbase.org/), a repository of pre-microRNAs of various organisms [4]. Based on experimental data, this database includes not only sequences, but also positions of miRNA on the 5′ and 3′ strand. Annotation and sequence data for each entry are displayed on the website, along with the proposed secondary structure model of the pre-miRNA.

4.1. WebLogo

Sequence analysis was performed using WebLogo [28], aimed to discover the most frequent nucleotide on each position of miRNA vicinity area. WebLogo version 2.8.2 [28] (https://weblogo. berkeley.edu/logo.cgi) produced diagrams showing the frequency of nucleotides at each analyzed position. The first position is marked as the closest one to miRNA. WebLogo version 3.0 [28] (http:// weblogo.threeplusone.com/create.cgi) was used to generate numerical values of nucleotide frequencies. WebLogo 2.8.2 [28] was used with the following settings for image format and size: *Image format* as eps (vector), and Logo Size per line equals to 18 × 5 cm. For advanced logo options, the settings were as follows: *Sequence Type* was automatic detection; *First Position Number* was 1; *Small Sample Correction* was true; *Frequency Plot* was true; *Logo Range* was none; *Multiline Logo (Symbols per Line)* was false. The advanced image options were set as follows: *Bitmap Resolution* at 96 pixels/inch (dpi); *Antialias Bitmaps* was set to true; *Title* was none; *Y-Axis Height* was none; *Show Y-Axis* was true; *Show X-Axis* was true; *Y-Axis Label* was none; *X-Axis Label* was none; *Show Error Bars* was false; *Boxed/Boxed Shrink Factor* was false; *Show Fine Print* was true; *Label Sequence Ends* was false; *Outline Symbols* was false; and *Y-Axis Tic Spacing* was 1 bit. Colors settings were selected as default. In the WebLogo 3.0 tool [28], we used following parameters: *Title* was none; *Output Format* was data (plain text); *Sequence type* was auto; *Logo size* was medium; *Stacks per Line* was 40; *Ignore lower case* was false; *Units* were probability; *First position number* was 1; *Logo range* was none; *Figure label* was none; *Scale stack widths* was true; *Composition* was auto; *Error bars* were false; *Show Sequence Ends labels* was false; *Version Fine Print* was true; *X-axis* was true; *Y-axis* was true; *Y-axis scale* was auto; *Y-axis tic spacing* was 1.0; and *Color Scheme* was auto.

4.2. Purine–Pyrimidine Patterns

The next phase of the study required changes in miRNA vicinity sequences. Adenine and guanine were represented as R (which denotes purines), while cytosine and uracil were represented as Y (which denotes any pyrimidine). These substitutions were applied by self-created script in Python language. Again, sequence patterns were searched in the modified sequences with using self-developed Python script.

4.3. ContextFold

In the second analytical step, the secondary structures were predicted via the ContextFold program [30]. This program, installed on a local computer, produces files which contain 2D structures defined in dot-bracket notation. In this format, each unpaired nucleotide (mismatch or gap) is represented as a single dot, and a paired nucleotide as an opening or closing bracket. The command used, *java-cp bin contextFold.app.Predict in: input_file.txt out:output_file.txt*, enabled prediction of the secondary structures for all RNA sequences in the input file, using the (default) supplied StHighCoHigh trained model, and saving the result to the output file [45].

4.4. RNApdbee

To facilitate further research, we used the RNApdbee webserver [32,34] (http://rnapdbee.cs.put. poznan.pl/) to convert dot-bracket representation into CT format. The latter data format describes the position of nucleotide in the sequence, nucleobase encoding, the position of the previous and next nucleotides in the sequence, and the index of the paired nucleotide. If the nucleotide is unpaired, the index equals 0. On the RNApdbee website, we chose the third mode of analysis (i.e., third tab page, selecting "(. . . .) → image"). After uploading the structures in dot-bracket notation, we selected the options to (1) identify the structural elements by treating pseudoknots as paired residues, and (2) visualize the secondary structure using the VARNA-based procedure. When the computation was finished, we downloaded the results in CT file format.

4.5. MotifSeeker

The secondary structures were examined by self-developed script named MotifSeeker. MotifSeeker reads CT files and additional information from the pre-miRNA id and its microRNA positions at the 5′ and 3′ ends. Next, the script searches for bulges and internal loops, providing information about the type of mismatch and its distance from miRNA.

4.6. RNAComposer

The last phase of our research involved the prediction of tertiary structures of RNA. We selected 10 secondary structures from each phylum, and used them to predict their 3D structures using RNAComposer (http://rnacomposer.cs.put.poznan.pl/), running it in batch mode [38,39]. RNAComposer allows us to automatically predict tertiary RNA structures, up to 500 nt per structure, based on their secondary structure in dot-bracket format. It is possible for the user to choose one of the six secondary structure prediction methods incorporated into the system. For our analysis, we set the *Select secondary structure prediction method* option to "true", and from the drop-down list we chose the ContextFold method [30]. The same can be done in the interactive mode of RNAComposer, where the user can either select the secondary structure prediction method by selecting it from drop-down list or by typing the method name in the next line after the sequence (no dot-bracket notation is required in this case), e.g.,:

```
#zma_MIR168a
>example
GAAGCCGCGCCGCCUCGGGCUCGCUUGGUGCAGAUCGGGACCCGCCGCCCGGCCGACGG
GACGGAUCCCGCCUUGCACCAAGUGAAUCGGAGCCGGCGGAGCGA
ContextFold
```

Since we have used the batch mode, we could generate more than one 3D structure per secondary structure input. However, we decided to generate a single 3D structure model, and the *Maximum number of generated 3D models* was set to 1.

4.7. PyMOL

The obtained 3D structures were processed in PyMOL [40]. PyMOL software enables molecular visualization, measurement, processing, and model comparison. We used it to align structures within each phylum, and to measure the RMSD values between them. RMSD (root mean square deviation) is one of the standard measures that calculates an average distance between the atoms.

4.8. BaRNAba

Finally, the baRNAba tool was applied to calculate eRMSD values, which refer to the distance considering only the relative positions and orientations of nucleobases [46]. The command applied for baRNAba tool was *./baRNAba –name output_file.txt ERMSD –pdb reference.pdb -f 1_structure.pdb 2_structure.pdb … 19_structure.pdb*.

Supplementary Materials: The following are available online. Table S1. Number of sequences extracted from miRBase website [4] distributed by phylum, clade, family and species.

Author Contributions: Marta Szachniuk conceived and supervised the study. Joanna Miskiewicz prepared the dataset, designed and carried the experiments. Both authors analyzed the results and participated in manuscript writing. Joanna Miskiewicz prepared the figures.

Funding: The authors acknowledge partial support from the National Science Center, Poland (grant 2016/23/B/ST6/03931), and Institute of Bioorganic Chemistry, Polish Academy of Sciences, within an intramural financing program.

Acknowledgments: This research was carried in the European Centre for Bioinformatics and Genomics, Poznan University of Technology, and the Institute of Bioorganic Chemistry PAS, Poland (granted HR Excellence in Research).

Conflicts of Interest: The authors declare that there is no conflict of interest regarding the publication of this paper.

References

1. Iorio, M.V.; Croce, C.M. microRNA involvement in human cancer. *Carcinogenesis* **2012**, *33*, 1126–1133. [CrossRef] [PubMed]
2. Tutar, Y. miRNA and cancer; computational and experimental approaches. *Curr. Pharm. Biotechnol.* **2014**, *15*, 429. [CrossRef] [PubMed]
3. Rogers, K.; Chen, X. Biogenesis, turnover, and mode of action of plant microRNAs. *Plant Cell* **2013**, *25*, 2383–2399. [CrossRef] [PubMed]
4. Kozomara, A.; Griffiths-Jones, S. miRBase: Annotating high confidence microRNAs using deep sequencing data. *Nucleic Acids Res.* **2014**, *42*, D68–D73. [CrossRef] [PubMed]
5. Stepien, A.; Knop, K.; Dolata, J.; Taube, M.; Bajczyk, M.; Barciszewska-Pacak, M.; Pacak, A.; Jarmolowski, A.; Szweykowska-Kulinska, Z. Posttranscriptional coordination of splicing and miRNA biogenesis in plants. *Wiley Interdiscip. Rev. RNA* **2017**, *8*, 1–23. [CrossRef] [PubMed]
6. Carthew, R.W.; Sontheimer, E.J. Origins and Mechanisms of miRNAs and siRNAs. *Cell* **2009**, *136*, 642–655. [CrossRef] [PubMed]
7. Bartel, D.P. MicroRNA Target Recognition and Regulatory Functions. *Cell* **2009**, *136*, 215–233. [CrossRef] [PubMed]
8. Axtell, M.J. Classification and comparison of small RNAs from plants. *Annu. Rev. Plant Biol.* **2013**, *64*, 137–159. [CrossRef] [PubMed]
9. Mickiewicz, A.; Rybarczyk, A.; Sarzynska, J.; Figlerowicz, M.; Blazewicz, J. AmiRNA Designer—New method of artificial miRNA design. *Acta Biochim. Pol.* **2016**, *63*, 71–77. [CrossRef] [PubMed]
10. Miskiewicz, J.; Tomczyk, K.; Mickiewicz, A.; Sarzynska, J.; Szachniuk, M. Bioinformatics Study of Structural Patterns in Plant MicroRNA Precursors. *BioMed Res. Int.* **2017**, *2017*, 6783010. [CrossRef] [PubMed]

11. Achkar, N.P.; Cambiagno, D.A.; Manavella, P.A. miRNA Biogenesis: A Dynamic Pathway. *Trends Plant Sci.* **2016**, *21*, 1034–1044. [CrossRef] [PubMed]

12. Cho, S.K.; Ryu, M.Y.; Shah, P.; Poulsen, C.P.; Yang, S.W. Post-Translational Regulation of miRNA Pathway Components, AGO1 and HYL1, in Plants. *Mol. Cells* **2016**, *39*, 581–586. [CrossRef] [PubMed]

13. Bartel, D.P. MicroRNAs: Genomics, Biogenesis, Mechanism, and Function. *Cell* **2004**, *116*, 281–297. [CrossRef]

14. Chávez Montes, R.A.; de Fátima Rosas-Cárdenas, F.; De Paoli, E.; Accerbi, M.; Rymarquis, L.A.; Mahalingam, G.; Marsch-Martínez, N.; Meyers, B.C.; Green, P.J.; de Folter, S. Sample sequencing of vascular plants demonstrates widespread conservation and divergence of microRNAs. *Nat. Commun.* **2014**, *5*, 3722. [CrossRef] [PubMed]

15. Tarver, J.E.; Donoghue, P.C.; Peterson, K.J. Do miRNAs have a deep evolutionary history? *BioEssays* **2012**, *34*, 857–866. [CrossRef] [PubMed]

16. Drusin, S.I.; Suarez, I.P.; Gauto, D.F.; Rasia, R.M.; Moreno, D.M. dsRNA-protein interactions studied by molecular dynamics techniques. Unravelling dsRNA recognition by DCL1. *Arch. Biochem. Biophys.* **2016**, *15*, 118–125. [CrossRef] [PubMed]

17. Dolata, J.; Bajczyk, M.; Bielewicz, D.; Niedojadlo, K.; Niedojadlo, J.; Pietrykowska, H.; Walczak, W.; Szweykowska-Kulinska, Z.; Jarmolowski, A. Salt stress Reveals a New Role for ARGONAUTE1 in miRNA Biogenesis at the Transcriptional and Posttranscriptional Levels. *Plant Physiol.* **2016**, *172*, 297–312. [CrossRef] [PubMed]

18. Conrad, T.; Orom, U.A. Insight into miRNA biogenesis with RNA sequencing. *Oncotarget* **2015**, *6*, 26546–26547. [CrossRef] [PubMed]

19. Zhu, H.; Zhou, Y.; Castillo-González, C.; Lu, A.; Ge, C.; Zhao, Y.T.; Duan, L.; Li, Z.; Axtell, M.J.; Wang, X.J.; et al. Bidirectional processing of pri-miRNAs with branched terminal loops by Arabidopsis Dicer-like1. *Nat. Struct. Mol. Biol.* **2013**, *20*, 1106–1115. [CrossRef] [PubMed]

20. Starega-Roslan, J.; Krol, J.; Koscianska, E.; Kozlowski, P.; Szlachcic, W.J.; Sobczak, K.; Krzyzosiak, W.J. Structural basis of microRNA length variety. *Nucleic Acids Res.* **2011**, *39*, 257–268. [CrossRef] [PubMed]

21. Voinnet, O. Origin, Biogenesis, and Activity of Plant MicroRNAs. *Cell* **2009**, *136*, 669–687. [CrossRef] [PubMed]

22. Mickiewicz, A.; Sarzynska, J.; Milostan, M.; Kurzynska-Kokorniak, A.; Rybarczyk, A.; Lukasiak, P.; Kulinski, T.; Figlerowicz, M.; Blazewicz, J. Modeling of the catalytic core of Arabidopsis thaliana Dicer-like 4 protein and its complex with double-stranded RNA. *Comput. Biol. Chem.* **2017**, *66*, 44–56. [CrossRef] [PubMed]

23. Beezhold, K.J.; Castranova, V.; Chen, F. Microprocessor of microRNAs: Regulation and potential for therapeutic intervention. *Mol. Cancer* **2010**, *9*, 1–9. [CrossRef] [PubMed]

24. Søkilde, R.; Newie, I.; Persson, H.; Borg, Å.; Rovira, C. Passenger strand loading in overexpression experiments using microRNA mimics. *RNA Biol.* **2015**, *12*, 787–791.

25. Zha, X.; Xia, Q.; Yuan, Y.A. Structural insights into small RNA sorting and mRNA target binding by Arabidopsis Argonaute Mid domains. *FEBS Lett.* **2012**, *586*, 3200–3207. [CrossRef] [PubMed]

26. Starega-Roslan, J.; Galka-Marciniak, P.; Krzyzosiak, W.J. Nucleotide sequence of miRNA precursor contributes to cleavage site selection by Dicer. *Nucleic Acids Res.* **2015**, *43*, 10939–10951. [CrossRef] [PubMed]

27. Flores-Jasso, C.F.; Arenas-Huertero, C.; Reyes, J.L.; Contreras-Cubas, C.; Covarrubias, A.; Vaca, L. First step in pre-miRNAs processing by human Dicer. *Acta Pharmacol. Sin.* **2009**, *30*, 1177–1185. [CrossRef] [PubMed]

28. Crooks, G.E.; Hon, G.; Chandonia, J.M.; Brenner, S.E. WebLogo: A sequence logo generator. *Genome Res.* **2004**, *14*, 1188–1190. [CrossRef] [PubMed]

29. Nucleotide Ambiguity Code (IUPAC). Available online: http://www.dnabaser.com/articles/IUPAC%20ambiguity%20codes.html (accessed on 1 February 2017).

30. Zakov, S.; Goldberg, Y.; Elhadad, M.; Ziv-Ukelson, M. Rich parameterization improves RNA structure prediction. *J. Comput. Biol.* **2011**, *18*, 1525–1542. [CrossRef] [PubMed]

31. Puton, T.; Kozlowski, L.P.; Rother, K.M.; Bujnicki, J.M. CompaRNA: A server for continuous benchmarking of automated methods for RNA secondary structure prediction. *Nucleic Acids Res.* **2013**, *41*, 4307–4323. [CrossRef] [PubMed]

32. Antczak, M.; Zok, T.; Popenda, M.; Lukasiak, P.; Adamiak, R.W.; Blazewicz, J.; Szachniuk, M. RNApdbee—A webserver to derive secondary structures from pdb files of knotted and unknotted RNAs. *Nucleic Acids Res.* **2014**, *42*, W368–W372. [CrossRef] [PubMed]

33. Rybarczyk, A.; Szostak, N.; Antczak, M.; Zok, T.; Popenda, M.; Adamiak, R.W.; Blazewicz, J.; Szachniuk, M. New in silico approach to assessing RNA secondary structures with non-canonical base pairs. *BMC Bioinform.* **2015**, *16*, 276. [CrossRef] [PubMed]

34. Zok, T.; Antczak, M.; Zurkowski, M.; Popenda, M.; Blazewicz, J.; Adamiak, R.W.; Szachniuk, M. RNApdbee 2.0: Multifunctional tool for RNA structure annotation. *Nucleic Acids Res.* **2018**, *46*. [CrossRef] [PubMed]

35. Wiedemann, J.; Milostan, M. StructAnalyzer—A tool for sequence versus structure similarity analysis. *Acta Biochim. Pol.* **2016**, *63*, 753–757. [CrossRef] [PubMed]

36. Wiedemann, J.; Zok, T.; Milostan, M.; Szachniuk, M. LCS-TA to identify similar fragments in RNA 3D structures. *BMC Bioinform.* **2017**, *18*, 456. [CrossRef] [PubMed]

37. Blazewicz, J.; Szachniuk, M.; Wojtowicz, A. RNA tertiary structure determination: NOE pathways construction by tabu search. *Bioinformatics* **2005**, *21*, 2356–2361. [CrossRef] [PubMed]

38. Antczak, M.; Popenda, M.; Zok, T.; Sarzynska, J.; Ratajczak, T.; Tomczyk, K.; Adamiak, R.W.; Szachniuk, M. New functionality of RNAComposer: An application to shape the axis of miR160 precursor structure. *Acta Biochim. Pol.* **2016**, *63*, 737–744. [CrossRef] [PubMed]

39. Purzycka, K.J.; Popenda, M.; Szachniuk, M.; Antczak, M.; Lukasiak, P.; Blazewicz, J.; Adamiak, R.W. Automated 3D RNA structure prediction using the RNAComposer method for riboswitches. *Methods Enzymol.* **2015**, *553*, 3–34. [PubMed]

40. *The PyMOL Molecular Graphics System*, version 1.8; Schrodinger, LLC: New York, NY, USA, 2015.

41. Bottaro, S.; Palma, F.D.; Bussi, G. The role of nucleobase interactions in RNA structure and dynamics. *Nucleic Acids Res.* **2014**, *42*, 13306–13314. [CrossRef] [PubMed]

42. Kufareva, I.; Abagya, R. Methods of protein structure comparison. *Methods Mol. Biol.* **2012**, *857*, 231–257. [PubMed]

43. Almeidaa, M.I.; Reisb, R.M.; Calin, G.A. MicroRNA history: Discovery, recent applications, and next frontiers. *Mutat. Res.* **2011**, *717*, 1–8. [CrossRef] [PubMed]

44. Varani, G. Twenty years of RNA: The discovery of microRNAs. *RNA* **2015**, *21*, 751–752. [CrossRef] [PubMed]

45. Context Fold 1.00. Available online: https://www.cs.bgu.ac.il/~negevcb/contextfold/readme.pdf (accessed on 1 February 2017).

46. eRMSD. Available online: https://plumed.github.io/doc-master/user-doc/html/_e_r_m_s_d.html (accessed on 1 February 2017).

Sample Availability: Samples of the compounds are not available from the authors.

molecules

MDPI

Article

ParaBTM: A Parallel Processing Framework for Biomedical Text Mining on Supercomputers

Yuting Xing [1], Chengkun Wu [1,*], Xi Yang [1], Wei Wang [1], En Zhu [1] and Jianping Yin [2]

[1] School of Computer Science, National University of Defense Technology, Changsha, Hunan 410073, China; xingyuting16@nudt.edu.cn (Y.X.); yangxi1016@nudt.edu.cn (X.Y.); g.webywang@gmail.com (W.W.); enzhu@nudt.edu.cn (E.Z.)

[2] School of Computer Science and Network Security, Dongguan University of Technology, Dongguan, Guangdong 523808, China; jpyin@dgut.edu.cn

* Correspondence: chengkun_wu@nudt.edu.cn; Tel.: +86-135-4964-2841

Received: 9 April 2018; Accepted: 25 April 2018; Published: 27 April 2018

Abstract: A prevailing way of extracting valuable information from biomedical literature is to apply text mining methods on unstructured texts. However, the massive amount of literature that needs to be analyzed poses a big data challenge to the processing efficiency of text mining. In this paper, we address this challenge by introducing parallel processing on a supercomputer. We developed paraBTM, a runnable framework that enables parallel text mining on the Tianhe-2 supercomputer. It employs a low-cost yet effective load balancing strategy to maximize the efficiency of parallel processing. We evaluated the performance of paraBTM on several datasets, utilizing three types of named entity recognition tasks as demonstration. Results show that, in most cases, the processing efficiency can be greatly improved with parallel processing, and the proposed load balancing strategy is simple and effective. In addition, our framework can be readily applied to other tasks of biomedical text mining besides NER.

Keywords: biomedical text mining; big data; Tianhe-2; parallel computing; load balancing

1. Introduction

With the rapid development of biotechnology, the amount of biomedical literature is growing exponentially. For instance, PubMed (https://www.ncbi.nlm.nih.gov/pubmed/), the most recognized biomedical literature database, indexes over 28 million entries for biomedical literature. Most of that information is presented in the form of unstructured texts. It is almost impossible for any domain expert to digest such a massive amount of information within a short period of time. Therefore, automated tools are essential for a systematic understanding of literature. To deal with the literature big data challenge, text mining methods are commonly applied to extract relevant knowledge from vast amounts of literature, and this has become a prominent trend in recent years [1].

Typical tasks of biomedical text mining include named entity recognition and relation extraction. One of the most fundamental tasks of biomedical text mining is named entity recognition (NER). Its task is to recognize target entities that represents key concepts from unstructured biomedical texts, such as proteins, genes, mutations, diseases, etc. There are some existing start-of-art biomedical tools that use text mining methods to identify some specific types of entities, such as mutations [2,3], genes [4,5], and diseases [6,7]. Most of these tools can achieve satisfactory recognition performance (F score over 80%) on standard datasets.

Relation extraction (RE) is a process that typically follows NER and aims to discover semantic connections between entities. Nowadays, there are a number of RE tools using different methods to identify biomedical entity interactions [8,9], such as drug–gene relationships [10–12], gene–disease

Molecules **2018**, *23*, 1028

relationships [9,13,14] and protein–protein interaction [15]. Some of them can achieve high F scores (over 80%) on several annotated datasets.

NER and RE are the preliminary steps in mining information from literature. With the uncovered facts, it is possible to construct a complex knowledge graph, which can assist new knowledge discovery and hypotheses generation. In order to achieve this goal, it is necessary to process as many articles as possible. However, text mining procedures are time consuming. BioContext, for instance, an integrated text mining system for large-scale extraction and contextualization of biomolecular events, took nearly 3 months to complete a full run of the system, which analyzed 20 million MEDLINE abstracts and several hundred thousand PMC open access full-texts using 100 concurrent processes [16]. In addition, some text mining tools, like GNormPlus [17] require a substantial amount of memory ($\geq$5 GB), due to the necessity of loading a large gene dictionary and complementary data structures. Consequently, commodity servers cannot fulfil the computation and storage demands of large-scale text mining. Cloud-based solutions in Map-Reduce mode can partially fulfil computational resource demands. However, practically speaking, many text mining components were written in different languages, and they are dependent on a complex collection of third-party libraries, which prevents them from being readily transplanted into a high-level framework, like Hadoop and Spark. In addition, we dived into the details of load balancing, which cannot be readily supported by Map-Reduce.

An alternative solution to address this computational challenge is to harness the power of high performance computers. High performance computers (HPC) like Tianhe-2 [18] represent high-end computing infrastructures that have traditionally been used to solve compute-intensive scientific and engineering problems. The system configuration of Tianhe-2 is listed in Table 1.

Table 1. The system configuration of Tianhe-2.

Items	Content
Manufacturer	NUDT
Cores	3,120,000
Memory	1,024,000 GB
CPU	Intel Xeon E5-2692v2 12 C 2.2 GHz
Interconnect	TH Express-2
Linpack Performance(Rmax)	33,862.7 TFlop/s
Theoretical Peak(Rpeak)	54,902.4 TFlop/s
HPCG [TFlop/s]	580.109
Operating System	Kylin Linux
MPI	MPICH2 with a customized GLEX channel

Although the software stack on Tianhe-2 is designed and optimized for compute-intensive tasks, its high-performance architecture does provide the capability and capacity of big data processing. Nonetheless, to employ Tianhe-2 for big data processing is not a trivial task, which requires expert knowledge of the system architecture and parallel programming. The programming model is MPI-based (message passing interface) [18], which adds an extra dimension of complexity to normal programming languages like C/C++, Python, Java, etc. Most existing text mining tools are implemented without parallel processing. Therefore, it is necessary to develop an enabling framework that can support parallel text mining without the need to rewrite the original code. In this paper, we develop a parallel processing framework for text mining on the Tianhe-2 supercomputer. The framework integrates text mining tools as plugins. It unifies the input–output stream, implements the parallel processing across multiple compute nodes using the MPI model, and it applies a carefully crafted load balancing strategy to improve the parallelization efficiency. Without a loss of generality, we demonstrate the effectiveness of our framework using multiple NER tools as the demonstration plugins, which can recognize genes, mutations and diseases appearing in biomedical literature. More sophisticated tools of biomedical text mining can be readily integrated into the framework. In the remaining of this paper, we will introduce how paraBTM works and evaluate its performance on Tianhe-2.

2. Results and Discussion

To verify the effectiveness of the parallel framework, we constructed a corpus named 60K, which consists of 60,000 randomly selected articles from PubMed. For NER plugins, we chose three state-of-the-art tools (GNormPlus [17], tmVar [2], DNorm [7]), developed by NCBI (National Centre for Biotechnology Information). We measured the performance in terms of the total processing time and the average processing time (across all processes), and the total time includes the time of initialization and the actual processing time of different plugins.

The 60 K corpus is presented in the NXML format, which is a standard format provided by NCBI. Titles, abstracts, and full-texts from NXML files are extracted and re-written in the PubTator format. All input and output files processed by paraBTM should follow the PubTator format and the PubTator format starts with:

<PMID>|t|<Title of the paper>
<PMID>|a|<The rest of the paper>

The output file will be appended with annotated information like named entities followed in a tab-separated way.

A basic fact is that the time overhead of text mining is not proportional to the number of input articles. We verified this via a single process run over several groups of randomly selected articles. The result is depicted in Figure 1. Here, different colors represent different processing plugins. Related numbers are also listed in Table 2.

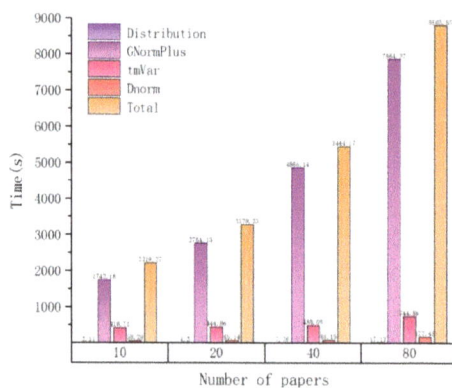

Figure 1. The time cost of processing different numbers of input articles in serial.

Table 2. Time distribution of processing different numbers of input articles in serial.

Number of Papers	Time(s)				
	Distribution	GNormPlus	tmVar	Dnorm	Total
10	2.11	1742.16	416.71	58.29	2219.27
20	4.2	2764.13	444.86	65.14	3278.33
40	7.76	4865.14	488.08	83.19	5444.17
80	17.17	7864.37	744.86	177.45	8803.85

As the number of input files increases, the time cost also increases but not linearly. For example, when the number of input articles is equal to 10, it takes about 36 min for tagging entities, and when the number of articles increases to 100, the spent time is about 3 h (180 min). This can be attributed to another important observation, that is, the total processing time is approximately proportional to the total input size (sum of file lengths as measured by number of characters), which is illustrated in

Figure 2 (size unit is MB, mega-bytes) and Table 3. The workload of each plugin can be better estimated by the total length of input files, which is the basis for our load balancing strategy in the following part.

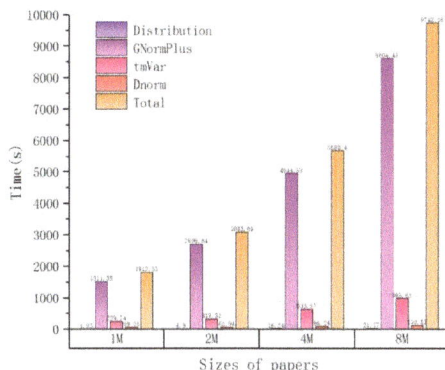

Figure 2. The time cost of processing different input sizes in serial.

Table 3. Time cost distribution of processing different input sizes in serial.

Sizes of Papers (M)	Time (s)				
	Distribution	GNormPlus	tmVar	Dnorm	Total
1	1.93	1511.35	239.74	59.51	1812.53
2	4.80	2696.64	319.31	62.94	3083.69
4	16.24	4944.35	633.57	86.24	5680.40
8	21.77	8604.43	985.63	130.33	9742.16

Figure 3 shows the time spent on paraBTM processing with different numbers of parallel processes on an input dataset of 16 MBs (including 175 articles) which is composed of articles randomly selected from the 60 K corpus. Parallel processing greatly reduces the processing time and different load-balancing strategies do affect the parallel efficiency. paraBTM costs about 500 s (under the Short-Board balancing strategy) when 64 processes are employed, which is around 1/16 the processing time of 2 processes. To note, each process needs to carry out initialization for every plugin, which means you cannot reduce the total processing time any further if the initialization time cost becomes the majority part.

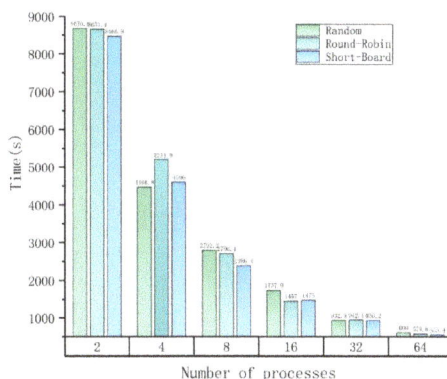

Figure 3. Effects of different load balancing strategies.

To profile different load balancing strategies, we summarize their effects under different parallel scales, as listed in Table 4. In all 6 test cases, the Short-Board strategy is the best in 4 cases and 2nd best in 2 remaining cases. We employ the load balancing efficiency (LBE) to quantify the effects of different strategies. Here, LBE is defined as:

$$LBE = AET/MET \tag{1}$$

Here, AET is the average execution time and MET is the maximum execution time. According to the above definition, the maxima of LBE is 1 (achieved if AET is equal to MET) and a greater LBE represents a better load balancing efficiency.

Table 4. Profiling of different load balancing strategies.

(a) Maximum times on different numbers of parallel processes with different strategies.

Number of Processes	Maximum Time among All Processes (s)		
	Random	Round-Robin	Short-Board
2	8676.60	8651.42	8466.89
4	4466.81	5211.96	4596.09
8	2792.29	2706.41	2386.45
16	1737.94	1457.01	1475.07
32	932.85	942.10	930.25
64	609.06	579.8	553.42

(b) Average times on different numbers of parallel processes with different strategies.

Number of Processes	Average Time of All Processes (s)		
	Random	Round-Robin	Short-Board
2	8513.10	8374.08	8392.46
4	4001.38	4535.42	4393.07
8	2287.19	2354.14	2242.42
16	1265.46	1174.12	1270.33
32	624.73	666.91	640.63
64	379.08	376.06	366.84

(c) Load balancing efficiencies on different numbers of parallel processes with different strategies.

Number of Processes	Efficiency (Average/Maximum)		
	Random	Round-Robin	Short-Board
2	0.98	0.97	0.99
4	0.90	0.87	0.96
8	0.82	0.87	0.94
16	0.73	0.81	0.86
32	0.67	0.71	0.69
64	0.62	0.65	0.66

Figure 4 shows that the Short-Board strategy exhibits the best LBE in almost all test cases. However, LBE values drop significantly when the number of parallel processes is greater than 16 in the 16M test set. The reason is that this test set contains only 175 articles, which means each process will only process two articles on average. If the input data set is big enough, the LBE will be maintained at a satisfactory level.

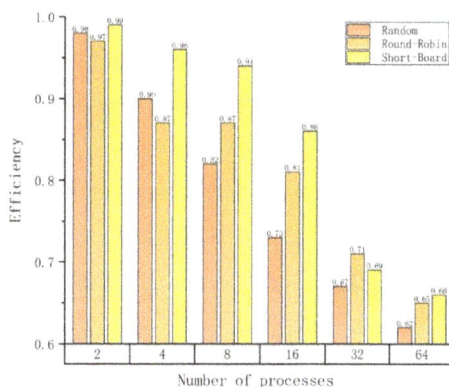

Figure 4. Load balancing efficiencies.

We also conducted an experiment on the whole 60 K corpus (61,078 articles). Table 5 shows that it took over 12 h to process 61,078 papers through three NER plugins (128 nodes under the Short-Board strategy, each node runs 5 processes). According to the results, we can see that parallelization greatly enhances the processing efficiency. To note, the speed-ups of different plugins differ as each plugin has its own characteristic computation and memory access patterns. To carry out a full-scale processing on the whole PMC-OA full-text dataset (over 1 million), it will take about 200 h if we only use 128 nodes. Fortunately, the computation capacity of Tianhe-2 is enormous, and we can reduce the total time down to several hours by harnessing the power of a few thousand nodes (over 16,000 available on Tianhe-2). We plan to carry out a full analysis on the whole PubMed dataset (the real large-scale biomedical texts) in the future. However, the cost of such a full run is currently beyond our funding support. We are currently in the application process of a bigger grant for this large-scale analysis. In our previous study, we have demonstrated that using text mining on a larger dataset does provide more comprehensive and insightful results compared with using a small dataset (say, can be handled by a few people) using thyroid cancer as a case study [19].

Table 5. The processing time of 61,078 papers running on 128 processes.

Number of Processes	Time (s)				
	Distribution	GNormPlus	tmVar2.0	Dnorm	Total
1	18,934.18	5,874,482.04	654,145.38	82,455.3	6,630,016.9
128	3643	23,733	16,214	233	43,823
Speed-up (x)	5.20	247.52	40.34	353.89	151.29

3. Materials and Methods

3.1. Data Sources and Storage

The biomedical literature has typical characteristics of large quantity, professional content, public resources, easy-accessibility, etc. Because of these characteristics, biomedical literature data has become one of the most noticeable data in biomedical field. For example, PubMed Central (PMC) is a free digital repository that archives publicly accessible articles. Until now, PMC has contained over 4.1 million references to full-text journal papers, covering a wide range of biomedical fields, and the literature data is stored in NXML format, from which we can extract some parts according to our interest.

However, most of the state-of-art NER tools do not support parallel processing, and it would take an enormous amount of time if we want to process the massive set of biomedical literature. One feasible solution is to harness the computing power provided by HPC systems by implementing a parallel NER processing framework. With this framework, text mining tools can be easily integrated into the framework and developers will not need to consider the details of parallel processing.

There are different levels of parallelism in text mining tasks. First, each input article is relatively independent; secondly, multiple sentences in each of the articles can be approximately regarded as independent. However, in practice, we usually use a single file as a processing unit, the reason is that many text mining tools spend a substantial amount of time to initialize on each processing pass. In addition, the memory size also limits the number of processes that can run in parallel on each computing node. For instance, on Tianhe-2 each node is equipped with 24 cores and 64 GB of memory, and the stable memory that users can control is about 50 GB (the operating system and other necessary tools need to use about 10 GB). The memory costs of a typical TmVar and gnormplus run for NER can be up to 5 GB and 10 GB. Therefore, at most 5 GNormPlus processes and 10 TmVar processes can run on one node. Figure 5 shows the implementation and deployment of a text mining system (paraBTM) in large-scale parallel environment.

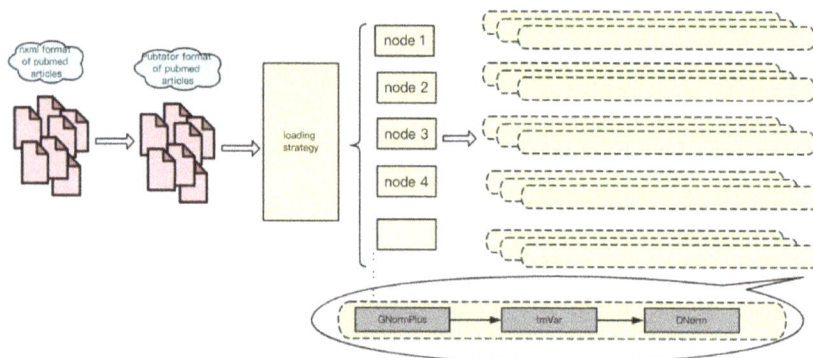

Figure 5. The implementation and deployment of paraBTM in large-scale parallel environment.

3.2. Parallel Processing

3.2.1. MPI-Based Multi-Node Computation

The message passing interface (MPI) is a standard model for parallel programming on HPCs. It is well established over 20 years, and has been implemented in different sorts of programming languages including C/C++ and Python. Our method can run on any supercomputer or cluster configured with MPI support. To note, different supercomputers might have different node configurations. When running on other platforms, the configuration (RAM, number of concurrent processes) might have to change accordingly.

In this work, we use MPI4PY (http://mpi4py.scipy.org/docs/) to implement the parallel processing. MPI4PY is a well-regarded, clean, and efficient implementation of MPI for Python. Our framework can simultaneously submit many jobs to cores distributed across computing nodes in Tianhe-2.

3.2.2. Load Balancing Strategy

A typical challenge in parallel computation is load unbalance, that is, workload is unevenly distributed among nodes, making some nodes very busy for a long time and others idle [20]. In this paper, we address this problem by designing an effective load balancing strategy.

Given a set F of files to be processed $F_0, F_1, F_2 \ldots F_{N-1}, F_n$, we initialize processes $P_0, P_1, P_2, \ldots \ldots, P_{n-1}, P_n$, and the number of processes is N, the problem is to allocate each file F_x to an appropriate process P_{rank}.

A naive solution is to randomly distribute target files into nodes. We can simply distribute files to by modulo operation rank $=\mathrm{mod}(P, \mathrm{size})$, P and size represent the position of the target file in the file list and number of processes respectively, and file F_p finally should be sent to P_{rank}. According to the formula, each file F to be processed is distributed to process $rank = P \% size$ in turn. As the files are arranged in a random order, this process is actually a simulation of random distribution. This is a naïve strategy and easy to implement. However, this strategy does not take into consideration the length of each file, and will very likely cause an unbalanced load distribution, which would detriment the overall parallel efficiency. For instance, if the total length of files assigned to one specific node is far larger than others, then the overall running time will be prolonged until this slowest node finishes. Figure 6 shows an example of the naïve random load balancing strategy.

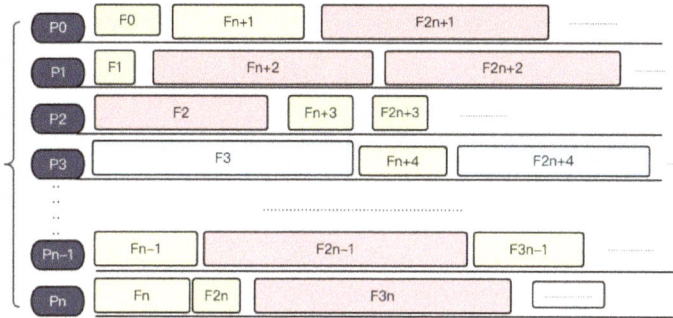

Figure 6. Random load balancing strategy.

A slightly more complex load balancing method is the round-robin (RR) method. Round-robin algorithm is a term that originally comes from the field of operating systems. Here, the general idea inspires us to mix small files with large files together into one process. After sorting files by size (see Figure 7), the system will assign files into processes in a snakelike way, making the size of files loaded in every process remains relatively balanced. Figure 8 shows an example of RR algorithm.

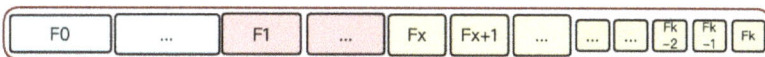

Figure 7. A sorted file list.

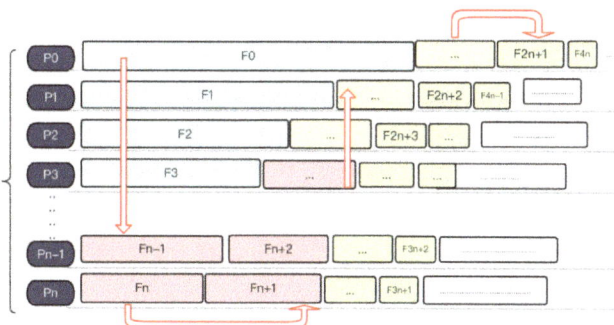

Figure 8. Round-robin algorithm.

The round-robin method also allocates the same number of files, and its serpentine way of load assignment ensures that the total size of the files in each process remains relatively balanced, since files were sorted by size in advance. However, in some circumstances, the lengths of input articles can be very biased, say, some files are extremely long while many others are short. In such cases, the RR method fails.

Instead of assignments based on the number of files, we proposed our "Short-Board" method. Firstly, the files that need to be processed are sorted in descending order according to the length of each file, and then files that need to be processed in the file list are sequentially fetched out and dispatched to the process whose current load is the smallest. Figure 9a–d shows an example of Short-Board algorithm. The pseudo code of Short-Board is shown in Figure 10.

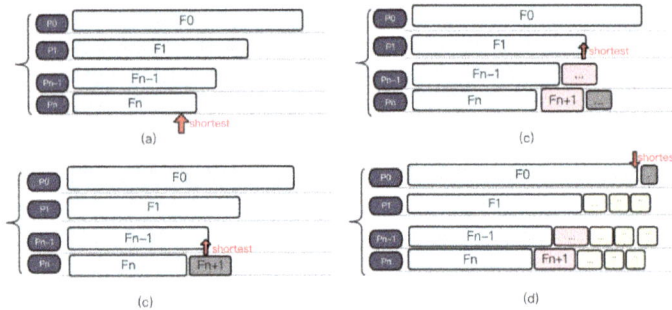

Figure 9. A demonstration of the Short-Board load balancing algorithm.

Loading Strategy: Short-Board
Input: sorted file information list by size in a descending order *Flist*   number of process *N*

Initialization:*ProcessList*(a list consist of processinfo)
            ProcessInfo[*index, capacity, subfilelist*]
            (information for each process, including the index in the
 processslist, the capacity, and the assigned subfile list)
            ReverseFlag(signal of whether the assignment list is reverse)
 	
 /* start the Short-Board Strategy */
 for each file ∈ Flist **do**
 	
            /* find the process which has smallest capacity */
            AssignIndex = FindMinCap(*ProcessList*)
            Use the AssignIndex to find the corresponding ProcessInfo in the
      ProcessList P[*index, capacity, subfilelist*]]
            P=ProcessList[AssignIndex]
            P[index]=AssignIndex
            P[capacity]=P[capacity] + filesize
            P[subfilelist] =P[subfilelist] ∪ fileindex
 	
 end for

Output: *ProcessList*

Figure 10. The pseudo code of Short-Board.

4. Conclusions

In this paper, we present paraBTM, a parallel framework for biomedical text mining developed on the Tianhe-2 supercomputer. It supports different types of components as plugins and its usage is straightforward. The parallel efficiency is guaranteed by a carefully devised load balancing strategy. We evaluated the performance of paraBTM on both small- and large-scale datasets. Experimental results validate that paraBTM effectively improve the processing speed of biomedical named entity recognition. On large scale of datasets, ParaBTM managed to process 60178 PubMed full-text articles in about 12 h. paraBTM is open-source and available at https://github.com/biotm/paraBTM.

Author Contributions: Y.X. and C.W. developed the algorithms and drafted the manuscript. They developed the codes of paraBTM together with X.Y. and W.W.; E.Z. and J.Y. proposed the idea of the work and revised the whole manuscript. All the authors have read and approve the manuscript.

Funding: This research was funded by the National Key R&D Program of China grant number 2018YFB1003203 and National Natural Science Foundation of China grant number 31501073, 61672528 and 61773392.

Acknowledgments: Special thanks to Zhuo Song and Gen Li (from GeneTalks Ltd., Beijing) for providing knowledge support on genomic variations.

Conflicts of Interest: The authors declare no conflict of interest.

References

1. Rebholz-Schuhmann, D.; Oellrich, A.; Hoehndorf, R. Text-mining solutions for biomedical research: Enabling integrative biology. *Nat. Rev. Genet.* **2012**, *13*, 1–11. [CrossRef] [PubMed]
2. Wei, C.-H.; Harris, B.R.; Kao, H.-Y.; Lu, Z. tmVar: A text mining approach for extracting sequence variants in biomedical literature. *Bioinformatics* **2013**, *29*, 1433–1439. [CrossRef] [PubMed]
3. Thomas, P.; Rocktäschel, T.; Hakenberg, J.; Lichtblau, Y.; Leser, U. SETH detects and normalizes genetic variants in text. *Bioinformatics* **2016**, *32*, 2883–2885. [CrossRef] [PubMed]
4. Wei, C.-H.; Kao, H.-Y. Cross-species gene normalization by species inference. *BMC Bioinform.* **2011**, *12* (Suppl. 8), S5. [CrossRef] [PubMed]
5. Pan, X.; Shen, H.B. OUGENE: A disease associated over-expressed and under-expressed gene database. *Sci. Bull.* **2016**, *61*, 752–754. [CrossRef]
6. Leaman, R.; Lu, Z. TaggerOne: Joint named entity recognition and normalization with semi-Markov Models. *Bioinformatics* **2016**, *32*, 2839–2846. [CrossRef] [PubMed]
7. Leaman, R.; Islamaj Dogan, R.; Lu, Z. DNorm: Disease name normalization with pairwise learning to rank. *Bioinformatics* **2013**, *29*, 2909–2917. [CrossRef] [PubMed]
8. Quan, C.; Wang, M.; Ren, F. An unsupervised text mining method for relation extraction from biomedical literature. *PLoS ONE* **2014**, *9*, e102039. [CrossRef] [PubMed]
9. Xu, D.; Zhang, M.; Xie, Y.; Wang, F.; Chen, M.; Zhu, K.Q.; Wei, J. DTMiner: Identification of potential disease targets through biomedical literature mining. *Bioinformatics* **2016**, *32*, 3619–3626. [CrossRef] [PubMed]
10. Xu, R.; Wang, Q. A knowledge-driven conditional approach to extract pharmacogenomics specific drug-gene relationships from free text. *J. Biomed. Inform.* **2012**, *45*, 827–834. [CrossRef] [PubMed]
11. Percha, B.; Garten, Y.; Altman, R.B. Discovery and explanation of drug-drug interactions via text mining. *Pac. Symp. Biocomput.* **2012**, 410–421. [CrossRef]
12. Segura-Bedmar, I.; Martínez, P.; de Pablo-Sánchez, C. Using a shallow linguistic kernel for drug-drug interaction extraction. *J. Biomed. Inform.* **2011**, *44*, 789–804. [CrossRef] [PubMed]
13. Bravo, À.; Piñero, J.; Queralt-Rosinach, N.; Rautschka, M.; Furlong, L.I. Extraction of relations between genes and diseases from text and large-scale data analysis: Implications for translational research. *BMC Bioinform.* **2015**, *16*, 55. [CrossRef] [PubMed]
14. Pletscher-Frankild, S.; Pallejà, A.; Tsafou, K.; Binder, J.X.; Jensen, L.J. DISEASES: Text mining and data integration of disease–gene associations. *Methods* **2015**, *74*, 83–89. [CrossRef] [PubMed]
15. Franceschini, A.; Szklarczyk, D.; Frankild, S.; Kuhn, M.; Simonovic, M.; Roth, A.; Lin, J.; Minguez, P.; Bork, P.; von Mering, C.; et al. STRING v9.1: Protein-protein interaction networks, with increased coverage and integration. *Nucleic Acids Res.* **2013**, *41*, 808–815. [CrossRef] [PubMed]

16. Gerner, M.; Sarafraz, F.; Nenadic, G.; Bergman, C.M. BioContext: An integrated text mining system for large-scale extraction and contextualisation of biomolecular events. *Bioinformatics* **2012**, *28*, 2154–2161. [CrossRef] [PubMed]

17. Wei, C.-H.; Kao, H.-Y.; Lu, Z. GNormPlus: An Integrative Approach for Tagging Genes, Gene Families, and Protein Domains. *BioMed Res. Int.* **2015**, *2015*, 918710–918717. [CrossRef] [PubMed]

18. Liao, X.; Xiao, L.; Yang, C.; Lu, Y. MilkyWay-2 supercomputer: System and application. *Front. Comput. Sci.* **2014**, *8*, 345–356. [CrossRef]

19. Wu, C.; Schwartz, J.M.; Brabant, G.; Nenadic, G. Molecular profiling of thyroid cancer subtypes using large-scale text mining. *BMC Med. Genom.* **2014**, *7* (Suppl. 3), S3. [CrossRef] [PubMed]

20. Kaur, S.; Kaur, G. A Review of Load Balancing Strategies for Distributed Systems. *IJCA* **2015**, *121*, 45–47. [CrossRef]

Sample Availability: Samples of the compounds are not available from the authors.

![molecules logo] *molecules*

MDPI

Article

To Decipher the *Mycoplasma hominis* Proteins Targeting into the Endoplasmic Reticulum and Their Implications in Prostate Cancer Etiology Using Next-Generation Sequencing Data

Mohammed Zakariah [1,†], Shahanavaj Khan [2,3,*,†], Anis Ahmad Chaudhary [4], Christian Rolfo [5], Mohamed Maher Ben Ismail [6] and Yousef Ajami Alotaibi [6]

1   Research Center, College of Computer and Information Science, King Saud University, Riyadh 11451, Saudi Arabia; mzakariah@ksu.edu.sa
2   Nanomedicine & Biotechnology Research Unit, Department of Pharmaceutics, College of Pharmacy, P.O. Box 2457, King Saud University, Riyadh 11451, Saudi Arabia
3   Department of Bioscience, Shri Ram Group of College (SRGC), Muzaffarnagar 251002, UP, India
4   Department of Pharmacology, College of Medicine, Al-Imam Mohammad Ibn Saud Islamic University, Riyadh 11451, Saudi Arabia; anis.chaudhary@gmail.com
5   Phase I-Early Clinical Trials Unit, Oncology Department, Antwerp University Hospital, "Centre for Oncological Research (CORE)", 2650 Edegem, Belgium; Christian.Rolfo@uza.be
6   Computer Science Department, College of Computer and Information Sciences, King Saud University, Riyadh 11451, Saudi Arabia; mbenismail@ksu.edu.sa (M.M.B.I.); yaalotaibi@ksu.edu.sa (Y.A.A.)
*   Correspondence: khan.shahanavaj@gmail.com; Tel.: +91-921-999-3262
†   These authors contributed equally to this work.

Academic Editors: Xiangxiang Zeng, Alfonso Rodríguez-Patón and Quan Zou
Received: 7 March 2018; Accepted: 18 April 2018; Published: 24 April 2018

Abstract: Cancer was initially considered a genetic disease. However, recent studies have revealed the connection between bacterial infections and growth of different types of cancer. The enteroinvasive strain of *Mycoplasma hominis* alters the normal behavior of host cells that may result in the growth of prostate cancer. The role of *M. hominis* in the growth and development of prostate cancer still remains unclear. The infection may regulate several factors that influence prostate cancer growth in susceptible individuals. The aim of this study was to predict *M. hominis* proteins targeted into the endoplasmic reticulum (ER) of the host cell, and their potential role in the induction of prostate cancer. From the whole proteome of *M. hominis*, 19 proteins were predicted to be targeted into the ER of host cells. The results of our study predict that several proteins of *M. hominis* may be targeted to the host cell ER, and possibly alter the normal pattern of protein folding. These predicted proteins can modify the normal function of the host cell. Thus, the intercellular infection of *M. hominis* in host cells may serve as a potential factor in prostate cancer etiology.

Keywords: prostate cancer; *Mycoplasma hominis*; endoplasmic reticulum; systems biology; protein targeting

1. Introduction

Bacterial infection is recognized to play a significant role in the progression and advancement of various forms of cancers, including prostate, lung, gastric, and colon cancer [1–3]. Recent study showed that prostate gland restrains a plethora of different strains of bacteria [4]. Bacterial dysbiosis, inflammation, and other factors are associated with the growth of prostate cancer, although the exact mechanisms involved in growth of cancer due to bacterial infection are not very clear. Mycoplasmas are bacteria that lack cell walls, and are causative agents of various diseases related to respiratory and urogenital tract among humans [5,6]. The dominant types of mycoplasmas found in the urogenital system of humans include *Mycoplasma genitalium*, *Ureaplasma urealyticum*, and *M. hominis*. The relationship between mycoplasmas and the human population was first detected in the 1960s [7]. Different studies have highlighted the connection between *M. hominis* infection and prostate cancer advancement [8–11]. This association was further supported by numerous studies on *M. hominis* infection and higher classification of prostate cancer. The cell cycle signal cascade, including DNA repair mechanisms and apoptosis, may be altered following mycoplasma infection [8]. Although a meta-analysis report has suggested a suspicious role of *M. hominis* in the growth of prostate cancer [12], various recent studies instead confirmed the involvement of different bacteria in the progression of different types of cancer [13–15]. Pathogenic bacteria and their subcellular constituents interact with several types of host cell receptors that may change the expression of various genes. These enigmatic alterations may affect the normal control and regulation of host cells [8,16]. Chronic inflammation and chronic infections (or both) are the cause of 20% of different forms of cancers in humans.

Induction of *pro-inflammatory cytokines* and reactive *oxygen species*, regulated by *chronic inflammation*, may promote nitration and chlorination of nucleic acids and proteins. *M. hominis* infection serves as a factor promoting the growth and development of prostate cancer, but several other causes are linked with the growth of prostate cancer [17–19]. Apart from chronic inflammation and mutations, different cyclomodulins have been associated with the growth of prostate cancer by the disturbance of homeostasis in *M. hominis*-infected cells. Some specific strains of bacteria have ability to produce different toxins known as cyclomodulins, that interfere in the host cell cycle, which suggests a potential association of different pathogenic bacteria with different type of cancers [20]. It has supposed that cyclomodulins have the capability to affect the usual growth cycle of the host cell, and are expected to grow as etiological aspects for *M. hominis*-mediated prostate cancer [8]. *M. hominis* strain is considered a usual Gram-negative pathogen. It is known to multiply and inhabit intracellularly during the progression of prostate cancer [21,22]. As *M. hominis* is colonized in the urogenital tract that comprises the prostate, *M. hominis* infection leads to some precise effects in the progression of prostate cancer. The mass collection of genomes for *M. hominis* revealed 715,165 base pairs and a G + C content of 26.94%.

Various strains of different bacteria are involved in the intracellular infection and duplication in specific host cells, wherein bacterial pathogens change the usual functioning of cells through the localization of their own proteins in different components of host cells, such as endoplasmic reticulum (ER), Golgi complex, mitochondria, nucleus, plasma membrane, and cytoplasm [13,21,22]. The complete genome of ATCC 27545 strain of *M. hominis* contains 563 open reading frames (ORFs) and encodes different enzymes and proteins. *M. hominis* has the capacity to naturally undergo intercellular replication, allowing the localization of numerous proteins within the host. The targeted proteins work as component of the host cell proteome. Hence, it is likely that several *M. hominis* proteins may possibly get localized within the host cell, due to the availability of signature sequence and evolutionary relatedness of proteins targeted within the cellular compartments of host cells. ER is an main compartment involved in proper protein folding, post-translation modification, translocation, and regulation of cellular homeostasis [23,24]. The unfolded protein response (UPR) is a highly conserved evolutionarily adaptive response that disrupts the ER physiology. UPR has been shown to be altered by different viruses and plays various roles during bacterial infection [25,26]. Both UPR and ER stress activation are involved in the growth and progression of various types of cancers [27]. The whole proteome of *M. hominis* may disturb the normal behavior of infected host cells and get involved in the development of prostate cancer. The main objective of the current work was to predict the protein localization of *M. hominis* in host cells and evaluate their role in the etiology of prostate cancer. We focused on ER proteins using bioinformatics tools and techniques and explored protein localization of *M. hominis* in the ER of host cells. We investigated the possible implication and relations of *M. hominis* proteome in the etiology of prostate cancer.

2. Results

2.1. Selection of the Whole Proteome Database of M. hominis

The whole protein sequence of ATCC-27545 strain of *M. hominis* was collected from UniProt, which has maximum proteins (563) with respect to other available strains. The unique selection of UniProt database was associated with its specific characteristics, including opulent, entirely classified, comprehensive, and accurately annotated sequences.

2.2. Prediction of the Subcellular Localization by cNLS Mapper

Nineteen proteins of *M. hominis* were predicted to be targeted in the ER with different NLSs (Figure 1). Figure 1 shows that among 19 ER targeting proteins, 15, 03, and 01 protein exhibit 0–3.0, 3.0–5.0, and 5.0–8.0 monopartite NLSs cutoff values respectively. Furthermore, the bipartite NLS was observed in 19 proteins and 06, 11, and 02 proteins exhibited 0–3.0, 3.0–5.0, and 5.0–8.0 cutoff values for NLSs, respectively. Different *M. hominis* proteins in a particular host cell may change the usual functioning of the host cell and promote the process of growth and development of cancer. cNLS (classical nuclear localization signals) mapper worked on amino acid sequence patterns, executed by three easy rules according to the NLSs classification [28]; these three rules are principally clusters of K and R basic amino acids and spaces between the clusters. cNLS mapper predicted nuclear localization signals for eukaryotic cells in *M. hominis* proteins. The literature of cNLS mapper has showed that the proteins with cut off value 8–10, 7–8, 3–5, and 1–2, were analyzed as targeted to the nucleus, targeted to both nucleus, partially targeted to nucleus and cytoplasm, and targeted to the cytoplasm, respectively. On the existence of multiple NLS sequence in same proteins, the elevated cut off value was documented.

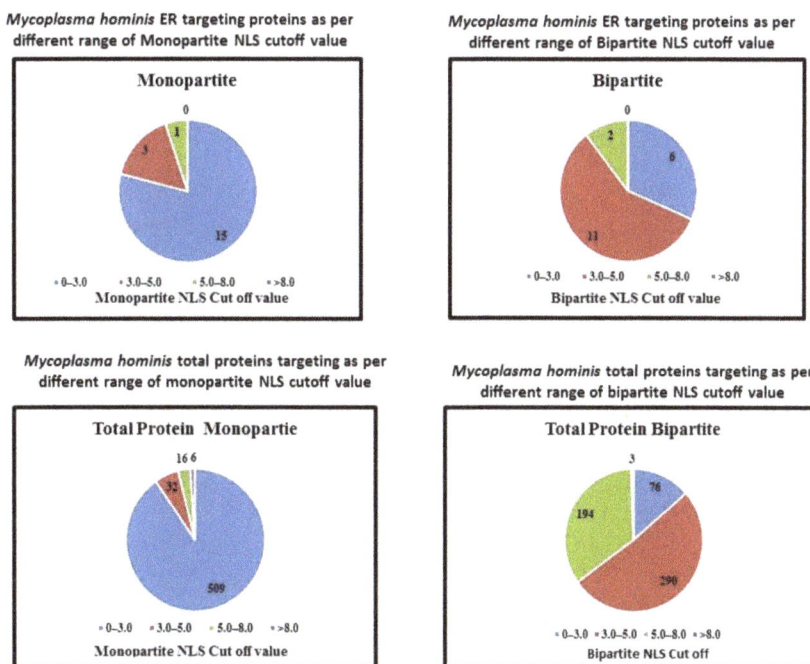

Figure 1. In silico analyses of the *Mycoplasma hominis* proteins localization in endoplasmic reticulum of host cells and their relationship with various NLS values.

2.3. Prediction of ER Localization in M. hominis Proteins Using Hum-Mploc 3.0

From the whole proteins (563) of *M. hominis*, only 19 were observed to be targeted into the ER of the host cell, as analyzed using Hum-mPLoc 3.0. The increase in cutoff values of NLS in monopartite and bipartite resulted in a decrease in ER localization (except 3.0–5.0 cutoff value of bipartite) (Table 1). Similarly, the increase in the molecular weight resulted in the decrease in the ER protein targeting of *M. hominis* (except for the range of 0–20 kDa). Proteins of 20–40 kDa molecular weights were observed mainly localized in the ER of the host cell (Table 2). Moreover, the outcome of values of isoelectric point (pI) failed to illustrate any constant pattern for ER localization of different proteins of *M. hominis* (Table 3). However, it has demonstrated that the bimodal character (pI and subcellular localization) in bacteria is likely to be a common property of proteomes, and is connected with the requirement of various pI values depending on subcellular protein localization [29].

Table 1. Computational prediction of *M. hominis* proteins targeted to the endoplasmic reticulum (ER) of host cells and their relation to all proteins with a similar NLS.

NLS	NLS Cutoff	Number of Proteins Targeting ER	Total Number of Proteins in This Range	Percentage
Monopartite NLS	0–3.0	15	509	2.94
	3.0–5.0	3	32	9.3
	5.0–8.0	1	16	6.25
	>8.0	0	6	0
Bipartite NLS	0–3.0	6	76	7.89
	3.0–5.0	11	290	3.79
	5.0–8.0	2	194	1.03
	>8.0	0	3	0

Table 2. Computational prediction of *Mycoplasma hominis* proteins targeted to ER of host cells and their relation to all proteins with similar molecular weight.

Molecular Weight	Number of Proteins Targeting to ER	Total Number of Proteins	Percentage
0–20 kDa	3	128	2.34
20–40 kDa	10	205	4.87
40–60 kDa	4	109	3.66
60–80 kDa	1	61	1.63
>80 kDa	1	60	1.66

Table 3. Computational prediction of *Mycoplasma hominis* proteins targeting to ER of host cells and their relation to all proteins with similar pI value.

Range of pI Value	Number of Proteins Targeting to ER	Total Number of Proteins	Percentage
3.0–5.0	0	25	0
5.0–6.0	1	105	0.95
6.0–7.0	0	68	0
7.0–8.0	0	35	0
8.0–9.0	2	109	1.83
9.0–10.0	12	181	6.62
10.0–11.0	4	33	12.12
12.0–13.0	0	7	0

The protein localization patterns of *M. hominis* in the ER of the host cell with diverse NLS values are shown in Figure 1. The patterns of protein localization in *M. hominis* for various ranges of molecular weights are shown in Figure 2. In the Figure 3, we illustrate the protein localization patterns of *M. hominis* in the ER of the host cell at various ranges of pI values. Various ER-targeting proteins are thought to be involved in different biochemical pathways. The details of the outcome of the ER-targeting *M. hominis* proteins with different roles are shown in Table 4. Various localized proteins could interfere in the regular growth behavior of host cells, thus leading to the alteration in the regular functioning of the host cell biochemical pathways. We have suggested a possible association of ER-targeting proteins of *M. hominis* in prostate cancer etiology. We have recently predicted the possible impacts of nucleus-, mitochondria-, and cytoplasm-targeting *M. hominis* proteins on the carcinogenesis of prostate cancer [10,30].

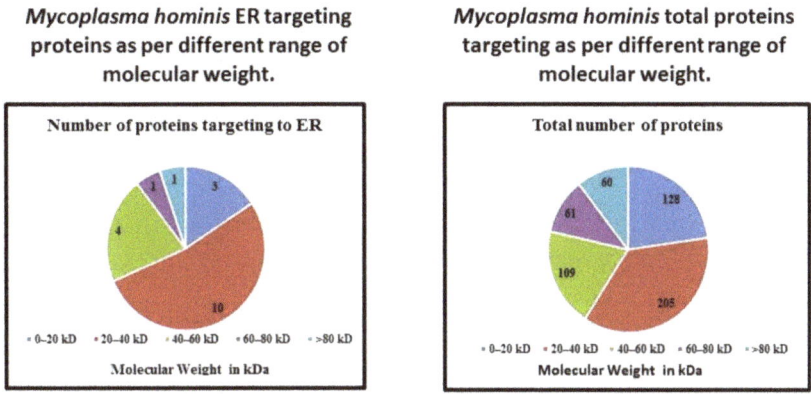

Figure 2. In silico analyses of the *Mycoplasma hominis* proteins localization in endoplasmic reticulum of host cells and their relation to proteins with different ranges of molecular weight.

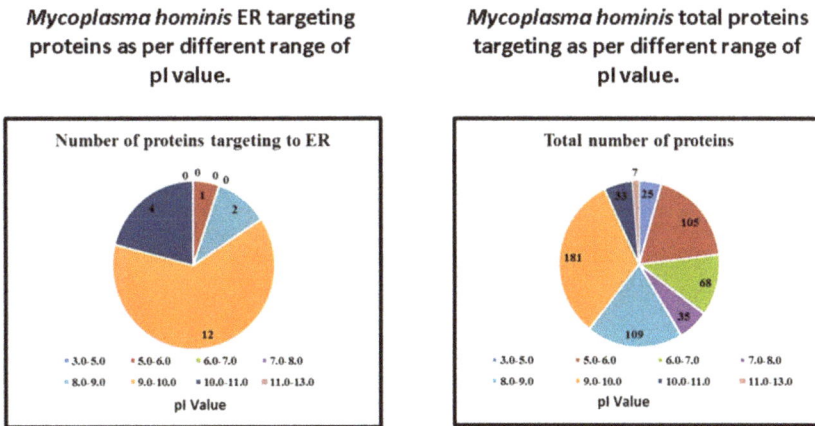

Figure 3. In silico analyses of the *Mycoplasma hominis* proteins localized in endoplasmic reticulum of host cells and their relation to proteins with different ranges of pI values.

Table 4. Details of *Mycoplasma hominis* proteins targeted to ER of host cells with their functions.

Accession Number	Protein Name	Function in Bacteria	Protein Existence	pI	Molecular Weight	NLS Mapper		Hum-mPLoc 3.0
						Monopartite	Bipartite	
A0A097NT54	Uncharacterized protein	Unknown	Protein predicted	10.17	13,450	0	22	Endoplasmic reticulum
A0A097NSU8	Uncharacterized protein	Unknown	Protein predicted	5.45	36,963	0	3.8	Endoplasmic reticulum
A0A097NTC6	Lipoprotein signal peptidase (EC 3.4.23.36)	Aspartic-type endopeptidase activity.	Protein inferred from homology	8.44	23,620	0	3.3	Endoplasmic reticulum
A0A097NTZ8	Cation transporting P-type ATPase (EC 3.6.3.8)	ATP binding, calcium-transporting ATPase activity, and metal ion binding activity.	Protein inferred from homology	8.56	107,148	2	6.1	Endoplasmic reticulum
A0A097NT87	Protein translocase subunit SecY	Intracellular protein transmembrane transport activity, protein transport activity hrough the Sec complex	Protein inferred from homology	9.51	54,926	6	3.9	Endoplasmic reticulum
A0A097NTV1	Membrane protein	Unknown	Protein predicted	9.54	25,294	0	5.5	Endoplasmic reticulum
A0A097NTR0	Prolipoprotein diacylglyceryl transferase (EC 2.4.99.-)	Phosphatidylglycerol-prolipoprotein diacylglyceryl transferase activity and involved in lipoprotein biosynthetic process.	Protein inferred from homology	9.56	35,944	0	3.8	Endoplasmic reticulum
A0A097NSJ5	Potassium transporter KtrB	Cation transmembrane transporter activity.	Protein predicted	9.6	58,434	0	3	Endoplasmic reticulum
A0A097NT10	ComEC/Rec2-related protein	Unknown	Protein predicted	9.65	54,059	0	4.4	Endoplasmic reticulum
A0A097NSI4	Uncharacterized protein	Unknown	Protein predicted	9.66	38,283	0	3.3	Endoplasmic reticulum
A0A097NSM2	Membrane protein	Unknown	Protein predicted	9.72	33,519	0	3.4	Endoplasmic reticulum
A0A097NSJ8	1-acyl-sn-glycerol-3-phosphate acyltransferase (EC 2.3.1.51)	1-acylglycerol-3-phosphate O-acyltransferase activity.	Protein predicted	9.79	31,034	0	4.7	Endoplasmic reticulum
A0A097NT14	Cobalt ABC transporter permease	Unknown	Protein predicted	9.85	35,998	5	4.4	Endoplasmic reticulum
A0A097NTJ8	ABC transporter permease	Transporter activity.	Protein inferred from homology	9.86	65,590	0	2.9	Endoplasmic reticulum
A0A097NTB8	Uncharacterized protein	Unknown	Protein predicted	9.9	57,701	0	5	Endoplasmic reticulum
A0A097NSN9	Uncharacterized protein	Unknown	Protein predicted	9.97	14,493	0	2.6	Endoplasmic reticulum
A0A097NSG1	Uncharacterized protein	Unknown	Protein predicted	10.1	17,661	4	2.6	Endoplasmic reticulum
A0A097NTR2	Membrane protein	Transporter activity.	Protein predicted	10.19	33,456	4	3.2	Endoplasmic reticulum
A0A097NT45	Membrane protein	Unknown	Protein predicted	10.47	20,769	0	2.5	Endoplasmic reticulum

3. Discussion

Prostate cancer is the frequently identified cancer and sixth leading reason of cancer-related death globally. The frequency of prostate disease has relentlessly expanded in Asian nations, including China, India, and Malaysia [19]. Cancer was initially considered as a genetic disease. Nevertheless, various studies have revealed the association between bacterial infections and progression of different cancer types. The infection may regulate different factors that influence prostate cancer progress in susceptible individuals. It has been reported that protein molecules of host cells face chronic and acute challenges for the maintenance of their integrity [31]. Proteostasis/protein homeostasis allows health and proper growth of eukaryotic cells. The proteostasis depends upon a complex network of proteases molecular, proteins, chaperones, and different regulatory factors [32]. Inadequacies in protein homeostasis or proteostasis direct to several cardiovascular, neurodegenerative, metabolic, and oncological disorders [31]. Different factors, such as chemical toxins, exogenous proteins, drugs, and environmental factors compromise proteostasis, which stimulates proteome stress, and cause different disorders [33,34]. Proteostasis in humans is regulated through quality control (QC) network of approximately 800 different proteins [35,36]. The ER is an important subcellular compartment responsible for the regulation of protein folding, post-translation modification, and protein translocation. Interruption in the environment of ER by pathological agents may cause alterations in glycosylation, DNA damage, nutrient deprivation, oxidative stress, calcium depletion, and energy fluctuation/disturbance, thereby resulting in ER stress and consequent accretion of misfolded or unfolded proteins in the ER lumen. These host cells must overcome perturbations of ER functions and stress for their survival. If ER stress is left unresolved, it may disturb the normal functioning of apoptosis [37]. The disturbance in apoptotic regulatory protein Bcl-2 function results in the increased transcription of p53 unregulated modulator of apoptosis (PUMA), Bcl2-like11 (BIM), BH3-only proteins, and NADPH oxidase activator (NOXA). ER stress promotes the interactions between Bax and PUMA, resulting in the release of cytochrome *c* and apoptosis activation by caspase-dependent modulation of p53 proteins. Many *M. hominis* proteins outmaneuver the host cell machinery, and possibly change the normal behavior of host cells, which may promote the growth of prostate cancer. The involvement of *M. hominis* in the progression of prostate cancer is not yet clear. The rationale of this study was to predict *M. hominis* proteins that are targeted into the ER of the host cell, and assess their potential roles in the growth and progression of prostate cancer.

Of the whole proteome, 19 proteins of *M. hominis* were expected to be targeted to the ER of host cells. These predicted proteins of *M. hominis* may modify the normal function of the host cell. The study of protein targeting into the host cell is very important to detect the progression and development of cancer, especially if the cancer growth is associated with the intracellular bacterial infection. The targeting of different bacterial proteins into host cell compartments, such as cytoplasm, mitochondria, and nucleus, has an important effect on the etiology of different cancer types [14,15,38]. The host cell gets affected by various types of bacterial proteins, which alter the usual development and normal behavior of the host cell. Infection is considered as a possible factor in the progression and development of different types of cancer, especially when it is linked with chronic inflammation that leads to cancer progression in various cases [39,40]. Several novel techniques could be developed to detect and treat cancer based on the study of chronic infection related to cancer and their mechanisms and activities that promote cancer [40]. The potential involvement of infection of *M. hominis* in prostate cancer and its evolution and progression depend on the estimation of its protein targeting into various sections of host cells. The protein-targeting ability of *M. hominis* may lead to several consequences related to prostate cancer etiology. Various *M. hominis* proteins targeted into host cells may disturb the behavior and functioning of infected cells [3,9].

Numerous advanced techniques have been developed for the analysis of protein localization; however, these may be inefficient, owing to their high cost and time-consuming protocols [41]. Several bioinformatics tools have been developed to calculate the subcellular targeting of proteins, thereby offering several advantages for investigational procedures [42–44]. The tools developed basically focus on the identity/alignment search or recognition of a particular sequence motif essential for particular protein localization [44]. The research work of protein targeting is also essential to infer the different functions of bacterial proteins within the host cell. The possible role of targeted proteins can be analyzed on the evidence of their relationships with different host proteins, whose role is already revealed. This information will act as a starting point for upcoming wet lab experiments.

Although nuclear-targeting proteins play a crucial role by controlling the normal functioning of host cells, other cell organelle-targeting proteins are also involved in the regulation of the normal behavior of host cells. At present, different predictors are developed for the prediction of specific motifs in protein sequences. Six NLS classes have been classified, wherein nuclear import proteins are transferred through α/β pathway of importin. The well-known dimer of α and β importin is present in the nuclear import receptor, wherein α importin works as a possible adapter protein and binds to cNLS that is identified either twice (bipartite NLS) or once (monopartite NLS), with highly basic stretches of amino acids [45]. The potential activity of NLS sequences changes within the identical class of NLS with altered sequence of NLS [46]. Hence, in the current study we particularly employed NLS mapper predictor in the present study to analyze NLS activity as an alternative to NLS sequence based on the contribution of every residue of amino acid in the NLSs. This could lead to more accurate prediction of results [44]. NLS predictor senses the activity of an NLS as a separate protein sequence, rather than the whole structural sequence of a particular native protein.

Hum-mPLoc 3.0 was used in the current in silico study for the analysis of *M. hominis* proteins targeted into the host subcellular compartment. Hum-mPLoc 3.0 tool is based on different Support Vector Machine (SVMs) method systematized in a decision tree, and predicts 12 different human subcellular localization, including ER, nucleus, mitochondrion, cytoplasm, Golgi apparatus, centriole, cytoskeleton, endosome, peroxisome, lysosome, extracellular region, and plasma membrane [47]. The tRNA threonylcarbamoyladenosine biosynthesis protein TsaE, membrane protein, lipoprotein signal peptidase, 1-acyl-*sn*-glycerol-3-phosphate acyltransferase, prolipoprotein diacylglyceryl transferase, cobalt ABC transporter permease, ComEC/Rec2-related protein, protein translocase subunit SecY, and potassium transporter KtrB are predicted to be targeted into the ER of the host cell, resulting in the alteration in the normal pattern of protein folding in ER.

4. Materials and Methods

4.1. Proteins and Prediction Analysis

The whole proteome of *M. hominis* comprising 563 proteins was selected for the prediction of proteins that are targeted into host cells. Computational predictions were used for the analysis of proteins targeted into the ER of host cells. Complete data were collected after the prediction analysis to predict implications of ER-targeting proteins in prostate cancer etiology.

4.2. Choice of Protein Database for M. hominis

We used the UniProt database (www.uniprot.org) to select the specific strain of *M. hominis*. The UniProt database is considered as a complete resource for the sequence of proteins and data annotation. This database was prepared with the combination of PIR database, Swiss-Prot, and TrEMBL activities [48], and is a collection of huge data with respect to *M. hominis* and its subcellular location, as described in Swiss-Prot/TrEMBL or PIR-PSD [49,50]. This database has two proteomes related to *M. hominis* strains, namely, ATCC-23114/PG21 and ATCC-27545 [51]. All *M. hominis* proteins of ATCC-27545 strain were used for the in silico analysis of the subcellular proteins targeted into host cells by using predictor cNLS mapper (Tsuruoka, Japan). Furthermore, Hum-mPLoc 3.0 predictor (Shanghai, China) was implicated to predict the possible location of ER-targeting *M. hominis* proteins in host cells.

4.3. Prediction of the Subcellular Localization by cNLS Mapper

Proteins targeted in various organelles of host cells were predicted using cNLS mapper [44]. Whole protein sequence of *M. hominis* was utilized for the analysis of monopartite and bipartite NLSs. The precise cNLS mapper cutoff values were 8–10, 1–2, 5–3, and 7–8, that were used to predict the localization of proteins in nucleus, cytoplasm, both cytoplasm and nucleus, and partially nucleus, respectively, as defined in the literature of cNLS. The values of monopartite and bipartite NLSs (basic amino acid stretches) were analyzed using the cNLS mapper in whole protein sequence of *M. hominis*. These basic stretches mediate binding of NLS to importin-α transport receptor, and this complex binds to importin β, through which a specific protein localizes to the nucleus. cNLS mapper detects contribution of each residue in NLS and predicts NLS activity, which is suggested to give more accurate prediction performance [44].

4.4. Prediction of ER Localization in M. hominis Proteins Using Hum-Mploc 3.0

Hum-mPLoc 3.0 was utilized to determine *M. hominis* protein localization in the ER and covered about 12 different human subcellular compartments. The protein sequences have showed multiview complementary features, i.e., peptide-based functional domains, context vocabulary annotation-based gene ontology, and amino acid residue-based statistical features, as most of the existing predictors to determine the subcellular targeting of human proteins are limited with unique location site. To prevail this barrier, Hum-mPLoc, a new ensemble classifier [47], was established and used for cases with multiple location sites. The predictor Hum-mPLoc is accessible generously by researchers from the web server at http://202.120.37.186/bioinf/hum-multi. This predictor has been employed to predict various human protein entries in Swiss-Prot database which do not have subcellular location annotations or are interpreted as "uncertain." Hum-mPLoc predictor may analyze the possible targeting of specific protein in three kingdoms, namely, plants, fungi, and animals. In the current research work, we determined proteins targeting in the ER of host cells using *M. hominis* proteins sequence as the query.

5. Conclusions

Protein homeostasis/proteostasis, is very crucial for viability and health of cells [33]. Alteration in proteostasis has been connected with the growth of many disorders, including cancer, which is considered as the most challenging disease of the current era [31]. Among different virulence aspects, various bacterial protein toxins that are somehow associated to the progression of various types of cancer have been the possible targets or markers for the management of cancer. The possible connection between *M. hominis* infection and risk of prostate cancer has gained attention in the past few years, but no detailed information is available. Here we predicted the connection between *M. hominis* infection and prostate cancer, and found that the intercellular *M. hominis* infection in host cells acts as a potential factor in prostate cancer etiology, owing to the accumulation of several *M. hominis* proteins in the ER of host cells. ER is an important organelle of eukaryotic cells involved in the regulation of secretory pathways, and release and storage of calcium. Misfolded proteins in ER cause ER stress through their accumulation and stimulation of UPR. ER stress and UPR activation are associated with the progression of different types of diseases in human, including different types of cancer. The present research work paves way for the analysis of the potential involvement of specific strain of *M. hominis* in prostate cancer.

Author Contributions: The project work was conceived by S.K. and Y.A.A., M.Z. performed bioinformatics analyses along with S.K., M.M.B.I., C.R. and A.A.C. were involved in protein sequencing. All authors were helped in writing and critical review of manuscript and approved the final draft of manuscript.

Acknowledgments: This project was supported by the Research Groups Program (Research Group number RG-1439-033), Deanship of Scientific Research, King Saud University, Riyadh, Saudi Arabia.

Conflicts of Interest: Authors confirm that they do not have any conflicts of interest related to current work.

References

1. Eaton, K.; Yang, W. Registered report: Intestinal inflammation targets cancer-inducing activity of the microbiota. *eLife* **2015**, *4*, e04186. [CrossRef] [PubMed]
2. Arthur, J.C.; Perez-Chanona, E.; Muhlbauer, M.; Tomkovich, S.; Uronis, J.M.; Fan, T.J.; Campbell, B.J.; Abujamel, T.; Dogan, B.; Rogers, A.B.; et al. Intestinal inflammation targets cancer-inducing activity of the microbiota. *Science* **2012**, *338*, 120–123. [CrossRef] [PubMed]
3. Barykova, Y.A.; Logunov, D.Y.; Shmarov, M.M.; Vinarov, A.Z.; Fiev, D.N.; Vinarova, N.A.; Rakovskaya, I.V.; Baker, P.S.; Shyshynova, I.; Stephenson, A.J.; et al. Association of *Mycoplasma hominis* infection with prostate cancer. *Oncotarget* **2011**, *2*, 289–297. [CrossRef] [PubMed]
4. Cavarretta, I.; Ferrarese, R.; Cazzaniga, W.; Saita, D.; Luciano, R.; Ceresola, E.R.; Locatelli, I.; Visconti, L.; Lavorgna, G.; Briganti, A.; et al. The Microbiome of the Prostate Tumor Microenvironment. *Eur. Urol.* **2017**, *72*, 625–631. [CrossRef] [PubMed]
5. Huang, S.; Li, J.Y.; Wu, J.; Meng, L.; Shou, C.C. Mycoplasma infections and different human carcinomas. *World J. Gastroenterol.* **2001**, *7*, 266–269. [CrossRef] [PubMed]
6. Razin, S.; Yogev, D.; Naot, Y. Molecular biology and pathogenicity of mycoplasmas. *Microbiol. Mol. Biol. Rev.* **1998**, *62*, 1094–1156. [PubMed]
7. Cimolai, N. Do mycoplasmas cause human cancer? *Can. J. Microbiol.* **2001**, *47*, 691–697. [CrossRef] [PubMed]
8. Feng, S.H.; Tsai, S.; Rodriguez, J.; Lo, S.C. Mycoplasmal infections prevent apoptosis and induce malignant transformation of interleukin-3-dependent 32D hematopoietic cells. *Mol. Cell. Biol.* **1999**, *19*, 7995–8002. [CrossRef] [PubMed]
9. Barykova Iu, A.; Shmarov, M.M.; Logunov, D.; Verkhovskaia, L.V.; Aliaev Iu, G.; Fiev, D.N.; Vinarov, A.Z.; Vinarova, N.A.; Rakovskaia, I.V.; Naroditskii, B.S.; et al. Identification of Mycoplasma in patients with suspected prostate cancer. *Zhurnal Mikrobiol. Epidemiol. Immunobiol.* **2010**, *4*, 81–85.
10. Khan, S.; Zakariah, M.; Rolfo, C.; Robrecht, L.; Palaniappan, S. Prediction of *Mycoplasma hominis* proteins targeting in mitochondria and cytoplasm of host cells and their implication in prostate cancer etiology. *Oncotarget* **2017**, *8*, 30830–30843. [CrossRef] [PubMed]
11. Lo, S.C.; Tsai, S. Mycoplasmas and human prostate cancer: An exciting but cautionary note. *Oncotarget* **2011**, *2*, 352–355. [PubMed]
12. Caini, S.; Gandini, S.; Dudas, M.; Bremer, V.; Severi, E.; Gherasim, A. Sexually transmitted infections and prostate cancer risk: A systematic review and meta-analysis. *Cancer Epidemiol.* **2014**, *38*, 329–338. [CrossRef] [PubMed]
13. Gagnaire, A.; Nadel, B.; Raoult, D.; Neefjes, J.; Gorvel, J.P. Collateral damage: Insights into bacterial mechanisms that predispose host cells to cancer. *Nat. Rev. Microbiol.* **2017**, *15*, 109–128. [CrossRef] [PubMed]
14. Khan, S.; Imran, A.; Khan, A.A.; Abul Kalam, M.; Alshamsan, A. Systems Biology Approaches for the Prediction of Possible Role of Chlamydia pneumoniae Proteins in the Etiology of Lung Cancer. *PLoS ONE* **2016**, *11*, e0148530. [CrossRef] [PubMed]
15. Khan, S. Potential role of Escherichia coli DNA mismatch repair proteins in colon cancer. *Crit. Rev. Oncol. Hematol.* **2015**, *96*, 475–482. [CrossRef] [PubMed]
16. Tsai, S.; Wear, D.J.; Shih, J.W.; Lo, S.C. Mycoplasmas and oncogenesis: Persistent infection and multistage malignant transformation. *Proc. Natl. Acad. Sci. USA* **1995**, *92*, 10197–10201. [CrossRef] [PubMed]
17. De Marzo, A.M.; Platz, E.A.; Sutcliffe, S.; Xu, J.; Gronberg, H.; Drake, C.G.; Nakai, Y.; Isaacs, W.B.; Nelson, W.G. Inflammation in prostate carcinogenesis. *Nat. Rev. Cancer* **2007**, *7*, 256–269. [CrossRef] [PubMed]
18. Sfanos, K.S.; De Marzo, A.M. Prostate cancer and inflammation: The evidence. *Histopathology* **2012**, *60*, 199–215. [CrossRef] [PubMed]
19. Ito, K. Prostate cancer in Asian men. *Nat. Rev. Urol.* **2014**, *11*, 197–212. [CrossRef] [PubMed]
20. Buc, E.; Dubois, D.; Sauvanet, P.; Raisch, J.; Delmas, J.; Darfeuille-Michaud, A.; Pezet, D.; Bonnet, R. High prevalence of mucosa-associated, *E. coli* producing cyclomodulin and genotoxin in colon cancer. *PLoS ONE* **2013**, *8*, e56964. [CrossRef] [PubMed]
21. Dessi, D.; Delogu, G.; Emonte, E.; Catania, M.R.; Fiori, P.L.; Rappelli, P. Long-term survival and intracellular replication of *Mycoplasma hominis* in Trichomonas vaginalis cells: Potential role of the protozoon in transmitting bacterial infection. *Infect. Immun.* **2005**, *73*, 1180–1186. [CrossRef] [PubMed]

22. Namiki, K.; Goodison, S.; Porvasnik, S.; Allan, R.W.; Iczkowski, K.A.; Urbanek, C.; Reyes, L.; Sakamoto, N.; Rosser, C.J. Persistent exposure to Mycoplasma induces malignant transformation of human prostate cells. *PLoS ONE* **2009**, *4*, e6872. [CrossRef] [PubMed]

23. Yadav, R.K.; Chae, S.W.; Kim, H.R.; Chae, H.J. Endoplasmic reticulum stress and cancer. *J. Cancer Prev.* **2014**, *19*, 75–88. [CrossRef] [PubMed]

24. Pluquet, O.; Pourtier, A.; Abbadie, C. The unfolded protein response and cellular senescence. A review in the theme: Cellular mechanisms of endoplasmic reticulum stress signaling in health and disease. *Am. J. Physiol. Cell Physiol.* **2015**, *308*, C415–C425. [CrossRef] [PubMed]

25. Dumartin, L.; Alrawashdeh, W.; Trabulo, S.M.; Radon, T.P.; Steiger, K.; Feakins, R.M.; di Magliano, M.P.; Heeschen, C.; Esposito, I.; Lemoine, N.R.; et al. ER stress protein AGR2 precedes and is involved in the regulation of pancreatic cancer initiation. *Oncogene* **2016**, *36*, 3094–3103. [CrossRef] [PubMed]

26. Celli, J.; Tsolis, R.M. Bacteria, the endoplasmic reticulum and the unfolded protein response: Friends or foes? *Nat. Rev. Microbiol.* **2015**, *13*, 71–82. [CrossRef] [PubMed]

27. Wang, M.; Kaufman, R.J. The impact of the endoplasmic reticulum protein-folding environment on cancer development. *Nat. Rev. Cancer* **2014**, *14*, 581–597. [CrossRef] [PubMed]

28. Hicks, G.R.; Raikhel, N.V. Protein import into the nucleus: An integrated view. *Annu. Rev. Cell Dev. Biol.* **1995**, *11*, 155–188. [CrossRef] [PubMed]

29. Schwartz, R.; Ting, C.S.; King, J. Whole proteome pI values correlate with subcellular localizations of proteins for organisms within the three domains of life. *Genome Res.* **2001**, *11*, 703–709. [CrossRef] [PubMed]

30. Khan, S.; Zakariah, M.; Palaniappan, S. Computational prediction of *Mycoplasma hominis* proteins targeting in nucleus of host cell and their implication in prostate cancer etiology. *Tumour Biol.* **2016**, *37*, 10805–10813. [CrossRef] [PubMed]

31. Balch, W.E.; Morimoto, R.I.; Dillin, A.; Kelly, J.W. Adapting proteostasis for disease intervention. *Science* **2008**, *319*, 916–919. [CrossRef] [PubMed]

32. Gupta, R.; Kasturi, P.; Bracher, A.; Loew, C.; Zheng, M.; Villella, A.; Garza, D.; Hartl, F.U.; Raychaudhuri, S. Firefly luciferase mutants as sensors of proteome stress. *Nat. Methods* **2011**, *8*, 879–884. [CrossRef] [PubMed]

33. Liu, Y.; Miao, K.; Li, Y.; Fares, M.; Chen, S.; Zhang, X. A HaloTag-Based Multicolor Fluorogenic Sensor Visualizes and Quantifies Proteome Stress in Live Cells Using Solvatochromic and Molecular Rotor-Based Fluorophores. *Biochemistry* **2018**. [CrossRef] [PubMed]

34. Liu, Y.; Fares, M.; Dunham, N.P.; Gao, Z.; Miao, K.; Jiang, X.; Bollinger, S.S.; Boal, A.K.; Zhang, X. AgHalo: A Facile Fluorogenic Sensor to Detect Drug-Induced Proteome Stress. *Angew. Chem. Int. Ed. Engl.* **2018**, *56*, 8672–8676. [CrossRef] [PubMed]

35. Hartl, F.U.; Bracher, A.; Hayer-Hartl, M. Molecular chaperones in protein folding and proteostasis. *Nature* **2011**, *475*, 324–332. [CrossRef] [PubMed]

36. Wood, R.J.; Ormsby, A.R.; Radwan, M.; Cox, D.; Sharma, A.; Vopel, T.; Ebbinghaus, S.; Oliveberg, M.; Reid, G.E.; Dickson, A.; et al. A biosensor-based framework to measure latent proteostasis capacity. *Nat. Commun.* **2018**, *9*, 287. [CrossRef] [PubMed]

37. Tabas, I.; Ron, D. Integrating the mechanisms of apoptosis induced by endoplasmic reticulum stress. *Nat. Cell Biol.* **2011**, *13*, 184–190. [CrossRef] [PubMed]

38. Khan, A.A. In silico prediction of escherichia coli proteins targeting the host cell nucleus, with special reference to their role in colon cancer etiology. *J. Comput. Biol.* **2014**, *21*, 466–475. [CrossRef] [PubMed]

39. Poutahidis, T.; Cappelle, K.; Levkovich, T.; Lee, C.W.; Doulberis, M.; Ge, Z.; Fox, J.G.; Horwitz, B.H.; Erdman, S.E. Pathogenic intestinal bacteria enhance prostate cancer development via systemic activation of immune cells in mice. *PLoS ONE* **2013**, *8*, e73933. [CrossRef] [PubMed]

40. Karin, M.; Lawrence, T.; Nizet, V. Innate immunity gone awry: Linking microbial infections to chronic inflammation and cancer. *Cell* **2006**, *124*, 823–835. [CrossRef] [PubMed]

41. Huh, W.K.; Falvo, J.V.; Gerke, L.C.; Carroll, A.S.; Howson, R.W.; Weissman, J.S.; O'Shea, E.K. Global analysis of protein localization in budding yeast. *Nature* **2003**, *425*, 686–691. [CrossRef] [PubMed]

42. Zhou, H.; Yang, Y.; Shen, H.B. Hum-mPLoc 3.0: Prediction enhancement of human protein subcellular localization through modeling the hidden correlations of gene ontology and functional domain features. *Bioinformatics* **2017**, *33*, 843–853. [CrossRef] [PubMed]

43. Kumar, M.; Raghava, G.P. Prediction of nuclear proteins using SVM and HMM models. *BMC Bioinform.* **2009**, *10*, 22. [CrossRef] [PubMed]

44. Kosugi, S.; Hasebe, M.; Tomita, M.; Yanagawa, H. Systematic identification of cell cycle-dependent yeast nucleocytoplasmic shuttling proteins by prediction of composite motifs. *Proc. Natl. Acad. Sci. USA* **2009**, *106*, 10171–10176. [CrossRef] [PubMed]

45. Lange, A.; Mills, R.E.; Lange, C.J.; Stewart, M.; Devine, S.E.; Corbett, A.H. Classical nuclear localization signals: Definition, function, and interaction with importin alpha. *J. Biol. Chem.* **2007**, *282*, 5101–5105. [CrossRef] [PubMed]

46. Kosugi, S.; Hasebe, M.; Entani, T.; Takayama, S.; Tomita, M.; Yanagawa, H. Design of peptide inhibitors for the importin alpha/beta nuclear import pathway by activity-based profiling. *Chem. Biol.* **2008**, *15*, 940–949. [CrossRef] [PubMed]

47. Shen, H.B.; Chou, K.C. Hum-mPLoc: An ensemble classifier for large-scale human protein subcellular location prediction by incorporating samples with multiple sites. *Biochem. Biophys. Res. Commun.* **2007**, *355*, 1006–1011. [CrossRef] [PubMed]

48. Apweiler, R.; Bairoch, A.; Wu, C.H.; Barker, W.C.; Boeckmann, B.; Ferro, S.; Gasteiger, E.; Huang, H.; Lopez, R.; Magrane, M.; et al. UniProt: The Universal Protein knowledgebase. *Nucleic Acids Res.* **2004**, *32*, D115–D119. [CrossRef] [PubMed]

49. Consortium, U. Activities at the Universal Protein Resource (UniProt). *Nucleic Acids Res.* **2014**, *42*, D191–D198.

50. Boeckmann, B.; Bairoch, A.; Apweiler, R.; Blatter, M.C.; Estreicher, A.; Gasteiger, E.; Martin, M.J.; Michoud, K.; O'Donovan, C.; Phan, I.; et al. The SWISS-PROT protein knowledgebase and its supplement TrEMBL in 2003. *Nucleic Acids Res.* **2003**, *31*, 365–370. [CrossRef] [PubMed]

51. Pereyre, S.; Sirand-Pugnet, P.; Beven, L.; Charron, A.; Renaudin, H.; Barre, A.; Avenaud, P.; Jacob, D.; Couloux, A.; Barbe, V.; et al. Life on arginine for *Mycoplasma hominis*: Clues from its minimal genome and comparison with other human urogenital mycoplasmas. *PLoS Genet.* **2009**, *5*, e1000677. [CrossRef] [PubMed]

Sample Availability: Not available.

![molecules logo] *molecules*

MDPI

Article

Causal Discovery Combining K2 with Brain Storm Optimization Algorithm

Yinghan Hong [1,2], **Zhifeng Hao** [1,3], **Guizhen Mai** [1,*], **Han Huang** [4] and **Arun Kumar Sangaiah** [5]

[1] School of Computer Science and Technology, Guangdong University of Technology, Guangzhou 510006, China; honyinghan@163.com (Y.H.); zfhao@gdut.edu.cn (Z.H.)
[2] School of Physics and Electronic Engineering, Hanshan Normal University, Chaozhou 521041, China
[3] School of Mathematics and Big Data, Foshan University, Foshan 528000, China
[4] School of Software Engineering, South China University of Technology, Guangzhou 510006, China; hhan@scut.edu.cn
[5] School of Computing Science and Engineering, Vellore Institute of Technology, Vellore-632014, Tamil Nadu, India; sarunkumar@vit.ac.in or arunkumarsangaiah@gmail.com
* Correspondence: mgz0323@126.com; Tel.: +86-20-3932-2277

Received: 12 May 2018; Accepted: 9 July 2018; Published: 16 July 2018

Abstract: Exploring and detecting the causal relations among variables have shown huge practical values in recent years, with numerous opportunities for scientific discovery, and have been commonly seen as the core of data science. Among all possible causal discovery methods, causal discovery based on a constraint approach could recover the causal structures from passive observational data in general cases, and had shown extensive prospects in numerous real world applications. However, when the graph was sufficiently large, it did not work well. To alleviate this problem, an improved causal structure learning algorithm named brain storm optimization (BSO), is presented in this paper, combining K2 with brain storm optimization (K2-BSO). Here BSO is used to search optimal topological order of nodes instead of graph space. This paper assumes that dataset is generated by conforming to a causal diagram in which each variable is generated from its parent based on a causal mechanism. We designed an elaborate distance function for clustering step in BSO according to the mechanism of K2. The graph space therefore was reduced to a smaller topological order space and the order space can be further reduced by an efficient clustering method. The experimental results on various real-world datasets showed our methods outperformed the traditional search and score methods and the state-of-the-art genetic algorithm-based methods.

Keywords: Bayesian causal model; causal direction learning; K2; brain storm optimization

---

1. Introduction

In recent years, the application of causal inference in bioinformatics has become more extensive, and plays a very important role in the development of this field. For instance, it is used for the discovery of the causal relationships between genes and the development of symptoms [1], and how to analyze the phenomenon of synthetic lethality [2,3] in biomedicine, which arises when a combination of mutations in two or more genes leads to cell death. Causal inference is different from the traditional statistical learning methods. The causal inference is the internal generative mechanism of the research data and the traditional statistical learning is the joint distribution of observation variables. The most significant difference between causality and relevance is whether or not to reflect the intrinsic relationship between the data. In scientific research, understanding the causal relationship of objects is crucial to predicting the laws of objects. Causal inference has already been applied in many fields, such as gene therapy, economic prediction, etc.

The problem of causal discovery or causal inference is generally formulated by a probabilistic graphical model where causal directions are represented by the directed edges [4]. In the causal inference algorithm, the techniques commonly used in local causality are conditional independent test (CI) method [5] and score & search method [4].

For example, Peter-Clack algorithms (PC algorithms) [5] determine causal relationships by finding out all the CIs in the given dataset, and the K2 algorithm [1] obtains the maximum score by searching for an optimal structure to discover causal relationships.

In general, a CI test method is used to detect the V-structure, and we can even infer the directed acyclic graph (DAG) [6] by the extension of the partial directed acyclic graph (PDAG). The accuracy of the above methods in causal inference is highly impacted by the number of the detected V-structures. In special cases, for example, without detecting the V-structure, the effect is poor. Therefore, the method cannot completely determine all edges and cannot distinguish the Markov equivalence classes, therefore often fails to uncover the true causal relationships contained in the given dataset if the number of equivalence classes is sufficiently large.

To distinguish causal direction in a non-experimental setting, some researchers recently resorted to using asymmetric relationships among variables under various hypothetical conditions. The additive noise model (ANM) proposed by Shimizu [7,8] is proved to be effective if the given data is generated by following linear non-Gaussian structural equation model. This method was later extended to nonlinear cases for continuous cases [9,10] as well as discrete cases [11,12].

Concretely, the existing ANM-based algorithms can be formulated as follows: assume there are two variables x and y satisfying a causal functional model $y = f(x) + \varepsilon$, where $f(*)$ is an arbitrary square-integrable function and ε is an independent noise of x. If the joint distribution $P(x,y)$ allows an ANM for one (forward) direction rather than the other one (backward), i.e., x cannot be obtained by a function of y plus an independent noise term, then the forward causal direction $x \rightarrow y$ is accepted as the true causal direction. The Post-Nonlinear (PNL) model [13] further extends ANM by making an additional function on the function $f(*)$ such that $y = g(f(x) + \varepsilon)$ with a bijective function $g: R \rightarrow R$. More recently, some researchers have aimed to detect the asymmetry from an information-geometric perspective [14–16]. We can see that these methods assume that reversible and deterministic mappings can get the random variables independently. According to the previous works, these methods are used to examine the asymmetry causality by different techniques, and effect in the low dimension is very good, but poor in the nonlinear high dimensional causal inference between variables.

There are also some hybrid algorithms such as the hybrid algorithm (HYA) [1] and three phases causal discovery method (TPCDM) algorithm [17], to some extent, are able to find the causal relationships of multidimensional networks. The additive noise method (ANM) differentiates the parent nodes and the child nodes in the HYA algorithm and also detects the relationship between the neighbor sets and the sink nodes in the TPCDM algorithm. However, the experimental results show that the effect of the methods above are not very accurate, because it is difficult to detect a one-to-many network structure by ANM methods [10,18–27].

We can see that all these methods for learning causal structure are unreliable, or the time complexity is so very high that we often cannot get the result in an acceptable time. In this situation, we resort to optimization algorithms.

Then, we study the optimization algorithms. Problems existing in many real worlds can be classified as optimization problems. The traditional optimization algorithm is based on a single point, such as gradient descent algorithm, which is a point that moves in the opposite direction of the gradient function. The traditional optimization algorithm mainly solves the problem of a single peak; it is easy to obtain the local optimal solution in the case of complex multiple modes and nonlinear problems.

In recent years, the swarm intelligence (SI) algorithm has been a topical research topic in solving the problem of multiple peaks. Swarm intelligence algorithms are used to solve problems by learning some life phenomena or natural phenomena in nature, which includes the characteristics of self-organization, self-learning and adaptability of natural life phenomena. Especially in 2011, a new

SI algorithm [28] called "Brain Storm Optimization" (also known as Brainstorm optimization, BSO) was proposed, which was inspired by human brainstorming activities. The paper demonstrates the ability of BSO to solve optimization problems by testing two basic functions. Based on the idea of human creative problem-solving, a new swarm intelligence algorithm, Shi's [9] optimization algorithm, was proposed. Unlike traditional swarm intelligence algorithms, such as ant colony optimization (ACO) and artificial bee colony (ABC), the BSO algorithm is the first one to solve the problem based on human creative thinking. Humans are the smartest animals in the world, and the BSO algorithm, which is inspired by their social behavior, is considered a promising method [9]. Shi [9,10] elaborated the thought and implementation process of BSO algorithm, and used the classical test function to simulate the BSO algorithm, and the results showed the effectiveness of BSO algorithm. However, there is still a problem of precocious maturity, and it is necessary to further study to optimize the algorithm itself, improve the effect of BSO algorithm [11].

In this study, we design an efficient method to support causal discovery by combining K2 with Brain Storm Optimization Algorithm (K2-BSO). We use the score returned by the K2 algorithm as the fitness function, and design an elaborate distance function for the clustering step in the BSO according to the mechanisms of K2. The graph space therefore was reduced to a smaller topological order space and the order space can be further reduced by an efficient clustering method. After a optimal causal order is returned by BSO, we run K2 to search for the optimal causal structure, and output the causal skeleton. In the case of high dimensions, the following methods are first used to process the skeleton. We split the causal skeleton into n (the number of variables in the skeleton) smaller sub-skeleton, and employ ANM to detect the causal directions between the target variables and all its parents from each causal skeleton. Consequently, we obtain a partial DAG (PDAG) w.r.t. each sub-skeleton. By merging all the PDAGs, the whole structure corresponding to the high dimensional causal network w.r.t. the given dataset is finally reconstructed. K2-BSO is designed for a certain problem, and the most different thing from other BSO methods should be the clustering procedure, since in the our design, we need to measure the distance between two node sequences in term of the corresponding orders instead of two sequences perset, therefore the existing clustering methodologies used in other BSO methods like those mentioned in [29–31] are not applicable for our case.

The rest of this paper is organized as follows: Section 2 briefly summarizes these definitions. Then we focus on the introduction to the basic concepts, algorithm flow and advantages and disadvantages of K2 and BSO algorithms in Section 3. The details of Causal Discovery combining K2 with Brain Storm Optimization Algorithm are discussed in Section 4. The correctness and performance characteristics of three algorithms are shown in the Section 5. Section 6 gives detailed experimental results. Finally, the conclusions are drawn in Section 7.

2. Definitions

In this section, we will introduce several basic definitions applied in our method. The concepts of the D-separation, V-structure and Additive noise model, which is described as follows:

A causal network can be expressed as a directed acyclic graph (DAG) $G_N = \{V_N, E_N\}$, in which $E_N = \{e_1, e_2, \ldots, e_n\}$ and $V_N = \{x_1, x_2, \ldots, x_n\}$ denote the sets of edges and nodes in G_N.

A. D-separation

Definition 1. *(d-Separation). Assume L is a path from x_i to x_j, and is blocked by a set of variables Z if one of the following conditions holds:*

(1) L either contains a chain, $x_i \leftarrow x_k \leftarrow x_j$, and $x_k \in Z$,

(2) or a fork, $x_i \leftarrow x_k \rightarrow x_j$, and $x_k \in Z$,

(3) or a collider, $x_i \rightarrow x_k \leftarrow x_j$, and $x_k \notin Z$, and no descendent of x_k is contained in Z.

We say a set Z separates two disjointed sets X_i and X_j ($X_i, X_j \subseteq V_D$) if Z blocks each path between X_i and X_j.

B. V-structure

Definition 2. *(V-Structure). Given three variables x, y, and z. If x and z are the parent nodes of y, and no other edge is existing between x and z, then x, z and y together form a V-structure. As shown in Figure* 1.

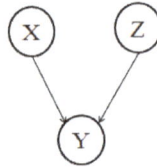

Figure 1. Illustration of a V-structure.

C. Additive noise model

Definition 3. *(Additive noise model (ANM for short)) ANM is represented by a collection of n equations* $S = \{S_1, S_2, \ldots S_n\}$: $S_i : x_i = f_i(x_{pa_{(i)}}) + \varepsilon_i$, $i = \{1, 2, \ldots, n\}$, *where* $x_{pa_{(i)}}$ *is the direct parent set of* x_i, *the noise terms* ε_i *are jointly independent, and are independent from* x_i.

It can be seen that the data-generating processes of X can be expressed as:

$$S_i : x_i = \varepsilon_i, i = \{1, 2, \ldots, k\} \text{ (the root nodes)}$$

$$S_j : x_j = f_j\left(x_{pa_{(j)}}\right) + \varepsilon_j, j = \{1, 2, \ldots, n-k\} \text{ (the other nodes)}$$

As shown aforementioned ANM provides a way for finding casualties by using the assumption of additional noise data generation process rather than satisfying Markov conditions.

3. The K2 and Brain Storm Optimization

In this section, we first introduce the K2 algorithm. Then, the basic concepts, algorithm flow and advantages and disadvantages of BSO algorithm are introduced in detail. All in all, the whole process of the K2 and Brain Storm Optimization can be described as follows.

3.1. The K2 Algorithm

K2 Algorithm, developed by Cooper and Herskovits in 1992, is a Bayesian Network Structure learning algorithm based on the score search method. It is a classical algorithm in the Bayesian Network Structure field with excellent learning performance [32].

As we all know, Bayesian Network Structure Learning aims to find the Bayesian Network Structure B_S which best connects with D through the analysis of data set D. That is the Bayesian Network Structure B_S with maximum posterior probability $P(B_S \mid D)$. Because $P(B_S \mid D) = P(B_S \mid D)/P(D)$ in which $P(D)$ is a constant, what we find is the network structure B_S that maximizes the probability $P(B_S \mid D)$, that is:

$$\max[P(B_S, D)] = c \prod_{i=1}^{n} \max\left[\prod_{j=1}^{q_1} \frac{(r_i - 1)!}{(N_{ij} + r_i - 1)!} \prod_{k=1}^{r_i} N_{ijk}!\right], \tag{1}$$

where c is the a priori probability $P(B_S \mid D)$ of each network structure, which is meant to be a constant because in the algorithm of K2, it is assumed that every network structure B_S has the same probability; n is the number of nodes; r_i is the number of values of node X_i; π_i is parent nodes set of node X_i; q_i is the number of configurations of π_i; N_{ijk} is the sample number of node X_i, which takes the value of k, and its parent set is the jth configuration in data set D; $N_{ij} = \sum_{k=1}^{r_i} N_{ijk}$.

As is showing above, K2 Algorithm uses Equation (1) as a score function to learn the Bayesian Network Structure. From Equation (1), the score function can be decomposed, that is, it can be seen as products of n local structures, which is made up of each node X_i, $i = 1, 2, \ldots, n$ and its corresponding parent nodes set. Then the following equation is derived:

$$g(X_i, \pi_i) = \sum_{j=1}^{q_i} \frac{(r_i - 1)!}{(N_{ij} + r_i - 1)!} \prod_{k=1}^{r_i} N_{ijk}! \tag{2}$$

$$G(B_S, D) = \sum_{i=1}^{n} g(X_i, \pi_i) \tag{3}$$

so we can maximize $G(B_S, D)$ if we maximize every local structure's scores $g(X_i, \pi_i)$, inevitably also maximizing the scores of the whole Bayesian Network Structure (Equation (1)). According to this idea, given nodes order ρ and the upper limits μ of each node's parent nodes, the K2 algorithm can use Greedy Searching to find each node's parent nodes in turn so as to finally construct a whole complete Bayesian Network. The concrete method is as follows: firstly, for every node X_i, $i = 1, 2, \ldots, n$, constantly choose the former nodes in former nodes' set from nodes order into parent set π_i of node X_i, making the score function $g(X_i, \pi_i)$ of π_i and X_i continuously increase. The above process cannot stop until function $g(X_i, \pi_i)$ does not increase any more when adding nodes. In that process, we need to limit that the parent node's number should be under μ.

As is known to all that the K2 algorithm has two prerequisites, given nodes order ρ and the upper limits μ of each node's parent nodes. With these two prerequisites, it can obtain a very good learning performance, but in most situations, we can't always meet the above prerequisites, causing difficulties in the application of the K2 algorithm.

3.2. Brain Storm Optimization

3.2.1. Brainstorming Algorithm Principle

Inspired by human behavior patterns, in 2011, a human brainstorming process was proposed for the first time by Shi et al., called Brainstorming Optimization Algorithm (BSO). Shi's article expounds the thought and realization process of BSO in detail, and simulates the BSO algorithm with classical test function, and the experimental results show that the BSO algorithm is effective. However, there are some deficiencies in the new algorithm, such as easily falling into local optima, resulting in premature convergence. Therefore, it is necessary to improve the BSO algorithm and optimize the algorithm so as to improve its effect [33–38].

The concept and theory of the basic BSO algorithm is derived from the simulation of the human brainstorming process. A brainstorming meeting needs a moderator, a number of owners to solve problems, and a group of parliamentarians with different backgrounds. Since parliamentarians have different backgrounds, different experiences and different ways of thinking, one problem will get different solutions. The moderator, in accordance with the four Rules of the Conference (see Table 1), presides over the meeting and gets solutions from as many as possible [38–43]. The algorithm needs a skilled host, with no or almost no problem-solving knowledge, so as not to lead host into bias, and also the host cannot engage in new ideas until all ideas are proposed. The host can divide it into K classes, and for each class, people can diversify their thinking and propose better solutions until they get the best solution. The BSO algorithm gets its inspiration from this model and then simulates the process. In the BSO algorithm, the feasible solution of each optimization problem is a quantity of information in the search space, all the information has an adaptive value which is determined by the function of optimization, and then the optimal information is iterated by clustering and learning all kinds of excellent information.

Table 1. Osborn's Original Rules for Idea Generation in a Brainstorming Process.

Rule 1	No bad ideas, every thought is good
Rule 2	Every thought has to be shared and recorded
Rule 3	Most ideas are based on existing ideas, and some ideas can and should be raised to generate new ideas
Rule 4	Try to produce more ideas

The brainstorming session procedure is as follows:

(A) Assemble as many parliamentarians with different backgrounds as possible;

(B) Get the solutions based on the brainstorming rules in Table 1;

(C) Choose a scheme as the best solution for the current problem from each of the problem-solving owners;

(D) Generate new schemes from the schemes selected in C according to the rules in Table 1

(E) Choose a solution from the idea of each problem-solving owner in D as the best solution for the current problem

(F) Randomly select a scheme as a clue to generate new schemes in the case of meeting the Rules in Table 1;

(G) Each problem-solving owner chooses a scheme from F to be the best solution for the current problem;

(H) Get the best solution that is desired by considering merging these programs.

3.2.2. BSO Algorithm Steps

The brainstorming algorithm is mainly composed of two modules: a clustering module and a learning module. In the clustering module, the algorithm uses the clustering method to gather the information into K classes, and the cluster center in each class is the optimal value. The algorithm is optimized by learning, also the information in each class is in parallel. Similarly the local search is promoted, and the algorithm jumps out of the local optimization through the cooperation between classes and the mutation operation, which promotes the global search. The convergence of the algorithm is ensured by the optimization process of cluster center, and the process of optimizing the information variation in the class ensures the diversity of the algorithm population. Each individual in the BSO algorithm represents a potential problem solution that is updated by the individual's evolution and fusion, a process similar to that of the human brainstorming process [44–46].

The implementation of BSO algorithm is simpler:

(1) Obtain the solution of n potential problems, then divide n individuals into M class by K-means method, the individual in each class is sorted by evaluating these n individuals, and the optimal individual is selected as the central point of the class;

(2) Randomly select the central individual of a class and determine whether it is replaced by a randomly generated individual according to the probability;

(3) to update the individual, the way is updated by the following four ways: (a) randomly select a class (the probability of selection is proportional to the number of individuals within the class), the random perturbation is added to the class center to produce a new individual; (b) randomly select a class (the probability of selection is proportional to the number of individuals within the class) and randomly select an individual in the selected class, plus a random perturbation to produce a new individual; (c) randomly selected two classes, the fusion of the class center and the random perturbation to produce a new individual; (d) randomly select two classes, randomly select an individual in each class, and then add a random perturbation to create a new individual. By adjusting the parameters to control the proportion of the above four ways to produce new individuals. After the new individual generation, compared with the original individual, the final

selection of the best one to retain to a new generation of individuals, repeat the above operation, the n individual to update each one, produced a new generation of n individuals.

This loops until the upper limit of the preset individual update algebra is reached. In the third step, the update of the individual has four ways to produce a new individual process; the selected amount of information plus a Gaussian random is worth the new amount of information, such as the following Equation (4):

$$X^d_{new} = X^d_{selected} + \varepsilon \times n(\mu, \sigma), \tag{4}$$

where X^d_{new} is the d dimension of the new information, $X^d_{selected}$ is the d dimension of the selected information, $n(\mu, \sigma)$ is the Gaussian function whose mean value is μ and variance is σ; ε is a weight coefficient which is described by Equation (5):

$$\xi = \log sig((0.5 \times max\_iteration - current\_iteration)/k) \times rand(), \tag{5}$$

where log*sig*() is a s-type logarithmic transfer function, and *max_iteration* is the maximum number of iterations, while *current_iteration* means the current number of iterations; k can change the slope of the function log*sig*(), *rand*() is the random value between (0,1).

4. The K2-BSO Method

In this section, the details of the K2-BSO method are given, we show that this method is able to discover causation combining K2 with the Brain Storm Optimization algorithm. All in all, the whole process of causality is deduced, which is described as follows:

4.1. Skeleton Learning Phase Based on K2-BSO

The Additive Noise Model (ANM) could find out the causal relationships correctly between variables in sparse causal networks, but this model would encounter multiple challenges when applied to high-dimensional complex network structures [12]. First of all, high-dimensional causal networks contain a large number of variables, and the causal relationships between them are very complex, so the algorithm requires the ability to quickly search. Causal relationship references based on the traversal method will face all possible network structures, which leads directly to the insufferable computational complexity, the storage space overflow and other problems. The K2 algorithm needs to satisfy two prerequisites, given nodes order and the upper limits of each node's parent nodes. However, it is difficult to make it in fact. What's more, the K2 algorithm is easy to fall into the local optimal solutions while the BSO algorithm could get rid of local optimizations. Therefore, the combination of the algorithm K2 and BSO can effectively solve the structural learning problem of causal network structure. As discussed in the previous section, there are three points we need to note:

(1) What needs to be optimized is the causal order that will highly affect the accuracy of K2. Generally, an input order approaching the actual topological order of the underling causal network will return the highest score and most similar causal structure.
(2) The fitness function is easy to be chosen, that is the score return by K2.
(2) The clustering method of BSO should be redesigned; all the distance function likes [46] cannot be directly applied to this case, as what we consider is the topological order. We design a new distance function like this:

Step I. Given two orders R_1 and R_2, for each variable in R_1, we find the same variable in R_2, assume it is v_1.
Step II. Consider n variables in front of v_1 in R_1, and m variables in front of v_1 in R_2, we calculate the number of the repeated variables in $n + m$ variables.
Step III. By literately sum up the repeated variables w.r.t. every variable in R_1 (or R_2), we get a number, and let this number as the distance between R_1 and R_2.

We note that, the clustering step is crucial to the BSO, as shown before, our distance function is designed based on the mechanism of K2, which will highly improve the clustering performance in BSO.

4.2. Direction Learning Phase

Algorithm 1 can obtain the skeleton of network returned by K2-BSO. Because the K2 can only examine a set of Markov equivalence classes rather than the realistic causal structure, we aim to detect the remaining directions of the output skeleton for distinguishing this equivalence in this section. Because of the existence of Markov equivalence classes, the structural learning methods are generally difficult to infer all causal direction. On the other hand, the ANM provides an effective way to learn causal direction in low-dimensional cases. Note that, we get the causal skeleton, then we can separate the causal skeleton S into n sub-skeletons $(S_i, \dots, S_n)$ which contain a target node X_i and all its neighbor nodes N_i according to S. In general, these sub-skeletons are generally low-dimensional and therefore can be solved by using ANM. The way to orient the edges of a skeleton in ANM method is described as follows:

Algorithm 1. Skeleton learning based on K2-BSO.

Input: dataset X, population size $|V|$.
Output: the skeleton w.r.t. X.
1: Randomly generate n potential causal order $R = R_1 \sim R_n$;
2: Cluster R into m clusters $C = C_1 \sim C_m$;
3: **For** each R_i
$Score_i = K2(R_i)$;
4: **End For**
5: Score = $Score_1 \sim Score_n$;
6: $R_{optimal}$ = BSO $(X, R, Score, C)$;
7: $G_{optimal}$ = K2($R_{optimal}$);
8: $X = G_{optimal}$;
9: **return** the causal skeleton X.

Firstly, consider a given dataset $X = \{X_1, X_2, \dots, X_n\}$ with index $V = \{1, 2, \dots, n\}$. X corresponds to an n-dimensional DAG $G = \{V, E\}$, where E represents the edges of V. Assume that X is generated by the following way: each variable $X_i \in X$ corresponds to one node $i \in V$ in G, and is determined by a causal function $X_i = f_i(x_{pa_{(i)}}) + \varepsilon_i$ in which f_i is nonlinear, $x_{pa_{(i)}}$ is the parent of x_i. The noise terms ε_i have a non-Gaussian distribution and are jointly independent.

In the issue of seeking out the causal direction, we aim to seek out all the parent nodes (contained in N_i) amount to each target X_i from S_i. On the basis of the mechanism of ANM, we denote the homologous remains between X_i and each candidate parent set C_{ik} as $X_i = f(C_{ik}) + \varepsilon_i$ by using GPR, and we test whether C_{ik} and ε are statistically independent. If they are independent we accept the model $C_{ik} \to X_i$; if not, we deny it. In this phase, we measure the independence by using the kernel-based conditional independence (KCI) test. The details of causal directions inference from a output causal skeleton is presented in Algorithm 2.

Algorithm 2. Learning causal direction from a sub-skeleton.

Input: sub-skeleton S_i and the corresponding target node X_i with all its neighbors N_i.
Output: the direction between X_i and (partial) N_i.
1: **For** each candidate parent set C_{ik};
2: fit X_i and C_{ik} to ANM;
3: **if** ε is independent of C_{ik} **then**
4:   accept $C_{ik} \to X_i$;
5: **end if**
6: **end for**

4.3. K2-BSO Framework (Algorithm 3)

We first present the details of the K2-BSO method:

Step 1. Learning the causal skeleton S by algorithm 1.

Step 2. Split S into n sub-skeleton $S_1, \ldots, S_n$ according to each node X_i contained in S.

Step 3. Perform Algorithm 2 for each sub-skeleton S_i.

Step 4. Merge all the partial results and output the final causal structure.

Algorithm 3. K2-BSO framework.

Input: Dataset X, threshold k
Output: Causal structure G.
1. Set Dimension X to n;
2. **if** $(n < k)$ **then**
3. $S = Algorithm\ 1(X); G = Algorithm\ 2(X,\ S)$;
4. **else**
5. Split S into n sub-skeleton $S_1, \ldots, S_n$ according to each node X_i contained in S;
6. **For** each S_i in S
7. $S_i = Algorithm\ 1(X_i); PDAG_i = Algorithm\ 2(X_i,\ S_i)$;
8. Merge all $PDAG_i$ to G;
9. **End for**
10. **end if**
11. **return** the final causal structure G.

---

5. The Correctness and Performance of the Algorithms

In this part, we analyze theoretically about the respective characteristics of the correctness and performance with the three algorithms (K2-Random, K2-BSO, K2-GA).

First, we discuss the K2-Random algorithm. It is a traditional method, and there is not much optimization process. The main process is: first step, randomly obtain p data sort, then sort the score from the top to the bottom and select the highest score. The second step is to continue to randomly obtain p data sort, found the highest score Tscore, until 10 consecutive times are the same highest score, and end the program; this method is very easy to enter the local optimization state, but the experimental result is unstable.

Second, we discuss the K2-BSO algorithm, which is the method proposed in this paper. It is better to avoid local optimization problems. The main process is: first step, randomly obtain p data sort, then sort the score from the top to the bottom and obtained m data sort by clustering method. The second step is to obtain m new subclasses by random perturbation about m subclasses by the BSO algorithm. Then we reevaluate the score until the score converges.

Third, we discuss the K2-GA algorithm. The main process is: The main process is: first step, randomly obtain p data sort, Then sort the score from the top to the bottom and select the highest score until the score converges. The second step is to obtain p new data sort by means of Genetic Algorithm (GA) method with randomly perturbation the highest ranking data. Then sort the score from the top to the bottom to obtain the highest score Tscore.

In summary, the first algorithm in time complexity is the best, but the accuracy rate is the lowest and unstable; the second algorithm and the third algorithm's time complexity are the same, especially with the increase of network dimensions, second algorithms tend to advance convergence faster than the third algorithms, and the accuracy of the second algorithms is better than the third algorithm. Next, we'll use real data to validate three algorithms in the next chapter.

6. Experiments

In this section, we evaluate our proposal on eight real-world datasets that cover a variety of applications including Small Networks (Asia, Sachs), Medium Networks (Child, Alarm),

Large Networks (Barley, Win95pts), and Very Large Networks (Pigs, MINUN) that cover a variety of applications, including, medicine (ASIA, SACHS, CHILD and ALARM), agricultural industry (BARLEY), system troubleshooting (WIN95PTS) and bioinformatics (PIGS and MUMIN) are available at "http://archive.ics.uci.edu/ml/datasets.html". The structural statistics of the eight networks are summarized in Table 2.

Table 2. Statistics on the network.

Network	Nodes	Edges	Avg Degree	Maximum in-Degree
ASIA	8	8	2	2
SACHS	11	17	3.09	3
CHILD	20	25	1.25	2
ALARM	37	46	2.49	4
BARLEY	48	84	3.5	4
WIN95PTS	76	112	2.95	7
PIGS	441	592	2.68	2
MUMIN	1041	1397	2.68	3

In this group of experiments, our proposed method is compared with other two mainstream causal discovery methods—K2-Random (Causal Discovery combining K2 with Random) method and K2-GA (Causal Discovery combining K2 with Genetic Algorithm) method. We evaluate these methods by different sample size at 250, 500, 1000, 2000, respectively. We use three criteria, Recall, Precision, and F1 to evaluate these methods, which are defined as follows:

$$Recall = (Inferred\ directions \cap Actual\ directions)/(Actual\ directions), \tag{9}$$

$$Precision = (Inferred\ directions \cap Actual\ directions)/(Inferred\ directions), \tag{10}$$

$$F1 = (2 \times Recall \times Precision)/(Recall + Precision) \tag{11}$$

Obviously Precision is the actual fraction of inferred causality with respect to a true graph. Similarly, Recall is the part of actual causality found by the algorithm. F1 is the organic combination of Precision and Recall which can serve as the accuracy standard for our algorithms.

The experimental environment is as follows:

(1) CPU of the physical host: CPU E5-2640 v3, 2.60 GHz (2-way 8-core);
(2) Platform belongs to the cloud platform version from Bingo Cloud: v6.2.4.161205143;
(3) Memory is 24 G.

As shown in Table 3, we can see that the K2-Random runs much faster than the other two algorithms. However, as showed in Figure 2, the accuracy of K2-Random is lowest, this means K2-Random easily falls into a local optimum. One can imagine that if we use K2-Random to test all possible causal orders detailed we can obtain the best score, but we usually cannot get the final result in an acceptable time, because the time complexity of such an exhaustive algorithm reaches the upper limit.

On the other hand, we can see that in the small networks, K2-GA runs faster than K2-BSO. However, as the size of the networks grows, the running time of K2-BSO increases slower than that consumed by K2-BSO, and the running time of the two methods tend to be very close. We can see that in the case of WIN95PTS, K2-BSO runs much faster than K2-GA. What is the most different between K2-BSO and K2-GA in the task is that K2-BSO performs a clustering step, which can greatly reduce the convergence time. Recall that, the clustering step in K2-BSO also costs time. Therefore, when the causal network spends more time in clustering step, K2-BSO is probably slower than K2-GA, while for a network to spend less time in the clustering step, theoretically K2-BSO runs much faster than K2-GA. Accordingly, the specific structure of a certain causal network weighs heavily on total time.

Table 3. Comparisons between three algorithms on execution time.

Dataset	Sample	K2-Random			K2-BSO			K2-GA		
		Best	Mean	Worst	Best	Mean	Worst	Best	Mean	Worst
ASIA	250	1.2555	1.749	2.6157	3.0185	4.5585	5.8543	2.8367	4.2504	5.5427
ASIA	500	1.085	1.4656	1.9224	3.2043	4.3753	6.3392	2.1125	4.8925	6.4698
ASIA	1000	1.5104	1.7994	2.2099	4.7037	5.0472	5.3491	3.1055	5.1146	8.2191
ASIA	2000	2.2676	2.4993	2.6629	3.3815	3.9774	4.3823	4.1878	5.506	8.1496
SACHS	250	4.5248	5.025	5.3196	6.2266	6.4963	6.6793	3.02	4.6432	7.501
SACHS	500	2.4084	3.6746	5.2433	8.7819	11.2107	15.6879	5.5198	8.1696	8.6039
SACHS	1000	2.7062	3.5759	4.0197	8.3591	11.7962	15.3562	5.4356	8.6128	12.8977
SACHS	2000	3.4966	4.6789	6.4649	9.3677	10.7807	11.5173	6.1085	6.1516	6.1762
CHILD	250	7.7092	9.117	12.5631	28.0133	29.961	31.325	21.1455	25.986	30.974
CHILD	500	8.9414	11.8924	14.1062	17.7665	28.9292	41.265	30.2484	34.1696	38.9523
CHILD	1000	10.059	17.5069	28.2669	33.4957	38.4505	43.9597	23.6723	24.608	25.3404
CHILD	2000	10.709	13.8898	18.3508	30.1317	58.6901	78.3929	21.3343	38.465	47.0478
ALARM	250	46.647	71.9888	86.3291	217.3178	266.3269	355.1756	147.9969	283.8744	371.7365
ALARM	500	94.125	103.2041	119.5914	227.3739	293.7428	377.54	133.6973	388.0371	519.749
ALARM	1000	62.140	92.5241	125.1668	157.303	247.1288	294.8618	144.2063	316.8503	475.2459
ALARM	2000	95.002	159.3802	232.1197	323.0716	394.6686	496.8733	219.1187	275.1111	345.3221
BARLEY	250	66.378	91.2914	136.5331	198.4244	325.1975	400.4625	160.181	205.9434	233.3933
BARLEY	500	75.057	99.7028	138.1832	304.5911	364.1937	368.7941	194.9004	358.4717	567.8317
BARLEY	1000	86.012	100.4193	116.8377	326.8585	370.2588	404.4023	255.2328	396.0264	505.5334
BARLEY	2000	96.159	103.1696	116.3037	478.3594	525.7576	549.7203	368.8762	810.5905	1.48×10^3
WIN95PTS	250	649.75	1.10×10^3	1.52×10^3	1.81×10^3	4.44×10^3	7.16×10^3	1.60×10^4	2.06×10^4	2.31×10^4
WIN95PTS	500	555.26	727.4609	819.2716	2.33×10^3	4.76×10^3	6.67×10^3	6.23×10^3	1.83×10^4	2.48×10^4
WIN95PTS	1000	693.01	746.7864	827.4684	2.73×10^3	4.38×10^3	6.00×10^3	2.01×10^4	2.57×10^4	3.19×10^4
WIN95PTS	2000	715.72	1.43×10^3	1.85×10^3	2.25×10^3	7.04×10^3	1.50×10^4	2.13×10^4	3.44×10^4	4.98×10^4
PIGS	250	1.87×10^4	2.52×10^4	3.96×10^4	2.60×10^5	3.89×10^5	4.85×10^5	1.45×10^5	2.48×10^5	3.77×10^5
PIGS	500	5.35×10^4	6.84×10^4	8.53×10^4	3.09×10^5	4.03×10^5	5.24×10^5	1.58×10^5	2.56×10^5	3.03×10^5
PIGS	1000	7.12×10^4	8.64×10^4	9.71×10^4	2.92×10^5	4.14×10^5	4.59×10^5	1.85×10^5	2.70×10^5	3.57×10^5
PIGS	2000	6.17×10^4	9.00×10^4	1.09×10^5	4.26×10^5	5.26×10^5	7.05×10^5	1.98×10^5	2.74×10^5	3.33×10^5
MINUN	250	1.98×10^5	2.70×10^5	4.33×10^5	1.71×10^6	2.70×10^6	3.46×10^6	4.54×10^5	7.74×10^5	1.09×10^6
MINUN	500	3.10×10^5	4.05×10^5	4.98×10^5	2.52×10^6	3.24×10^6	4.17×10^6	5.45×10^5	9.00×10^5	1.07×10^6
MINUN	1000	3.39×10^5	4.14×10^5	4.72×10^5	2.43×10^6	3.41×10^6	3.88×10^6	8.19×10^5	1.17×10^6	1.57×10^6
MINUN	2000	2.93×10^5	4.23×10^5	5.06×10^5	2.93×10^6	3.67×10^6	4.99×10^6	9.81×10^5	1.35×10^6	1.65×10^6

As shown in Figure 2, K2-BSO achieves the better score in the majority of cases, which means that the clustering step can not only improve the convergence speed on the basis of the number of iterations, but also prevent K2-BSO from falling into local optima. Even the largest network PIGS shows that the F1 score is 2% better than K2-GA.

Figure 2 also shows the main trends of the indexes (Recall (R), Precision (P), and F1) of the three algorithms (K2-R, K2-BSO, K2-GA), with different samples [250,500,1000,2000] in eight datasets, including ASIA, ALARM, SACHS, BARLEY, CHILD, Win95pt, PIGS and MINUN. The blue line 'o:' represents the numerical trend of the Recall of K2-Random; the blue line 'o–' indicates the numerical trend of the Precision of K2- Random; the blue line '*—' indicates the numerical trend of the F1 of K2-Random; The red line 'o:' represents the numerical trend of the Recall of K2-BSO; the red line 'o–' indicates the numerical trend of Precision of K2-BSO; the red line '*—' indicates the numerical trend of F1 of K2-BSO; The green line 'o:' represents numerical trend of the Recall of K2-GA; the green line "o–" indicates the numerical trend of the Precision of K2-GA, and the blue line "*—" indicates the numerical trend of the F1 of K2-GA.

Figure 2a shows the curves of the three methods (K2-R, K2-BSO, K2-GA) with different samples in the data set ASIA. It can be seen that the red curve basically goes above the green one and the blue one, which means that K2-BSO's indexes R, P, F1 are higher than that of K2-R and K2-GA, thus proves that K2-BSO algorithm is better than K2-R algorithm and K2-GA algorithm.

Figure 2b–e show the curves of the three algorithms (K2-R, K2-BSO, K2-GA) with different samples in data sets SACHS, CHILD, ALARM and BARLEY. It can be observed that the results are similar to that in Figure 2a, that K2- BSO's indexes R,P,F1 are higher than that of K2-R and K2- GA, thus also proves that K2-BSO algorithm is better than K2-R algorithm and K2-GA algorithm. Figure 2f–g shows the curves of the three methods (K2-R, K2-BSO, K2-GA) with different samples in data set WIN95PTS, PIGS and MINUN. WIN95PTS is a 76-dimensional network, PIGS is a 441-dimensional network and MINUN is a 1041-dimensional network, so they belongs to the high dimensional networks. We can see from Figure 2f that the curve of the blue value is the lowest; with sample 500 and 2000, the value of the green curve is slightly higher than that of the red curve, while on the whole, the red curve goes above the green. However, when we refer to Table 3, it is obvious that the execution time of K2-BSO is much less than that of K2-GA, which means the K2-BSO is better than the other algorithms in this network. On the other hand, Figure 2g shows that the curves are slightly different from the form's results, the Recall of the three methods grows with the increase of sample size while the Precision reduces with the increase of sample size.

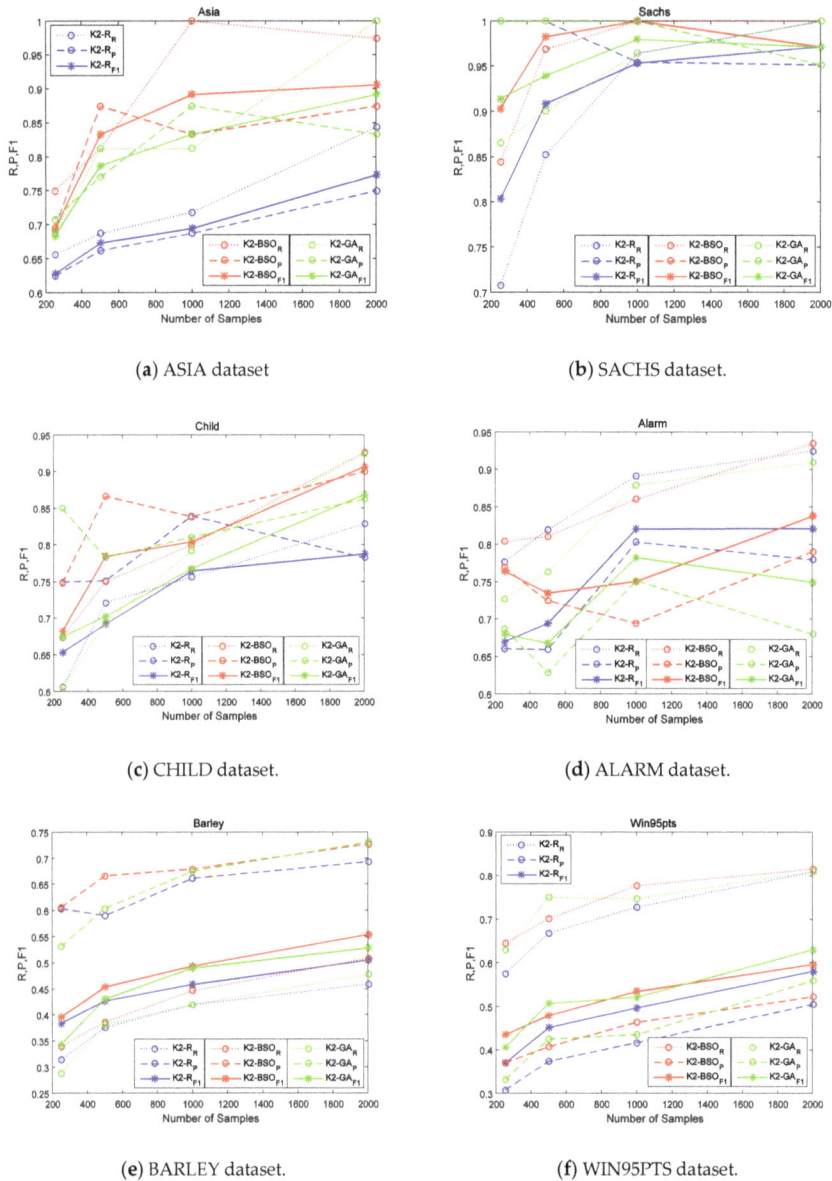

(**a**) ASIA dataset

(**b**) SACHS dataset.

(**c**) CHILD dataset.

(**d**) ALARM dataset.

(**e**) BARLEY dataset.

(**f**) WIN95PTS dataset.

Figure 2. *Cont.*

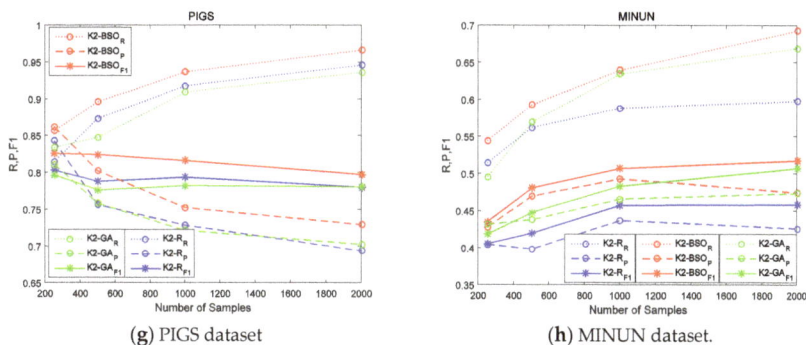

(g) PIGS dataset.   (h) MINUN dataset.

Figure 2. R, p & F1 of the K2-R, K2-BSO and K2-GA with eight dataset: (**a**) ASIA dataset; (**b**) SACHS dataset; (**c**) CHILD dataset; (**d**) ALARM dataset; (**e**) BARLEY dataset; (**f**) Win95pt dataset; (**g**) PIGS dataset; (**h**) MINUN dataset.

The reason for such a difference is that PIGS is a genetic network, that is, PIGS has a very complex structure where some nodes connect with many neighboring nodes, for example, the maximum degree is 41 (maximum in-degree is 2). Accordingly it is difficult for the subroutine K2 to remove these in-direct causal relationships. Even so, it can be seen that the F1 score of K2-BSO is still 2% better than those of K2-R and K2-GA. Figure 2h shows that even in the case of MINUN network of more than 1000 dimensionality, K2-BSO works much better than K2-GA and K2-R. These results demonstrate that our method K2-BSO is much reliable than the counterparts in more complexity and higher-dimensional cases, and also shows that K2-BSO is able to learn the causal structure from a dataset with hundreds of variables. In summary, K2-BSO performs better than K2-GA if the accuracy and execution time are combined, so in our future work, we will continue to perfect the K2-BSO algorithm, making it adapt to high-dimensional network accuracy problems at the cost of some appropriate execution time.

7. Conclusions

To reduce the search space of graphs is important in causal relationship discovery; however, the existing methods show inefficiency for large scale causal networks. In this work, an improved causal structure learning algorithm combining K2 with brain storm optimization (BSO) called K2-BSO is presented to alleviate this problem. In contrast to other evolutionary algorithms based on the search and score methods, K2-BSO has two significant advantages, (1) K2-BSO searches optimal topological order of nodes instead of graph space. The order space should be much smaller than the whole graph space. In this phase, an elaborate distance function is introduced for clustering nodes' orders in BSO based on the mechanism of K2. The graph space therefore is reduced to a smaller topological order space that can be further reduced by an efficient clustering method. (2) Our method is designed through the following split and merge strategy, the original dataset is split into a set of subdata sets in the first place. The BSO will run on these subdata sets to recover the corresponding substructures. Here we further use additive noise model approach to rectify the direction of the erroneous orientation or the side without direction. We eventually merge all these substructures and obtain the entire structure of the graph. The experimental results on various causal networks showed our method could outperform the traditional search and score method and the state-of-the-art genetic algorithm-based method.

Author Contributions: Y.H. and Z.H. designed the study. Y.H., Z.H., G.M., H.H. and A.K.S. performed the study. Y.H. and G.M. wrote the manuscript. G.M. was the principal investigator and corresponding author.

Funding: This research was supported in part by NSFC-Guangdong Joint Found (U1501254), Natural Science Foundation of China (61472089), (61473331), (61502108), Science and Technology Planning Project of Guangdong Province, China (2015A030401101), (2015B090922014), (2017A040405063), (2017B030307002),

Guangdong High-level personnel of special support program (2014TQ01X664) and International Cooperation Project of Guangzhou (201807010047).

Conflicts of Interest: The authors declare that they have no competing interests.

References

1. Hao, Z.; Huang, J.; Cai, R.; Wen, W. A hybrid approach for large scale causality discovery. In *Emerging Intelligent Computing Technology and Applications, Proceedings of the 8th International Conference, ICIC 2012, Huangshan, China, 25–29 July 2012*; Springer: Berlin, Germany, 2013; Volume 375, pp. 1–6.
2. Li, X.J.; Mishra, S.K.; Wu, M.; Zhang, F.; Zheng, J. Syn-lethality: An integrative knowledge base of synthetic lethality towards discovery of selective anticancer therapies. *BioMed Res. Int.* **2014**, *2014*, 196034. [CrossRef] [PubMed]
3. Wu, M.; Li, X.; Zhang, F.; Li, X.; Kwoh, C.K.; Zheng, J. In silico prediction of synthetic lethality by meta-analysis of genetic interactions, functions, and pathways in yeast and human cancer. *Cancer Inform.* **2014**, *13*, 71–80. [CrossRef] [PubMed]
4. Pearl, J. *Causality*; Cambridge University Press: Cambridge, UK, 2009.
5. Spirtes, P.; Glymour, C.N.; Scheines, R.; Heckerman, D.; Meek, C.; Cooper, G.; Richardson, T. *Causation, Prediction, and Search*; Springer: New York, NY, USA, 1993; pp. 272–273.
6. Chickering, D.M. Learning equivalence classes of bayesian-network structures. *J. Mach. Learn. Res.* **2002**, *2*, 150–157.
7. Shimizu, S.; Hoyer, P.O.; Hyvärinen, A.; Kerminen, A. A linear non-gaussian acyclic model for causal discovery. *J. Mach. Learn. Res.* **2006**, *7*, 2003–2030.
8. Shimizu, S.; Inazumi, T.; Sogawa, Y.; Hyvärinen, A.; Kawahara, Y.; Washio, T.; Hoyer, P.O.; Bollen, K. Directlingam: A direct method for learning a linear non-gaussian structural equation model. *J. Mach. Learn. Res.* **2011**, *2*, 1225–1248.
9. Hoyer, P.O.; Janzing, D.; Mooij, J.M.; Peters, J.; Schölkopf, B. Nonlinear causal discovery with additive noise models. In Proceedings of the International Conference on Neural Information Processing Systems, Vancouver, BC, Canada, 8–10 December 2008; pp. 689–696.
10. Peters, J.; Mooij, J.M.; Janzing, D.; Schölkopf, B. Causal discovery with continuous additive noise models. *J. Mach. Learn. Res.* **2013**, *15*, 2009–2053.
11. Peters, J.; Janzing, D.; Scholkopf, B.; Teh, Y.W.; Titterington, M. Identifying cause and effect on discrete data using additive noise models. In Proceedings of the International Conference on Artificial Intelligence and Statistics, Sardinia, Italy, 13–15 May 2010; pp. 597–604.
12. Peters, J.; Janzing, D.; Scholkopf, B. Causal inference on discrete data using additive noise models. *IEEE Trans. Pattern Anal. Mach. Intell.* **2011**, *33*, 2436–2450. [CrossRef] [PubMed]
13. Zhang, K.; Hyvärinen, A. Distinguishing causes from effects using nonlinear acyclic causal models. In Proceedings of the 2008th International Conference on Causality: Objectives and Assessment, Vancouver, BC, Canada, 8–10 December 2008; Volume 6, pp. 157–164.
14. Daniusis, P.; Janzing, D.; Mooij, J.; Zscheischler, J.; Steudel, B.; Zhang, K.; Schölkopf, B. Inferring deterministic causal relations. In Proceedings of the Conference on UAI, Catalina Island, CA, USA, 8–11 July 2010; pp. 143–150.
15. Janzing, D.; Mooij, J.; Zhang, K.; Lemeire, J.; Zscheischler, J.; Daniušis, P.; Steudel, B.; Scholkopf, B. Information-geometric approach to inferring causal directions. *Artif. Intell.* **2012**, *182–183*, 1–31. [CrossRef]
16. Janzing, D.; Steudel, B.; Shajarisales, N.; Scholkopf, B. Justifying information-geometric causal inference. In *Measures of Complexity*; Springer: Cham, Switzerland, 2015; pp. 253–265.
17. Chen, W.; Hao, Z.; Cai, R.; Zhang, X.; Hu, Y.; Liu, M. Multiple-cause discovery combined with structure learning for high-dimensional discrete data and application to stock prediction. *Soft Comput.* **2016**, *20*, 4575–4588. [CrossRef]
18. Cai, R.; Zhang, Z.; Hao, Z. Causal gene identification using combinatorial v-structure search. *Neural Netw.* **2013**, *43*, 63–71. [CrossRef] [PubMed]
19. Cai, R.; Zhang, Z.; Hao, Z. SADA: A General Framework to Support Robust Causation Discovery. In Proceedings of the 30th International Conference on Machine Learning (ICML), Atlanta, GA, USA, 16–21 June 2013; Volume 28, pp. 208–216.

20. Cai, R.; Zhang, Z.; Hao, Z.; Winslett, M. Understanding Social Causalities Behind Human Action Sequences. *IEEE Trans. Neural Netw. Learn. Syst.* **2016**, *28*, 1801–1813. [CrossRef] [PubMed]
21. Cai, R.; Zhang, Z.; Hao, Z. BASSUM: A Bayesian semi-supervised method for classification feature selection. *Pattern Recognit.* **2011**, *44*, 811–820. [CrossRef]
22. Mooij, J.; Janzing, D.; Peters, J.; Scholkopf, B. Regression by dependence minimization and its application to causal inference in additive noise models. In Proceedings of the 26th Annual International Conference on Machine Learning, Montreal, QC, Canada, 14–18 June 2009; pp. 745–752.
23. Zhang, K.; Peters, J.; Janzing, D.; Scholkopf, B. Kernel-based conditional independence test and application in causal discovery. *Comput. Sci.* **2012**, *6*, 895–907.
24. Cheng, S.; Qin, Q.; Chen, J.; Shi, Y. Brain storm optimization algorithm: A review. *Artif. Intell. Rev.* **2016**, *46*, 445–458. [CrossRef]
25. Hong, Y.H.; Liu, Z.S.; Mai, G.Z. An efficient algorithm for large-scale causal discovery. *Soft Comput.* **2016**, *21*, 7381–7391. [CrossRef]
26. Hong, Y.H. Fast causal network skeleton learning algorithm. *J. Nanjing Univ. Sci. Technol.* **2016**, *40*, 315–321.
27. Hong, Y.H.; Mai, G.Z.; Liu, Z.S. Learning tree network based on mutual information. *Metall. Min. Ind.* **2015**, *12*, 146–151.
28. Duan, H.; Li, S.; Shi, Y. Predator–prey brain storm optimization for DC brushless motor. *IEEE Trans. Mag.* **2013**, *49*, 5336–5340. [CrossRef]
29. Shi, Y. Brain storm optimization algorithm. In Proceedings of the International Conference in Swarm Intelligence, Chongqing, China, 12–15 June 2011; Springer: Berlin/Heidelberg, Germany, 2011; pp. 303–309.
30. Zhan, Z.H.; Zhang, J.; Shi, Y.H.; Liu, H.L. A modified brain storm optimization. In Proceedings of the 2012 IEEE Congress on Evolutionary Computation, Brisbane, QLD, Australia, 10–15 June 2012; pp. 1–8.
31. Xue, J.; Wu, Y.; Shi, Y.; Cheng, S. Brain storm optimization algorithm for multi-objective optimization problems. In *Advances in Swarm Intelligence*; Springer: Berlin/Heidelberg, Germany, 2012; pp. 513–519.
32. Cooper, G.F.; Herskovits, E. A bayesian method for the induction of probabilistic networks from data. *Mach. Learn.* **1992**, *9*, 309–347. [CrossRef]
33. Shi, Y. An optimization algorithm based on brainstorming process. In *Emerging Research on Swarm Intelligence and Algorithm Optimization*; Information Science Reference: Hershey, Pennsylvania, 2015; pp. 1–35.
34. Zhou, D.; Shi, Y.; Cheng, S. Brain storm optimization algorithm with modified step-size and individual generation. *Adv. Swarm Intell.* **2012**, *7331*, 243–252.
35. Sun, C.; Duan, H.; Shi, Y. Optimal satellite formation reconfiguration based on closed-loop brain storm optimization. *IEEE Comput. Intell. Mag.* **2013**, *8*, 39–51. [CrossRef]
36. Jadhav, H.T.; Sharma, U.; Patel, J.; Roy, R. Brain storm optimization algorithm based economic dispatch considering wind power. In Proceedings of the 2012 IEEE International Conference on Power and Energy (PECon), Parit Raja, Malaysia, 2–5 December 2012; pp. 588–593.
37. Qiu, H.; Duan, H. Receding horizon control for multiple UAV formation flight based on modified brain storm optimization. *Nonlinear Dyn.* **2014**, *78*, 1973–1988. [CrossRef]
38. Shi, Y.; Xue, J.; Wu, Y. Multi-objective optimization based on brain storm optimization algorithm. *Int. J. Swarm Intell. Res.* **2013**, *4*, 1–21. [CrossRef]
39. Shi, Y. Brain storm optimization algorithm in objective space. In Proceedings of the 2015 IEEE Congress on Evolutionary Computation (CEC), Sendai, Japan, 25–28 May 2015; pp. 1227–1234.
40. Yang, Z.; Shi, Y. Brain storm optimization with chaotic operation. In Proceedings of the 2015 Seventh International Conference on Advanced Computational Intelligence (ICACI), Wuyi, China, 27–29 March 2015; pp. 111–115.
41. Yang, Y.; Shi, Y.; Xia, S. Advanced discussion mechanism-based brain storm optimization algorithm. *Soft Comput.* **2015**, *19*, 2997–3007. [CrossRef]
42. Jia, Z.; Duan, H.; Shi, Y. Hybrid brain storm optimisation and simulated annealing algorithm for continuous optimisation problems. *Int. J. Bio-Inspired Comput.* **2016**, *8*, 109–121. [CrossRef]
43. Cheng, S.; Shi, Y.; Qin, Q.; Gao, S. Solution clustering analysis in brain storm optimization algorithm. In Proceedings of the 2013 IEEE Symposium on Swarm Intelligence (SIS), Singapore, 16–19 April 2013; pp. 111–118.
44. Cheng, S.; Shi, Y.; Qin, Q.; Zhang, Q.; Bai, R. Population diversity maintenance in brain storm optimization algorithm. *J. Artif. Intell. Soft Comput. Res.* **2014**, *4*, 83–97. [CrossRef]

45. Cheng, S.; Shi, Y.; Qin, Q.; Ting, T.O.; Bai, R. Maintaining population diversity in brain storm optimization algorithm. In Proceedings of the 2014 IEEE Congress on Evolutionary Computation (CEC), Beijing, China, 6–11 July 2014; pp. 3230–3237.
46. Georgiou, D.N.; Karakasidis, T.E.; Nieto, J.J.; Torres, A. A study of entropy/clarity of genetic sequences using metric spaces and fuzzy sets. *J. Theor. Biol.* **2010**, *267*, 95–105. [CrossRef] [PubMed]

Sample Availability: Samples of the compounds are not available from the authors.

![molecules logo] *molecules*

MDPI

Article

A Resolution-Free Parallel Algorithm for Image Edge Detection within the Framework of Enzymatic Numerical P Systems

Jianying Yuan [1,2,3], Dequan Guo [1,2,3], Gexiang Zhang [4,5,*], Prithwineel Paul [5], Ming Zhu [2] and Qiang Yang [4]

1   The Postdoctoral Station at Xihua University Based on Collaboration Innovation Center of Sichuan Automotive Key Parts, Xihua University, Chengdu 610039, China; yuanjy@cuit.edu.cn (J.Y.); guodq@cuit.edu.cn (D.G.)
2   School of Control Engineering, Chengdu University of Information Technology, Chengdu 610225, China; zhuming@126.com
3   School of Aeronautics and Astronautics, University of Electronic Science and Technology, Chengdu 610054, China
4   Robotics Research Center, Xihua University, Chengdu 610039, China; qiangychd@126.com
5   School of Electrical Engineering, Southwest Jiaotong University, Chengdu 610031, China; prithwineelpaul@gmail.com
*   Correspondence: zhgxdylan@126.com; Tel.: +86-028-8772-9583

Received: 3 February 2019; Accepted: 23 March 2019; Published: 29 March 2019

Abstract: Image edge detection is a fundamental problem in image processing and computer vision, particularly in the area of feature extraction. However, the time complexity increases squarely with the increase of image resolution in conventional serial computing mode. This results in being unbearably time consuming when dealing with a large amount of image data. In this paper, a novel resolution free parallel implementation algorithm for gradient based edge detection, namely EDENP, is proposed. The key point of our method is the introduction of an enzymatic numerical P system (ENPS) to design the parallel computing algorithm for image processing for the first time. The proposed algorithm is based on a cell-like P system with a nested membrane structure containing four membranes. The start and stop of the system is controlled by the variables in the skin membrane. The calculation of edge detection is performed in the inner three membranes in a parallel way. The performance and efficiency of this algorithm are evaluated on the CUDA platform. The main advantage of EDENP is that the time complexity of $O(1)$ can be achieved regardless of image resolution theoretically.

Keywords: membrane computing; edge detection; enzymatic numerical P system; resolution free

1. Introduction

In recent decades, image processing technology has experienced dramatic growth and widespread applications. Nearly no area escapes impact in some way by digital image processing. Normally, digital image processing includes three main levels, i.e., low-level, mid-level and high-level processing [1]. As one of the most basic operators in low-level image processing, edge detection can preserve the important structural properties of an image while significantly reducing the amount of data. This excellent property makes it a basic tool for many high-level image processing algorithms and is extensively applied in target tracking [2], image compression [3], and object recognition [4]. An edge can be defined as points in a digital image at which the image brightness changes sharply or has discontinuities. This phenomenon may be caused by depth discontinuous, illumination changes, or intrinsic texture properties of objects. In various edge detection algorithms, the gradient based method is a type of classic edge detection approach with the merit of simple theory and good performance. However, as convolution calculation (i.e., a classic neighborhood computing in image processing) [5] is involved in this kind of algorithm, the time complexity increases squarely with the increase of image resolution. So it is difficult to deal with images with large resolution, such as remote sensing images, medical images, etc., in real time processing.

In order to achieve real-time calculation of high resolution images, many researchers have put much effort into this problem and several methods have been proposed. Generally, there are two main categories of resolutions. The first type of resolution concerns computational algorithms. In this kind of method, an elaboratively computational algorithm is usually designed to reduce the computational complexity. For the template matching problem, integral image [6] and dual-bound algorithm [7] are two classical approaches to speed up the computation. In [8], Fast LDA feature extraction is present, where steepest descent and conjugate direction methods are combined to optimize the step size in each iteration. In [9], common orthogonal basis extraction is proposed to extract a common basis of collection of matrices. The second category is based on hardware with parallel architecture, such as Graphics Processing Unit (GPU) [10–12] and Field Programmable Gate Array (FPGA) [13,14]. GPU uses hundreds of parallel processor cores executing tens of thousands of parallel threads to rapidly solve large problems having substantial inherent parallelism. However, with the shrinking volume of chips, semiconductor technology begins to reach its physical limits, which means the performance of conventional computing technique based on silicon chip integrated circuit microprocessors will be difficult to improve further [15]. Under this background, some scholars have turned their attention to non-traditional computing, such as quantum computing [16], DNA computing [17] and membrane computing (MC) [18]. MC is a new active branch of natural computing that simulates the function and structure of living cells and tissues, abstracting their biochemical reactions and material exchanges [19]. One of the most prominent features of MC is its capability of generating exponential growth space over a polynomial time, which makes it a promising method to resolve the conflict between the ever-increasing amount of data in the image processing field and the backward computing power of conventional computer [20]. In recent years, image edge detection and image segmentation [21–24], image smoothing [25], obtaining homology groups of 2D images [26,27], counting cells [28], Enzymatic numerical P systems image thinning [29] and corner detection [30] in MC framework have been vividly studied. In the previous literature about MC and image processing, much work is based on tissue-like P systems. However, when designing a parallel implementation program of an existing image processing algorithm, it is difficult to realize the mathematical formula in "tissue-like P systems language". The reasons for this are as follows. First, the data type of an image is an integer between 0 and 255. When design image processing algorithm uses tissue-like P systems, the image data should be coded to symbolic variables and those symbolic variables need to be decoded to integer for display as the algorithm finished. Second, most image processing algorithms are composed of several steps in determined logical order, which means variables in the membrane system need to be calculated in a deterministic way, rather than in a random manner. Since the rules in tissue-like P systems are implemented randomly, it is difficult to control the execution orders of different rules.

In order to overcome the above shortcomings when tissue-like P systems are combined to image processing, we make the first attempt to introduce enzymatic numerical P system (ENPS) to image processing. Concretely, a parallel algorithm for gradient based edge detection algorithm is designed and tested in the framework of ENPS. Besides the features described in [25], ENPS has another two good properties which make it particularly appropriate for image processing. One is that numerical variables and numerical expressions can be used directly in ENPS. Thus, image data can be directly operated without the additional encoding and decoding process. Another important characteristic is that enzymatic variables can control the execution orders of multiple rules in ENPS, i.e., the algorithms with complex logical steps can be designed easily.

The main contribution of this paper is that a parallel algorithm for image edge detection in the framework of ENPS, namely EDENP, is designed. The significant advantage of EDENP is that it can achieve the time complexity of $O(1)$ theoretically, no matter how large the image resolution is.

Moreover, the performance is equivalent to the performance run on the serial computing platform. This is very important for real projects, because most of the classical image processing algorithms have been widely proven to be effective in practical engineering, so the designed parallel implementation algorithm can be directly applied to the real image processing project without the need to perform large-scale testing. To the best of our knowledge, it is the first time to bridge problems from image processing with ENPS.

The rest of this paper is structured as follows. Section 2 introduces the definition, characteristics and applications of MC and ENPS. The problem statement is elaborated in Section 3. Section 4 discusses the EDENP algorithm in detail. The experiments and results are presented in Section 5. Conclusions are drawn in Section 6.

2. MC and ENPS

MC is a young biocomputing model proposed by Gh.Păun in 2000 [19]. The computational devices in MC are called P systems. Generally, a P system includes three ingredients: (i) the membrane structure; (ii) multisets of objects; (iii) rules of a bio-chemical inspiration. The multisets of objects are placed in the membrane, and evolved according to given rules which are usually applied in a synchronous non-deterministic maximally parallel manner. Since being proposed, MC has received great attention from scientists in many fields [31–33]. In the past 20 years, both the theory [32,34–37] and application [31,38–41] of MC have been greatly developed, and many different classes of P systems have been investigated. According to the way in which membranes are structured, there are three major types of P systems, i.e., cell-like [19], tissue-like [42] and spiking neural P systems [43,44]. Enzymatic numerical P system comes from numerical P system (NPS). NPS is a new special research branch of cell-like P systems, proposed by the founder of MC, Gh.Păun in 2006 [45]. In NPS, multisets of objects associated to membranes are sets of numerical variables, and the evolutionary rules are composed of a production function and a repartition protocol [46–48]. The most common widely application area of NPS is robot controller design [49–52]. Although NPS can deal with numerical variables, it can only execute one production function per membrane at a time. When there are multiple production functions per membrane, one is selected randomly. This limits its application in some situations where the rules should be executed deterministically. In order to solve this problem and expand the application of NPS, ENPS is put forward [24]. It is extended from NPS by introducing enzyme-like variables which can make rules run deterministically [53]. The standard form of ENPS is defined as follows:

$$\Pi = \Big(m, H, \mu, \big(Var_1, E_1, \mathrm{Pr}_1, Var_1(0)\big), \ldots, \big(Var_m, E_m, \mathrm{Pr}_m, Var_m(0)\big)\Big).$$

where:

1. m is the number of membranes used.
2. H is an alphabet that contains m symbols, and $H = \{1, 2, \ldots, m\}$.
3. μ is the membrane structure.
4. Var_i is the set of variables from membrane i and $Var_i(0)$ are the initial values for these variables.
5. Pr_i is the set of rules in membrane i, composed of a production function and a repartition protocol. A typical rule is as follows.

$$F_{l,i}(y_{1,i}, \ldots, y_{k_l,i})\big|_{e_{j,i}} \to c_{l,1}|v_1 + c_{l,2}|v_2 + \ldots + c_{l,n_i}|v_{n_i},$$

where $e_{j,i}$ is a variable from Var_i different from $y_{1,i}, \ldots, y_{k_l,i}$ and $v_1, v_2, \ldots, v_{n_i}$. The rule can be executed at a time t only if $e_{j,i} > \min\{y_{1,i}(t), y_{2,i}(t), \ldots y_{k_l,i}(t)\}$. From the definition of ENPS, it is clear that with enzymes-like variables, the system can control multiple production functions to run in parallel in the same membrane deterministically [54]. Hence, it can overcome the disadvantages of traditional NPS that only run one rule nondeterministically at a time in a membrane. The ENPS with deterministic, parallel execution model has already been proved to be Turing universal [55,56]. In [57], it is shown that any ENPS working in all-parallel mode or one parallel model can be simulated by an equivalent one-membrane ENPS working in the same mode. Since the proposal of ENPS, this model has been successfully applied in a wide range of domains, such as robot control [58], big data field [59], and sequential minimal optimization [60] fields. In this paper, ENPS is used to solve the problem of gradient based image edge detection.

3. Problem Statement

Edges generally occur in areas where the brightness of the image changes dramatically. These changes can be described by image gradients. Usually, a pair of convolution masks are used to estimate the gradients in the x and y directions, respectively, as shown in Equations (1)–(3), where ($Sobel_x$, $Sobel_y$), ($Prew_x$, $Prew_y$), (Rob_x, Rob_y) are three classic pairs of convolution masks. In this paper, we take Sobel operator as an example of gradient based edge detection (GBED). When the masks are sliding over the image, a square of pixels are operated. Then both directional gradients and absolute gradient magnitudes of image are computed, as shown in Equations (4) and (5), where I is the image, (g_x, g_y) are gradients in x and y direction respectively, $g_{i,j}$ is the absolute gradient magnitude of a pixel with coordinate (i, j), $2 \leq i, j \leq n - 1$ for image with resolution of $n \times n$.

$$Sobel_x = \begin{bmatrix} -1 & 0 & 1 \\ -2 & 0 & 2 \\ -1 & 0 & 1 \end{bmatrix}; \quad Sobel_y = \begin{bmatrix} 1 & 2 & 1 \\ 0 & 0 & 0 \\ -1 & -2 & -1 \end{bmatrix} \tag{1}$$

$$Prew_x = \begin{bmatrix} -1 & 0 & 1 \\ -1 & 0 & 1 \\ -1 & 0 & 1 \end{bmatrix}; \quad Prew_y = \begin{bmatrix} 1 & 1 & 1 \\ 0 & 0 & 0 \\ -1 & -1 & -1 \end{bmatrix} \tag{2}$$

$$Rob_x = \begin{bmatrix} -1 & 0 \\ 0 & 1 \end{bmatrix}; \quad Rob_y = \begin{bmatrix} 0 & -1 \\ 1 & 0 \end{bmatrix} \tag{3}$$

$$g_x = Sobel_x * I; \quad g_y = Sobel_y * I; \tag{4}$$

$$g_{i,j} = \sqrt{g_{x_{i,j}}{}^2 + g_{y_{i,j}}{}^2} \tag{5}$$

When the gradient magnitude $g_{i,j}$ is computed, the difference between it and a predefined threshold θ is used to judge whether this pixel is an edge pixel or not, as presented in Equation (6), where $d_{i,j}$ is the difference. More concretely, if $d_{i,j}$ is greater than or equal to 0, then the pixel is assumed as an edge point, otherwise, it is a background point, as shown in Equation (7). It is worth noting that in real application, before thresholding, the gradient image should be filtered by "non-maximum suppression" for getting more real edges. In this paper, in order to simplify the algorithm, this step is ignored.

$$d_{i,j} = g_{i,j} - \theta \tag{6}$$

$$edg_{i,j} = \begin{cases} 1 & \text{if } (d_{i,j} \geq 0) \\ 0 & \text{if } (d_{i,j} < 0) \end{cases} \tag{7}$$

The program pseudo code of GBED run on conventional serial computer is illustrated in Algorithm 1, where the initial value of $edg_{i,j}$ is set to 0. From Algorithm 1, it can be deduced that the computational complexity is $O(n^2)$ because two loops are involved. When n becomes large, the calculations are very time-consuming under the serial computing platform.

Algorithm 1 The pseudo code of GBED

Input: $I(n * n)$
Output: $edg(n * n)$
1: **for** $i = 2 : n - 2$ **do**

2:   **for** $j = 2 : n - 1$ **do**

3:     Computing $gx_{i,j}$
4:     Computing $gy_{i,j}$
5:     Computing $g_{i,j}$
6:     Computing $d_{i,j}$
7:     Computing $edg_{i,j}$
8:   **end for**
9: **end for**

In order to reduce the calculation time complexity, we attempt to introduce an enzymatic numerical P system to design a high parallel computing algorithm for edge detection. The details of how to design the algorithm will be given in the next section.

4. The EDENP Algorithm

This section starts with the mathematical model of EDENP followed by the detailed description of EDENP. The execution process and resources needed are discussed lastly.

4.1. Mathematical Model of EDENP

From Section 3, we know that the GBED algorithm contains four steps for a certain pixel in an image. In EDENP, the four steps will be executed in a cell-like P system under the control of enzyme variables, as illustrated in Figure 1. The initialization of variables, start and stop of the system will be controlled in the skin membrane. The directional gradients estimation will be completed in membrane 1. The absolute gradient magnitude estimation will take place in membrane 2. Membrane 3 is responsible for computing the image edge. The corresponding membrane structure is illustrated in Figure 2.

The mathematical expression of EDENP is as follows, and

$$\Pi = \left(m, H, \mu, \left(Var_1, E_1, \text{Pr}_1, Var_1(0) \right), ..., \left(Var_4, E_4, \text{Pr}_4, Var_4(0) \right) \right),$$

where

1. $m = 4$.
2. $H = \{1, 2, 3, 4\}$.
3. $u = [[[[\ ]_1]_2]_3]_4$.
4. $Var_1 = \{gx_{i,j}, gy_{i,j}\}$, $Var_2 = g_{i,j}$, $Var_3 = \{ed_1, ed_2, ed_3, E_{i,j}, E_{D_{i,j}}\}$, $Var_4 = \{x_{i,j}, edg_{i,j}, \theta, e_{1,1}, E_D\}$.

    $x_{i,j} (1 \leq i, j \leq n)$, are the gray value of pixel with coordinate of (i, j) on the source image plane.

    $edg_{i,j} (1 \leq i, j \leq n)$, are the corresponding edge points of the source image with initial value 0.

    $\theta[threshold]$, is a numerical variable which is used as the threshold value for edge detection, and the value of threshold should be predefined.

    $gx_{i,j} (1 \leq i, j \leq n)$, are the horizontal derivative approximations at each pixel.

    $gy_{i,j} (1 \leq i, j \leq n)$, are the vertical derivative approximations at each pixel.

    $g_{i,j} (1 \leq i, j \leq n)$, are the gradient magnitude approximations at each pixel.

$ed_1[0]$, is a numerical variable with initial value 0, which is used as the background value of the edge image.

$ed_2[1]$, is a numerical variable with initial value 1, which is used as the edge point value of the edge image.

$ed_3[-256]$, is a numerical variable with initial value -256, which is used as a intermediate variable.

5. E_k is a set of enzyme variables from membrane k, i.e., $E_1 = [\ ], E_2 = [\ ], E_3 = \{E_{i,j}, ED_{i,j}\}, E_4 = \{e_{1,1}, E_D\}$.

6. Pr_k is the set of programs (rules) in membrane k, composed of a production function and a repartition protocol.

$Pr_{1,CE_{i,j}}$:
$$(|x_{i,j+2} + 2x_{i+1,j+2} + x_{i+2,j+2} - x_{i,j} - 2x_{i+1,j} - x_{i+2,j}|)|e_{1,1} \rightarrow 1|gx_{i,j}, (2 \leq i, j \leq n - 2),$$

$Pr_{2,CE_{i,j}}$:
$$(|x_{i,j} + 2x_{i,j+1} + x_{i,j+2} - x_{i+2,j} - 2x_{i+2,j+1} - x_{i+2,j+2}|)|e_{1,1} \rightarrow 1|gy_{i,j}, (2 \leq i, j \leq n - 2),$$

$Pr_{3,CE_{1,i}} : 0|e_{1,1} \rightarrow |gx_{1,i}; (1 \leq i \leq n),$

$Pr_{4,CE_{n,i}} : 0|e_{1,1} \rightarrow |gx_{n,i}; (1 \leq i \leq n),$

$Pr_{5,CE_{i,1}} : 0|e_{1,1} \rightarrow |gx_{i,1}; (2 \leq i \leq n - 1),$

$Pr_{6,CE_{i,n}} : 0|e_{1,1} \rightarrow |gx_{i,n}; (2 \leq i \leq n - 1),$

$Pr_{7,CE_{1,i}} : 0|e_{1,1} \rightarrow |gy_{1,i}; (1 \leq i \leq n),$

$Pr_{8,CE_{n,i}} : 0|e_{1,1} \rightarrow |gy_{n,i}; (1 \leq i \leq n),$

$Pr_{9,CE_{i,1}} : 0|e_{1,1} \rightarrow |gy_{i,1}; (2 \leq i \leq n - 1),$

$Pr_{10,CE_{i,n}} : 0|e_{1,1} \rightarrow |gy_{i,n}; (2 \leq i \leq n - 1).$

Those rules are used to execute Formula (1). The enzyme in $Pr_{1,CE_{i,j}} \sim Pr_{10,CE_{i,j}}$ must exist in enough amount so that the rules can be activated. Specifically, if the value of the enzyme $e_{1,1}$ is greater than variable $x_{i,j}(1 \leq i, j \leq n)$, then rules $Pr_{1,CE_{i,j}} \sim Pr_{10,CE_{i,j}}$ are effective. Since variable $x_{i,j}$ is the gray value of image, the maximum value is 255. So, the initial value of $e_{1,1}$ is set to 256, such that the condition modeled by rule $Pr_{1,CE_{i,j}} \sim Pr_{10,CE_{i,j}}$ are satisfied. It is important to note that the number of rules are $n \times n$, and all the rules are executed in parallel.

7. $Pr_{21,CE_{i,j}} : (\sqrt{gx_{i,j}^2 + gy_{i,j}^2})|e_{1,1} \rightarrow 1|g_{i,j}; 1 \leq i, j \leq n$

$Pr_{21,CE_{i,j}}$ are the rules which are executed by Formula (5). Hence, after executing $Pr_{1,CE_{i,j}} \sim Pr_{10,CE_{i,j}}$, the value of the variables $gx_{i,j}, gy_{i,j}$ are obtained. The maximum value of $gx_{i,j}$ and $gy_{i,j}$ is 255, and the enzyme $e_{1,1}$ is 256. So the condition of execution for rules $Pr_{21,CE_{i,j}}$ is satisfied. Hence, all $n \times n$ rules are executed concurrently.

8. $Pr_{31,CE_{i,j}} : (2*(g_{i,j} - \theta))| \rightarrow 1|g_{i,j}+1|E_{i,j}; 1 \leq i, j \leq n$

$Pr_{31,CE_{i,j}}$ are the rules which compute $d_{i,j}$ in Formula (6). After executing $Pr_{31,CE_{i,j}}$, the value of $d_{i,j}$ are obtained, which is equal to variables $g_{i,j}$ and $E_{i,j}$ in rule $Pr_{31,CE_{i,j}}$.

9. $Pr_{32,CE_{i,j}} : (ed_1 + 2 * ed_2)|E_{i,j} \rightarrow 1|edg_{i,j} + 1|ED_{i,j}; Pr_{33,CE_{i,j}} : (0 * ed_1 + 0 * ed_3)|E_{i,j} \rightarrow 1|edg_{i,j} + 1|ED_{i,j}; 1 \leq i, j \leq n.$

$Pr_{32,CE_{i,j}}$ and $Pr_{33,CE_{i,j}}$ are rules for computing edge value as Formula (7). If $E_{i,j}$ is greater than or equal to 0, then $Pr_{32,CE_{i,j}}$ and $Pr_{33,CE_{i,j}}$ are executed. Because ed_1 is 0, and ed_3 is -256, so $E_{i,j} \geq min(ed_1, ed_2)$ and $E_{i,j} \geq min(ed_1, ed_3)$. The execution condition of $Pr_{32,CE_{i,j}}$ and $Pr_{33,CE_{i,j}}$ is satisfied. If $d_{i,j} < 0$, only $Pr_{33,CE_{i,j}}$ will be executed. Because $E_{i,j} \geq min(ed_1, ed_3)$ and $E_{i,j} < min(ed_1, ed_2)$, only the execution condition of $Pr_{33,CE_{i,j}}$ can be satisfied. After executing $Pr_{32,CE_{i,j}}$ and $Pr_{33,CE_{i,j}}$, variables $edg_{i,j}$ will be set to 1 if $d_{i,j} \geq 0$ and every variable $ED_{i,j}$ will be assigned.

10. $\mathrm{Pr}_{main} : (0 * E_{D_{1,1}} + 0 * E_{D_{1,2}} + \ldots + 0 * E_{D_{m,n}} + 1)| \rightarrow 1|E_D$

   Pr_{main} is a rule contained in membrane 4, which controls the stop condition of the P system. For pixel (i, j), if all the enzyme variables $E_{D_{i,j}}$ are assigned, the condition for Pr_{main} is meet. Enzyme variable E_D is set to 1 by rule Pr_{main}, and the system stops running.

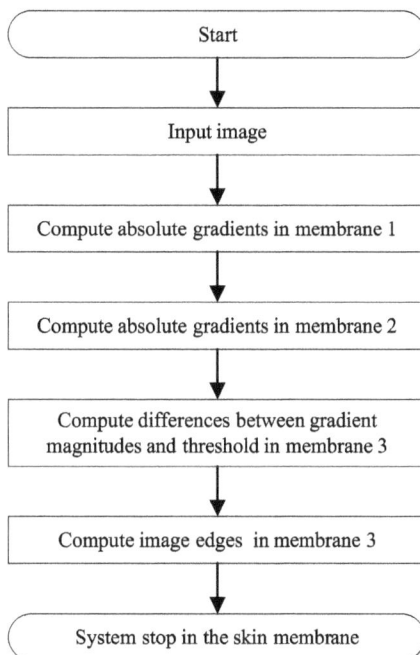

Figure 1. The flowchart of EDENP.

4.2. The Structure and Execution Processes of EDENP

As shown in Figure 2, the structure of EDENP includes four membranes. The system begins to start when the input variables $x_{i,j}$ representing the gray value of source image at location (i, j) appear in the skin membrane. The whole process includes five steps.

Step 1: Horizontal and vertical derivative approximations of every pixel are computed in membrane 1 by using rules of $\mathrm{Pr}_{1,CE_{i,j}} \sim \mathrm{Pr}_{10,CE_{i,j}}$ in a parallel manner. When the directional gradients are computed, membrane 2 will be activated.

Step 2: The gradient magnitude of all the pixels are obtained at the same time with rules of $\mathrm{Pr}_{21,CE_{i,j}}$ in membrane 2.

Step 3: The comparisons between the gradient magnitudes of all pixels and the predefined threshold are executed by rules of $\mathrm{Pr}_{31,CE_{i,j}}$ in membrane 3.

Step 4: The edge pixels are detected and marked with 1, while the background pixels are marked with 0 by rules of $\mathrm{Pr}_{32,CE_{i,j}}$ and $\mathrm{Pr}_{33,CE_{i,j}}$ in membrane 3.

Step 5: The system stop condition is satisfied and the system stops working by rules of Pr_{main} in membrane 4.

So as described above, only five steps are needed in the proposed algorithm for images with arbitrary resolution. Since we do not change the mathematical model of Sobel based edge detection, the detection result by our proposed method is the same as if run on a serial computing platform.

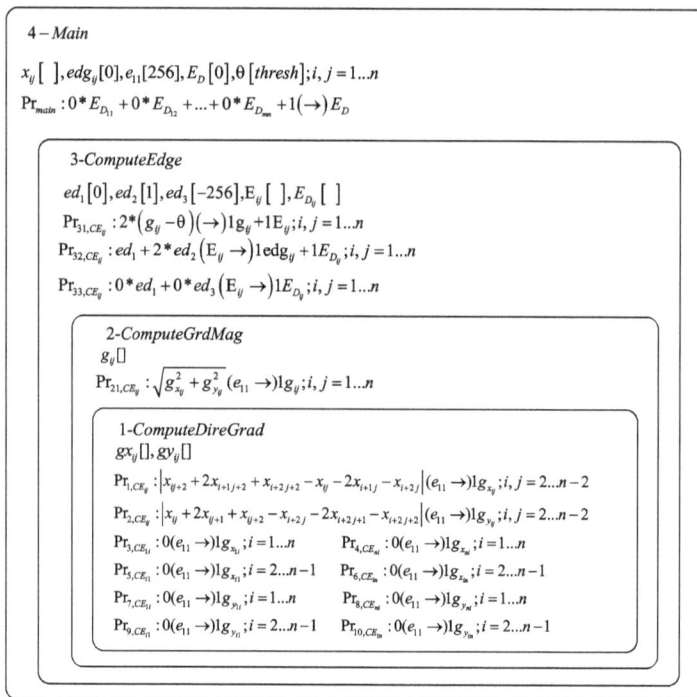

Figure 2. The structure of EDENP.

4.3. Complexity and Resources Analysis

Taking into account that the size of the input data is $n \times n$, and the image is a gray image. The amount of resources needed is illustrated in Table 1. From Table 1, we can see that there are $(7n^2 + 6)$ variables, including $(2n^2 + 2)$ enzymatic variables and $(5n^2 + 4)$ numerical variables. $(6n^2 + 4)$ rules are involved in this system. The total storage space is 1 cell with $(13n^2 + 10)$ molecules. So the space complexity is $O(n^2)$ theoretically. The time complexity is $O(1)$ because the number of execution steps is 5, which implies the computational efficiency is constant for images under arbitrary resolutions.

From the above analysis, we can see that the core of the proposed algorithm is to use space to replace time to obtain high-performance parallel computing, which is exactly the prominent characteristic of MC. Since molecules are used as storage units in a real biological computer, huge storage space can be utilised when this algorithm is implemented on it. So we think the proposed parallel algorithm is effective for images with high resolutions, at least at a theoretical level.

Table 1. Complexity and resources needed for EDENP.

Term	Necessary Resources
Initial number of cells	1
Number of enzymatic variables	$2n^2 + 2$
Number of numerical variables	$5n^2 + 4$
Number of rules	$6n^2 + 4$
Execution steps	5

5. Experiments and Results

In this section, both the performance and efficiency of our proposed EDENP algorithm are evaluated. Since there is no hardware implementation of MC systems at present, the only way to test the behaviors of the designed P systems is to simulate them in conventional computers. In this paper, a parallel computing architecture, Compute Unified Device Architecture (CUDA), is used as the simulating platform, as it has been reported in literature [24,61]. The parameters of the platform on which our experiments are carried out are illustrated in Table 2. The threshold θ for all the experiments is set to 0.2.

Table 2. Parameters the computer used.

Term	Parameters
CPU model	Intel(R) Core(TM) i7-7700HQ
cache memory	8 MB, 16-Way, 64 byte lines
main memory	16 GB (2* DDR4 2400MHz)
hard disc	SSD, SK hynix SC308 SATA 128GB, 600 Mbps; MQ01ABD100, 1TB
GPU model	Nvidia GeForce GTX 1050 Ti (4 GB)
execution steps	5

5.1. Performance Evaluation

Two case studies are considered to evaluate the performance of the proposed method for different types of images. Since the proposed algorithm is in the framework of MC, edge detection methods based on tissue-like P systems [21,24] are chosen as comparison methods. Algorithms in the literature [21,24] are sketched and implemented on a CPU platform using the MATLAB program.

5.1.1. Qualitative Evaluation

Case study 1 is considered to evaluate the performance of the three algorithms for images with rich textures. Four images named *rice, cameraman, mri,* and *AT3_lm4_01* randomly collected from the MATLAB Image Tool Box are used as testing samples in this experiment, as shown in Figure 3a,e,i,m. Figure 3b–d,f–h,j–l,n–p show the detailed qualitative edge detection results of the three algorithms for the four images. It can be clearly observed from Figure 3b,f,j,n, that the contours of the objects can be detected, but meanwhile the noise in the background is also detected, which will make the following image processing, such as object recognition, more difficult to deal with. The results by reference [21] are shown in Figure 3c,g,k,o. It can be seen that there are too many small edges, and the main outlines of the targets can hardly be found even by human eyes. The results of EDENP are illustrated in Figure 3d,h,l,p, from which we can see that not only the main contours of objects can be detected successfully, but also the noise is well suppressed.

Case study 2 is used to test the performance of the three methods for images with less texture, in which images named *toyobjects, circbw, text, testpart1* randomly selected from MATLAB Image Tool Box are used as testing image samples. In image *toyobjects*, each object has a constant gray value, while the other three images are binary images. Like in Case 1, the detected edge results by the three approaches are shown in Figure 4. Figure 4b,f,j,n clearly show that there are many discontinuous edges when using algorithm in reference [24], while the other two methods can detect the edges completely. When comparing the thickness of the edges, it is obvious to see that the method in reference [21] can achieve the thinnest edges, then the EDENP method, and the edges detected by [24] is the thickest. Although the method in [21] can obtain the finest edges, those edges often have burrs, as shown in Figure 5. Figure 5a,e are the whole edge image of *toyobjects* and *circbw*. Figure 5b–d,f–h are the local enlargement of areas in pink rectangles in Figure 5a,e. Areas marked in green in Figure 5b,f are some examples of discontinuous edges by [24]. When comparing Figure 5c,g with Figure 5d,h, it is clear that edges by EDENP are much smoother than by algorithm [21].

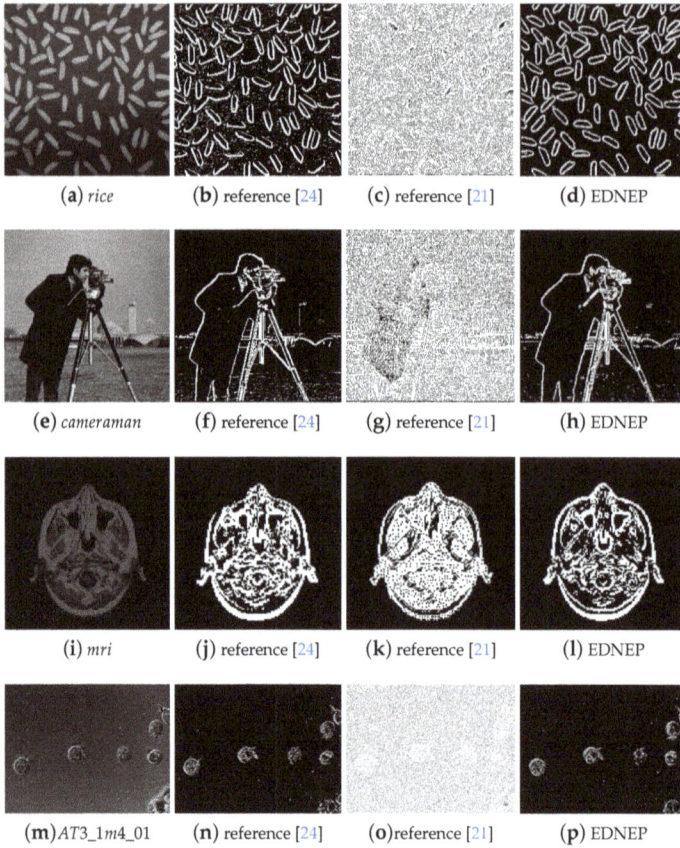

(**a**) *rice* (**b**) reference [24] (**c**) reference [21] (**d**) EDNEP

(**e**) *cameraman* (**f**) reference [24] (**g**) reference [21] (**h**) EDNEP

(**i**) *mri* (**j**) reference [24] (**k**) reference [21] (**l**) EDNEP

(**m**)*AT3_1m4_01* (**n**) reference [24] (**o**)reference [21] (**p**) EDNEP

Figure 3. Edge detection results of images with rich texture (the first column: the source gray images; the second to the last column: results by using methods in [21,24] and EDENP respectively).

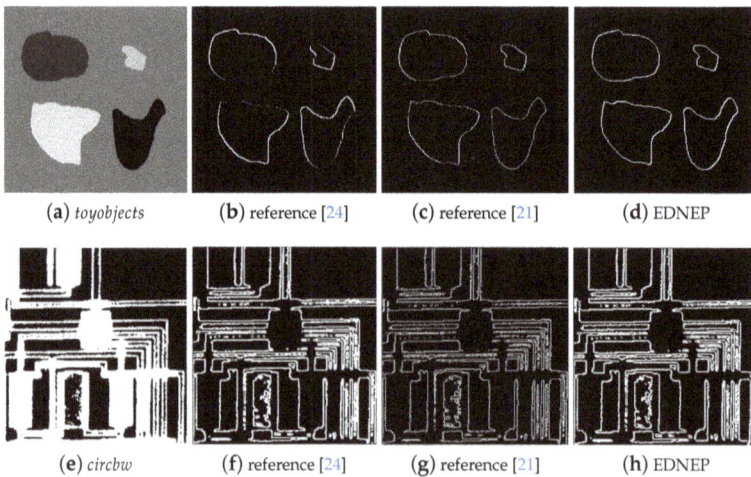

(**a**) *toyobjects* (**b**) reference [24] (**c**) reference [21] (**d**) EDNEP

(**e**) *circbw* (**f**) reference [24] (**g**) reference [21] (**h**) EDNEP

Figure 4. *Cont.*

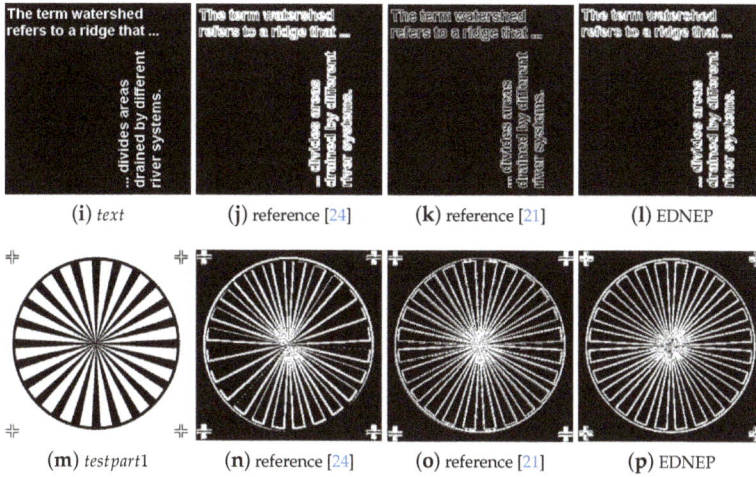

Figure 4. Edge detection results of images with less texture (the first column: the source gray images; the second to the last column: results by using methods in [21,24] and EDENP, respectively).

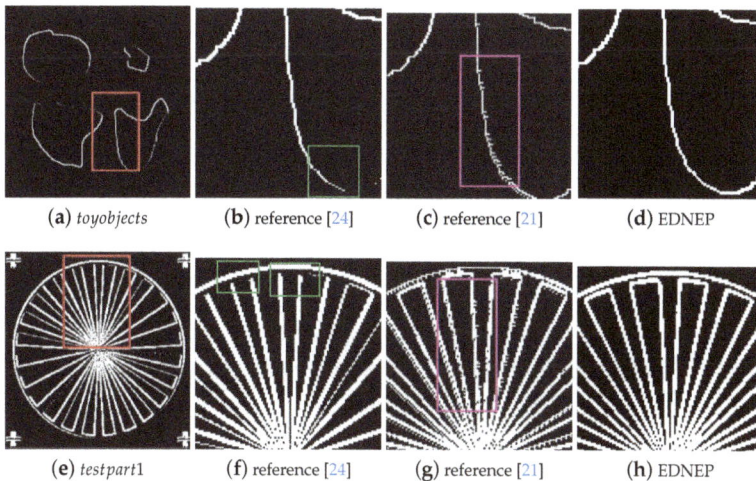

Figure 5. Edge detection results of *toyobjects* and *testpart*1 (the first column: the edge image; the second to the last columns: the local edge image enlarged by using methods in [21,24] and EDENP respectively).

5.1.2. Quantitative Evaluation

The confidence degree of the edge image is one of the most used indexes for evaluating the authenticity of the edge pixels. In general, the greater the edge confidence degree is, the more reliable the edges are. In this paper, we use this index to evaluate the performance of the edge detection algorithm quantitatively, whose mathematical definition is presented in reference [62].

Table 3 provides the comparison results of the three methods in terms of edge confidence degree. It can be seen from Table 3 that the EDENP method has the highest edge confidence degree for images with both high and low texture, which means edges detected by EDENP have less false edges.

Through the above quantitative and qualitative results, it can be deduced that the method in reference [21] is nearly invalid for grayscale images with rich texture. For images with less textures, this method can get the fine edges of the objects. However, the edges are not smooth in some cases because of the false burr edge points. The approach in [24] cannot get the whole contours of the objects due to the discontinuous edges detected for images with both rich and less rich textures. The EDENP algorithm has the highest performance and can obtain clear, continuous, and authentic edges of images with both rich and less rich textures.

Table 3. The edge confidence degree.

	Reference [24]	Reference [21]	EDENP
rice	0.75	0.56	**0.84**
cameraman	0.66	0.32	**0.74**
mri	0.63	0.56	**0.68**
AT3_lm4_01	0.44	0.12	**0.5**
toyobjects	0.85	0.76	**0.86**
circbw	0.94	0.93	**0.95**
text	0.93	0.90	**0.94**
testpart1	0.81	0.79	**0.86**

In this paper, only edge detection methods in the framework of MC are chosen for a comparison. From the above experimental results, we can see that the proposed algorithm has better performance compared with the existing tissue-like based edge detection methods. The fundamental reason for this is that with the help of "enzyme variables" in ENPS, the rules can be controlled flexibly, thus the existing Sobel edge detection algorithm can be programmed in "membrane computing language" easily.

5.2. Efficiency Evaluation

To better describe the computation efficiency of EDENP, a speedup ratio is defined as the elapsed time of algorithm on CPU platform divided by running time on GPU platform. The running times of images with different resolutions under GPU and CPU platform and corresponding speedup ratios for one image (*camera*) are illustrated in Table 4. From Table 4, we can see, although the computation times of EDENP are independent of resolutions theoretically, it takes different times to execute the EDENP algorithm for the same image at different resolutions. The reason for this is that the programs do not run on real bio-computers. Table 5 gives the speedup ratios results of the other seven images. It can be found that the lowest speedup is 53, and the maximum speedup can reach up to 262. It is obvious that the computing power of the proposed algorithm is much superior compared with the traditional algorithm implemented on CPU platform.

Table 4. Elapsed time of images with different resolution (*cameraman*).

Image Resolution	256^2	384^2	512^2	768^2	1024^2	2048^2	Platform
Elapsed time(ms)	0.014	0.03	0.05	0.12	0.23	0.86	GPU
Elapsed time(ms)	3.5	9.1	4.4	9.6	41.9	72.8	CPU
Speedup ratio	250	303	88	80	182	130	

Table 5. The speedup ratio of seven images.

Image Resolution	256^2	384^2	512^2	768^2	1024^2	2048^2
rice	79	121	79	101	136	82
mri	60	80	77	62	73	66
AT3_lm4_01	80	90	102	172	71	75
toyobjects	187	162	163	81	182	62
circbw	193	213	262	210	176	66
text	167	180	194	118	57	65
testpart1	53	76	100	161	87	64

6. Conclusions

Membrane computing is a new branch of natural computing, and its amazing storage space and high parallel computing characteristics are very suitable for big data processing. Among various membrane systems, the ENPS can directly deal with numeric variables, and the enzyme variables can flexibly control the execution orders of different rules. In this paper, we attempt to apply ENPS to image processing, and take Sobel edge detection as an example. Compared with the previous works which are based on tissue-like P systems, the advantage of the proposed method is that it does not need to encode and decode the image data, and it is easy to write the program for algorithms with complex execution orders in "membrane computing language". The limitation of the proposed algorithm mainly has two aspects. One is that the execution of the algorithm is based on real biological computers. However, there are no universal biological computers at present, so it is difficult to evaluate the real computing efficiency of the proposed algorithm. The other shortage is that the space complexity is $O(n^2)$, which means large storage space is needed for the proposed algorithm. In future research, we will simulate the algorithm on FPGA hardware and try to combine the ENPS with other, more complex image processing algorithms.

Supplementary Materials: The following are available online.

Author Contributions: The research structure was conceived and designed by J.Y. and G.Z.; D.G., M.Z. and Q.Y. wrote the program and performed the experiments; J.Y. and P.P. wrote the paper and analyzed the data; G.Z. made revisions to the final manuscript. The final manuscript was read and corrected by all authors.

Funding: This work was partially supported by the National Natural Science Foundation of China, under Grant #162300410079, #61672437, #61702428; the Sichuan Science and Technology Program China, under Grant #2018GZ0385, #2017GZ0431, #2018GZ0245, #2018GZ0185, #2018GZ0086, #2018GZ0095, #2019YFG0188; Sichuan education department Program China, under Grant #17ZB0095, #17ZB0090; the Talent Import Fund of CUIT China, under Grant #KYTZ201633; the Chengdu Science and Technology Program China, under Grant #2017-GH02-00049-HZ, #2018-YF05-00981-GX and the New Generation Artificial Intelligence Science and Technology Major Project of Sichuan Province China, under Grant #2018GZDZX0043.

Acknowledgments: The authors would like to thank the anonymous reviewers for their valuable suggestions on improving this paper.

Conflicts of Interest: The authors declare no conflict of interest.

References

1. Zheng, S.; Yuille, A.; Tu, Z. Detecting object boundaries using low-, mid-, and high-level information. *Comput. Vis. Image Underst.* **2010**, *114*, 1055–1067. [CrossRef]
2. Zhao, L.; Zhao, Q.; Liu, H.; Lv, P.; Gu, D. Structural sparse representation-based semi-supervised learning and edge detection proposal for visual tracking. *Vis. Comput.* **2017**, *33*, 1169–1184. [CrossRef]
3. Hua, S.; Wen, H. Moment-preserving edge detection and its application to image data compression. *Opt. Eng.* **2013**, *32*, 1596–1608. [CrossRef]
4. Satpathy, A.; Jiang, X.; Eng, H.L. LBP-based edge-texture features for object recognition. *IEEE Trans. Image Process.* **2014**, *23*, 1953–1964. [CrossRef]
5. Saif, J.; Hammad, M.; Alqubati, I. Gradient based image edge detection. *Int. J. Eng. Technol.* **2016**, *8*, 153–156. [CrossRef]
6. Jung, J.; Lee, H.; Lee, J.; Park, D. A novel template matching scheme for fast full-Search boosted by an integral image. *IEEE Signal Proc. Lett.* **2010**, *17*, 107–110. [CrossRef]
7. Schweitzer, H.; Deng, R.; Anderson, R.F. A dual-bound algorithm for very fast and exact template matching. *IEEE Trans. Pattern Anal. Mach. Intell.* **2011**, *33*, 459–470. [CrossRef] [PubMed]
8. Ghassabeh, Y.A.; Rudzicz, F.; Moghaddam, H.A. Fast incremental LDA feature extraction. *Pattern Recognit.* **2015**, *48*, 1999–2012. [CrossRef]
9. Zhou, G.; Cichocki, A.; Zhang, Y.; Mandic, D.P. Group component analysis for multiblock data: common and individual feature extraction. *IEEE Trans. Neural Netw.* **2016**, *27*, 2426–2439. [CrossRef]
10. Herout, A.; Josth, R.; Juranek, R.; Havel, J.; Hradis, M.; Zemcik, P. Real-time object detection on CUDA. *J. Real-Time Image Process.* **2011**, *6*, 159–170. [CrossRef]

11. Jiang B. Real-time multi-resolution edge detection with pattern analysis on graphics processing unit. *J. Real-Time Image Process.* **2018**, *14*, 293–321. [CrossRef]
12. Zuo H.; Zhang Q.; Yong, X.; Zhao, R. Fast sobel edge detection algorithm based on GPU. *Opto-Electron. Eng.* **2009**, *36*, 8–12. [CrossRef]
13. Jiang, J.; Liu, C.; Ling, S.R. An FPGA implementation for real-time edge detection. *J. Real-Time Image Process.* **2018**, *15*, 787–797. [CrossRef]
14. Nausheen, N.; Seal, A.; Khanna, P.; Halder, S. A FPGA based implementation of Sobel edge detection. *Microprocess Microsy* **2018**, *56*, 84–91. [CrossRef]
15. Paolo,B.; Sipp, D. Regulation: Sell help not hope. *Nature* **2014**, *510*, 336–337. [CrossRef]
16. Jiang, N.; Dang, Y.; Wang, J. Quantum image matching. *Quantum Inf. Process.* **2016**, *15*, 3543–3572. [CrossRef]
17. Tsaftaris, S.A.; Katsaggelos, A.K.; Pappas, T.N.; Papoutsakis E.T. How can DNA computing be applied to digital signal processing?. *IEEE Signal Process. Mag.* **2004**, *21*, 57–61. [CrossRef]
18. Díaz-Pernil, D.; Gutierrez-Naranjo, M; Peng, H. Membrane computing and image processing: A short survey. *J. Membr. Comput.* **2019**. [CrossRef]
19. Păun, Gh. Computing with membranes. *J. Comput. Syst. Sci.* **2000**, *61*, 108–143. [CrossRef]
20. Alsalibi, B.; Venkat, I.; Subramanian, K.; Lebai, L.; De, W. The impact of bio-inspired approaches toward the advancement of face recognition. *ACM Comput. Surv.* **2015**, *48*, 1–33. [CrossRef]
21. Christinalh, A.; Díaz-Pernil, D.; Real, P. Region-based segmentation of 2D and 3D images with tissue-like P systems. *Pattern Recognit. Lett.* **2011** *32*, 2206–2212. [CrossRef]
22. Díaz-Pernil, D.; Gutiérrez-Naranjo, M.; Molina-Abril, H.; Real, P. Designing a new software tool for digital imagery based on P systems. *Nat. Comput.* **2012**, *11*, 381–386. [CrossRef]
23. Carnero, J.; Díaz-Pernil, D.; Molina-Abril, H.; Real, P. Image segmentation inspired by cellular models using hardware programming. In Prceedings of the 3rd International Workshop onComputational Topology in Image Context, Chipiona, Spain, 10–12 November 2010; Volume 1, pp. 143–150.
24. Díaz-Pernil, D.; Berciano, A.; Peña-Cantillana, F.; Gutiérrez-Naranjo, M. Segmenting images with gradient-based edge detection using membrane computing. *Pattern Recognit. Lett.* **2013**, *34*. 846–855. [CrossRef]
25. Peña-Cantillana, F.; Díaz-Pernil, D.; Christinal, H.; Gutiérrez-Naranjo, M. Implementation on CUDA of the smoothing problem with tissue-like P systems. *Int. J. Nat. Comput. Res.* **2011**, *2*, 25–34. [CrossRef]
26. Alsalibi, B.; Venkat, I.; Subramanian, K.; Christinal, H. A bio-inspired software for homology groups of 2D digital images. In Proceedings of the Asian Conference on Membrane Computing (ACMC), Coimbatore, India, 18–19 September 2014; IEEE Computer Society: Washington, DC, USA, 2014. [CrossRef]
27. Díaz-Pernil, D.; Christinal, H.; Gutiérrez-Naranjo, M.; Real, P. Using membrane computing for effective homology. *Appl. Algebr. Eng. Commun.* **2012** *23*, 233–249. [CrossRef]
28. Ardelean, I.; Díaz-Pernil, D.; Gutiérrez-Naranjo, M.; Peña-Cantillana, F.; Reina-Molina, R.; Sarchizian, I. Counting cells with tissue-like P systems. In Proceedings of the Tenth Brainstorming Week on Membrane Computing, Seville, Spain, 30 January–3 February 2012; Fénix Editora: Seville, Spain, 2012.
29. Reina-Molina, R.; Díaz-Pernil, D.; Gutiérrez-Naranjo, M. Cell complexes and membrane computing for thinning 2D and 3D images. In Proceedings of the Tenth Brainstorming Week on Membrane Computing, Seville, Spain, 30 January–3 February 2012; Fénix Editora: Seville, Spain, 2012.
30. Berciano, A.; Díaz-Pernil, D.; Christinal, H.; Venkat, I. First steps for a corner detection using membrane computing. In Proceedings of the Asian Conference on Membrane Computing, Coimbatore, India, 18–19 September 2014; IEEE Computer Society: Washington, DC, USA, 2014.
31. Enguix, G. Preliminaries about some possible applications of P systems in linguistics. *Lect. Notes Comput. Sci.* **2002**, *2597*, 74–89. [CrossRef]
32. Cabarle, F.; de la Cruz, R.; Zhang, X.; Jiang, M.; Liu, X.; Zeng, X. On string languages generated by spiking neural P systems with structural plasticity. *IEEE Trans. Nanobiosci.* **2018**, *17*, 560–566. [CrossRef]
33. Song, T.; Zeng, X.; Zheng, P.; Jiang, M.; Rodríguez-Patón, A. A parallel workflow pattern modelling using spiking neural P systems with colored spikes. *IEEE Trans. Nanobiosci.* **2018**, *17*, 474–484. [CrossRef] [PubMed]
34. Mayne, R.; Phillips, N.; Adamatzky, A. Towards experimental P-systems using multivesicular liposomes. *J. Membr. Comput.* **2019**. [CrossRef]
35. Mitrana, V. Polarization: a new communication protocol in networks of bio-inspired processors. *J. Membr. Comput.* **2019**, published online. [CrossRef]

36. Pan, L.; Păun, Gh.; Zhang, G.; Neri, F. Spiking Neural P Systems with Communication on Request. *Int. J. Neural Syst.* **2017**, *27*, 1750042. [CrossRef]

37. Orellana-Martín, D.; Valencia-Cabrera, L.; Riscos-Núñez, A.; Pérez-Jiménez, M.J. P systems with proteins: A new frontier when membrane division disappears. *J. Membr. Comput.* **2019**. [CrossRef]

38. Sánchez-Karhunen, E.; Valencia-Cabrera, L. Modelling complex market interactions using PDP systems. *J. Membr. Comput.* **2019**. [CrossRef]

39. Zhang, G.; Rong, H.; Neri, F.; Pérez-Jiménez, M.J. An optimization spiking neural P system for approximately solving combinatorial optimization problems. *Int. J. Neural Syst.* **2014**, *24*, 1440006. [CrossRef] [PubMed]

40. Zhang, G.; Cheng, J.; Gheorghe, M.; Meng, Q. A hybrid approach based on differential evolution and tissue membrane systems for solving constrained manufacturing parameter optimization problems. *Appl. Soft Comput.* **2013**, *13*, 1528–1542. [CrossRef]

41. Wang, T.; Zhang, G.; Zhao, J.; He, Z.; Wang, J.; Pérez-Jiménez, M.J. Fault diagnosis of electric power systems based on fuzzy reasoning spiking neural P systems. *IEEE Trans. Power Syst.* **2015**, *30*, 1182–1194. [CrossRef]

42. Martín-Vide, C.; Păun, Gh.; Pazos, J.; Rodríguez-Patón, A. Tissue P systems. *Theor. Comput. Sci.* **2003**, *296*, 295–326. [CrossRef]

43. Ionescu, M.; Păun, Gh.; Yokomori, T. Spiking neural P systems. *Fund. Inform.* **2006**, *71*, 279–308. [CrossRef]

44. Zeng, X.; Pan, L.Q; Pérez-Jiménez, M.J. Small universal simple spiking neural P systems with weights. *Sci. China Inf. Sci.* **2014**, *57*, 1–11. [CrossRef]

45. Păun, G.; Păun, R. Membrane computing and economics: numerical P systems. *Fund. Inform.* **2006**, *73*, 213–227.

46. Zhang, Z.; Wu, T.; Paun, A.; Pan, L. Numerical P systems with migrating variables. *Theor. Comput. Sci.* **2016**, *641*, 85–108. [CrossRef]

47. Pan, L.; Zhang, Z.; Wu, T.; Xu, J. Numerical P systems with production thresholds. *Theor. Comput. Sci.* **2017**, *673*, 30–41. [CrossRef]

48. Zhang, Z.; Pan, L. Numerical P systems with thresholds. *Int. J. Comput. Commun.* **2017**, *11*, 292–304. [CrossRef]

49. Buiu, C.; Vasile, C.; Arsene, O. Development of membrane controllers for mobile robots. *Inform. Sci.* **2012**, *187*, 33–51. [CrossRef]

50. Wang, X.; Zhang, G.; Neri, F.; Jiang T.; Zhao, J.; Gheorghe, M.; Ipate, F.; Lefticaru, R. Design and implementation of membrane controllers for trajectory tracking of nonholonomic wheeled mobile robots. *Integr. Comput.-Aid Eng.* **2016**, *23*, 15–30. [CrossRef]

51. Zhang, G.; Gheorghe, M.; Pérez-Jiminez, M.J. *Real-Life Applications with Membrane Computing*; Springer International Publishing AG: Cham, Switzerland, 2017; pp. 130–141.

52. Mahalingam, K.; Rama, R.; Sureshkumar, W. Robot motion planning inside a grid using membrane computing. *Int. J. Imaging Robot.* **2017**, *17*, 33–51.

53. Pavel, A.; Arsene, O.; Buiu, C. Enzymatic numerical P systems—A new class of membrane computing systems. In Proceedings of the Fifth International Conference on Bio-Inspired Computing: Theories and Applications, Changsha, China, 23–26 September 2010; IEEE Computer Society: Washington, DC, USA, 2010. [CrossRef]

54. Zhang, Z.; Wu, T.; Paun, A.; Pan, L. Universal enzymatic numerical P systems with small number of enzymatic variables, *Sci. China Inf. Sci.* **2018**, *61*, 38–49. [CrossRef]

55. Vasile, C.; Pavel, A.; Dumitrache, I.; Păun, Gh. On the power of enzymatic numerical P system. *ACTA Inform.* **2012**, *49*, 395–412. [CrossRef]

56. Vasile, C.; Pavel, A.; Dumitrache, I. Universality of enzymatic numerical P systems. *Int. J. Comput. Math.* **2013**, *90*, 869–879. [CrossRef]

57. Leporati, A.; Porreca, A.; Zandron, C.; Mauri, G. Improved universality results for parallel enzymatic numerical P systems. *Int. J. Unconv. Comput.* **2013**, *9*, 385–404.

58. Pavel, A.; Buiu, C. Using enzymatic numerical P systems for modeling mobile robot controllers. *Nat. Comput.* **2012**, *11*, 387–393. [CrossRef]

59. Li, W.; Yang, J.; Zhang, J. Handling big data field with enzymatic numerical P System. *J. Sichuan Univ. Nat. Sci. Ed.* **2013**, *45*, 96–104.

60. Pang S.C.; Ding, T.; Rodriguez-Paton, A.; Song, T.; Zheng, P. A parallel bioinspired framework for numerical calculations using enzymatic P system with an enzymatic environment. *IEEE Access* **2018**, *6*, 65548–65556. [CrossRef]

61. Cecilia, J.; García, J.; Guerrero, G.; Martínet-del-Amor, M.; Pérez-Hurtado, I.; Pérez-Jiménez, M. Simulation of P systems with active membranes on CUDA. *Brief Bioinform.* **2009**, *11*, 313–322. [CrossRef] [PubMed]

62. Wang, J.; Bi, J.; Wang, L.; Wang, X. A non-reference evaluation method for edge detection of wear particles in ferrograph images. *Mech. Syst. Signal Process.* **2018**, *100*, 863–876. [CrossRef]

![molecules logo] **molecules**

MDPI

Article

Small Universal Bacteria and Plasmid Computing Systems

Xun Wang [1], Pan Zheng [2], Tongmao Ma [1] and Tao Song [1,3,*]

[1]   College of Computer and Communication Engineering, China University of Petroleum, Qingdao 266580, China; wangsyun@upc.edu.cn (X.W.); matongmao@163.com (T.M.)
[2]   Department of Accounting and Information Systems, University of Canterbury, Christchurch 8041, New Zealand; panzheng@ieee.org
[3]   Departamento de Inteligencia Artificial, Universidad Politcnica de Madrid (UPM), Campus de Montegancedo, 28660 Boadilla del Monte, Spain
*   Correspondence: tsong@upc.edu.cn or t.song@upm.es; Tel.: +86-150-532-587-69

Received: 25 April 2018; Accepted: 21 May 2018; Published: 29 May 2018

Abstract: Bacterial computing is a known candidate in natural computing, the aim being to construct "bacterial computers" for solving complex problems. In this paper, a new kind of bacterial computing system, named the bacteria and plasmid computing system (BP system), is proposed. We investigate the computational power of BP systems with finite numbers of bacteria and plasmids. Specifically, it is obtained in a constructive way that a BP system with 2 bacteria and 34 plasmids is Turing universal. The results provide a theoretical cornerstone to construct powerful bacterial computers and demonstrate a concept of paradigms using a "reasonable" number of bacteria and plasmids for such devices.

Keywords: bacterial computing; bacteria and plasmid system; Turing universality; recursively enumerable function

1. Introduction

In cell biology, bacteria, despite their simplicity, contain a well-developed cell structure that is responsible for some of their unique biological structures and pathogenicity. The bacterial DNA resides inside the bacterial cytoplasm, for which transfer of cellular information, transcription, and DNA replication occurs within the same compartment [1,2]. Along with chromosomal DNA, most bacteria also contain small independent pieces of DNA called plasmids, which can be conveniently obtained and released by a bacterium to act as a gene delivery vehicle between bacteria in the form of horizontal gene transfer [3].

Bacterial computing was coined with the purpose of building biological machines, which are developed to solve real-life engineering and science problems [4]. Practically, bacterial computing proves mechanisms and the possibility of using bacteria for solving problems in vivo. If an individual bacterium can perform computation work as a computer, this envisions a way to build millions of computers in vivo. These "computers", combined together, can perform complicated computing tasks with efficient communication via plasmids. Using such conjugation, DNA molecules, acting as information carriers, can be transmitted from one cell to another. On the basis of the communication, information in one bacteria can be moved to another and can be used for further information processing [5,6].

Bacterial computing models belong to the field of bio-computing models, such as DNA computing models [7–9] and membrane computing models [10–12]. Because of the computational intelligence and parallel information processing strategy in biological systems, most of the bio-computing models have been proven to have the desired computational power. Most of these can do what a Turing machine can do (see, e.g., [13–19]). The proposed bacterial computing models can provide powerful computing models at the theoretical level but a lack of practical results. Current bacterial computing models are designed for solving certain specific biological applications, such as bacteria signal pathway detecting, but give no result for computing power analysis.

In general bacterial computing models, information to be processed is encoded by DNA sequences, and conjugation is the tool for communicating among bacteria. The biological process is shown in Figure 1.

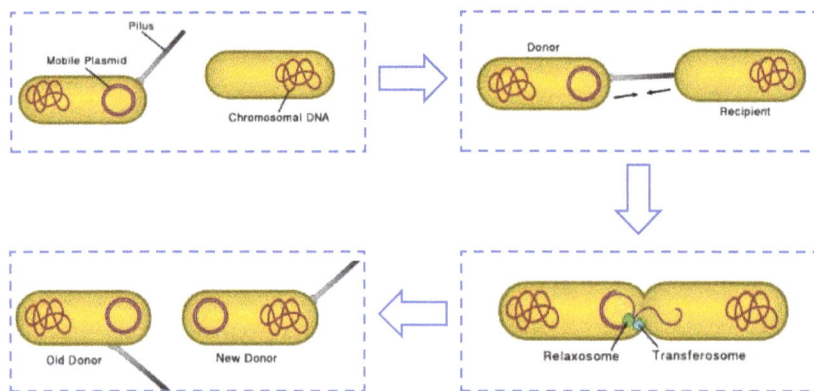

Figure 1. Bacteria conjugation from biological point of view.

Looking for small universal computing devices, such as small universal Turing machines [20,21], small universal register machines [22], small universal cellular automata [23], small universal circular Post machines [24], and so on, is a natural and well-investigated topic in computer science. Recently, this topic started to be considered also in the framework of bio-computing models [25–31].

In this work, we focus on designing small universal bacteria and plasmid computing systems (BP systems); that is, we construct Turing universal BP systems with finite numbers of bacteria and plasmids. Specifically, we demonstrate that a BP system with 2 bacteria and 34 plasmids is universal for computing recursively enumerable functions and families of sets of natural numbers. In the universality proofs, 2 bacteria are sufficient, as in [32], but the numbers of plasmids needed are reduced to about 10 from a possible infinite number. The results provide a theoretical cornerstone to construct powerful "bacterial computers" and demonstrate a concept of paradigms using a "reasonable" number of bacteria and plasmids for these devices.

2. The Bacteria and Plasmid System

In this work, as for automata in automata theory, the BP system is formally designed and defined. In general, the system is composed of three main components:

– a set of bacteria;
– a set of plasmids;
– a set of evolution rules in each bacterium, including conjugation rules and gene-editing (inserting/deleting) rules.

The evolution rules are in the form of productions in formal language theory, which are used to process and communicate information among bacteria. Such a system is proven to be powerful for a number of computing devices; that is, they can compute the sets of natural numbers that are Turing computable. However, in the universality proof, the number of plasmids involved is not limited. It is possible to use an infinite number of plasmids for information processing and exchanging. Such a feature is acceptable (as for the infinite tape in Turing machines) in mathematic theory but is not feasible with the biological facts.

A BP system of degree m is a construct of the following form:

$$\Pi = (O, b_1, b_2, \ldots, b_m, P, b_{out}), \text{where the following are true.}$$

- $O = \{g_1, g_2, \ldots, g_n\}$ is a set of genes in the chromosomal DNA of bacteria.
- $P = P_{crispr} \cup P_{temp} \cup \{p_{null}\}$ is a set of plasmids.

  - Plasmids in P_{crispr} are of the form $(cas9, gRNA_{g_i}^{\alpha})$ with $\alpha \in \{insert, delete\}$, which is used for cutting specific genes.
  - Plasmids in P_{temp} are of the form $(gRNA_{g_i}^{template})$, which takes templates of genes to be inserted.
  - Plasmid p_{null} is of the form (Pro_{Rap}^{Rel}) for bacteria conjugation.

- Variables $b_1, b_2, \ldots, b_m$ are m bacteria of the form $b_i = (w_i, R_i)$, where

  - w_i is a set of genes over O initially placed in bacterium b_i;
  - R_i is a set of rules in bacterium b_i of the following forms:

    (1) **Conjugation rule** is of the form $(ATP\text{-}P_c, b_i/b_j, ATP\text{-}P_c')$, by which ATP in bacterium b_i is consumed and a set of plasmids $P_c' \subseteq P$ associated with ATP is transmitted into bacterium b_j.

    (2) **CRISPR/Cas9 gene inserting rule** is of the form $p_i p_{s_i} \times (g_j, g_k)$, where $p_i \in P_{crispr}$, $\alpha = insert$, $p_{s_i} \in P_{temp}$, and g_j and g_k are two neighboring genes. The insertion is operated if and only if g_j and g_k are neighboring genes and plasmids $p_i p_{s_i}$ are present in the bacterium.

    (3) **CRISPR/Cas9 gene deleting rule** is of the form $p_i \times (g_j, g_k)$ with $p_i p_{null} \in P_{crispr}$, $\alpha = detele$, and g_j and g_k being two neighboring genes. The rule can be used if and only there exists gene g_i placed between the two neighboring genes.

- Variable b_{out} is the output bacterium.

It is possible to have more than one enabled conjugation rule at a certain moment in a bacterium, but only one is non-deterministically chosen for use. This is due to the biological fact that ATP can support the transmission of one plasmid but not all of the plasmids. If a bacterium has more than one CRISPR/Cas9 operating rule associated with a certain common plasmid, only one of the rules is non-deterministically chosen for use; if the enabled CRISPR/Cas9 operating rules are associated with different plasmids, all of them will be used to edit the related genes.

The configuration of the system is described by chromosomal DNA encoding the information in each bacterium. Thus, the initial configuration is $\langle (w_1, w_2, \ldots, w_m) \rangle$. Using the conjugation and CRISPR/Cas9 rules defined above, we can define the transitions among configurations. Any sequence of transitions starting from the initial configuration is called a computation. A computation is called successful if it reaches a halting configuration, that is, no rule can be used in any bacterium. The computational result is encoded by the chromosomal DNA in bacterium b_{out} when the system halts, where $b_{out} \in \{b_1, b_2, \ldots, b_m\}$ denotes the output bacterium. There are several ways to encode numbers by the chromosomal DNA. We use the number of genes in the chromosomal DNA to encode different numbers computed by the system.

The set of numbers computed by system Π is denoted by $N(\Pi)$. We denote by $NBP(bact_j, plas_k)$ the family of sets of numbers computed/generated by BP systems with m bacteria and k plasmids (if no limit is imposed on the values of parameters m and k, then the notation is replaced by $*$).

We need an input bacterium to receive genetic signals in the form of short DNA segments from the environment or certain bacteria, as well as an output bacteria, with which the system can compute functions. The input bacterium is denoted by b_{in} with $b_{in} \in \{b_1, b_2, \ldots, b_m\}$. Input bacterium b_{in} can read/receive information from the environment, where information is encoded by DNA segments or a string of genes. When a BP system has both input and output bacteria, it starts by reading/receiving information from the environment through input bacterium b_{in}. After reading the input information, the system starts its computation by using the conjugation and CRISPR/Cas9 gene inserting/deleting rules; it then finally halts. The computational result is stored in the output bacterium b_{out} encoded by a number of certain genes.

Mathematically, if the input information is x, which is encoded by DNA segments composed of x genes, when the system halts, bacterium b_{out} holds y genes. It is said that the BP system can compute the function $f(x) = y$. In general, if the inputs are $x_1, x_2, \ldots, x_n$ in the form of DNA strands containing x_i copies of gene g_i with $i = 1, 2, \ldots, n$, when the system halts, we obtain the computational result y in the output bacterium in the form of y copies of genes. The system is said to compute the function $f(x_1, x_2, \ldots, x_n) = y$.

3. Universality Results

In this section, we construct two small universal BP systems. Specifically, we construct a Turing universal BP system with 2 bacteria and 34 plasmids to compute recursively enumerable functions. As a natural-number computing device, a universal BP system with 2 bacteria and 34 plasmids is achieved.

In the following universality proofs, the notion of a register machine is used. A register machine is a construct of the form $M = (m, H, l_0, l_h, R)$, where m is the number of registers, H is the set of instruction labels, l_0 is the start label, l_h is the halt label (assigned to instruction HALT), and R is the set of instructions; each label from H labels only one instruction from R, thus precisely identifying it. The instructions are of the following forms:

- $l_i : (\mathrm{ADD}(r), l_j, l_k)$ (add 1 to register r and then go to one of the instructions with labels l_j and l_k);
- $l_i : (\mathrm{SUB}(r), l_j, l_k)$ (if register r is non-zero, then subtract 1 from it, and go to the instruction with label l_j; otherwise, go to the instruction with label l_k);
- $l_h : \mathrm{HALT}$ (the halt instruction).

A register machine M generates a set $N(M)$ of numbers in the following way: it starts with all registers being empty (i.e., storing the number zero) and then applies the instruction with label l_0; it continues to apply instructions as indicated by the labels (and made possible by the contents of registers). If the register machine finally reaches the halt instruction, then the number n present in specified register 0 at that time is said to be generated by M. If the computation does not halt, then no number is generated. It is known (e.g., see [33]) that register machines generate all sets of numbers that are Turing computable.

A register machine can also compute functions. In [22], register machines are proposed for computing functions, with the universality defined as follows: Let $\varphi_x(y)$ be a fixed admissible enumeration of the unary partial recursive functions. A register machine M is said to be universal if there is a recursive function g such that for all natural numbers x and y, it holds $\varphi_x(y) = M(g(x), y)$; that is, with input $g(x)$ and y introduced in registers 1 and 2, the result $\varphi_x(y)$ is obtained in register 0 when M halts.

A specific universal register machine M_u shown in Figure 2 is used here, which was modified by a universal register machine from [22]. Specifically, the universal register machine from [22] contains a separate check for zero of register 6 of the form $l_8 : (\mathrm{SUB}(6), l_0, l_{10})$; this instruction was replaced in M_u by $l_8 : (\mathrm{SUB}(6), l_9, l_0)$, $l_9 : (\mathrm{ADD}(6), l_{10})$ (see Figure 2). Therefore, in the modified universal register machine, there are 8 registers (numbered from 0 to 7) and 23 instructions (hence 23 labels), the last instruction being the halting instruction. The input numbers are introduced in registers 1 and 2, and the result is obtained in register 0.

$$l_0 : (\text{SUB}(1), l_1, l_2), \qquad\qquad l_1 : (\text{ADD}(7), l_0),$$
$$l_2 : (\text{ADD}(6), l_3), \qquad\qquad l_3 : (\text{SUB}(5), l_2, l_4),$$
$$l_4 : (\text{SUB}(6), l_5, l_3), \qquad\qquad l_5 : (\text{ADD}(5), l_6),$$
$$l_6 : (\text{SUB}(7), l_7, l_8), \qquad\qquad l_7 : (\text{ADD}(1), l_4),$$
$$l_8 : (\text{SUB}(6), l_9, l_0), \qquad\qquad l_9 : (\text{ADD}(6), l_{10}),$$
$$l_{10} : (\text{SUB}(4), l_0, l_{11}), \qquad\qquad l_{11} : (\text{SUB}(5), l_{12}, l_{13}),$$
$$l_{12} : (\text{SUB}(5), l_{14}, l_{15}), \qquad\qquad l_{13} : (\text{SUB}(2), l_{18}, l_{19}),$$
$$l_{14} : (\text{SUB}(5), l_{16}, l_{17}), \qquad\qquad l_{15} : (\text{SUB}(3), l_{18}, l_{20}),$$
$$l_{16} : (\text{ADD}(4), l_{11}), \qquad\qquad l_{17} : (\text{ADD}(2), l_{21}),$$
$$l_{18}: (\text{SUB}(4), l_0, l_h), \qquad\qquad l_{19} : (\text{SUB}(0), l_0, l_{18}),$$
$$l_{20} : (\text{ADD}(0), l_0), \qquad\qquad l_{21} : (\text{ADD}(3), l_{18}),$$
$$l_h : \text{HALT}$$

Figure 2. The universal register machine for computing Turing-computable functions [22].

3.1. A Small Universal BP System as Function Computing Device

Theorem 1. *There exists a Turing universal BP system with 2 bacteria and 34 plasmids that can compute Turing-computable recursively enumerable functions.*

Proof. To this aim, we construct a BP system Π with 2 bacteria and 34 plasmids to simulate the register machine M_u shown in Figure 2. The system Π is of the following form:

$$\Pi = (O, b_1, b_2, P, b_{in}, b_{out}), \text{where the following are true.}$$

- $O = \{g_0, g_1, \ldots, g_7, g_m\}$ is set of genes in chromosomal DNA of bacteria.
- $P = P_{crispr} \cup P_{temp} \cup \{p_{null}\}$ is a set of plasmids shown in Table 1, where

  - $P_{crispr} = \{p_1, p_2, \ldots, p_{22}, p_h\}$, whose elements associated with the labels of instructions are used for gene cutting;
  - $P_{temp} = \{p_{s_i} \mid i = 1, 2, 5, 7, 9, 16, 17, 20, 21\}$ are plasmids taking templates of genes to be inserted, which are used for simulating ADD instructions;
  - plasmid p_{null} for bacteria conjugation is used for simulating SUB instructions.

- $b_1 = (w_1, R_1)$, where $w_1 = \lambda$, meaning no initial chromosomal DNA is placed in bacteria b_1; the set of rules R_1 is shown in Table 2.
- $b_2 = (w_2, R_2)$, where $w_2 = g_0 g_m g_1 g_m g_2 g_m g_3 g_m g_4 g_m g_5 g_m g_6 g_m g_7 g_m$, indicating the initially placed chromosomal DNA in bacterium b_2; the set of rules R_2 is shown in Table 2.
- $b_{in} = b_{out} = b_2$, which means bacterium b_2 can read signals from the environment, and when the system halts, the computational result is stored in bacterium b_2.

In general, for each add instruction l_i acting on register $r \in \{0, 1, 2, 3, 4, 5, 6, 7\}$, plasmids $p_i = (cas9, gRNA_{g_r}^{insert})$ and $p_{s_i} = (gRNA_{g_r}^{template})$ are associated; for any SUB instruction l_i acting on register $r \in \{0, 1, 2, 3, 4, 5, 6, 7\}$, a plasmid $p_i = (cas9, gRNA_{g_r}^{delete})$ is associated in system Π. The numbers stored in register r are encoded by the number of copies of gene g_r with $r \in \{0, 1, 2, 3, 4, 5, 6, 7\}$ in chromosomal DNA of bacterium b_2. Specifically, if the number stored in register r is $n \geq 0$, then bacterium b_2 contains $n + 1$ copies of gene g_r.

During the simulation of register machine M_u by system Π, when bacterium b_1 holds a pair of plasmids $p_i p_{s_i}$ (respectively $p_i p_{null}$) and ATP, the system starts to simulate an ADD instruction (respectively a SUB instruction) l_i of M_u: plasmids $p_i p_{s_i}$ (respectively $p_i p_{null}$) are transmitted to bacterium b_2 by the conjugation rule; then one copy of gene g_r between neighboring genes g_r and g_m is inserted (respectively deleted) to simulate increasing (respectively decreasing) the number in register r by 1; after this, bacterium b_2 sends ATP and plasmids $p_j p_{null}$ to bacterium b_1 if the proceeding instruction l_j is a SUB instruction or plasmids $p_j p_{s_j}$ if the proceeding l_j is an ADD instruction.

Table 1. Plasmids in system Π.

Plasmid	Forms of Plasmids	Plasmid	Forms of Plasmids
p_0	$p_0 = (cas9, gRNA_{g_1}^{delete})$	p_{16}	$p_{16} = (cas9, gRNA_{g_4}^{insert})$
p_1	$p_1 = (cas9, gRNA_{g_7}^{insert})$	p_{17}	$p_{17} = (cas9, gRNA_{g_2}^{insert})$
p_2	$p_2 = (cas9, gRNA_{g_6}^{insert})$	p_{18}	$p_{18} = (cas9, gRNA_{g_4}^{delete})$
p_3	$p_3 = (cas9, gRNA_{g_5}^{delete})$	p_{19}	$p_{19} = (cas9, gRNA_{g_3}^{delete})$
p_4	$p_4 = (cas9, gRNA_{g_6}^{delete})$	p_{20}	$p_{20} = (cas9, gRNA_{g_0}^{insert})$
p_5	$p_5 = (cas9, gRNA_{g_5}^{insert})$	p_{21}	$p_{21} = (cas9, gRNA_{g_3}^{insert})$
p_6	$p_6 = (cas9, gRNA_{g_7}^{delete})$	p_h	$p_h = (cas9, gRNA_{g_h}^{delete})$
p_7	$p_7 = (cas9, gRNA_{g_1}^{insert})$	p_{s_1}	$p_{s_1} = (gRNA_{g_7}^{template})$
p_8	$p_8 = (cas9, gRNA_{g_6}^{delete})$	p_{s_2}	$p_{s_2} = (gRNA_{g_6}^{template})$
p_9	$p_9 = (cas9, gRNA_{g_6}^{insert})$	p_{s_5}	$p_{s_5} = (gRNA_{g_5}^{template})$
p_{10}	$p_{10} = (cas9, gRNA_{g_4}^{delete})$	p_{s_7}	$p_{s_7} = (gRNA_{g_1}^{template})$
p_{11}	$p_{11} = (cas9, gRNA_{g_5}^{delete})$	p_{s_9}	$p_{s_9} = (gRNA_{g_6}^{template})$
p_{12}	$p_{12} = (cas9, gRNA_{g_5}^{delete})$	$p_{s_{16}}$	$p_{s_{16}} = (gRNA_{g_4}^{template})$
p_{13}	$p_{13} = (cas9, gRNA_{g_2}^{delete})$	$p_{s_{17}}$	$p_{s_{17}} = (gRNA_{g_2}^{template})$
p_{14}	$p_{14} = (cas9, gRNA_{g_5}^{delete})$	$p_{s_{20}}$	$p_{s_{20}} = (gRNA_{g_0}^{template})$
p_{15}	$p_{15} = (cas9, gRNA_{g_3}^{delete})$	$p_{s_{21}}$	$p_{s_{21}} = (gRNA_{g_3}^{template})$
p_{null}	(Pro_{Rap}^{Rel})	p_{s_h}	$p_{s_h} = (Pro_{Rap}^{Rel})$

Initially, there is no chromosomal DNA initially placed in bacterium b_1, but bacterium b_2 has genes $w_2 = g_0 g_m g_1 g_m g_2 g_m g_3 g_m g_4 g_m g_5 g_m g_6 g_m g_7 g_m$. At the beginning, the system receives $g(x)$ copies of gene g_1 and y copies of gene g_2 from the environment through input bacterium b_2, which simulates the numbers $g(x)$ and y being introduced in registers 1 and 2 for register machine M_u. In this way, the chromosomal DNA of bacterium b_2 becomes

$$g_0 g_m g_1^{g(x)+1} g_m g_2^{y+1} g_m g_3 g_m g_4 g_m g_5 g_m g_6 g_m g_7 g_m.$$

Once completing the reading of information from the environment, a pair of plasmids $p_0 p_{s_0}$ and one unit of ATP is placed in bacterium b_1 to trigger the computation; meanwhile no plasmid or ATP is initially contained in bacterium b_2. The transition of system Π by reading input signals encoded by $g(x)$ copies of genes g_1 and y copies of gene g_2 through input bacterium b_2 is shown in Figure 3.

In what follows, we explain how system Π simulates ADD instructions and SUB instructions and outputs the computational result.

Simulating the ADD instruction: $l_i : (DD(r), l_j)$.

We assume at a certain moment that system Π starts to simulate an ADD instruction l_i of M_u, acting on register $r \in \{0, 1, 2, \ldots, 7\}$. At that moment, bacterium b_1 holds two plasmids $p_i p_{s_i}$ and ATP, such that the conjugation rule $(ATP\text{-}p_i p_{s_i}, b_1/b_2, ATP\text{-}p_i p_{s_i})$ is used. By using the conjugation rule, plasmids $p_i p_{s_i}$ and ATP are transmitted to bacterium b_2. In system Π, plasmids p_i and p_{s_i} are associated with the ADD instruction l_i, where plasmid p_i is of the form $p_i = (cas9, gRNA_{g_r}^{insert})$ for cutting a certain site of chromosomal DNA, and p_{s_i} is of the form $p_i = (gRNA_{g_r}^{template})$ carrying the gene to be inserted.

In bacterium b_2, the CRISPR/Cas9 inserting rule $p_i \times (g_r, g_m)$ is used to insert gene g_r between neighboring genes g_r and g_m. In this way, the number of gene g_r of bacterium b_2 is increased by 1, which simulates the number in register r being increased by 1. We note that there is a unique position at which gene g_r can be inserted with the context of neighboring g_r and g_m.

Table 2. Rules in each bacterium of system Π.

Sim.	Rules	Bac.
l_0	$(\text{ATP-}p_0 p_{null}, b_1/b_2, \text{ATP-}p_0 p_{null})$	b_1
	$p_0 \times (g_1, g_m), (\text{ATP-}p_{null}, b_2/b_1, \text{ATP-}p_1 p_{s_1}), (\text{ATP-}p_0 p_{null}, b_2/b_1, \text{ATP-}p_2 p_{s_2})$	b_2
l_1	$(\text{ATP-}p_1 p_{s_1}, b_1/b_2, \text{ATP-}p_1 p_{s_1})$	b_1
	$p_1 \times (g_7, g_m), (\text{ATP-}p_{s_1}, b_2/b_1, \text{ATP-}p_0 p_{null})$	b_2
l_2	$(\text{ATP-}p_2 p_{s_2}, b_1/b_2, \text{ATP-}p_2 p_{s_2})$	b_1
	$p_2 \times (g_6, g_m), (\text{ATP-}p_{s_2}, b_2/b_1, \text{ATP-}p_3 p_{null})$	b_2
l_3	$(\text{ATP-}p_3 p_{null}, b_1/b_2, \text{ATP-}p_3 p_{null})$	b_1
	$p_3 \times (g_5, g_m), (\text{ATP-}p_{null}, b_2/b_1, \text{ATP-}p_2 p_{s_2}), (\text{ATP-}p_3 p_{null}, b_2/b_1, \text{ATP-}p_4 p_{null})$	b_2
l_4	$(\text{ATP-}p_4 p_{null}, b_1/b_2, \text{ATP-}p_4 p_{null})$	b_1
	$p_4 \times (g_6, g_m), (\text{ATP-}p_{null}, b_2/b_1, \text{ATP-}p_5 p_{s_5}), (\text{ATP-}p_4 p_{null}, b_2/b_1, \text{ATP-}p_3 p_{null})$	b_2
l_5	$(\text{ATP-}p_5 p_{s_5}, b_1/b_2, \text{ATP-}p_5 p_{s_5})$	b_1
	$p_1 \times (g_5, g_m), (\text{ATP-}p_{s_5}, b_2/b_1, \text{ATP-}p_6 p_{null})$	b_2
l_6	$(\text{ATP-}p_6 p_{null}, b_1/b_2, \text{ATP-}p_6 p_{null})$	b_1
	$p_6 \times (g_7, g_m), (\text{ATP-}p_{null}, b_2/b_1, \text{ATP-}p_7 p_{s_7}), (\text{ATP-}p_6 p_{null}, b_2/b_1, \text{ATP-}p_8 p_{null})$	b_2
l_7	$(\text{ATP-}p_7 p_{s_7}, b_1/b_2, \text{ATP-}p_7 p_{s_7})$	b_1
	$p_7 \times (g_1, g_m), (\text{ATP-}p_4 p_{null}, b_2/b_1, \text{ATP-}p_4 p_{null})$	b_2
l_8	$(\text{ATP-}p_8 p_{null}, b_1/b_2, \text{ATP-}p_8 p_{null})$	b_1
	$p_8 \times (g_6, g_m), (\text{ATP-}p_{null}, b_2/b_1, \text{ATP-}p_9 p_{s_9}), (\text{ATP-}p_0 p_{null}, b_2/b_1, \text{ATP-}p_0 p_{null})$	b_2
l_9	$(\text{ATP-}p_9 p_{s_9}, b_1/b_2, \text{ATP-}p_9 p_{s_9})$	b_1
	$p_9 \times (g_6, g_m), (\text{ATP-}p_{10} p_{null}, b_2/b_1, \text{ATP-}p_{10} p_{null})$	b_2
l_{10}	$(\text{ATP-}p_{10} p_{null}, b_1/b_2, \text{ATP-}p_{10} p_{null})$	b_1
	$p_{10} \times (g_4, g_m), (\text{ATP-}p_0 p_{null}, b_2/b_1, \text{ATP-}p_0 p_{null}), (\text{ATP-}p_{10} p_{null}, b_2/b_1, \text{ATP-}p_{11} p_{null})$	b_2
l_{11}	$(\text{ATP-}p_{10} p_{null}, b_1/b_2, \text{ATP-}p_{11} p_{null})$	b_1
	$p_{11} \times (g_5, g_m), (\text{ATP-}p_{null}, b_2/b_1, \text{ATP-}p_{12} p_{null}), (\text{ATP-}p_{11} p_{null}, b_2/b_1, \text{ATP-}p_{13} p_{null})$	b_2
l_{12}	$(\text{ATP-}p_{12} p_{null}, b_1/b_2, \text{ATP-}p_{12} p_{null})$	b_1
	$p_{12} \times (g_5, g_m), (\text{ATP-}p_{null}, b_2/b_1, \text{ATP-}p_{14} p_{null}), (\text{ATP-}p_{12} p_{null}, b_2/b_1, \text{ATP-}p_{15} p_{null})$	b_2
l_{13}	$(\text{ATP-}p_{13} p_{null}, b_1/b_2, \text{ATP-}p_{13} p_{null})$	b_1
	$p_{13} \times (g_2, g_m), (\text{ATP-}p_{null}, b_2/b_1, \text{ATP-}p_{18} p_{null}), (\text{ATP-}p_{13} p_{null}, b_2/b_1, \text{ATP-}p_{19} p_{null})$	b_2
l_{14}	$(\text{ATP-}p_{14} p_{null}, b_1/b_2, \text{ATP-}p_{14} p_{null})$	b_1
	$p_{14} \times (g_5, g_m), (\text{ATP-}p_{null}, b_2/b_1, \text{ATP-}p_{16} p_{s_{16}}), (\text{ATP-}p_{14} p_{null}, b_2/b_1, \text{ATP-}p_{17} p_{s_{17}})$	b_2
l_{15}	$(\text{ATP-}p_{15} p_{null}, b_1/b_2, \text{ATP-}p_{15} p_{null})$	b_1
	$p_{15} \times (g_3, g_m), (\text{ATP-}p_{null}, b_2/b_1, \text{ATP-}p_{18} p_{null}), (\text{ATP-}p_{15}, b_2/b_1, \text{ATP-}p_{20} p_{s_{20}})$	b_2
l_{16}	$(\text{ATP-}p_{16} p_{s_{16}}, b_1/b_2, \text{ATP-}p_{16} p_{s_{16}})$	b_1
	$p_{16} \times (g_4, g_m), (\text{ATP-}p_{s_{16}}, b_2/b_1, \text{ATP-}p_{11} p_{null})$	b_2
l_{17}	$(\text{ATP-}p_{17} p_{s_{17}}, b_1/b_2, \text{ATP-}p_{17} p_{s_{17}})$	b_1
	$p_{17} \times (g_2, g_m), (\text{ATP-}p_{s_{17}}, b_2/b_1, \text{ATP-}p_{21} p_{s_{21}})$	b_2
l_{18}	$(\text{ATP-}p_{18} p_{null}, b_1/b_2, \text{ATP-}p_{18} p_{null})$	b_1
	$p_{18} \times (g_4, g_m), (\text{ATP-}p_{null}, b_2/b_1, \text{ATP-}p_0 p_{null}), (\text{ATP-}p_{18} p_{null}, b_2/b_1, \text{ATP-}p_h p_{s_h})$	b_2
l_{19}	$(\text{ATP-}p_{19} p_{null}, b_1/b_2, \text{ATP-}p_{19} p_{null})$	b_1
	$p_{19} \times (g_3, g_m), (\text{ATP-}p_{null}, b_2/b_1, \text{ATP-}p_0 p_{null}), (\text{ATP-}p_{15}, b_2/b_1, \text{ATP-}p_{18} p_{null})$	b_2
l_{20}	$(\text{ATP-}p_{20} p_{s_{20}}, b_1/b_2, \text{ATP-}p_{20} p_{s_{20}})$	b_1
	$p_{20} \times (g_0, g_m), (\text{ATP-}p_{s_{20}}, b_2/b_1, \text{ATP-}p_0 p_{null})$	b_2
l_{21}	$(\text{ATP-}p_{21} p_{s_{21}}, b_1/b_2, \text{ATP-}p_{21} p_{s_{21}})$	b_1
	$p_0 \times (g_3, g_m), (\text{ATP-}p_{s_{21}}, b_2/b_1, \text{ATP-}p_{18} p_{null})$	b_2
l_h	$(\text{ATP-}p_h p_{s_h}, b_1/b_2, \text{ATP-}p_h p_{s_h})$	

By using the CRISPR/Cas9 inserting rule, plasmid p_i is consumed, and plasmid p_{s_i} and ATP remain in bacterium b_2. The conjugation rule in bacterium b_2 is designed by the operation of the proceeding instruction l_j. One of the following two cases occurs in bacterium b_2.

- If instruction l_j is an ADD instruction, then bacterium b_2 has the conjugation rule $(\text{ATP-}p_{si}, b_2/b_1, \text{ATP-}p_j p_{s_j})$. By using the rule, plasmids $p_j p_{s_j}$ and ATP are conjugated to bacterium b_1. In this case, system Π starts to simulate the proceeding ADD instruction l_j.
- If instruction l_j is a SUB instruction, then bacterium b_2 has the conjugation rule $(\text{ATP-}p_{si}, b_2/b_1, \text{ATP-}p_j p_{null})$, by which plasmids $p_j p_{null}$ and ATP are transmitted to bacterium b_1. In this case, system Π starts to simulate the proceeding SUB instruction l_j.

Therefore, system Π can correctly simulate the ADD instruction of M_u. The system starts from bacterium b_1 having plasmid $p_i p_{s_i}$ and ATP, which are transmitted to bacterium b_2 by the conjugation rule. In bacterium b_2, the number of gene g_r in chromosomal DNA is increased by 1 using the CRISPR/Cas9 gene inserting rule, and plasmids $p_j p_{s_j}$ (if the proceeding instruction l_j is an ADD instruction) or $p_j p_{null}$ (if the proceeding instruction l_j is a SUB instruction) are transmitted to bacterium b_1, which means that system Π starts to simulate instruction l_j.

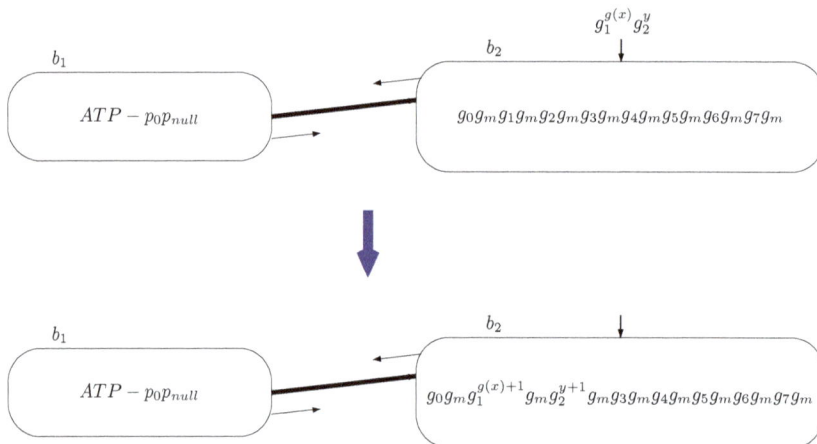

Figure 3. The transition of system Π by reading input information encoded by $g(x)$ copies of genes g_1 and y copies of gene g_2 through input bacterium b_2.

Simulating the SUB instruction: $l_i : (\text{SUB(r)}, l_j, l_k)$.

We suppose at a certain computation step that system Π has to simulate a SUB instruction $l_i :$ $(\text{SUB(r)}, l_j, l_k)$. For any SUB instruction l_i, plasmid p_i of the form $p_i = (cas9, gRNA_{g_r}^{delete})$ is associated in system Π. In bacterium b_1, there are plasmids $p_i p_{null}$ and ATP such that the conjugation rule $(\text{ATP-}p_i p_{null}, b_1/b_2, \text{ATP-}p_i p_{null})$ can be used. In bacterium b_2, it has the following two cases.

- If there is at least one gene g_r existing between neighboring genes g_r and g_m in chromosomal DNA of bacterium b_2 (corresponding to the case that the number stored in register r is $n > 0$), then the CRISPR/Cas9 deleting rule $p_i \times (g_1, g_m)$ is used to delete one copy of gene g_r from chromosomal DNA. This simulates the number stored in register r being decreased by 1. By consuming plasmid p_i, bacterium b_2 retains plasmid p_{null} and ATP such that a conjugation rule $(\text{ATP-}p_{null}, b_2/b_1, \text{ATP-}p_j p_{s_j})$ or $(\text{ATP-}p_{null}, b_2/b_1, \text{ATP-}p_j p_{null})$ is used, which depends on whether the proceeding instruction would be an ADD or a SUB instruction. In this way, plasmids $p_j p_{s_j}$ or $p_j p_{null}$) and ATP are transmitted to bacterium b_1. The system starts to simulate instruction l_j.

- If there is no gene g_r existing between neighboring genes g_r and g_m in chromosomal DNA of bacterium b_2 (corresponding to the case that the number stored in register r is 0), then the CRISPR/Cas9 deleting rule $p_i \times (g_1, g_m)$ cannot be used, but a conjugation rule $(\text{ATP-}p_i p_{null}, b_2/b_1, \text{ATP-}p_k p_{s_k})$ or $(\text{ATP-}p_i p_{null}, b_2/b_1, \text{ATP-}p_k p_{null})$ is able to be used. Plasmids $(p_k p_{s_k}$ or $p_k p_{null})$ and ATP are conjugated to bacterium b_1, which means the system starts to simulate instruction l_k.

We note that when plasmids $p_i p_{null}$ are conjugated to bacterium b_2 from bacterium b_1, it may happen that both the CRISPR/Cas9 deleting rule $p_i \times (g_1, g_m)$ and $(\text{ATP-}p_i p_{null}, b_2/b_1, \text{ATP-}p_k p_{s_k})$ (or $(\text{ATP-}p_i p_{null}, b_2/b_1, \text{ATP-}p_k p_{null}))$ can be used. In this case, the CRISPR/Cas9 deleting rule $p_i \times (g_1, g_m)$ will be applied because of the fact that it has priority over the plasmid transferring rule.

The simulation of a SUB instruction is correct: System Π starts from bacterium b_1 having plasmid $p_i p_{null}$ and ATP and ends with plasmid $p_j p_{s_j}$ or $p_j p_{null}$ and ATP (if the number stored in register r is $n > 0$) to start the simulation of instruction l_j; otherwise it ends with plasmid $p_k p_{s_k}$ or $p_k p_{null}$ and ATP (if the number stored in register r is 0) to start the simulation of instruction l_k.

Simulating the halt instruction: l_h : HALT.

When register machine M_u reaches the halt instruction l_h : HALT, the computation of register machine M_u halts. At that moment, bacterium b_1 in system Π holds plasmids $p_h p_{s_h}$ and ATP, and the conjugation rule $(\text{ATP-}p_h p_{sh}, b_1/b_2, \text{ATP-}p_h p_{s_h})$ can be used. By using the rule, plasmids $p_h p_{s_h}$ and ATP are transmitted to bacterium b_2; no gene can be edited by plasmid p_h, and no rule can be used. Hence, the computation of system Π finally halts.

The number of gene g_0 in chromosomal DNA of bacterium b_2 encodes the number stored in register 0 of M_u. If the number stored in register 0 is $n > 0$, then there are $n + 1$ copies of gene g_0 in chromosomal DNA of bacterium b_2. The computational result can be obtained by counting the number of gene g_0 in chromosomal DNA of bacterium b_2.

From the above description of system Π and its work, it is clear that system Π can simulate each computation of M_u. We can check that the constructed system Π has

- 2 bacterium for conjugation with each other;
- 22 plasmids p_i for the 22 ADD and SUB instructions with $i = 0, 1, 2, \ldots 21$;
- 9 plasmids p_{s_i} for 9 ADD instructions with $i = 1, 2, 5, 7, 9, 16, 17, 20, 21$;
- 1 plasmid p_{null} for the 13 SUB instructions;
- 2 plasmids p_h and p_{s_h} for the HALT instruction;
- 8 genes g_i for encoding numbers in registers i with $i = 0, 1, 2 \ldots 7$;
- 1 gene g_m for separating gene g_i in chromosomal DNA.

This gives, in total, 2 bacteria, 34 plasmids, and 9 genes.
This concludes the proof. $\square$

3.2. A Small Universal BP System as a Number Generator

In this section, we construct a small universal BP system as a number generator. A BP system Π_u is universal if, given a fixed admissible enumeration of the unary partial recursive functions $(\varphi_0, \varphi_1, \ldots)$, there is a recursive function g such that for each natural number x, whenever we input the number $g(x)$ in Π_u, the set of numbers generated by the system is equal to $\{n \in N | \varphi_x(n) \text{ is defined}\}$. In other words, after introducing the "code" $g(x)$ of the partial recursive function φ_x in the form of $g(x)$ copies of certain genes in chromosomal DNA of the input bacterium, the BP system generates all numbers n for which $\varphi_x(n)$ is defined.

System Π_u has the same topological structure, plasmids, and evolution rules as system Π constructed in Section 3.1, but the input bacterium is b_2 and the output bacterium is b_1. Differently from the universal computing devices considered in Section 3.1, the strategy to simulate a universal register machine as a number generator is as follows.

Step 1. The output bacterium b_1 initially has n copies of gene g_m.

Step 2. System Π_u starts by loading $g(x)$ copies of gene g_1 and n copies of gene g_2 in the input bacterium b_2.

Step 3. The computation of Π_u is activated by using plasmid $p_0 p_{null}$ to simulate the register machine M_u from Figure 2, with $g(x)$ stored in register 1, and number n stored in register 2.

If the computation in register machine M_u halts, instruction σ_{l_h} can finally be activated. To simulate register machine M_u reaching the HALT instruction, system Π_u holds plasmids $p_h p_{s_h}$ and transmits them to bacterium b_2. After this, system Π_u halts, as no rule can be used in bacterium b_2. When the system halts, the number of gene g_m in the output bacterium b_1 is the computational result, which is exactly the number n. Hence, the number n can be computed/generated by system Π_u.

The difference between systems Π and Π_u is the loading input information process. The initial configuration and transition of system Π_u by reading input signals encoded by $g(x)$ copies of genes g_1 and n copies of gene g_2 through input bacterium b_2 are shown in Figure 4.

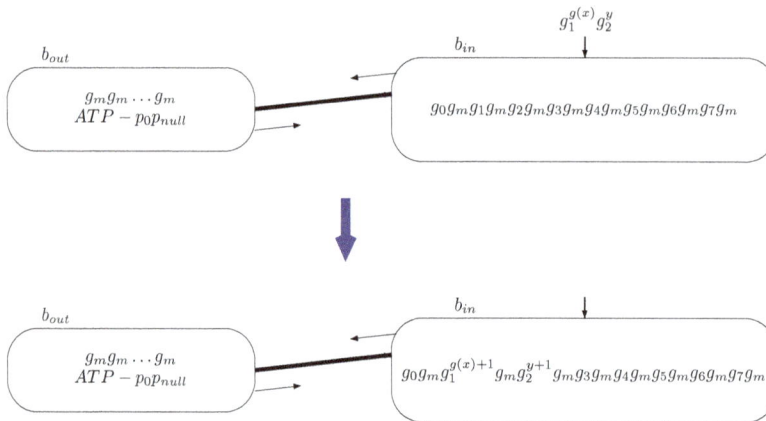

Figure 4. The initial configuration and transition of system Π_u by reading input information encoded by $g(x)$ copies of genes g_1 and n copies of gene g_2 through input bacterium b_2.

We can check that the constructed system Π_u has

- 2 bacterium for the conjugation with each other;
- 22 plasmids p_i for the 22 ADD and SUB instructions with $i = 0, 1, 2, \ldots 21$;
- 9 plasmids p_{s_i} for 9 ADD instructions with $i = 1, 2, 5, 7, 9, 16, 17, 20, 21$;
- 1 plasmid p_{null} for the 13 SUB instructions;
- 2 plasmids p_h and p_{s_h} for the HALT instruction;
- 8 genes g_i for encoding numbers in registers i with $i = 0, 1, 2 \ldots 7$;
- 1 gene g_m for separating gene g_i in chromosomal DNA.

This gives, in total, 2 bacteria, 34 plasmids, and 9 genes.
Therefore, we have the following theorem.

Theorem 2. *There is a Turing universal BP system with 2 bacteria and 34 plasmids that can compute a Turing-computable set of natural numbers.*

4. Conclusions

In this work, we construct two small universal BP systems. Specifically, it is obtained that a BP system with 2 bacteria, 34 plasmids, and 9 genes is universal for both computing recursively enumerable functions and computing/generating a family of sets of natural numbers. It is obtained that 34 plasmids are sufficient for constructing Turing universal BP systems. This provides theoretical support as well as paradigms using a reasonable number of bacteria and plasmids to construct powerful bacterial computers.

Following the research line, finding smaller universal BP systems deserves further research. A possible way to slightly decrease the number of plasmids used in small universal BP systems is using code optimization, exploiting some particularities of the register machine M_u. For example, as considered in [25], for the sequence of two consecutive ADD instructions $l_{17}: (ADD(2), l_{21})$ and $l_{21}: (ADD(3), l_{18})$, without any other instruction addressing the label l_{21}, the two ADD modules can be combined. However, a challenging problem regards what the minimum size of a universal BP system is—in other words, what the borderline between universality and non-universality is. Characterization of universality by BP systems is expected. A balance between the number of bacteria and plasmids in universal BP systems can be considered, that is, using more bacteria to reduce the number of plasmids.

It is worth developing the applications of BP systems. Bio-inspiring computing models perform well in computations, particularly in solving computational complex problems in feasible time [34–36]. It is of interest to use BP systems to solve computationally hard problems. Some specific applications using BP systems would be of interest to researchers from biological fields.

In artificial intelligence, there are many bio-inspired algorithms (see, e.g., [37,38]). It is worth designing bacteria-computing-inspired algorithms or introducing bacteria computing operators in classical algorithms. Additionally, it would be meaningful to construct powerful bacterial computers or computing devices in biological labs.

Author Contributions: Conceptualization, X.W. and T.S.; Methodology, X.W.; Software, T.M.; Validation, P.Z., T.S. and X.W.

Acknowledgments: This work was supported by the National Natural Science Foundation of China (Grant Nos. 61502535, 61572522, 61572523, 61672033, and 61672248), PetroChina Innovation Foundation (2016D-5007-0305), Key Research and Development Program of Shandong Province (No. 2017GGX10147), Natural Science Foundation of Shandong Province (No. ZR2017MF004), Talent introduction project of China University of Petroleum (No. 2017010054), Research Project TIN2016-81079-R (AEI/FEDER, Spain-EU) and Grant 2016-T2/TIC-2024 from Talento-Comunidad de Madrid, Project TIN2016-81079-R (MINECO AEI/FEDER, Spain-EU), and the InGEMICS-CM Project (B2017/BMD-3691, FSE/FEDER, Comunidad de Madrid-EU).

Conflicts of Interest: The authors declare no conflict of interest.

References

1. Gitai, Z. The new bacterial cell biology: Moving parts and subcellular architecture. *Cell* **2005**, *120*, 577–586. [CrossRef] [PubMed]
2. Goldstein, E.; Drlica, K. Regulation of bacterial dna supercoiling: Plasmid linking numbers vary with growth temperature. *Proc. Natl. Acad. Sci. USA* **1984**, *81*, 4046–4050. [CrossRef] [PubMed]
3. Summers, D. *The Biology of Plasmids*; John Wiley & Sons: Hoboken, NJ, USA, 2009.
4. Poet, J.L.; Campbell, A.M.; Eckdahl, T.T.; Heyer, L.J. Bacterial computing. *XRDS Crossroads ACM Mag. Stud.* **2010**, *17*, 10–15. [CrossRef]
5. Gupta, V.; Irimia, J.; Pau, I.; Rodríguez-Patón, A. Bioblocks: Programming protocols in biology made easier. *ACS Synth. Biol.* **2017**, *6*, 1230–1232. [CrossRef] [PubMed]

6. Gutierrez, M.; Gregorio-Godoy, P.; del Pulgar, G.P.; Munoz, L.E.; Sáez, S.; Rodríguez-Patón, A. A new improved and extended version of the multicell bacterial simulator gro. *ACS Synth. Biol.* **2017**, 6, 1496–1508. [CrossRef] [PubMed]

7. Adleman, L.M. Molecular computation of solutions to combinatorial problems. *Sciences* **1994**, *266*, 1021–1024. [CrossRef]

8. Carell, T. Molecular computing: Dna as a logic operator. *Nature* **2011**, *469*, 45–46. [CrossRef] [PubMed]

9. Xu, J. Probe machine. *IEEE Trans. Neural Netw. Learn. Syst.* **2016**, *27*, 1405–1416. [CrossRef] [PubMed]

10. Păun, G.; Rozenberg, G.; Salomaa, A. *The Oxford Handbook of Membrane Computing*; Oxford University Press: Oxford, UK, 2010.

11. Păun, G.; Rozenberg, G. A guide to membrane computing. *Theor. Comput. Sci.* **2002**, *287*, 73–100. [CrossRef]

12. Păun, G. Computing with membranes. *J. Comput. Syst. Sci.* **2000**, *61*, 108–143. [CrossRef]

13. Freund, R.; Kari, L.; Păun, G. DNA computing based on splicing: The existence of universal computers. *Theory Comput. Syst.* **1999**, *32*, 69–112. [CrossRef]

14. Kari, L.; Păun, G.; Rozenberg, G.; Salomaa, A.; Yu, S. DNA computing, sticker systems, and universality. *Acta Inform.* **1998**, *35*, 401–420. [CrossRef]

15. Martın-Vide, C.; Păun, G.; Pazos, J.; Rodrıguez-Patón, A. Tissue P systems. *Theor. Comput. Sci.* **2003**, *296*, 295–326. [CrossRef]

16. Ionescu, M.; Păun, G.; Yokomori, T. Spiking neural P systems. *Fundam. Inform.* **2006**, *71*, 279–308.

17. Song, T.; Pan, L.; Păun, G. Asynchronous spiking neural p systems with local synchronization. *Inform. Sci.* **2013**, *219*, 197–207. [CrossRef]

18. Ezziane, Z. DNA computing: Applications and challenges. *Nanotechnology* **2005**, *17*, R27. [CrossRef]

19. Chen, X.; Pérez-Jiménez, M.J.; Valencia-Cabrera, L.; Wang, B.; Zeng, X. Computing with viruses. *Theor. Comput. Sci.* **2016**, *623*, 146–159. [CrossRef]

20. Rogozhin, Y. Small universal turing machines. *Theor. Comput. Sci.* **1996**, *168*, 215–240. [CrossRef]

21. Baiocchi, C. Three small universal turing machines. In *Machines, Computations, and Universality*; Springer: Berlin/Heidelberg, Germany, 2001; pp. 1–10.

22. Korec, I. Small universal register machines. *Theor. Comput. Sci.* **1996**, *168*, 267–301. [CrossRef]

23. Iirgen Albert, J.; Culik, K., II. A simple universal cellular automaton and its one-way and totalistic version. *Complex Syst.* **1987**, *1*, 1–16.

24. Kudlek, M.; Yu, R. Small universal circular post machines. *Comput. Sci. J. Mold.* **2001**, *9*, 25.

25. Păun, A.; Păun, G. Small universal spiking neural P systems. *BioSystems* **2007**, *90*, 48–60. [CrossRef] [PubMed]

26. Zhang, X.; Zeng, X.; Pan, L. Smaller universal spiking neural P systems. *Fundam. Inform.* **2008**, *87*, 117–136.

27. Pan, L.; Zeng, X. A note on small universal spiking neural P systems. *Lect. Notes Comput. Sci.* **2010**, *5957*, 436–447.

28. Păun, A.; Sidoroff, M. Sequentiality induced by spike number in SNP systems: Small universal machines. In *Membrane Computing*; Springer: Berlin/Heidelberg, Germany, 2012; pp. 333–345.

29. Song, T.; Jiang, Y.; Shi, X.; Zeng, X. Small universal spiking neural P systems with anti-spikes. *J. Comput. Theor. Nanosci.* **2013**, *10*, 999–1006. [CrossRef]

30. Song, T.; Xu, J.; Pan, L. On the universality and non-universality of spiking neural P systems with rules on synapses. *IEEE Trans. Nanobiosci.* **2015**, *14*, 960–966. [CrossRef] [PubMed]

31. Song, T.; Pan, L. Spiking neural P systems with request rules. *Neurocomputing* **2016**, *193*, 193–200. [CrossRef]

32. Song, T.; Rodríguez-Patón, A.; Gutiérrez, M.; Pan, Z. Computing with bacteria conjugation and crispr/cas9 gene editing operations. *Sci. Rep.* **2018**, sumitted.

33. Minsky, M.L. *Computation: Finite and Infinite Machines*; Prentice-Hall, Inc.: Upper Saddle River, NJ, USA, 1967.

34. Valencia-Cabrera, L.; Orellana-Martín, D.; Martínez-del Amor, M.A.; Riscos-Núnez, A.; Pérez-Jiménez, M.J. Computational efficiency of minimal cooperation and distribution in polarizationless p systems with active membranes. *Fundam. Inform.* **2017**, *153*, 147–172. [CrossRef]

35. Song, B.; Pérez-Jiménez, M.J.; Pan, L. An efficient time-free solution to qsat problem using p systems with proteins on membranes. *Inf. Comput.* **2017**, *256*, 287–299. [CrossRef]

36. Macías-Ramos, L.F.; Pérez-Jiménez, M.J.; Riscos-Nú nez, A.; Valencia-Cabrera, L. Membrane fission versus cell division: When membrane proliferation is not enough. *Theor. Comput. Sci.* **2015**, *608*, 57–65. [CrossRef]
37. Ma, X.; Sun, F.; Li, H.; He, B. Neural-network-based sliding-mode control for multiple rigid-body attitude tracking with inertial information completely unknown. *Inf. Sci.* **2017**, *400*, 91–104. [CrossRef]
38. Alsaeedan, W.; Menai, M.E.B.; Al-Ahmadi, S. A hybrid genetic-ant colony optimization algorithm for the word sense disambiguation problem. *Inf. Sci.* **2017**, *417*, 20–38. [CrossRef]

Sample Availability: Samples of the compounds are not available.

![molecules](molecules logo) *molecules*

MDPI

Article

8-Bit Adder and Subtractor with Domain Label Based on DNA Strand Displacement

Weixuan Han [1,2] and Changjun Zhou [1,3,*]

1   College of Mathematics and Computer Science, Zhejiang Normal University, Jinhua 321004, China;
    m18340852618@163.com
2   College of Nuclear Science and Engineering, Sanmen Institute of technicians, Sanmen 317100, China
3   Key Laboratory of Advanced Design and Intelligent Computing (Dalian University) Ministry of Education,
    Dalian 116622, China
*   Correspondence: zhou-chang231@163.com

Academic Editor: Xiangxiang Zeng
Received: 16 October 2018; Accepted: 13 November 2018; Published: 15 November 2018

Abstract: DNA strand displacement, which plays a fundamental role in DNA computing, has been widely applied to many biological computing problems, including biological logic circuits. However, there are many biological cascade logic circuits with domain labels based on DNA strand displacement that have not yet been designed. Thus, in this paper, cascade 8-bit adder/subtractor with a domain label is designed based on DNA strand displacement; domain t and domain f represent signal 1 and signal 0, respectively, instead of domain t and domain f are applied to representing signal 1 and signal 0 respectively instead of high concentration and low concentration high concentration and low concentration. Basic logic gates, an amplification gate, a fan-out gate and a reporter gate are correspondingly reconstructed as domain label gates. The simulation results of Visual DSD show the feasibility and accuracy of the logic calculation model of the adder/subtractor designed in this paper. It is a useful exploration that may expand the application of the molecular logic circuit.

Keywords: DNA strand displacement; cascade; 8-bit adder/subtractor; domain label

---

1. Introduction

In recent years, biological computing has become a new hotspot due to DNA molecules having the advantages of parallelism, low energy consumption, and high storability in dealing with massive information; therefore, DNA nanotechnology stands out DNA nanotechnology has potential applications in biological calculations. A range of information circuits and bio-computing models have been implemented in DNA by using strand displacement. Examples include DNA strand displacement reactions [1,2], molecular motors [3–5], catalytic signal amplification circuits [6–8], and biological logic circuits [9–11], as well as computing with membranes, bacteria, conjugation and RNA computing [12–16]. As a new technique in the field of self-assembled DNA, the DNA strand displacement reaction has been widely used in the field of molecular computing. DNA molecular circuits corresponding to different logic gates have been designed on the basis of the DNA self-assembly calculation principle. When the DNA signal strand is input into the molecular logic circuit, molecular logic gates with different molecular concentration ratios are mixed, and the molecular logic circuit outputs the signal strand through intermolecular specific hybridization and the DNA strand displacement reaction. In 2006, Seeling designed the AND Gate, OR Gage, NOT Gate signal amplifier and signal feedback using single-stranded nucleic acids as the input and output signals based on DNA strand displacement [17]. However, NOT Gate output is unstable due to its single-stranded input. In 2011, Qian designed a simple Seesaw logic gate, four-bit square root biological logic circuit and avoided the problem of NOT Gate output instability via the use of dual-rail [18,19]. However, dual-rail logic is not suitable for large-scale cascaded molecular logic circuits. In 2013, Zhang proposed and verified the logical "AND" gate and "OR" gate [20]. In 2014, Guo designed multiple types of logic gate based on a single g-quadrupled DNA strand [21]. In 2015, Wang used strand displacement to achieve the multi-bit adder design [22]. In 2016, Lakin presented a framework for the development of adaptive molecular circuits using buffered DNA strand displacement networks and designed supervised learning in adaptive DNA strand displacement networks [23]. In 2017, Sun presented a one-bit half adder-half subtractor logical operation based on DNA strand displacement [24].

Although Winfree [19] solved the instability caused by NOT Gate and then designed many stable biological logic circuits, dual-rail logic brought many new problems. The scale of a dual-rail logic circuit is two times that of a single-rail logic circuit, which increases the material cost, complexity, and difficulty of designing the logic circuit. The most fundamental problem of the instability caused by NOT Gate and the scale of the dual-rail logic circuit is that concentration is applied to the presentation of logic 1 and logic 0. The concentration of a reactant has an important effect on the reaction rate, because the reaction rate of high concentration reactants is faster than that of low concentration reactants, and a reaction works from an area of high concentration to an area of low concentration under the same circumstances [25]. It appears that changing the conditions of the NOT Gate of a single-rail logic circuit goes against the design of a large-scale single-rail logic circuit.

In this paper, DNA signals are marked with domain labels based on the freedom of DNA hybridization and the high sensitivity of domain labels, and then logic gates with domain labels are constructed by redesigning the special DNA structure, where logic value 1 and logic value 0 are represented by domain t and domain f, respectively, which solves the instability issue of NOT Gate in dual-rail logic. Based on this, the first domain label, cascade 8-bit adder/subtractor is designed. The innovation of this paper is that the design of the molecular domain label is used in the logic gates, where logical results are detected through the labeling of the domain label with a fluorescent label in the reaction solution. This was not seen in the previous design of the molecular logic circuit, and this paper broadens the range of input signals with DNA molecules to construct the logic circuit. The detection method of the logic gate model has high sensitivity and simple operation. It has less stringent requirements for base mismatches, reducing the impact of hybrid competition in the experimental results to a certain extent. In addition, the domain label cascade 8-bit adder/subtractor can be used to design large scale biochemical circuits to allow good encapsulation.

This paper is arranged as follows: the development of DNA molecule logic circuits is introduced in the first part; the background of DNA strand displacement and logic gates is presented in the second part; a brief method for building domain label logic gates is presented in the third part; the simulation of the domain label cascade 8-bit adder/subtractor by Visual DSD is presented in the fourth part; and the fifth part presents the conclusions of this paper.

2. Backgrounds

2.1. DNA Strand Displacement Reaction

Utilizes the characteristics of the free energy of the molecular hybridization system to stabilize the state and control or induce downstream strand displacement reactions by changing the sequence and length of the input signal. Intuitively, DNA strand displacement is the process of replacing a shorter hybridization region with a longer, double-stranded hybridization region. The process of this is shown in Figure 1 [26].

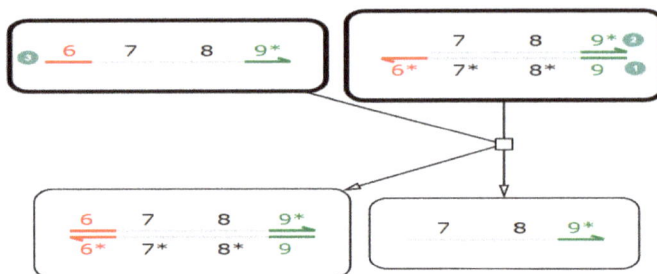

Figure 1. DNA strand displacement.

The process of the reversible DNA strand displacement reaction is shown in Figure 1. Firstly, two partially complementary DNA strands are joined together (the 1-strand is longer than the 2-strand), Secondly, the 3-strand is added to the solution at room temperature (the sequence of the 3-strand is completely complementary to the 1-strand). Thirdly, the specific recognition region is first combined with a single strand of the 1-strand. In order to achieve the most stable state, the binding sites of the 1-strand and 2-strand are gradually occupied by the 3-strand, and finally, the 3-strand completely replaces the 2-strand. As the DNA strand displacement reaction with highly specific identification sites can start in parallel and realize a multi-level nested trigger, it has developed rapidly in recent years and has become a hotspot in the field of molecular computing.

2.2. DNA Logic Module with Domain Label

On the basis of the existing DNA logic model, the domain label is used to realize the operation of the domain label logic module which can react spontaneously at room temperature. Domain t and domain f respectively represent signal 1 and the signal 0, which correspond to the regions of high concentration and low concentration in the solution, which are shown in Figure 2 [27].

As can be seen from Figure 2, a domain label DNA signal strand consists of a left domain and a right domain, where the right domain is responsible for passing the logic signal to the downstream logic gate, and the left domain is responsible for receiving the upstream DNA signal. Therefore, the (f, t) and (t, t) strands are known as domain t and correspond to signal 1 while the (f, f) and (t, f) strands are known as domain f and correspond to signal 0.

Figure 2. Domain f and domain t.

The AND Gate with a domain label, the OR Gate with a domain label and the NOT Gate with a domain label are the most elementary logic modules, the logic gates of which are shown in Figure 3. The AND Gate with a domain label and the OR Gate with a domain label are made up of three DNA double strands. The NOT Gate with a domain label consists of two DNA double strands. In Figure 3a, AND Gates with domain labels are denoted by $G_{ms,f}$, $G_{ns,f}$ and $G_{mns,t}$ respectively. In Figure 3b, OR Gates with domain labels are denoted by $G_{ms,t}$, $G_{ns,t}$ and $G_{mns,f}$ respectively In Figure 3c, the two DNA double strands are the same, except for the locations of t and f of the NOT Gate with a domain label.

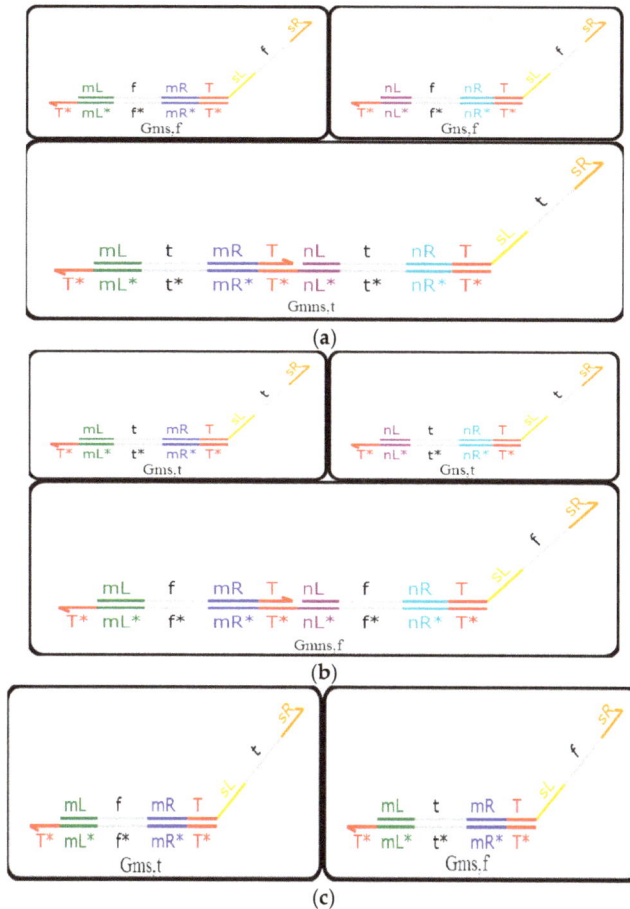

Figure 3. (a) The AND Gate module with a domain label; (b) the OR Gate module with a domain label; (c) the NOT Gate module with a domain label.

Whether a DNA single strand represents logic 1 or logic 0 depends on its presence (domain t~logic value 1, domain f~logic value 0). When two strands with domain t are input to the AND Gate with a domain label at the same time, the DNA molecule reaction depends on $G_{mns,t}$, and finally, it outputs a strand with domain t; in other cases, the DNA molecule reaction depends on o, $G_{ms,f}$ or $G_{ns,f}$, and finally, it outputs a strand with domain f. When two strands with domain f are input to the OR Gate with a domain label at the same time, the DNA molecule reaction depends on $G_{mns,f}$, and finally, it outputs a strand with domain f; in other cases, the DNA molecule reaction depends on $G_{ms,t}$ or $G_{ns,t}$, and finally, it outputs a strand with domain t. As for the NOR Gate, the situation is much simpler. When a strand with domain f is input to the NOR Gate with a domain label, the DNA molecule reaction depends on $G_{ms,t}$, and finally, it outputs a strand with domain t; otherwise, the DNA molecule reaction depends on $G_{ms,f}$, and finally, it outputs a strand with domain f.

2.3. Mapping

K operation: Let A be a non-empty set. The Cartesian product $A^K = A \times A \times A \times \cdots \times A$ to A mapping f is called the N operation on the set A. In addition, each element in $\left| A^K \right|$ has $|A|$ possible correspondences of $|A|$, so from A^K to A, it has $|A|^{|A|^K}$ possible mappings, which is called m, m = $m = |A|^{|A|^K}$. K = 1 is called 1-input mapping. K = 2 is called 2-input mapping. If A = $[0, 1]$, n = 1, then $m = |A|^{|A|^K} = 2^{2^1} = 4$. n = 2, then m = $|A|^{|A|^K} = 2^{2^2} = 16$, and so on. If each N-input mapping module is a mapping from $A^K = \{t, f\}^K$ to $A^K = \{t, f\}^K$, there are a total of $|A|^{|A|^K} = 2^{2^K}$ mappings, so K-input mapping has 2^{2^K} modules.

3. Methods

In this paper, the logic circuit domain label and double-dual are briefly compared in terms of the stability of the molecular reaction process. In the double-dual logic circuit, the Seesaw module contains threshold gates which react with the upstream DNA signal strands irreversibly. The next step of the molecular reaction can be carried out only the threshold gate is completely consumed, so theoretically, the smaller the concentration of the threshold is, the faster the speed of reaction is. The significance of the threshold is to distinguish between the high and low concentrations of the reaction process so that the logical values 1 and 0 are correctly expressed. The double-dual logic circuit corresponds to logic 1 and logic 0 via the high concentration and low concentration, respectively. In general, a DNA signal strand with a unit concentration of 0.9~1 (1 unit concentration of 10,000 nM in this paper) represents logic value 1, and a DNA signal strand with a unit density of 0~0.1 represents logic value 0. It is worth noting that the threshold can correctly express the logical value at the average concentration value, but the molecular reaction is extremely unstable at the threshold concentrations of 0.1~0.2 units and 0.8~0.9 units. In such cases, the threshold cannot strictly distinguish between high concentration and low concentration in the reaction process, resulting in the output of the wrong signal, while logic circuits with domain t and domain f are determined throughout the reaction process and will not output the wrong signal, thereby avoiding the instability of the molecular circuit in the reaction process. On this basis, the paper constructed an N-mapping module with a domain label, an amplification gate with a domain label, a fan-out gate with a domain label, and a reporter gate with a domain label. This formed the basis for constructing the 8-bit addition and subtraction with a domain label.

3.1. N-Mapping Module with Domain Label

The 1-input mapping module consists of strands with the domain label {T^*} [mL^ f mR^ T^] <nL^ n1 nR> and {T^*} [mL^ t mR^ T^] <nL^ n2 nR>, which note (f, n1), (t, n2). As shown in Figure 4a, there are 4 corresponding mapping modules (n1 = t, n2 = f; n1 = f, n2 = t; n1 = f, n2 = f; n1 = t, n2 = t). It is noteworthy that, when n1 = t, n2 = f, the DNA strands output the opposite domain label signal, and the mapping module implements the NOT gate function. The reaction is shown in Figure 4b. Generally, this function is commonly used in domain label 1 input module mapping.

As shown in Figure 4a, although the 1-mapping module consists of two DNA strands with the same structure and initial concentration, the DNA signal strand actually reacts with one of them, and the reaction is accomplished within only one step. The 1-mapping module shown in Figure 4b implements the NOT operation, in which case n1 = t, n2 = f, and it transforms logic value 1, which is represented by <m2L^ f m2R^ T^mL^ t mR^>, to logic value 0, which is represented by <mL^ t mR^ T^ nL^ f nR^>. The remaining double strands, {T ^*} [mL ^ f mR ^ T ^] <nL ^ t nR>, follow the same principle. A single strand containing domain labels <m2L^ t m2R^ T^ mL^ f mR^> is input, and this reacts with {T ^*} [mL ^ f mR ^ T ^] <nL ^ n1 nR> and outputs logic value 1, which is represented by <mL^ f mR^ T^ nL^ t nR>.

Therefore, when double strands with domain labels n1 and n2 are inverted (n1 = t, n2 = f; n1 = f, n2 = t), the NOT operations are performed with each of them at the same time, which makes the reaction proceed simultaneously and reduces the time consumed by the reaction. The advantages of the modules are demonstrated here.

The K-input mapping module consists of 2^K strands with domain labels, namely, {T^*} [m1L^ f m1R^ T^]: ... :[mkL^ f mkR^ T^] <nL^ n1 nR>, {T^*} [m1L^ f m1R^ T^]: ... :[mkL^ t mkR^ T^] <nL^ n2 nR>, ... , and {T^*} [m1L^ t m1R^ T^]: ... :[mkL^ t mkR^ T^] <nL^ n2^K nR>, Among them (ff, ... f, n1), (f ... f, t, n2),

..., and (t ... t, t, n2^K)., n1, n2 ... n2^K ∈ {t, f}. There are 2$^{2^K}$ corresponding mapping modules, and similar to the 1-input mapping module, the K-input mapping module can also perform different logic operations under k inputs and produces one output. In this paper, we describe 1, 2, and 3-input mapping modules.

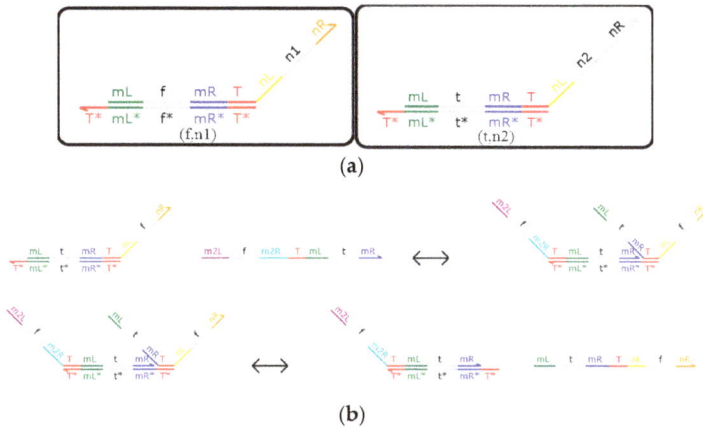

(a)

(b)

Figure 4. (**a**) 1-Mapping module.; (**b**) Reaction of the 1-mapping module.

The 2-input mapping module consists of strands with domain labels {T^*} [mL^ f mR^ T^]: [nL^ f nR^ T^] <hL^ n1 hR^>, {T^*} [mL^ f mR^ T^]: [nL^ t nR^ T^] <hL^ n2 hR^>, {T^*} [mL^ t mR^ T^]: [nL^ f nR^ T^] <hL^ n3 hR^>, and {T^*} [mL^ t mR^ T^]: [nL^ t nR^ T^] <hL^ n4 hR^> Among them (ff, n1), (ft, n2), (tt, n3), and (tf, n4). As shown in Table 1, there are 16 corresponding mapping modules, namely, (ff, n1)~(ff, n1); (ff, n1)~(ft, n2); (ff, n1)~(tf, n3); (ff, n1)~(tt, n4); (ft, n2)~(ft, n2); (ft, n2)~(ff, n1); (ft, n2)~(tft, n3); (ft, n2)~(tt, n4); (tf, n3)~(tf, n3); (tf, n3)~(ff, n1); (tf, n3)~(ft, n2); (tf, n3)~(tt, n4); (tt, n4)~(tt, n4); (tt, n4)~(ff, n1); (tt, n4)~(ft, n2); (tt, n4)~(tf, n3). It is noteworthy that, when n1 = n2 = n3 = f, n4 = t, only two single strands need to be input, then the AND gate logic can be realized; when n1 = f, n2 = n3 = n4 = f = t, the input of two single strands can implement OR gate logic; when n1 = n4 = f, n2 = n4 = t, only two single strands need to be inputted for XOR gate logic to be realized, which is called a 2-input 1-output mapping module. Generally, this function is commonly used for 2-input module mapping.

Table 1. Double strands of the 2-input mapping module.

Identification	DNA Strands	Simple Note
n1 ∈ {t, f}	{T*} [mL^ f mR^ T^] : [nL^ f nR^ T^] < hL^ n1 hR >	(ff, n1)
n2 ∈ {t, f}	{T*} [mL^ f mR^ T^] : [nL^ t nR^ T^] < hL^ n2 hR >	(ft, n2)
n3 ∈ {t, f}	{T*} [mL^ t mR^ T^] : [nL^ f nR^ T^] < hL^ n3 hR >	(tf, n3)
n4 ∈ {t, f}	{T*} [mL^ t mR^ T^] : [nL^ t nR^ T^] < hL^ n4 hR >	(tt, n4)

Therefore, when n1 = n4 = f, n2 = n4 = t, XOR logic can be realized, which is used as an n-bit adder with a domain label. Generally, the reactions proceed simultaneously, reducing the time consumed by the reaction. The advantages of the modules are demonstrated again.

3.2. Amplification Gate with Domain Label

DNA signals will be attenuated during the reaction process, and lower concentrations will affect the rate of the DNA reaction and the final detection accuracy, so an amplification gate with a domain label was designed in this paper, consisting of two DNA double strands with domain labels and two DNA single strands with domain labels, namely, {T^*} [mL^ f mR^ T^] <nL^ f nR^>, {T^*} [mL^ t mR^ T^] <nL^ t nR^>, and <mL^ f mR^ T^ iL^ jiR^>, <mL^ t mR^ T^ iL^ jiR^>, which corresponds to amplifier strand-3, amplifier strand-4, fuel strand-1, and fuel strand-2. As shown in Figure 5, in the amplifier, the two single strands are of the same concentration and act as fuel during the DNA reaction. Of course, the two double strands also have the same concentration, and the total concentration of the DNA single strand with a domain label is always set to two times that of double strands with a domain label.

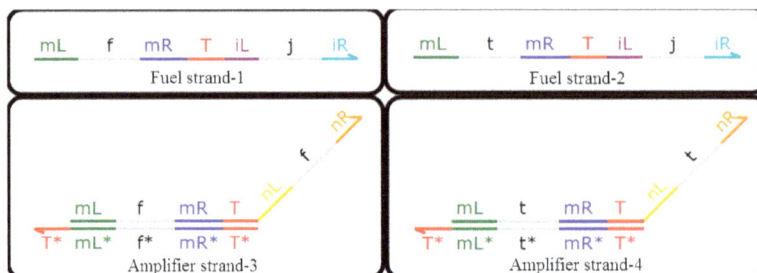

Figure 5. Amplification gate with a domain label.

It is noteworthy that the domain label is a long domain so only two DNA strands with domain labels are involved in the reaction, amplifying the concentration of the upstream DNA signal strand with a domain label to a set value (one-unit concentration in this paper). Namely, fuel strand-2 and amplifier strand-4 will react with the upstream DNA signal strand that represents logic value 1, and the other two strands do not participate in the reaction. Similarly, fuel strand-1 and amplifier strand-3 will react with the upstream DNA signal strand that represents logic value 0, and the other two strands do not participate in the reaction. In summary, one and only one of the above situations occurs when the DNA system is being reacted. More specifically, for DNA signal strands with a domain label that represents a logical value of 0, the DNA single strands <mL^ t mR^ T^ nL^ t nR^> cannot appear in the output strands. For DNA signal strands with a domain label that represents a logical value of 1, it is also unlikely that the DNA single strands <mL^ f mR^ T^ nL^ fnR^> will appear in the output strands. The reaction between the amplification gate with a domain label and the DNA signal strand with a domain label is shown in Figure 6. Compared with the amplifier of DNA dual-track logic circuit, the results of domain-labeled DNA reaction systems have better certainty.

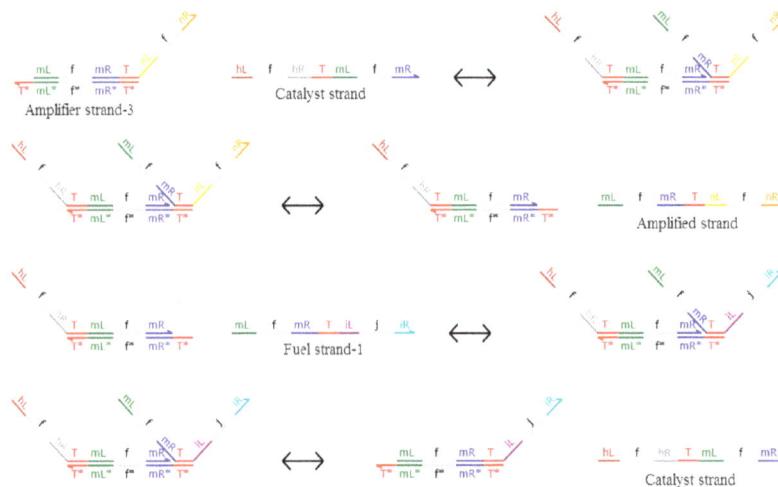

Figure 6. Reaction between fuel strand-1and amplifier strand-3.

Figure 6 shows fuel strand-1 and amplifier strand-3 as the reaction strands. <mL^ f mR^ T^ nL^ f nR^> as the output strands, which are called amplified strands and have a logic value of 0; <hL^ f hR^ T^ mL^ f mR^> as the catalytic strand; and other strands as the middle process reaction strands. When fuel strand-1 and amplifier strand-3 participate in the reaction, fuel strand-2 and amplifier strand-4 are invalid. Similarly, when fuel strand-2 and amplifier strand-4 participate in the reaction, fuel strand-1 and amplifier strand-3 are invalid, and the logic value of the output strand is 1.

3.3. Fan-Out Gate with Domain Label

The functions of the fan-out gate with a domain label and the fan-out gate with a dual-rail logic circuit are the same principle, so they can transform the DNA signal strand into several DNA signal strands representing the same logic signal (the specific quantity can be set). The N fan-out gate's function is to convert a domain-labeled DNA signal strand into an identical domain-labeled DNA signal strands, with a concentration N times that of the original signal strand. Through the fan-out gate with domain label conversion, the output strands can react with different encapsulated logic modules which are relatively independently packaged. This ensures that the logic inside the DNA system is encapsulated, as are the combined DNA systems, and the entire DNA system is at a steady state. The 2-fan out gate with a domain label is shown in Figure 7.

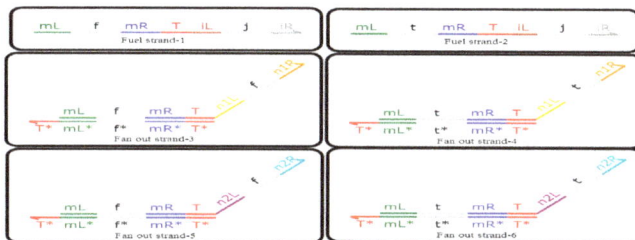

Figure 7. 2-Fan out gate with a domain label.

It can be seen from Figure 7 that the 2-fan out gate with a domain label is composed of the domain-labeled DNA double-stranded DNA and the domain-labeled DNA single-stranded DNA, that is, fan out strand-1, fan out strand-2, fan out strand-3, fan out strand-4, fan out strand-5, fan out strand-6, fuel strand-1, and fuel strand-2. The reactions are shown in Figure 8.

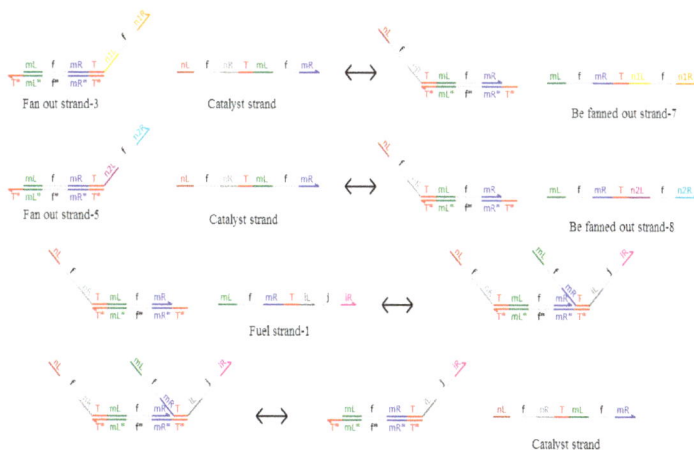

Figure 8. Reactions among fan out strand-3, fan out strand-5, and fuel strand-1.

In the reaction shown in Figure 8, the signal strand with domain label <nL^ f nR^ T^ mL^ f mR^> is equivalent to the catalyst. The reactions of fan out strand-3, fan out strand-5, and fuel strand-1 represent be the final outputs of fan out strand-7 and fan out strand-8, which have a logic value of 0. Similarly, when fan out strand-4, fan out strand-6, and fuel strand-2 participate in the reaction, strands <mL^ t mR^ T^ n1L^ t n1R^> and <mL^ t mR^ T^ n2L^ t n2R^> are the final outputs, and the logic value of the output strands is 1. The concentration of the output signal stand is determined by the concentration of the domain-labeled DNA double strands.

3.4. Reporter Gate with Domain Label

In this paper, the experimental results are tested via the reporter gate with a domain label. The detection gate is composed of fluorophore and quenchers, which converts the DNA signal strands into fluorescent signal strands. Fluorescent signal strands are released when the reporter gate with a domain label reacts with DNA signal strands, and the fluorescent signal strands do not react with other logic gates. As shown in Figure 9, the reaction between the reporter gate with a domain label and the DNA signal strand with a domain label is shown in Figure 10.

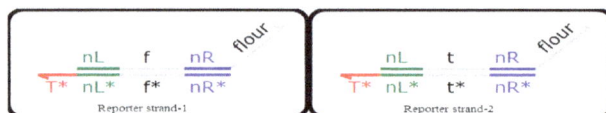

Figure 9. Reporter gate with a domain label.

As can be seen in Figure 9, the reporter gate with a domain label consists of reporter strand-1 and reporter strand-2, which respectively convert the corresponding domain-labeled DNA signal strands into fluorophore signal strands for detection.

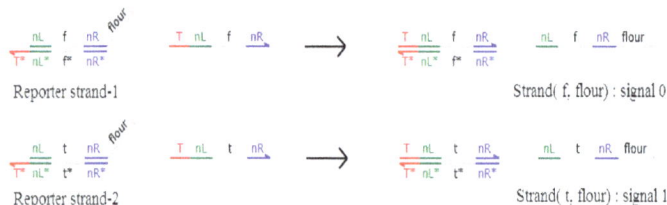

Figure 10. Domain label signal detection reaction.

In Figure 10, the logic value of the DNA signal strand can be judged by detecting strand (f, flour) or (t, flour), namely, the logic value of strand (t, flour) is 1, and the logic value of strand (f, flour) is 0.

4. Simulation

4.1. One-Bit Full Adder with Domain Label

The one-bit full adder takes the adjacent low-order carry into account. The one-bit full adder has DNA input strands denoted by c1 (low-order carry), x1, and x2, respectively, and has two DNA output strands denoted by y (sum) and c2 (carry). It can implement the addition of three binary logic values and simulate an electronic one-bit full adder. The logic circuit is shown in Figure 11a. The logic circuit of the one-bit full adder with a domain label is shown in Figure 11b.

(a)

(b)

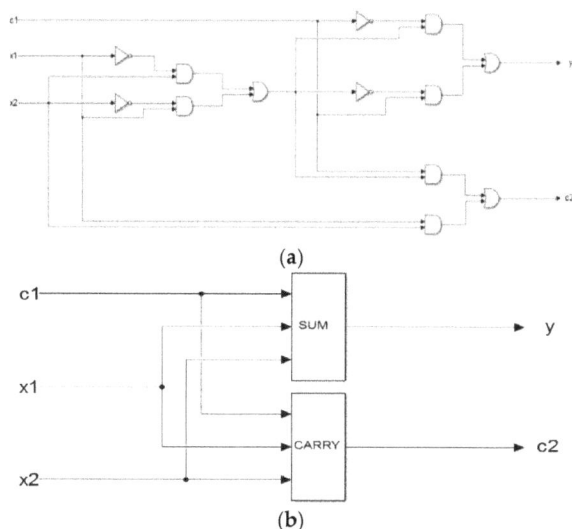

Figure 11. (a). Single-rail one-bit full adder; (b). One-bit full adder with a domain label.

The single-rail one-bit full adder consisting of six AND gates, three OR gates, and four NOT gates, as shown in Figure 11a, has six layers. It operates on three inputs and produces two outputs.

4.2. Simulation of the 1-Bit Full Adder with a Domain Label.

The 1-bit binary adder constructed by two 3-input mapping modules, which consists of 1 XOR gates with domain labels, 3 amplification gates with domain labels, one 2-fan out gate with a domain label, and reporter gates with domain labels. It can implement an addition between the 1-bit binary A_0 and 1-bit binary B_0, and finally, outputs a 2-bit binary number, S_1S_0 (S_1 is the output carry-bit).

The reporter gates with domain labels are <S1L^ _ S1R^ fluor> and <S2L^ _ S2R^ fluor>, which correspond to the summation-bit and carry-bit respectively. Specifically, when the 1-bit binary numbers are $A_0 = 1$ and $B_0 = 0$, the corresponding results are $S_0 = 1$ and $S_1 = 0$. Specifically, the output strands are <S1L^ t S1R^ fluor> and <S2L^ f S2R^ fluor>. When the 1-bit binary numbers are $A_0 = 1$ and $B_0 = 1$, the corresponding results are $S_0 = 0$ and $S_1 = 1$. Specifically, the output strands are <S1L^ f S1R^ fluor> and <S2L^ t S2R^ fluor>. When the 1-bit binary numbers are $A_0 = 0$ and $B_0 = 1$, the corresponding results are $S_0 = 1$ and $S_1 = 0$. Specifically, the output strands are <S1L^ t S1R^ fluor> and <S2L^ f S2R^ fluor>. When the 1-bit binary numbers are $A_0 = 0$ and $B_0 = 0$, the corresponding results are $S_0 = 0$ and $S_1 = 0$. Specifically, the output strands are <S1L^ f S1R^ fluor> and <S2L^ f S2R^ fluor>. The simulation results are shown in Figure 12.

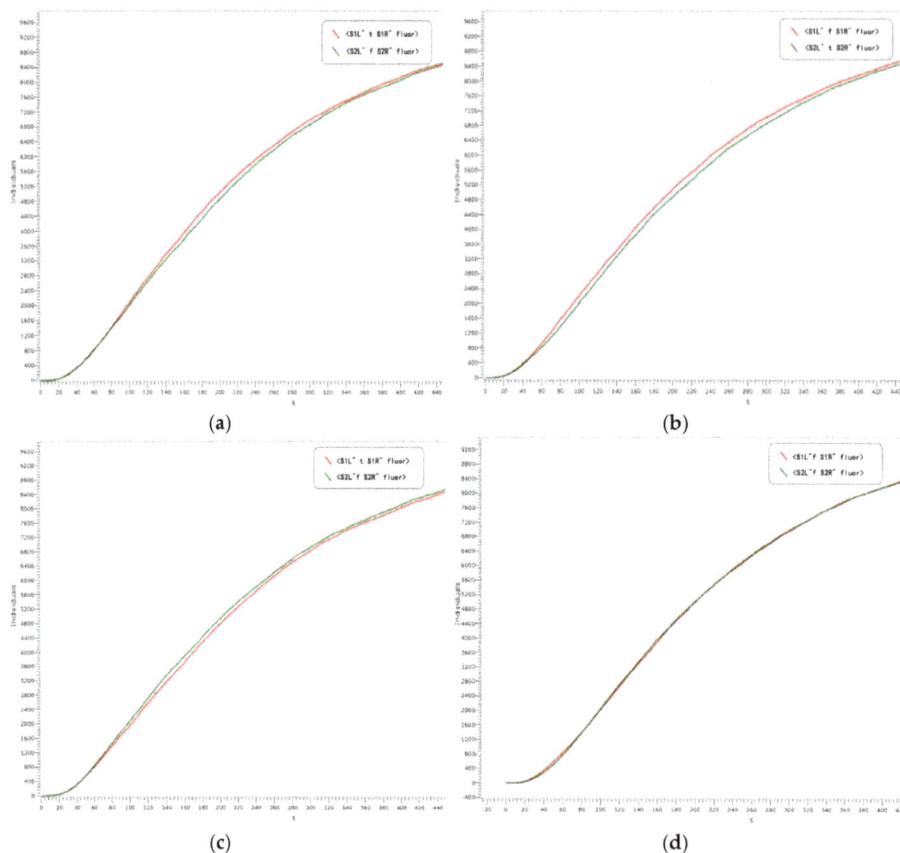

(a)   (b)

(c)   (d)

Figure 12. (**a**): $A_0 = 1, B_0 = 0, S_0 = 1, S_1 = 0$; (**b**): $A_0 = 1, B_0 = 1, S_0 = 0, S_1 = 1$; (**c**): $A_0 = 0, B_0 = 1, S_0 = 1, S_1 = 0$; (**d**): $A_0 = 0, B_0 = 0, S_0 = 0, S_1 = 0$.

From Figure 12, the following conclusions can be drawn. Firstly, the logical values of the reaction results in Figure 12a–d are correct, in accordance with the logical operation of binary summation, indicating that the adder with a domain label is feasible and it has a high accuracy. Secondly, the entire reaction curve is smooth and the reaction process is very stable, which indicating the stability of the adder with a domain label is improved, so the reaction state can be determined. Thirdly, the reaction reaches a state of equilibrium in about 540 s and the sensitivity is higher. In summary, the accuracy, stability, and sensitivity responses of the adder with a domain label satisfy our experimental requirements, thus providing a new perspective for the construction of other bio-circuits.

To further verify the advantages of the adder with a domain label, we simulate the double-rail 1-bit adder when the threshold concentration in the double-rail logic is at the extreme edge, as shown in Figure 13.

In Figure 13, the reporter gates are <S60L S60 S60R Fluor01> (SM^1), <S50L S50 S50R Fluor00> (SM^0), <S55L S55 S55R Fluor11> (CY^1), and <S58L S58 S58R Fluor10> (CY^0), which correspond, respectively, to $S_0^1 = 1$, $S_0^0 = 1$, $S_1^1 = 1, S_1^0 = 1$. The 1-bit binary numbers A and B_0 are converted into A_0^0, A_0^1, B_0^0, B_0^1. Their DNA input strands correspond to <S4L^ S4 S4R^ T^ S5L^ S5 S5R^>, <S6L^ S6 S6R^ T^ S7L^ S7 S7R^>, <S8L^ S8 S8R^ T^ S9L^ S9 S9R^>, and <S10L^ S10 S10R^ T^ S11L^ S11 S11R^>.

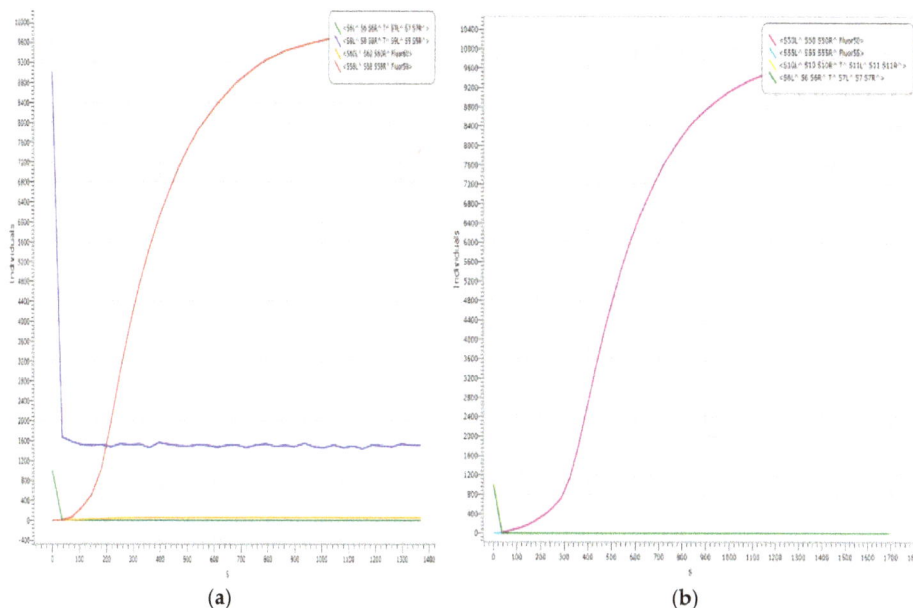

(a)   (b)

Figure 13. (a): $A_0^1 = 1, B_0^0 = 1, S_0^1 = 1, S_1^0 = 1$; (b): $A_0^1 = 1, B_0^1 = 1, S_0^1 = 0, S_1^1 = 1$.

In theory, when the 1-bit binary numbers are $A_0^1 = 1$ and $B_0^0 = 1$, the corresponding results are $S_0^1 = 1$ and $S_1^0 = 0$; specifically, when the input strands are <S6L^ S6 S6R^ T^ S7L^ S7 S7R^> ((A_0^1) A_0^1 and <S8L^ S8 S8R^ T^ S9L^ S9 S9R^> (B_0^0), the output strands are <S60L S60 S60R Fluor01> (SM^1) and <S58L S58 S58R Fluor10> (CY^0). In other words, the logic values of them are both 1. When the 1-bit binary numbers are $A_0^1 = 1$ and $B_0^0 = 1$ the corresponding results are $S_0^0 = 1$ and $S_1^1 = 1$. Specifically, when the input strands are <S6L^ S6 S6R^ T^ S7L^ S7 S7R^> (A_0^1) and <S10L^ S10 S10R^ T^ S11L^ S11 S11R^> (B_0^1), the output strands are <S50L S50 S50R Fluor00> (SM^0) and <S55L S55 S55R Fluor11> (CY^1), which both have logic values of 1.

However, the actual logic operations in Figure 13a,b are wrong. Specifically, in Figure 13a, when the input strands are <S6L^ S6 S6R^ T^ S7L^ S7 S7R^> (A_0^1) and <S8L^ S8 S8R^ T^ S9L^ S9 S9R^> (B_0^0), the logic value of output strand <S58L S58 S58R Fluor10> (CY^0) is 1, while the logic value of output strand <S60L S60 S60R Fluor01> (SM^1) is 0, which goes against the binary logic algorithms. In Figure 13b, when the input strands are <S6L^ S6 S6R^ T^ S7L^ S7 S7R^> (A_0^1) and <S10L^ S10 S10R^ T^ S11L^ S11 S11R^> (B_0^1), the logic value of output strand <S50L S50 S50R Fluor00> (SM^0) is 1, while the logic value of output strand<S55L S55 S55R Fluor11> (CY^1) is 0, which goes against the binary logic algorithms.

The reason for the logic error in the reactions shown in Figure 13 is the threshold concentration in the double-rail logic Seesaw gate. In general, a DNA signal strand with a unit concentration of 0.9 to 1 represents logic value 1, and a DNA signal strand with a 0 to 0.1 unit density represents a logic value of 0. However, the threshold cannot strictly distinguish between high and low concentrations when its concentration is at 0.1~0.2 units or 0.8~0.9 units, resulting in the wrong signal being output (Figure 13). And the logic circuit with domain t and domain f is determined throughout the entire reaction process (Figure 11), which provides the potential for molecular automation, such as DNA 4 × 4 multiplier operations, n-bit addition, and so on.

4.3. Simulation of DNA 4 × 4 Multiplier with Domain Label

Figure 14a shows a domain-labeled binary DNA 4 × 4 multiplier based on DNA strand permutation, which consists of 16 domain-labeled AND gates, four one-half adders (without detection gates), and eight one-bit full adders (without detection gates). Eight domain-labeled detection gates are composed. Of course, domain-labeled amplifiers can also be added at other desired locations. Since the domain-labeled fluorescent signal chain cannot react with other logic gates just to facilitate detection, the removal of the detection gate does not affect the result of the reaction, so detection gates are removed from semi-adder and full adder, and only the A domain-labeled detection gate is added to the final output of the multiplier to facilitate the detection of the eight domain tag output signal strands.

Figure 14b is a simulation of a binary DNA 4 × 4 multiplier, simulated with 1111 × 1111 = 11100001 as an example. Table 2 shows the logical values of the DNA input and output strands.

Table 2. Logic values of DNA signal strands (1111 × 1111 = 11100001).

DNA Strands with Domain Labels	Input/Output	Logic Value
<A0L^ t A0R^ T^ A0L^ t A0R^>	A0	1
<A1L^ t A1R^ T^ A1L^ t A1R^>	A1	1
<A2L^ t A2R^ T^ A2L^ t A2R^>	A2	1
<A3L^ t A3R^ T^ A3L^ t A3R^>	A3	1
<B0L^ t B0R^ T^ B0L^ t B0R^>	B0	1
<B1L^ t B1R^ T^ B1L^ t B1R^>	B1	1
<B2L^ t B2R^ T^ B2L^ t B2R^>	B2	1
<B3L^ t B3R^ T^ B3L^ t B3R^>	B3	1
<J0L^ t J0R^ fluor>	S0	1
<J3L^ f J3R^ fluor>	S1	0
<J10L^ f J10R^ fluor>	S2	0
<J20L^ f J20R^ fluor>	S3	0
<J29L^ f J29R^ fluor>	S4	0
<J35L^ t J35R^ fluor>	S5	1
<J38L^ t J38R^ fluor>	S6	1
<J39L^ t J39R^ fluor>	S7	1

The concentration of the eight domain-labeled DNA signal strands rapidly decreases to less than 0.1 times the unit concentration within the initial 60 s, and then slowly decreases to almost zero over the remaining time period; the concentration of the domain-labeled fluorescent signal chain rapidly rose to 0.9 times the unit concentration or more within 2100 s, and then it slowly rose until it was very close to a concentration of 1 unit. The domain-labeled binary DNA 4 × 4 multiplier can realize the multiplication of 4-bit binary number and 4-bit binary number. The whole reaction is very stable, which indicates that the designed domain-labeled binary DNA 4 × 4 multiplier has good stability and encapsulation, which further demonstrates that the AND gate, OR gate, NOT gate, amplifier, fan-out gate, and detection gate with the domain label have good stability and encapsulation, which lays the foundation for the realization of DNA computers.

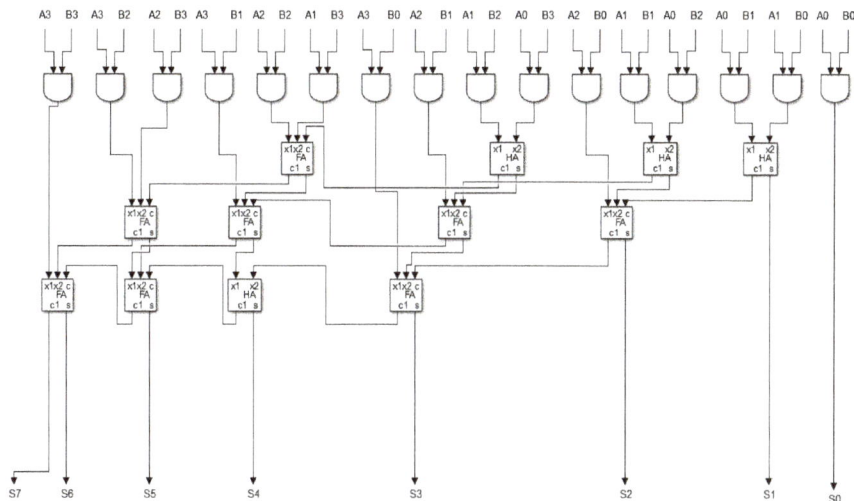

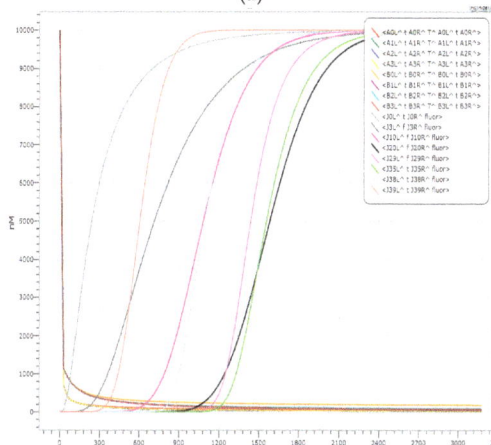

Figure 14. (**a**): Binary DNA 4 × 4 multiplier with a domain label; (**b**): Simulation of DNA 4 × 4 multiplier with a domain label.

4.4. Simulation of 8-Bit Binary Adder/Subtractor with Domain Label

Similar to the 1-bit full adder, the 8-bit adder takes the adjacent low-order carry into account. It is constructed by 1-mapping modules, 2-mapping modules, and 3-mapping modules, and consists of eight XOR gates with a domain label, eight one-bit full adders with a domain label, 24 amplification gates with a domain label and one 9-fan out gate with a domain label. It can implement an adder or subtractor between the 8-bit binary $A_7A_6A_5A_4A_3A_2A_1A_0$ and the 8-bit binary $B_7B_6B_5B_4B_3B_2B_1B_0$, and finally, outputs a 9-bit binary number $S_8S_7S_6S_5S_4S_3S_2S_1S_0$ (S_8 is the output carry-bit). It has 16 input DNA strands with domain labels, nine DNA output strands with domain labels, and one DNA switch strand with a domain label (denoted by A#S) which decides whether to implement an 8-bit adder or 8-bit subtractor (when A#S = 0, the DNA 8-bit adder is used, when A#S = 1, the DNA 8-bit subtractor is used). The logical values corresponding to the DNA input and output strands with domain labels are shown in Table 1 (take 00101101 − 10010110 = 010010111 as an example).

The single logic circuit discussed below has eight inputs and eight outputs. A_i, B_i and C_i represent the ith input and the ith output respectively. The logic function expressions of the single-rail logic circuit are as follows:

$$S_i = A_i(A_0, A_1, \ldots, A_7) + B_i(B_0, B_1, \ldots, B_7)$$
$$= f_i(A_0, A_1, \ldots, A_7, \overline{A}_0, \overline{A}_1, \ldots, \overline{A}_7) + g_i(B_0, B_1, \ldots, B_7, \overline{B}_0, \overline{B}_1, \ldots, \overline{B}_7)$$

In Table 3, the decimal number 77 minus the decimal number 150 is equal to the negative decimal number 73. Based on the binary complement operation, the logic expression is $77 + (2^8 - 150) = 77 + 106 = 183$, and the decimal number 183 corresponding to the binary number is 10110111. Obviously, the complement of the negative decimal number 73 is the binary number 10110111. So, the 8-bit binary adder/subtractor correctly calculates the result of subtracting two 8-bit binary numbers, namely, $00101101 - 10010110 = 010010111$. The simulation of it is shown in Figure 15.

Table 3. Logic values of DNA signal strands ($00101101 - 10010110 = 010010111$).

DNA Strands with Domain Label	Input/Output	Logic Value
<S2L^ t S2R^ T^ S2L^ t S2R^>	A_0	1
<S3L^ f S3R^ T^ S3L^ f S3R^>	A_1	0
<S4L^ t S4R^ T^ S4L^ t S4R^>	A_2	1
<S5L^ t S5R^ T^ S5L^ t S5R^>	A_3	1
<S6L^ f S6R^ T^ S6L^ f S6R^>	A_4	0
<S7L^ t S7R^ T^ S7L^ t S7R^>	A_5	1
<S8L^ f S8R^ T^ S8L^ f S8R^>	A_6	0
<S9L^ f S9R^ T^ S9L^ f S9R^>	A_7	0
<S10L^ f S10R^ T^ S10L^ f S10R^>	B_0	0
<S11L^ t S11R^ T^ S11L^ t S11R^>	B_1	1
<S12L^ t S12R^ T^ S12L^ t S12R^>	B_2	1
<S13L^ f S13R^ T^ S13L^ f S13R^>	B_3	0
<S14L^ t S14R^ T^ S14L^ t S14R^>	B_4	1
<S15L^ f S15R^ T^ S15L^ f S15R^>	B_5	0
<S16L^ f S16R^ T^ S16L^ f S16R^>	B_6	0
<S17L^ t S17R^ T^ S17L^ t S17R^>	B_7	1
<S0L^ t S0R^ T^ S0L^ t S0R^>	A#S	1
<S69L^ t S69R^ fluor>	S_0	1
<S109L^ t S109R^ fluor>	S_1	1
<S149L^ t S149R^ fluor>	S_2	1
<S189L^ f S189R^ fluor>	S_3	0
<S229L^ t S229R^ fluor>	S_4	1
<S269L^ f S269R^ fluor>	S_5	0
<S309L^ f S309R^ fluor>	S_6	0
<S349L^ t S349R^ fluor>	S_7	1
<S350L^ f S350R^ fluor>	S_8	0

In Figure 15, the function of the 8-bit binary subtractor with a domain label is realized; in the input 8-bit binary code, the meaning of 00101101 is 77, and the meaning of 10010110 is 150. That is, the reaction input $00101101 - 10010110$ will output 010010111. If we input strands with the domain label <S2L^ t S2R^ T^ S2L^ t S2R^> ... <S9L^ f S9R^ T^ S9L^ f S9R^> and strands with the domain label <S10L^ f S10R^ T^ S10L^ f S10R^> ... <S17L^ t S17R^ T^ S17L^ t S17R^> at the same time, we do not forget the strands <S0L^ t S0R^ T^ S0L^ f S0R^> and <S0L^ t S0R^ T^ S0L^ t S0R^>, which determine the addition and subtraction of the reaction. When the reaction input <S0L^ t S0R^ T^ S0L^ t S0R^>, the reaction is a subtraction, and it outputs strands with domain labels <S69L^ t S69R^ fluor>, <S109L^ t S109R^ fluor>, <S149L^ t S149R^ fluor>, <S189L^ f S189R^ fluor>, <S229L^ t S229R^ fluor>, <S269L^ f S269R^ fluor>, <S309L^ f S309R^ fluor>, <S349L^ t S349R^ fluor>, and <S350L^ f S350R^ fluor>. Otherwise, the reaction performs an addition operation. The entire reaction correctly calculates the subtraction of two 8-bit binary numbers, and the entire reaction occurs quickly and orderly.

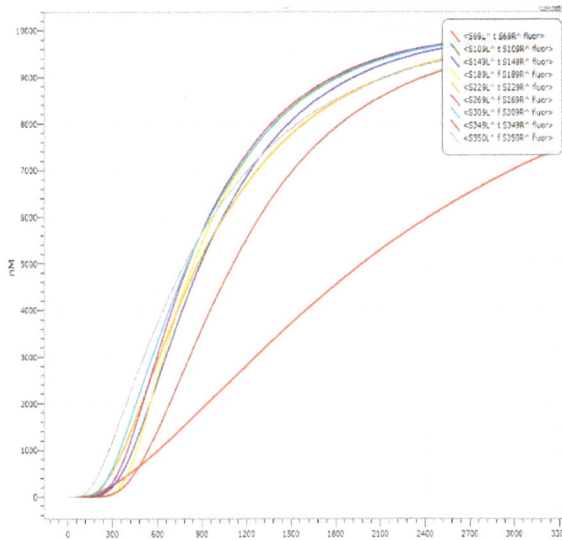

Figure 15. Simulation of 00101101 − 10010110 = 010010111.

4.5. Summary

The logic circuit constructed by a 1-mapping module, 2-mapping module and 3-mapping module has a lower design complexity, a shorter calculation time, and higher stability than the dual-rail logic circuit. Regarding the complexity of the design, firstly, the 1-mapping module and the 2-mapping module are combined to simulate the correctness of the one-bit full adder with the domain label, and then, the n-bit stable logic circuit is constructed. The fewer layers used, the higher the parallelism is, and the shorter the computing time is. Regarding the stability, logic values of 1 and 0 in the dual-rail logic circuit are represented by a higher concentration and a lower concentration, respectively; however, thresholding strands may not distinguish DNA signal strands with higher concentrations from DNA signal strands with lower concentrations, which introduces an error signal into the DNA reaction. Logic values of 1 and 0 in the logic circuit constructed by mapping modules are represented by the domain labels t and f. This is deterministic and does not make DNA reaction produce an error signal. In addition, the standard deviation of the computation time of the logic circuit constructed by mapping modules is far less than that of the dual-rail logic circuit, indicating that mapping modules possess stability and logic circuits constructed by them are more stable.

5. Conclusions

Basic logic gates, an amplification gate, a fan-out gate, a reporter gate with a domain label (domains t and f), and N-mapping modules with domain labels were designed in this paper. The mapping modules included a 1-mapping module, a 2-mapping module, a 3-mapping module and an N-mapping module according to how many inputs they operated on. DNA logic circuits constructed with a 1-mapping module, a 2-mapping module, a 3-mapping module and an N-mapping module were shown to possess a lower design complexity, fewer layers, higher parallelism, higher stability, and shorter time complexity, which was verified through a comparison with a one-bit full adder with a domain label. A DNA 8-bit adder/subtractor was designed with mapping modules; this could be applied to design more stable and faster DNA computers in the future, so that more and more NP-complete problems can be solved with shorter time complexity.

Author Contributions: Conceptualization, C.Z.; Methodology, C.Z. and W.H.; Software, W.H.; Validation, C.Z. and W.H.

Funding: This research received no external funding.

Acknowledgments: This work is supported by the National Natural Science Foundation of China (Nos. 61672121, 61425002, 61751203, 61772100, 61672467, 61702070, 61572093), Program for Changjiang Scholars and Innovative Research Team in University (No. IRT_15R07), the Dalian Outstanding Young Science and Technology Talent Support Program (No.2017RJ08).

Conflicts of Interest: The authors declare no conflict of interest.

References

1. Ramezani, H.; Jed, H. DNA strand displacement reaction for programmable release of biomolecules. *Chem. Commun.* **2015**, *51*, 8307–8310. [CrossRef] [PubMed]
2. Zhang, D.Y.; Seelig, G. Dynamic DNA nanotechnology using strand-displacement reactions. *Nat. Chem.* **2011**, *3*, 103–114. [CrossRef] [PubMed]
3. Qian, L.; Winfree, E.; Bruck, J. Neural network computation with DNA strands displacement cascades. *Nature* **2011**, *475*, 368–372. [CrossRef] [PubMed]
4. Keller, N.; Grimes, S.; Jardine, P.J.; Smith, D.E. Single DNA molecule jamming and history-dependent dynamics during motor-driven viral packaging. *Nat. Phys.* **2016**, *12*, 757–761. [CrossRef] [PubMed]
5. Miyazono, Y.; Endo, M.; Ueda, T.; Sugiyama, H.; Harada, Y.; Tadakuma, H. 1M1524 constructing DNA-kinesin hybrid-nanomachine using the DNA-tile scaffold. *Biophysics* **2017**, *51*, 64–65. [CrossRef]
6. Prokup, A.; Hemphill, J.; Liu, Q.; Deiters, A. Optically controlled signal amplification for DNA computation. *ACS Synth. Biol.* **2015**, *4*, 1064–1069. [CrossRef] [PubMed]
7. Chen, S.X.; Seelig, G. An engineered kinetic amplification mechanism for SNV discrimination by DNA hybridization probes. *Chem. Soc.* **2016**, *138*, 5076–5086. [CrossRef] [PubMed]
8. Chen, X.; Ren, H.; Dong, Y. A unimolecular multifunctional DNA cascaded logic circuit and signal amplifier based on Hg^{2+} and Ag^+. *J. Comput. Theor. Nanosci.* **2016**, *13*, 4083–4087. [CrossRef]
9. Wang, Y.; Tian, G.; Hou, H.; Ye, M.; Cui, G. Simple logic computation based on the DNA strand displacement. *J. Comput. Theor. Nanosci.* **2014**, *11*, 1975–1982. [CrossRef]
10. Zhang, X.; Zhang, W.; Zhao, T.; Wang, Y.; Cui, G. Design of logic circuits based on combinatorial displacement of DNA strands. *J. Comput. Theor. Nanosci.* **2015**, *12*, 1161–1164. [CrossRef]
11. Wang, Z.; Cai, Z.; Sun, Z.; Ai, J.; Wang, Y.; Cui, G. Research of molecule logic circuit based on DNA strand displacement reaction. *J. Comput. Theor. Nanosci.* **2016**, *13*, 7684–7691. [CrossRef]
12. Song, T.; Liu, X.; Zhao, Y.; Zhang, X. Spiking neuralp systems with white hole neurons. *IEEE Trans. Nanobioscience* **2016**, *15*, 666–673. [CrossRef] [PubMed]
13. Song, T.; Zheng, P.; Wong, D.; Wang, X. Design of logic gates using spiking neural p systems with homogeneous neurons and astrocytes-like control. *Inf. Sci.* **2016**, *372*, 380–391. [CrossRef]
14. Song, T.; Rodríguez-Patón, A.; Zheng, P.; Zeng, X. Spiking neural p systems with colored spikes. *IEEE Trans. Cogn. Develop. Syst.* **2018**. [CrossRef]
15. Li, X.; Hong, L.; Song, T.; Rodríguez-Patón, A.; Chen, C.; Zhao, H.; Shi, X. Highly biocompatible drug-delivery systems based on DNA nanotechnology. *J. Biomed. Nanotech.* **2017**, *13*, 747–757. [CrossRef]
16. Wang, X.; Zheng, P.; Ma, T.; Song, T. Computing with bacteria conjugation: Small universal system. *Moleculer* **2018**, *23*. [CrossRef]
17. Seelig, G.; Soloveichik, D.; Winfree, E.; Zhang, D. Enzyme-free nucleic acid logic circuits. *Science* **2006**, *314*, 1585–1588. [CrossRef] [PubMed]
18. Qian, L.; Winfree, E. A simple DNA gate motif for synthesizing large-scale circuits. *J. R. Soc. Interface* **2011**, *8*, 1281–1297. [CrossRef] [PubMed]
19. Qian, L.; Winfree, E. Scaling up digital circuit computation with DNA strand displacement cascades. *Science* **2011**, *332*, 1196–1200. [CrossRef] [PubMed]
20. Zhang, C.; Ma, L.N.; Dong, Y.F.; Yang, J.; Xu, J. Molecular logic computing model based on DNA self-assembly strand branch migration. *Chin. Sci. Bull.* **2013**, *58*, 32–38. [CrossRef]
21. Guo, Y.; Zhou, L.; Xu, L.; Zhou, X.; Hu, J.; Pei, R. Multiple types of logic gates based on a single g-quadruplex DNA strand. *Sci. Rep.* **2014**, *4*, 7315–7322. [CrossRef] [PubMed]

22. Wang, Z.; Wu, Y.; Tian, G.; Wang, Y.; Cui, G. The application researchon multi-digit logic operation based on DNA strand displacement. *J. Comput. Theor. Nanosci.* **2015**, *12*, 1252–1257. [CrossRef]

23. Lakin, M.R.; Stefanovic, D. Supervised learning in adaptive DNA strand displacement networks. *ACS Synth. Biol.* **2016**, *5*, 885–914. [CrossRef] [PubMed]

24. Sun, J.; Li, X.; Cui, G.; Wang, Y. One-bit half adder-half subtractor logical operation based on the DNA strand displacement. *J. Nanoelectron. Optoe.* **2017**, *12*, 375–380. [CrossRef]

25. Levine, R.D. *Molecular Reaction Dynamics*; Cambridge University Press: Cambridge, UK, 2009.

26. Song, T.; Garg, S.; Mokhtar, R.; Bui, H.; Reif, J. Analog computation by DNA strand displacement circuits. *ACS Synth. Biol.* **2016**, *5*, 898–937. [CrossRef] [PubMed]

27. Yang, Q.; Zhou, C.; Zhang, Q. Logic gates designed with domain label based on DNA strand displacement. *Adv. Swarm Intell.* **2016**, *9712*, 244–255.

Sample Availability: Samples of the compounds are not available.

molecules

MDPI

Article

Correcting Errors in Image Encryption Based on DNA Coding

Bin Wang [1],*, Yingjie Xie [2], Shihua Zhou [1], Xuedong Zheng [1] and Changjun Zhou [3],*

[1] Key Laboratory of Advanced Design and Intelligent Computing, Dalian University, Ministry of Education, Dalian 116622, China; shihuajo@gmail.com (S.Z.); xuedongzheng@163.com (X.Z.)

[2] Applied Technology College, Dalian Ocean University, Dalian 116300, China; yingjieying@163.com

[3] College of Mathematics, Physics and Information Engineering, Zhejiang Normal University, Jinhua 321004, China

* Correspondence: wangbinpaper@gmail.com (B.W.); zhou-chang231@163.com (C.Z.); Tel.: +86-0411-8740-2106 (B.W. & C.Z.)

Received: 25 June 2018; Accepted: 24 July 2018; Published: 27 July 2018

Abstract: As a primary method, image encryption is widely used to protect the security of image information. In recent years, image encryption pays attention to the combination with DNA computing. In this work, we propose a novel method to correct errors in image encryption, which results from the uncertainty of DNA computing. DNA coding is the key step for DNA computing that could decrease the similarity of DNA sequences in DNA computing as well as correct errors from the process of image encryption and decryption. The experimental results show our method could be used to correct errors in image encryption based on DNA coding.

Keywords: image encryption; chaotic map; DNA coding; Hamming distance

1. Introduction

With wide usage of multimedia technologies and excessive spread of internet, the awareness of protecting information, especially image information, is heightened day by day. As we known, encrypting technology can usually be used to protect the security of image information. In image encryption, chaotic maps are usually employed to encrypt image, because they have the features of ergodicity, sensitivity to initial conditions, control parameters and so on [1–7]. Chen et al. proposed a novel 3D cat maps to design a real-time secure symmetric encryption scheme [1]. Lian et al. first analyzed the parameter sensitivity of standard map and proposed an improved standard map to encrypt image [3]. Wong et al. proposed a fast algorithm of image encryption, where the overall encryption time was reduced as fewer rounds were required [2]. Zhang et al. proposed a new image encryption algorithm based on the spatiotemporal chaos of the mixed linear-nonlinear coupled map lattices [7]. Wang et al. combined genetic recombination with hyper-chaotic system to design a novel image encryption and experiment results proved that the proposed algorithm was effective for image encryption [8]. Zhang et al. analyzed different kinds of permutation algorithms and proposed a new cryptosystem to address these drawbacks [5]. In the recent past, although these methods have made some progress, they lack the capability of parallel computing.

Inspired by the biological character of DNA sequences, such as parallel computing, low-energy and so on, DNA computing and DNA coding are widely used to encrypt image [6,9–17]. Zhang et al. combined DNA sequence addition operation with chaotic map to design a novel image encryption scheme [9]. The experimental results shown that the proposed scheme could achieve good encryption and resist some kind attacks. In Ref. [10], the authors transformed DNA sequences into its base pair for random times to confuse the pixels, generate the new keys according to the plain image and the common keys. Wei et al. further utilized DNA sequence addition operation and Chen's hyper-chaotic map to encrypt a color image [11]. Due to some disadvantages in One-Time-Pad (OTP) algorithm, the author used logistic chaotic map as an input of OTP algorithm and proposed an interesting encryption algorithm based on a chaotic selection between original message DNA strands and OTP DNA strands [12]. In Ref. [13], the authors used genetic algorithm to determine the best masks, which result from DNA and logistic map functions. Ozkaynak et al. broke a previous cryptosystem and proposed an improved image encryption algorithm [14]. Rehman et al. utilized whole set of DNA complementary rules dynamically and employed DNA addition operation to encrypt image [15]. Song and Qiao proposed a novel image encryption scheme based on DNA encoding and spatiotemporal chaos, which was of high key sensitivity and large key space [16]. In Ref. [17], DNA coding combined with an improved 1D chaotic systems to design image encryption. Kulsoom et al. employed an entire set of DNA complementary rules along with 1D chaotic maps to design an image encryption algorithm [6]. Wang et al. proposed a new chaotic image encryption scheme based on Josephus traversing and mixed chaotic map [18]. Parvaz and Zarebnia defined a combination chaotic system and studied its properties [19].

DNA computing was addressed to solve the seven-point Hamiltonian path problem by Adleman in 1994 [20]. Along with the development of research, there are a large number of applications about DNA computing, such as DNA logic gates [21], neural network [22], cryptography [4], data storage [23], image watermarking [24] and so on. Hybridization reaction is the key operation for DNA sequences and influences the reliability of DNA computing. However, the false hybridization is unavoidable because of the limit of biological technology, result from false positive and false negative. The lack of similarity between DNA sequences could result in false positive and generating hybridization reaction between two unmatched DNA sequences. The mistake in the biochemical operation result in false negative in which two matched DNA sequences did not hybridize each other [25]. Chai et al. encoded plain image by DNA matrix and permuted the image with a new wave-based permutation scheme [26]. In Ref. [27], DNA sequence operation combining with one-way coupled-map lattices was to structure a robust and lossless color image encryption algorithm and the three gray-level components of plain-image were converted into three DNA matrices and performed XOR operation twice. Designing DNA coding could obtain high quality DNA sequences which satisfy some constraints, such as Hamming distances, GC content and so on, to decrease the similarity between DNA sequences [28,29]. Inspired by Hybridization reaction is the kernel for DNA computing and influences the reliability of DNA computing. However, the false hybridization is unavoidable because of the limit of biological technology, result from false positive and false negative. The lack of similarity between DNA sequences could result in false positive and generating hybridization reaction between two unmatched DNA sequences. The mistake in the biochemical operation result in false negative in which two matched DNA sequences did not hybridize each other [25]. Designing DNA coding could obtain high quality DNA sequences which satisfy some constraints, such as Hamming distance, GC content and so on, to decrease the similarity between DNA sequences [28,29]. Inspired by communication theory, Hamming code can be used to correct errors. For example, d is the Hamming distance between two strings and then the bits of correcting errors are equal to $\left\lfloor \frac{d-1}{2} \right\rfloor$. So, in this paper, we introduce Hamming distance to decrease the similarity between DNA sequences as well as correct errors from hybridization reaction. Furthermore, to improve the accuracy of DNA computing, the constraints of DNA coding are used to decrease the generation of false positive. Finally, the experimental results show that the number of pixels change rate (NPCR) has achieved 99.57% and the unified average changing intensity (UACI) has achieved 32.38%. The proposed method could effectively correct the encrypted image contained 1000 errors and improve the accuracy of hybridization reaction.

2. Methods

2.1. DNA Coding

Hamming distance is widely used to design DNA coding. It is the number of positions at which the corresponding symbols are different when two strings have the equal length [30]. In the alphabet $\Sigma = \{A, C, G, T\}$, there exists a set S with length n and size of $|S| = 4^n$. A subset $C \subseteq S$ and let u, v any two codes in the C satisfy [31]:

$$\tau(u, v) \geq d \tag{1}$$

d is a positive integer, τ is the constraint criteria (or criterion) for designing DNA coding. In this paper, τ is denoted as the Hamming distance.

2.1.1. Sequences-Sequence Hamming Distance (SS)

Sequences-sequence Hamming Distance [31]: for the DNA sequences u, v with given length n (written from the 5' to the 3' end), the Hamming distance between u and v is denoted as $H(u, v)$. The minimal $H(u_i, v_j)$ in all DNA sequences is denoted as $SS(u_i)$ and it should not be less than parameter d,

$$SS(u_i) = \min_{1 \leq j \leq n, j \neq i} \{H(u_i, v_j)\} \geq d \tag{2}$$

2.1.2. Sequences-Complementarity Hamming Distance (SC)

Sequences-complementarity Hamming Distance [31]: for the DNA sequences u, v with given length n (written from the 5' to the 3' end), $H(u, v^C)$ denotes the Hamming distance between u and v^C, where v^C is the complementary sequence of v. For example, $v = $ ACTG, then $v^C = $ CAGT. The minimal $H(u_i, v_j^C)$ in all DNA sequences is denoted as $SS(u_i)$ and it should not be less than parameter d,

$$SC(u_i) = \min_{1 \leq j \leq n, j \neq i} \left\{ H\left(u_i, v_j^C\right) \right\} \geq d \tag{3}$$

2.1.3. GC Content

In order to approximate the thermodynamic properties of DNA sequences, GC content constraint is used to combine with distance constraint, such as Hamming distance. The percentage of G or C bases within each DNA is denoted as GC content. In this paper, GC content is equal to 50%. The GC content is denoted as follows:

$$GC\_content = Num\_gc/n \times 100\% \tag{4}$$

2.1.4. DNA Coding Rule

Adenine (A), Cytosine (C), Guanine (G) and Thymine (T) are the four elements that make up the whole DNA sequence. When paring, the principle of complementary base pairing is observed, namely A with T and C with G [32]. There is a complementary relationship between 0 and 1 in the binary bit. Similarly, there is a complementary relationship between 01 and 10 as well as 00 and 11. In the previous works, the authors converted binary message to DNA sequences based on the DNA coding rule in Table 1 [6,9,13–15,17,23,33,34]. There are eight DNA coding methods to convert binary message to DNA sequences [9,24]. For example, the pixel 65 is firstly transformed into binary bit 01000001 and then 01000001 transformed into DNA sequence ACCA for the first rule.

Table 1. DNA coding list [9,24].

	1	2	3	4	5	6	7	8
A	01	01	00	00	10	10	11	11
T	10	10	11	11	01	01	00	00
C	00	11	01	10	00	11	01	10
G	11	00	10	01	11	00	10	01

From the Table 1, the information only is simple transformed between binary and DNA sequence. It does not consider the characters of DNA sequences, especially specific hybridization. So, in this paper, we designed DNA coding that satisfied three constraints above to encrypt image and use this DNA coding to correct the errors.

2.2. New DNA Coding Rule for Correcting Errors

As shown above, the DNA sequences used to encode pixels and chaotic orbits should satisfy these constraints to decrease the similarity between DNA sequences and correct the errors. In our previous work, we proposed a dynamic genetic algorithm to design DNA sequence sets which satisfy the combinational constraints [35]. DNA sequence set denotes that any pair of DNA sequences in this set satisfies the combinational constraints. In this paper, we use the DNA coding $A_4^{SS+SC+GC}(8,3) = 336$, namely the length is equal to 8 and Hamming distance is equal to 3. We randomly select 256 elements from this set to encode the pixels between 0 and 255. In this paper, we denote these 256 DNA sequences as DNA coding rule. Table 2 lists the first 50 DNA coding rule to encrypt image. The whole $A_4^{SS+SC+GC}(8,3)$ and DNA coding rule are shown in the Supplement.

Table 2. The first 50 DNA coding rule.

Pixel	DNA Coding	Pixel	DNA Coding	Pixel	DNA Coding	Pixel	DNA Coding	Pixel	DNA Coding
0	ATCATGCC	1	CTCGATCA	2	GCTCTTCT	3	AGTGGGAT	4	ACTCTCTG
5	AATCTGCG	6	ACTCACGT	7	CTTCCAAC	8	GCTTCTAG	9	TAGGAGGT
10	GATCGACT	11	TAACGCTG	12	TAAGCGGA	13	CTGTGATC	14	CCCTAATC
15	TGGAAGGA	16	TACTACCG	17	CTTATGGG	18	TCAGCAAG	19	CGACTTCT
20	AGTGTCGA	21	TGCGATTC	22	CAACGACA	23	GATCTGTC	24	GCCAACTA
25	ATGAGGGA	26	TAGAACGG	27	CCGTAACA	28	TAGACTGC	29	GCTGGATT
30	GTGAGTCA	31	TCATGGAC	32	ACCACTAC	33	TCCTAAGG	34	GGCTAAAG
35	CCAACTGA	36	TCGTCTTG	37	TTGGGAAC	38	AATAGCCC	39	CTGTCGAA
40	CCCCATAT	41	AACCTCTC	42	GGTTTACG	43	GCAGAAGA	44	TAGAGGAG
45	GAAAGGGA	46	ATCGACGA	47	GCAAGTAC	48	TCAGACAC	49	CTTGGTTG

2.3. Process of Encrypting and Decrypting Image Based on DNA Coding

2.3.1. Encrypting Image

Recently, there are some works on cryptanalysis of encrypting schemes based on chaotic map and DNA coding [36–38]. In this paper, in order to improving the security of our encrypting scheme, two logistic maps with different parameters and initial values are chosen to generate pseudorandom sequence. The different parameters and initial values for the Equation (5) are denoted as $\mu_1, \mu_2, x_1^1(0)$ and $x_1^2(0)$, respectively, where $\mu_1, \mu_2 \in [3.9, 4]$ and $x_1^1(0), x_1^2(0) \in (0, 1)$.

$$x_{i+1} = \mu x_i(1 - x_i) \tag{5}$$

The detailed of encrypting image is described as follows:

Step 1. The key with 16 elements is randomly generated as the initial key and the initial key is implemented XOR operation with every pixel value of the plain image. The result of XOR operation is regard as the relating key;

Step 2. According to initial condition of logistic maps, namely two parameters μ_1, μ_2 and two initial value $x_1^1(0)$, $x_1^2(0)$, the relating key is evenly dividing relating key into four parts. These logistic maps are to iterate for 100 times to get rid of the transient effect of chaotic systems;

Step 3. The logistic maps are continuingly iterated base on the number of pixels, namely one map for the half number and the pseudorandom sequence consists of the logistic chaotic orbits;

Step 4. In order to permute the plain image, the chaotic orbits are sorted in ascending order. This operation (permutation) only changes the location of pixels of plain image;

Step 5. The XOR operation is implemented between the pixels of the permuted image and the pseudorandom sequence from the logistic maps. This operation (diffusion) only changes the value of pixels of digital image;

Step 6. According to the new DNA coding rule, the encrypted image is encoded by DNA coding;

Step 7. Outputting the encrypted image.

The flowchart of encrypting image is illustrated in Figure 1.

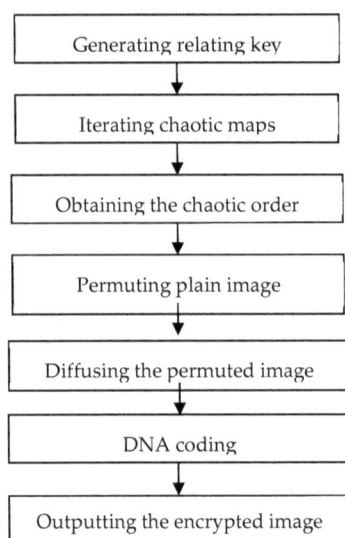

Figure 1. The flowchart of encrypting image.

2.3.2. Decrypting Image

The decryption process is similar to that of encryption procedure in the reversed order. It can be briefly stated as follows:

Step 1. According to the same relating key, the chaotic maps are to iterate for 100 times to get rid of the transient effect;

Step 2. The chaotic orbits are regenerated based on the same parameters and initial values as well as the encryption process;

Step 3. Decoding the cipher image based on the DNA coding rule;

Step 4. The XOR operation is implemented between the pixels of the cipher image and the pseudorandom sequence from the logistic maps and the permuted image is recovered;

Step 5. According to the order of chaotic sequences, the plain image is recovered from the permuted image;

Step 6. Outputting the plain image.

Note that the permutation–diffusion architecture is widely used into image encryption based on chaotic map and DNA coding. So, the whole architecture of the proposed method is the permutation–diffusion. However, the previous works are mainly to convert the pixel value into an 8-bit binary sequence and then perform a simple one-to-one correspondence between the binary (or ASCII codes) and the DNA sequence without the function of error correction. For example, the binary sequence of the pixel value 1 is 00000001 and the corresponding DNA sequence is AAAC (A for 00, C for 01, G for 10 and T for 11). In this paper, a DNA coding scheme with the function of error correction is proposed, where the pixel value of image is directly corresponded to a piece of DNA sequence with the function of error correction.

3. Experiment and Simulation

In order to resist the brute-force attack, the key space must be large enough for a secure image cryptosystem. 16 elements make up the key in our paper, $key = \{x_i\}, i = 1, 2, \ldots, 16, x_i \in [0, 255]$. It is sufficiently large to ensure the security of digital image when the key space reaches to $2^{128} \approx 3.4 \times 10^{38}$. All the following experiment have the same size for key space.

3.1. Key Sensitivity

The test of key sensitivity can be stated as follows:

Step 1. Generating the key 123456789012345 and using this key to encrypt the test images;
Step 2. Generating another key—123456789012346—with a slight difference and using this key to encrypt the same test image;
Step 3. Calculating the difference between different cipher images.

From the results, although the two different keys are only slightly different—by one bit—the cipher image with the key 123456789012345 is 99.63% different from the cipher image with the key 123456789012346. Figure 2 shows the results of test image Lena. For the same keys of Cameraman, there is 99.59% difference shown in Figure 3. There is 99.55% difference for Boat shown in Figure 4.

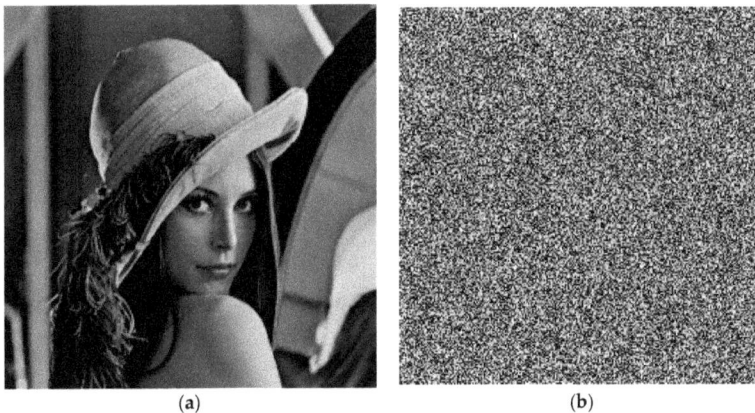

(a)   (b)

Figure 2. *Cont.*

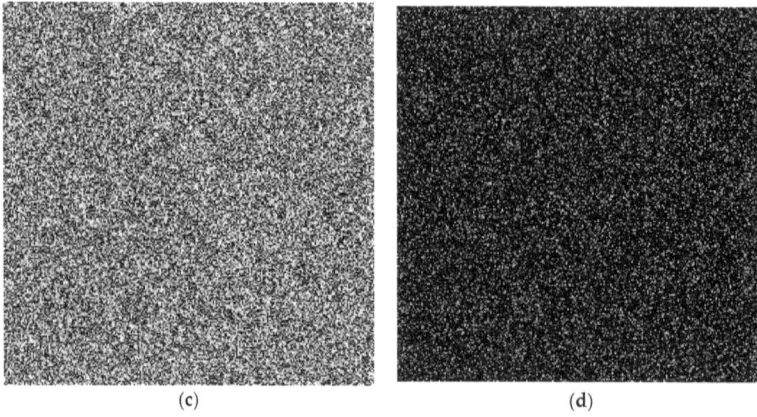

Figure 2. Key sensitivity for Lena. (a) Plain-image of Lena; (b) Encrypted image by key: 123456789012345; (c) Encrypted image by key: 123456789012346; (d) Difference image.

Figure 3. Key sensitivity for Cameraman. (a) Plain-image of Cameraman; (b) Encrypted image by key: 123456789012345; (c) Encrypted image by key: 123456789012346; (d) Difference image.

Figure 4. Key sensitivity for Boat. (**a**) Plain-image of Boat; (**b**) Encrypted image by key: 123456789012345; (**c**) Encrypted image by key: 123456789012346; (**d**) Difference image.

3.2. Statistical Analysis

The statistical characteristics of digital image can be exploited to attack the encryption system. The correlation of two adjacent pixels, as one of statistical characteristics of digital image, is the main aspect of statistical attack. 1000 pairs of adjacent pixels are respectively selected from vertical pixels, horizontal pixels and diagonal pixels. The correlation coefficient of each pair is calculated by the following formulas [1]:

$$\text{cov}(x,y) = E\{(x - E(x))(y - E(y))\} \qquad (6)$$

$$r_{xy} = \frac{\text{cov}(x,y)}{\sqrt{D(x)}\sqrt{D(y)}} \qquad (7)$$

where x and y are grey-scale values of two adjacent pixels in the image. As digital image consists of discrete pixels, we adapt the following discrete formulas for calculating the correlation:

$$E(x) = \frac{1}{N}\sum_{i=1}^{N} x_i \qquad (8)$$

$$D(x) = \frac{1}{N}\sum_{i=1}^{N} (x_i - E(x))^2 \qquad (9)$$

$$\mathrm{cov}(x,y) = \frac{1}{N}\sum_{i=1}^{N}\{(x_i - E(x))(y_i - E(y))\} \tag{10}$$

Table 3 shows the results of horizontal, vertical and diagonal directions. The values outside the brackets indicate the correlation between two adjacent pixels for three different plaintext images and the correlation between the cipher text images is indicated in the brackets. From the experimental results, the proposed algorithm greatly reduces the correlation between pixel values of horizontally, vertically and diagonally adjacent images and improves the ability to resist statistical attacks.

Table 3. The correlation coefficient of adjacent pixels.

	Horizontal	Vertical	Diagonal
Lena	0.9727(0.0073)	0.9481(0.0058)	0.9250(−0.0091)
Cameraman	0.9561(−0.0053)	0.9213(−0.0062)	0.9145(−0.0059)
Boat	0.9334(0.0006)	0.9249(0.0009)	0.8891(−0.0002)

3.3. Differential Attack

Number of pixels change rate (NPCR) and Unified average changing intensity (UACI) are the common quantitative criteria for image cryptosystem to evaluate the property of resisting differential attack.

The NPCR and UACI are defined as follows [39–41]:

$$NPCR = \frac{\sum\limits_{i,j} D(i,j)}{W \times H} \times 100\% \tag{11}$$

$$UACI = \frac{1}{W \times H}\left[\sum_{i,j}\frac{|C_1(i,j) - C_2(i,j))|}{255}\right] \times 100\% \tag{12}$$

where C_1 and C_2 denotes two different cipher images. These cipher images only have one pixel difference. $C_1(i,j)$ and $C_2(i,j)$ respectively denote the pixel values at the same point (i,j) of C_1 and C_2; H and W are respectively the height and width of the image; $C_1(i,j)$ and $C_2(i,j)$ determine the value of $D(i,j)$, namely, if $C_1(i,j) = C_2(i,j)$ then $D(i,j) = 0$ otherwise, $D(i,j) = 1$.

The comparing results of NPCR and UACI list in the Table 4, where the image cryptosystem adapts the permutation—diffusion architecture with only one round.

The average of ten trials for our method is listed the table. According to the comparison, it shows that our method has higher security.

Table 4. The value of NPCR and UACI for Lena.

	NPCR	UACI
Proposed algorithm	99.57%	32.38%
Wang's work [42]	44.27%	14.874%
Gupta's work [43]	99.62%	17.30%

4. Correcting Errors

In this chapter, we simulated the process of correcting errors. First, we encode the cipher image to DNA sequences and randomly change 1000 bases. Each DNA sequence encoded pixel only change one base. Figure 5 shows the effect of correcting errors. Figure 5a shows the encrypted image contain 1000 errors. Figure 5b shows the image after correcting errors by Hamming code. Figure 5c shows the difference between Figure 5a and Figure 5b. Figure 5d shows the decrypted image after correcting errors. The experimental results express that the proposed method could effectively correct the errors and improve the accuracy of hybridization reaction.

Note that if the changed DNA sequence does not match the according to the DNA coding rule, we compulsively set this DNA sequence correspond to the pixel 255. For example, the pixel 16 match DNA sequence GCCTATCT according to DNA coding rule. If the third base is changed, namely GCGTATCT, there will be no pixel match this changed DNA sequence. So, we set GCGTATCT to correspond to pixel 255.

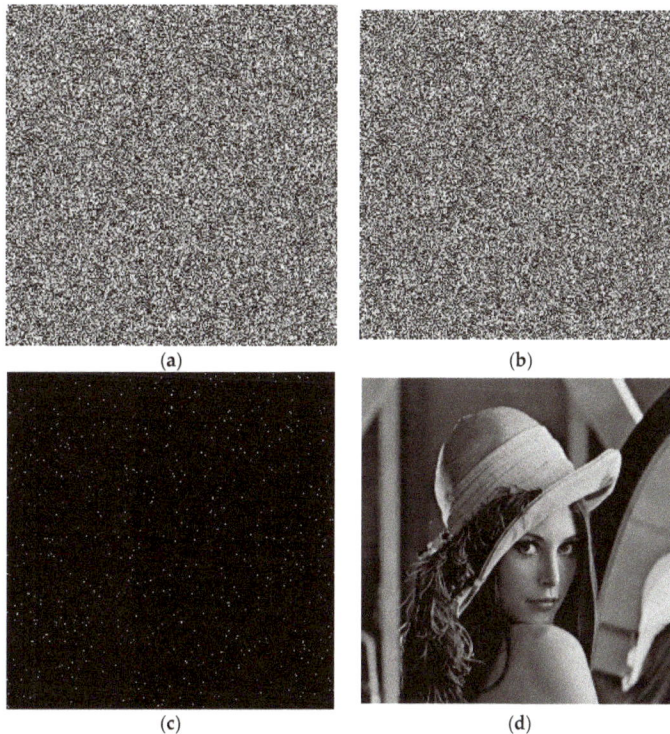

Figure 5. Correcting errors. (**a**) Containing errors image; (**b**) image after correcting errors; (**c**) Difference image; (**d**) Decrypted image after correcting errors.

5. Conclusions

In this paper, in order to improve the accuracy of DNA computing, we propose a novel method which could decrease the similarity of DNA sequences in DNA computing as well as correct errors from the process of image encryption and decryption. We first analyze the characteristic of DNA hybridization reaction and introduce the combinatorial constraints, namely Sequences-sequence Hamming Distance, Sequences-complementarity Hamming Distance and GC content, to design DNA coding. Then we use the chaotic map to generate pseudo-random sequences and encrypt the plain image by the permuting-diffusing architecture. Finally, we propose a novel DNA coding rule to encode the encrypted image. The experimental results show our method could be used to correct errors in image encryption based on DNA coding.

Bio-inspired computing models, such as membrane computing models [44–50], may provide intelligent methods for Image Encryption. As well, DNA coding strategies can provide biological ways in solving chemical information processing problems.

Supplementary Materials: The supplementary materials are available online.

Author Contributions: Conceptualization, B.W.; Methodology, B.W.; Validation, B.W. and C.Z.; Formal Analysis, X.Z.; Investigation, Y.X.; Data Curation, S.Z.; Writing-Original Draft Preparation, B.W.; Writing-Review & Editing, B.W.; Project Administration, B.W.; Funding Acquisition, B.W.

Funding: This work is supported by the National Natural Science Foundation of China (Nos. 61425002, 61751203, 61772100, 61702070, 61672121, 61572093), Program for Changjiang Scholars and Innovative Research Team in University (No. IRT_15R07), the Program for Liaoning Innovative Research Team in University (No. LT2015002), the Basic Research Program of the Key Lab in Liaoning Province Educational Department (No. LZ2015004).

Conflicts of Interest: The authors declare no conflict of interest.

References

1. Chen, G.; Mao, Y.; Chui, C.K. A symmetric image encryption scheme based on 3D chaotic cat maps. *Chaos Solitons Fractals* **2004**, *21*, 749–761. [CrossRef]
2. Wong, K.-W.; Kwok, B.S.-H.; Law, W.-S. A fast image encryption scheme based on chaotic standard map. *Phys. Lett. A* **2008**, *372*, 2645–2652. [CrossRef]
3. Lian, S.; Sun, J.; Wang, Z. A block cipher based on a suitable use of the chaotic standard map. *Chaos Solitons Fractals* **2005**, *26*, 117–129. [CrossRef]
4. Chang, W.L.; Guo, M.Y.; Ho, M.S.H. Fast parallel molecular algorithms for DNA-based computation: Factoring integers. *IEEE Trans. Nanobiosci.* **2005**, *4*, 149–163. [CrossRef]
5. Zhang, W.; Yu, H.; Zhao, Y.L.; Zhu, Z.L. Image encryption based on three-dimensional bit matrix permutation. *Signal Process.* **2016**, *118*, 36–50. [CrossRef]
6. Kulsoom, A.; Xiao, D.; Ur, R.A.; Abbas, S.A. An efficient and noise resistive selective image encryption scheme for gray images based on chaotic maps and DNA complementary rules. *Multimedia Tools Appl.* **2016**, *75*, 1–23. [CrossRef]
7. Zhang, Y.Q.; Wang, X.Y. A symmetric image encryption algorithm based on mixed linear-nonlinear coupled map lattice. *Inf. Sci.* **2014**, *273*, 329–351. [CrossRef]
8. Wang, X.Y.; Zhang, H.L. A novel image encryption algorithm based on genetic recombination and hyper-chaotic systems. *Nonlinear Dyn.* **2016**, *83*, 333–346. [CrossRef]
9. Zhang, Q.; Guo, L.; Wei, X.P. Image encryption using DNA addition combining with chaotic maps. *Math. Comput. Model.* **2010**, *52*, 2028–2035. [CrossRef]
10. Liu, H.J.; Wang, X.Y.; Kadir, A. Image encryption using DNA complementary rule and chaotic maps. *Appl. Soft Comput.* **2012**, *12*, 1457–1466. [CrossRef]
11. Wei, X.P.; Guo, L.; Zhang, Q.; Zhang, J.X.; Lian, S.G. A novel color image encryption algorithm based on DNA sequence operation and hyper-chaotic system. *J. Syst. Softw.* **2012**, *85*, 290–299. [CrossRef]
12. Babaei, M. A novel text and image encryption method based on chaos theory and DNA computing. *Nat. Comput.* **2013**, *12*, 101–107. [CrossRef]
13. Enayatifar, R.; Abdullah, A.H.; Isnin, I.F. Chaos-based image encryption using a hybrid genetic algorithm and a DNA sequence. *Opt. Lasers Eng.* **2014**, *56*, 83–93. [CrossRef]
14. Ozkaynak, F.; Yavuz, S. Analysis and improvement of a novel image fusion encryption algorithm based on DNA sequence operation and hyper-chaotic system. *Nonlinear Dyn.* **2014**, *78*, 1311–1320. [CrossRef]
15. Rehman, A.U.; Liao, X.F.; Kulsoom, A.; Abbas, S.A. Selective encryption for gray images based on chaos and DNA complementary rules. *Multimedia Tools Appl.* **2015**, *74*, 4655–4677. [CrossRef]
16. Song, C.Y.; Qiao, Y.L. A Novel Image Encryption Algorithm Based on DNA Encoding and Spatiotemporal Chaos. *Entropy* **2015**, *17*, 6954–6968. [CrossRef]
17. Wu, X.J.; Kan, H.B.; Kurths, J. A new color image encryption scheme based on DNA sequences and multiple improved 1D chaotic maps. *Appl. Soft Comput.* **2015**, *37*, 24–39. [CrossRef]
18. Wang, X.Y.; Zhu, X.Q.; Zhang, Y.Q. An Image Encryption Algorithm Based on Josephus Traversing and Mixed Chaotic Map. *IEEE Access* **2018**, *6*, 23733–23746. [CrossRef]
19. Parvaz, R.; Zarebnia, M. A combination chaotic system and application in color image encryption. *Opt. Laser Technol.* **2018**, *101*, 30–41. [CrossRef]
20. Adleman, L.M. Molecular computation of solutions to combinatorial problems. *Science* **1994**, *266*, 1021–1024. [CrossRef] [PubMed]

21. Li, T.; Wang, E.K.; Dong, S.J. Potassium-Lead-Switched G-Quadruplexes: A New Class of DNA Logic Gates. *J. Am. Chem. Soc.* **2009**, *131*, 15082–15083. [CrossRef] [PubMed]
22. Qian, L.; Winfree, E.; Bruck, J. Neural network computation with DNA strand displacement cascades. *Nature* **2011**, *475*, 368–372. [CrossRef] [PubMed]
23. Scudellari, M. Inner Workings: DNA for data storage and computing. *Proc. Natl. Acad. Sci. USA* **2015**, *112*, 15771–15772. [CrossRef] [PubMed]
24. Wang, B.; Zhou, S.; Zheng, X.; Zhou, C.; Dong, J.; Zhao, L. Image watermarking using chaotic map and DNA coding. *Opt.-Int. J. Light Electron Opt.* **2015**, *126*, 4846–4851. [CrossRef]
25. Zhang, Q.; Wang, B.; Wei, X.; Fang, X.; Zhou, C. DNA word set design based on minimum free energy. *IEEE Trans. NanoBiosci.* **2010**, *9*, 273–277. [CrossRef] [PubMed]
26. Chai, X.L.; Chen, Y.R.; Broyde, L. A novel chaos-based image encryption algorithm using DNA sequence operations. *Opt. Lasers Eng.* **2017**, *88*, 197–213. [CrossRef]
27. Wu, X.J.; Kurths, J.; Kan, H.B. A robust and lossless DNA encryption scheme for color images. *Multimedia Tools Appl.* **2018**, *77*, 12349–12376. [CrossRef]
28. Marathe, A.; Condon, A.E.; Corn, R.M. On combinatorial DNA word design. *J. Comput. Biol.* **2001**, *8*, 201–219. [CrossRef] [PubMed]
29. Shin, S.-Y.; Lee, I.-H.; Kim, D.; Zhang, B.-T. Multiobjective evolutionary optimization of DNA sequences for reliable DNA computing. *IEEE Trans. Evol. Comput.* **2005**, *9*, 143–158. [CrossRef]
30. Hamming, R.W. Error detecting and error correcting codes. *Bell Syst. Tech. J.* **1950**, *29*, 147–160. [CrossRef]
31. Zhang, Q.; Wang, B.; Wei, X.P.; Zhou, C.J. A Novel Constraint for Thermodynamically Designing DNA Sequences. *PLoS ONE* **2013**, *8*. [CrossRef] [PubMed]
32. Watson, J.D.; Crick, F.H.C. Molecular structure of nucleic acids—A structure for deoxyribose nucleic acid. *Nature* **1953**, *171*, 737–738. [CrossRef] [PubMed]
33. Zhang, Q.; Guo, L.; Wei, X.P. A novel image fusion encryption algorithm based on DNA sequence operation and hyper-chaotic system. *Optik* **2013**, *124*, 3596–3600. [CrossRef]
34. Kracht, D.; Schober, S. Insertion and deletion correcting DNA barcodes based on watermarks. *BMC Bioinform.* **2015**, *16*, 50. [CrossRef] [PubMed]
35. Wang, B.; Zhang, Q.; Zhang, R.; Xu, C.X. Improved Lower Bounds for DNA Coding. *J. Comput. Theor. Nanosci.* **2010**, *7*, 638–641. [CrossRef]
36. Su, X.; Li, W.; Hu, H. Cryptanalysis of a chaos-based image encryption scheme combining DNA coding and entropy. *Multimedia Tools Appl.* **2017**, *76*, 14021–14033. [CrossRef]
37. Zhang, Y. Breaking a RGB image encryption algorithm based on DNA encoding and chaos map. *Int. J. Inf. Secur.* **2014**, *1*, 22–28.
38. Liu, Y.; Tang, J.; Xie, T. Cryptanalyzing a RGB image encryption algorithm based on DNA encoding and chaos map. *Opt. Laser Technol.* **2014**, *60*, 111–115. [CrossRef]
39. Kwok, H.; Tang, W.K.S. A fast image encryption system based on chaotic maps with finite precision representation. *Chaos Solitons Fractals* **2007**, *32*, 1518–1529. [CrossRef]
40. Peng, J.; Zhang, D.; Liao, X. A digital image encryption algorithm based on hyper-chaotic cellular neural network. *Fundam. Inf.* **2009**, *90*, 269–282.
41. Wang, B.; Zheng, X.; Zhou, S.; Zhou, C.; Wei, X.; Zhang, Q.; Che, C. Encrypting the compressed image by chaotic map and arithmetic coding. *Opt.-Int. J. Light Electron Opt.* **2014**, *125*, 6117–6122. [CrossRef]
42. Wang, Y.; Wong, K.W.; Liao, X.; Chen, G. A new chaos-based fast image encryption algorithm. *Appl. Soft Comput.* **2011**, *11*, 514–522. [CrossRef]
43. Gupta, K.; Silakari, S. Novel Approach for fast Compressed Hybrid color image Cryptosystem. *Adv. Eng. Softw.* **2012**, *49*, 29–42. [CrossRef]
44. Yuan, S.; Deng, G.; Feng, Q.; Zheng, P.; Song, T. Multi-objective evolutionary algorithm based on decomposition for energy-aware scheduling in heterogeneous computing systems. *J. Univ. Comput. Sci.* **2017**, *23*, 636–651.
45. Wang, X.; Gong, F.; Zheng, P. On the computational power of spiking neural P systems with self-organization. *Sci. Rep.* **2016**. [CrossRef] [PubMed]
46. Song, T.; Wong, P.Z.D.M.; Wang, X. Design of logic gates using spiking neural p systems with homogeneous neurons and astrocytes-like control. *Inf. Sci.* **2016**, *372*, 380–391. [CrossRef]

47. Song, T.; Rodríguez-Patón, A.; Zheng, P.; Zeng, X. Spiking neural p systems with colored spikes. *IEEE Trans. Cognit. Dev. Syst.* **2018**. [CrossRef]
48. Wang, X.; Zheng, P.; Ma, T.; Song, T. Computing with bacteria conjugation: Small universal systems. *Moleculer* **2018**, *23*, 1307. [CrossRef] [PubMed]
49. Zhang, L.; Yuan, S.; Feng, L.; Guo, B.; Qin, J.; Xu, B.; Lollar, C.; Sun, D.; Zhou, H. Pore-Environment Engineering with Multiple Metal Sites in Rare-Earth Porphyrinic Metal-Organic Frameworks. *Angew. Chem.* **2018**. [CrossRef]
50. Zhang, M.; Xin, X.; Xiao, Z. A multi-aromatic hydrocarbon unit induced hydrophobic metal–organic framework for efficient C2/C1 hydrocarbon and oil/water separation. *J. Mater. Chem.* **2017**, *5*, 1168–1175. [CrossRef]

Sample Availability: Samples of the images and DNA sequences are available from the authors.

MDPI

St. Alban-Anlage 66

4052 Basel

Switzerland

Tel. +41 61 683 77 34

Fax +41 61 302 89 18

www.mdpi.com

Molecules Editorial Office

E-mail: molecules@mdpi.com

www.mdpi.com/journal/molecules

www.ingramcontent.com/pod-product-compliance
Lightning Source LLC
Chambersburg PA
CBHW051708210326
41597CB00032B/5407